Readings in Conservation Ecology

Contributors

P.L. Ames
A.H. Baumhover
A.M. Beeton
A.I. Bischoff
B.A. Bitter
F.H. Bormann
C.C. Bradley
L.R. Brown
R.C. Bushland
L.R. Clark
O.B. Cope
B.J. Copeland
A.T. Cringan
R.F. Dasmann
J.J. Davis
H.F. Dorn
R.G. Downes
F.H. Dudley
W.T. Edmondson
L. Ellison
F.C. Evans
R.F. Foster
P.W. Geier
A.J. Graham
K.S. Hagen
W.C. Hanson
G.H. Hepting
D.E. Hopkins
G.B. Huffaker
R.R. Humphrey
E.G. Hunt
G.E. Hutchinson
P.A. Isaacson
J.O. Ivie
D. Kirk
C.E. Kennett
A. Leopold
A.S. Leopold
G.E. Likens
W.J. McConnell
L.A. Mehrhoff
W.D. New
E.P. Odum
O.T. Oglesby
D. Pimentel
D.H. Pimlott
J.H.A. Ruzicka
F. Sargent, II
C.N. Sawyer
R.F. Smith
V.M. Stern
E.C. Stone
R.D. Taber
J. O'G. Tatton
O.C. Taylor
M.D. Thomas
C.R. Thompson
R. van den Bosch
G.M. Van Dyne
G.M. Woodwell
C.F. Wurster, Jr.

Readings in Conservation Ecology

edited by
George W. Cox
SAN DIEGO STATE COLLEGE

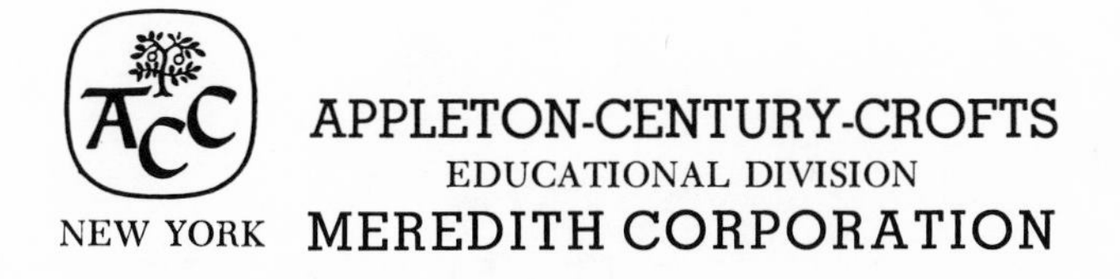

APPLETON-CENTURY-CROFTS
EDUCATIONAL DIVISION
NEW YORK MEREDITH CORPORATION

679-2

Library of Congress Card Number: 69-18685

PRINTED IN THE UNITED STATES OF AMERICA

390-21340-3

Editor's Introduction

The most important trend in the conservation movement today is the growing realization that conservation policy and practice must be based on sound ecological theory and must be directed toward the management of entire ecological systems. This trend has given rise to the field of conservation ecology, the aim of which is perpetuation of the economic and esthetic values of ecological systems through maintenance of their functional integrity. Recognition of the need for such an approach to conservation has developed slowly from observations of the failures and shortcomings of practices directed toward management of individual resources or individual species, and from the growing evidence that man's utilization of his environment has brought about major, yet often inconspicuous, changes in the functioning of many ecological systems. The theory and technology for this approach to conservation have been provided largely by the field of ecology, which in recent years has increasingly emphasized study of the dynamics of ecological systems. The growing acceptance of this approach is evidenced by the frequency with which papers concerned with a systems approach to diverse conservation problems are now appearing in major scientific journals. That this approach is still young, however, is attested to by the scarcity of challenging textbooks dealing with conservation ecology. This last fact has served as the immediate stimulus for the development of this set of readings. Hopefully this collection of papers will prove useful and stimulating to students and workers in this new field.

Conservation ecology is based on recognition of the existence of functional environmental units known as ecosystems. An ecosystem is a unit consisting of groups of organisms together with their physical environment. The components of an ecosystem, both biotic and abiotic, show a high degree of interpendence as a result of processes involving the exchange of energy and specific chemical nutrients. An example of a specific ecosystem type is a lake, with the various plant, animal, and decomposer organisms representing the biotic components, and the water mass and bottom sediments representing two of the major abiotic components. Ecoystems are thus dynamic in nature, and exhibit a structure which reflects the pattern of interchange of nutrients and energy. Although some ecosystems may approach a steady-state condition in which the structure and function of the system change little through time, most ecosystems exhibit more or less rapid change in these features. Since these systems are dynamic, active management by man is usually necessary for the perpetuation of some desired condition. For ecosystems not in a steady-state condition, or for ecosystems influenced by broad environmental changes caused by man, simple protection from direct exploitation, without active management, may rapidly lead to the loss of important ecosystem values. The goal of conservation ecology is thus to develop an understanding of the functioning of world ecosystems and to devise ecologically sound techniques for their management.

Inherent in this approach to conservation is the realization that man is one of the interrelated components of most ecological systems. This relationship implies that advancing human technology, which results in the utilization of environmental resources in an increasingly intensive fashion, increases the complexity of man's interrelationships with natural ecological systems, rather than making him more independent of them. It also demands that direct consideration be given to the implications of human population behavior and the effects of human technology in an overall survey of the field of conservation ecology.

The readings in this collection are organized around the concept of the ecosystem. The collection is introduced by three papers. The first, by Francis C. Evans, concisely presents the concept of the ecosystem while the second, by Eugene P. Odum, considers in greater detail the relationships between ecosystem structure and function. The third, by George M. Van Dyne, discusses the systems approach to the study of ecosystem problems. The remaining papers, concerned with specific problems in conservation ecology, are organized into two groups representing the two major aspects of ecosystem function: the flow of energy and the cycling of nutrients and other materials among ecosystem components. Included in the section relating to problems of energy flow are papers dealing with management or control of specific biotic components of ecosystems. Within this section papers are grouped under the topics of (1) management of wildlife populations, (2) control of undesirable species, (3) vegetation management, and (4) human populations and food resources. The common feature of these problems is the fact that management is concerned with the regulation of energy flow through one or more of the biotic components of an ecosystem.

The section relating to problems of cycling of nutrients and other materials includes papers dealing with the ecological effects of various types of environmental pollution or disturbance of natural patterns of nutrient cycling. This section is introduced with a general paper by G. E. Hutchinson on the management of nutrient cycles. The topics under which the papers in this section are grouped consist of the major pollution problems of (1) pesticides, (2) radioactive materials, (3) air pollution, (4) eutrophication of aquatic ecosystems, and problems of sustained use management of (5) aquatic ecosystems, and water resources, and (6) terrestrial ecosystems and soils resources. The common feature of these problems is the fact that management is concerned with regulation of the loss or gain of nutrients and other chemical substances by ecosystems and with the patterns of movement of such materials among ecosystem components. The collection is drawn together with a paper by Aldo Leopold, written over 30 years ago, but still pertinent to the problem of development of social and political attitudes necessary for the implementation of ecologically sound management practices.

This set of readings is designed to complement a one-semester course in conservation as presented from an ecological viewpoint. The specific organization of such a course can vary greatly, depending on the philosophy of the instructor and the background of the students to whom the course is directed. At San Diego State College, where the course is available both to undergraduate majors in biology and to nonmajors who have had an introductory biology course, the conservation course is introduced by a series of lectures on principles of ecology relating to ecosystem function. This material includes principles of energy flow and nutrient cycling in ecological systems, principles of population growth and regulation, and principles relating to biotic succession and stability in communities and ecosystems. Following this introduction, specific problems in conservation ecology, corresponding to the topics under which papers are grouped in this collection, are considered in relation to these principles.

The diversity in the nature of the papers included in this collection allows them to play a variable role in such a course. Some of the papers deal with basic theory of management, some summarize the state of knowledge on particular problems, and others deal with specific studies of individual problems. A few of the papers may be most useful as substitutes for textbook coverage of certain problems. Others are probably most useful as examples of the application of an ecological approach to specific problems. One of the most important functions of many of the papers may be, however, as background reading on which class discussions of particular problems may be based. The complexity of many of these problems and the variety of ecological considerations involved in the development of appropriate management procedures often may best be explored through discussions drawing upon the ideas of many individuals.

I wish to express my gratitude to members of the ecology group at San Diego State College who have all, directly or indirectly, contributed to my interest and awareness in the field of conservation ecology. I am especially indebted to Boyd D. Collier, William E. Hazen, Albert W. Johnson, Herbert Melchior, and Phillip C. Miller for specific suggestions regarding the inclusion of papers. Finally, I wish to thank the individual authors and publishers for their permission to reprint the papers included in this collection.

G.W.C.

Contents

Part IV Outlook for the Future

I

Ecosystem approach to conservation problems

1

Ecosystem as the basic unit in ecology

Francis C. Evans

The term *ecosystem* was proposed by Tansley (1) as a name for the interaction system comprising living things together with their nonliving habitat. Tansley regarded the ecosystem as including "not only the organism-complex, but also the whole complex of physical factors forming what we call the environment." He thus applied the term specifically to that level of biological organization represented by such units as the community and the biome. I here suggest that it is logically appropriate and desirable to extend the application of the concept and the term to include organization levels other than that of the community.

In its fundamental aspects, an ecosystem involves the circulation, transformation, and accumulation of energy and matter through the medium of living things and their activities. Photosynthesis, decomposition, herbivority, predation, parasitism, and other symbiotic activities are among the principal biological processes responsible for the transport and storage of materials and energy, and the interactions of the organisms engaged in these activities provide the pathways of distribution. The food-chain is an example of such a pathway. In the nonliving part of the ecosystem, circulation of energy and matter is completed by such physical processes as evaporation and precipitation, erosion and deposition. The ecologist, then, is primarily concerned with the quantities of matter and energy that pass through a given ecosystem and with the rates at which they do so. Of almost equal importance, however, are the kinds of organisms that are present in any particular ecosystem and the roles that they occupy in its

Reprinted with permission from *Science*, 123:1127–1128 (22 June 1956).

structure and organization. Thus, both quantitative and qualitative aspects need to be considered in the description and comparison of ecosystems.

Ecosystems are further characterized by a multiplicity of regulatory mechanisms which, in limiting the numbers of organisms present and in influencing their physiology and behavior, control the quantities and rates of movement of both matter and energy. Processes of growth and reproduction, agencies of mortality (physical as well as biological), patterns of immigration and emigration, and habits of adaptive significance are among the more important groups of regulatory mechanisms. In the absence of such mechanisms, no ecosystem could continue to persist and maintain its identity.

The assemblage of plants and animals visualized by Tansley as an integral part of the ecosystem usually consists of numerous species, each represented by a population of individual organisms. However, each population can be regarded as an entity in its own right, interacting with its environment (which may include other organisms as well as physical features of the habitat) to form a system of lower rank that likewise involves the distribution of matter and energy. In turn, each individual animal or plant, together with its particular microenvironment, constitutes a system of still lower rank. Or we may wish to take a world view of life and look upon the biosphere with its total environment as a gigantic ecosystem. Regardless of the level on which life is examined, the ecosystem concept can appropriately be applied. The ecosystem thus stands as a basic unit of ecology, a unit that is as important to this field of natural science as the species is to taxonomy and systematics. In any given case, the particular level on which the ecosystem is being studied can be specified with a qualifying adjective—for example, community ecosystem, population ecosystem, and so forth.

All ranks of ecosystems are open systems, not closed ones. Energy and matter continually escape from them in the course of the processes of life, and they must be replaced if the system is to continue to function. The pathways of loss and replacement of matter and energy frequently connect one ecosystem with another, and therefore it is often difficult to determine the limits of a given ecosystem. This has led some ecologists to reject the ecosystem concept as unrealistic and of little use in description or analysis. One is reminded, however, of the fact that it is also difficult, if not impossible, to delimit a species from its ancestral or derivative species or from both; yet this does not destroy the value of the concept. The ecosystem concept may indeed be more useful when it is employed in relation to the community than to the population or individual, for its limits may be more easily determined on that level. Nevertheless, its application to all levels seems fully justified.

The concept of the ecosystem has been described under many names, among them those of *microcosm* (2), *naturkomplex* (3), *holocoen* (4) and *biosystem* (5). Tansley's term seems most successfully to convey its meaning and has in fact been accepted by a large number of present-day ecologists. I hope that it will eventually be adopted universally and that its application will be expanded beyond its original use to include other levels of biological organization. Recognition of the ecosystem as the basic unit in ecology would be helpful in focussing attention upon the truly fundamental aspects of this rapidly developing science.

REFERENCES

1. A. G. Tansley, *Ecology* 16, 296 (1935).
2. S. A. Forbes, *Bull. Peoria Sci. Assoc.* (1887).
3. E. Markus, *Sitzber. Naturforsch. Ges. Univ. Tartu* 32, 79 (1926).
4. K. Friederichs, *Die Grundfragen und Gesetzmässigheiten der land-und forstwirtschaftlichen zoologie.* (Parey, Berlin, 1930).
5. K. Thienemann, *Arch. Hydrobiol.* 35, 267 (1939).

2

Relationships between structure and function in the ecosystem

Eugene P. Odum

The topic I wish to discuss with you today is: Relationships between structure and function in the ecosystem. As you know ecology is often defined as: The study of interrelationships between organisms and environment. I feel that this conventional definition is not suitable; it is too vague and too broad. Personally, I prefer to define ecology as: The study of the structure and function of ecosystems. Or we might say in a less technical way: The study of structure and function of nature.

By structure we mean: (1) The composition of the biological community including species, numbers, biomass, life history and distribution in space of populations; (2) the quantity and distribution of the abiotic (nonliving) materials such as nutrients, water, etc.; (3) the range, or gradient, of conditions of existence such as temperature, light, etc. Dividing ecological structure into these three divisions is, of course, arbitrary but I believe convenient for actual study of both aquatic and terrestrial situations.

By function we mean: (1) The rate of biological energy flow through the ecosystem, that is, the rates of production and the rates of respiration of the populations and the community; (2) the rate of material or nutrient cycling, that is, the biogeochemical cycles; (3) biological or ecological regulation including both regulation of organisms by environment (as, for

Address given at the 9th Annual Meeting of the Ecological Society of Japan, April 4, 1962. Reprinted with modification from the *Japanese Journal of Ecology*, 12(3):108–118.

example, in photoperiodism) and regulation of environment by organisms (as, for example, in nitrogen fixation by microorganisms). Again, dividing ecological function into these three divisions is arbitrary but convenient for study.

Until recently ecologists have been largely concerned with structure, or what we might call the descriptive approach. They were content to describe the conditions of existence and the standing crop of organisms and materials. In recent years equal emphasis is being placed on the functional approach as indicated by the increasing number of studies on productivity and biological regulation. Also the use of experimental methods, both in the field and in the laboratory, has increased. Today, there exists a very serious gap between the descriptive and the functional approach. It is very important that we bring together these two schools of ecology. I should like to present some suggestions for bridging this gap.

The main features of the structure of a terrestrial and an aquatic ecosystem may be illustrated by comparing an open water community, such as might be found at sea or in a large lake, with a land community such as a forest. In our discussion we shall consider these two types as models for the extremes in a gradient of communities which occur in our biosphere. Thus, such ecosystems as estuaries, marshes, shallow lakes, grasslands and agricultural croplands will have a community structure intermediate between the open water and forest types.

Both aquatic and terrestrial community types have several structural features in common. Both must have the same three necessary biological components: (1) producers or green plants capable of fixing light energy (i.e., autotrophs); (2) animals or macro-consumers which consume particulate organic matter (i.e., phagotrophs); and (3) microorganism decomposers which dissolve organic matter releasing nutrients (i.e., osmotrophs). Both ecosystems must be supplied with the same vital materials such as nitrogen, phosphorus, trace minerals, etc. Both ecosystems are regulated and limited by the same conditions of existence such as light and temperature. Finally, the arrangement of biological units in vertical space is basically the same in the two contrasting types of ecosystems. Both have two strata, an autotrophic stratum above and a heterotrophic stratum below. The photosynthetic machinery is concentrated in the upper stratum or photic zone where light is available, while the consumer-nutrient regenerating machinery is concentrated largely below the photic zone. It is important to emphasize that while the vertical extent or thickness of communities varies greatly (especially in water), light energy comes into the ecosystem on a horizontal surface basis which is everywhere the same. Thus, different ecosystems should be compared on a square meter basis, not on a cubic or volume basis.

On the other hand, aquatic and terrestrial ecosystems differ in structure in several important ways. Species composition is, of course, completely different; the roles of producers, consumers and decomposers are carried out by taxonomically different organisms which have become adapted through evolution. Trophic structure also differs in that land plants tend to be large in size but few in number while the autotrophs of open water ecosystems (i.e., phytoplankton) are small in size but very numerous. In general, autotrophic biomass is much greater than heterotrophic biomass on land, while the reverse is often true in the sea. Perhaps the most important difference is the following: The matrix, or supporting framework, of the community is largely physical in aquatic ecosystems, but more strongly biological on land. That is to say, the community itself is important as a habitat on land, but not so important in water.

Now, we may ask: How do these similarities and differences in structure affect ecological function?

One important aspect of function is shown in Figure 2–1 which compares energy flow in an aquatic and a terrestrial ecosystem. The lower diagram is an energy flow model for a marine community; the upper diagram is a comparable model for a forest. The boxes represent the average standing crop biomass of organisms to be expected; the light gray boxes are the autotrophs, the darker boxes are the heterotrophs. Three trophic levels are shown: (1) Producers, the phytoplankton of the sea and the leaves of the forest trees; (2) primary consumers (herbivores, etc.); and (3) secondary consumers (carnivores). The pipes or flow channels represent the energy flow through the ecosystems beginning with the incoming solar energy and passing through the successive trophic levels. At each transfer a large part of the energy is dissipated in respiration and passes out of the system as heat. The amount of energy remaining after three steps is so small that it can be ignored in so far as the energetics of the community are concerned. However, tertiary consumers ("top carnivores") can be important as regulators; that is, predation may have an important effect on energy flow at the herbivore level.

All numbers in the diagrams are in terms of large or Kilogram Cal-

---

Fig. 2–1. Energy flow models for two contrasting types of ecosystems, an open water marine ecosystem and a terrestrial forest

Standing crop biomass (in terms of KCal./M^2) and trophic structure are shown by means of shaded rectangles. Energy flows in terms of KCal/M^2/day (average annual rate) are shown by means of the unshaded flow channels. The aquatic system is characterized by a small biomass structure (hence the habitat is largely physical) while the forest has a very large biomass structure (hence the habitat is strongly biological). In both types of systems the energy of net primary production passes along two major pathways or food chains: (1) the grazing food chain (upper sequence in the water column or vegetation), and (2) the detritus food chain (lower sequence in sediments or soil).

The marine diagram is based on work of Riley and Harvey, the forest diagram on the work of Ovington and unpublished data from research at the University of Georgia. In some cases figures are hypothetical since no complete study has yet been made of any ecosystem. Hence, the diagrams should be considered as "working models" which do not represent any one situation.

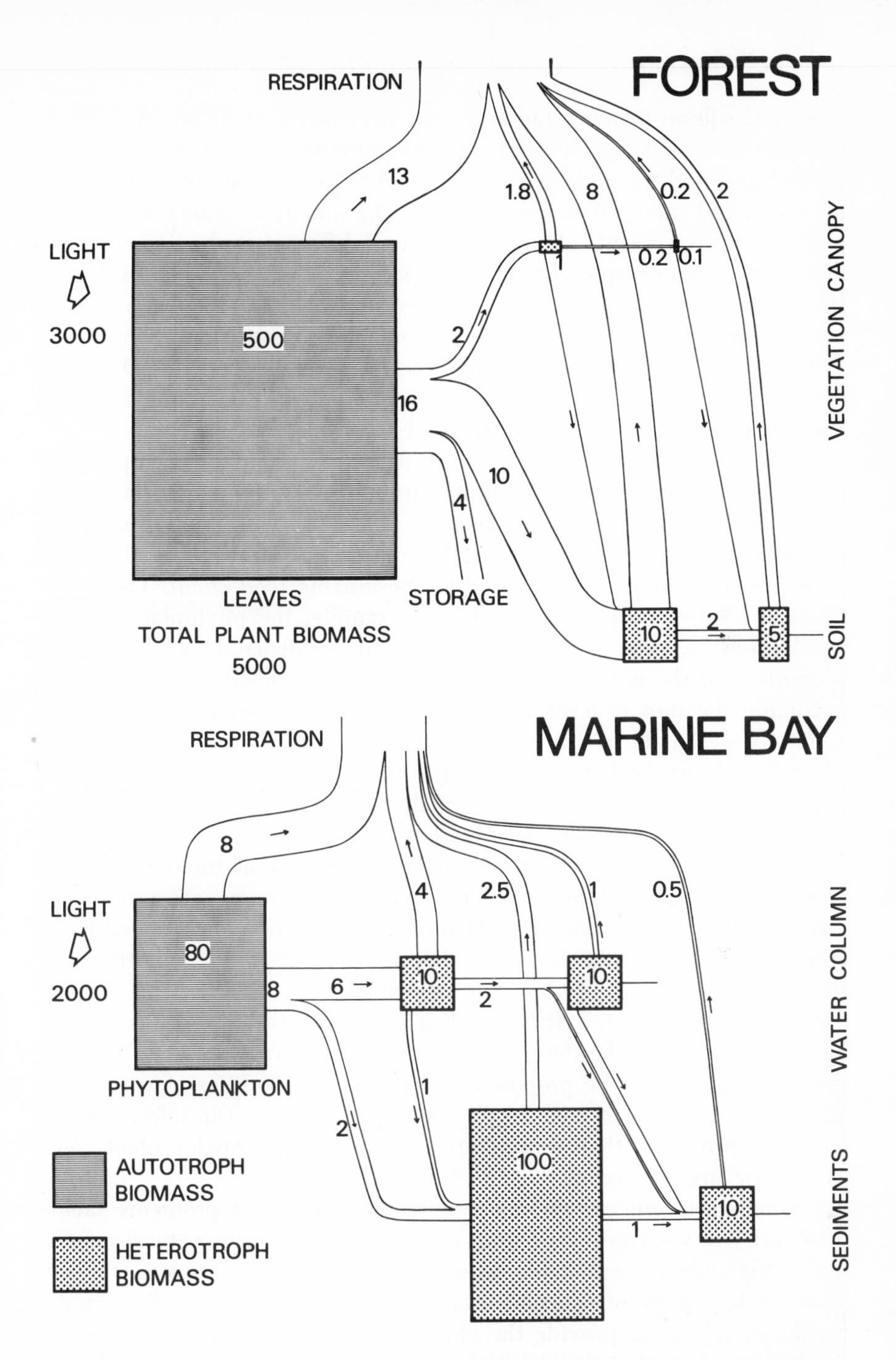

STANDING CROP BIOMASS IN KILOGRAM CALORIES / SQUARE METER

ENERGY FLOW IN KILOGRAM CALORIES / M^2 / DAY

ories and square meters; standing crop is in terms of KCal./M^2; energy flow is in terms of KCal./M^2/day. The diagrams are drawn so that the areas of the boxes and the pipes are proportional to the magnitude of the standing crops and energy flows respectively. The quantities shown are a composite of measurements obtained in several different studies; some of the figures for higher trophic levels are hypothetical since complete information is not yet available for any one ecosystem. The marine community is particularly based on the work of Gordon Riley (Long Island Sound) and H.W. Harvey (English Channel), and the forest on the work of J.D. Ovington (pine forest) and unpublished data on terrestrial communities from our research group at the University of Georgia.

The autotrophic-heterotrophic stratification, which we emphasized as a universal feature of community structure, results in two basic food chains as shown in both diagrams (Fig. 2–1). The consumption of living plants by herbivores which live in the autotrophic stratum together with their predators may be considered as the *grazing food chain.* This is the classical food chain of ecology, as, for example, the phytoplankton-zooplankton-fish sequence or the grass-rabbit-fox sequence. However, a large proportion of the net production may not be consumed until dead, thus becoming the start of a rather different energy flow which we may conveniently designate as the *detritus food chain.* This energy flow takes place largely in the heterotrophic stratum. As shown in Figure 2–1 the detritus energy flow takes place chiefly in the sediments of water systems, and in the litter and soil of land systems.

Ecologists have too often overlooked the fact that the detritus food chain is the more important energy pathway in many ecosystems. As shown in Figure 2–1 a larger portion of net production is estimated to be consumed by grazers in the marine bay than in the forest; nine-tenths of the net production of the forest is estimated to be consumed as detritus (dead leaves, wood, etc.). It is not clear whether this difference is a direct or indirect result of the difference in community structure. One tentative generalization might be proposed as follows: communities of small, rapidly growing producers such as phytoplankton or grass can tolerate heavier grazing pressure than communities of large, slow-growing plants such as trees or large seaweeds.

Grazing is one of the most important practical problems facing mankind; yet we know very little about the situation in natural ecosystems. Well-ordered and stable ecosystems seem to have numerous mechanisms which prevent excessive grazing of the living plants. Sometimes, predators appear to provide the chief regulation; sometimes weather or life history characteristics (limited generation time or limited number of generations of herbivores) appear to exercise control. Unfortunately, man with his cattle, sheep and goats often fails to provide such regulation with result that overgrazing and declining productivity are apparent in large areas of the world, especially in grasslands. A study of the division of energy flow between grazing and detritus pathways in stable natural ecosystems can provide a guide for man's utilization of grasslands, forests, the sea, etc.

The energy flow diagrams, as shown in Figure 2–1, reemphasize the difference in biomass as mentioned previously. Autotrophic biomass is very large and envelops or encloses the whole community in the forest; such extensive biological structure buffers and modifies physical factors such as temperature and moisture. In contrast, the aquatic community stands naked or exposed to the direct action of physical factors. In the marine situation the animal biomass often exceeds the plant biomass, and sessile animals (oysters, barnacles, etc.) instead of plants often provide some protection or habitat for other organisms.

Despite the large difference in relative size of standing crops in the two extreme types of ecosystems, the actual energy flow may be of the same order of magnitude if light and available nutrients are similar. In Figure 2–1 we have shown the available light (absorbed light) and the resulting net production as being somewhat lower in the marine community, but this may not always be true. Thus, 80 KCals of phytoplankton may have a net production almost as large as 5000 KCals of trees (or 500 KCals of green leaves). Therefore, productivity is not proportional to the size of the standing crop except in special cases involving annual plants (as in some agriculture). Unfortunately, many ecologists confuse productivity and standing crop. The relation between structure and function in this case depends on the size and rate of metabolism (and rate of turnover) of the organisms.

To summarize, we see that biological structure influences the pattern of energy flow, particularly the fate of net production and the relative importance of grazers and detritus consumers. However, total energy flow is less affected by structure, and is thus less variable than standing crop. A functional homeostasis has been evolved in nature despite the wide range in species structure and in biomass structure.

So far we have dealt with structure in relation to one aspect of function of the entire ecosystem. Now let us turn to structure and function at the population level and consider a second major aspect of function, namely, the cycling of nutrients. As an example I shall review the work of Dr. Edward J. Kuenzler at the University of Georgia Marine Institute on Sapelo Island. The study concerned a species of mussel of the genus *Modiolus* in the intertidal salt marshes. There are similar species of filter-feeding mollusca in the intertidal zone in all parts of the northern hemisphere.

First, we shall take a look at the salt marsh ecosystem and the distribution of the species in the marsh. The mussels live partly buried in the sediments and attached to the stems and rhizomes of the marsh grass,

Spartina alterniflora. Individuals are grouped into colonies (clumped distribution), but the colonies are widely scattered over the marsh. Numbers average 8/M² for the entire marsh and 32/M² in the most favorable parts of the marsh. Biomass in terms of ash-free dry weight averages 11.5 gms/M². When the tide covers the colonies the valves partly open and the animals begin to pump large quantities of water.

Figure 2–2 illustrates the role of the mussel population in phosphorus cycling and energy flow according to Dr. Kuenzler's data. Each day the population removes a large part of the phosphorus from the water, especially the particulate fraction. Most of this does not actually pass through the body but is sedimented in the form of pseudofeces which fall on the sediments. Thus, the mussel makes large quantities of phosphorus avail-

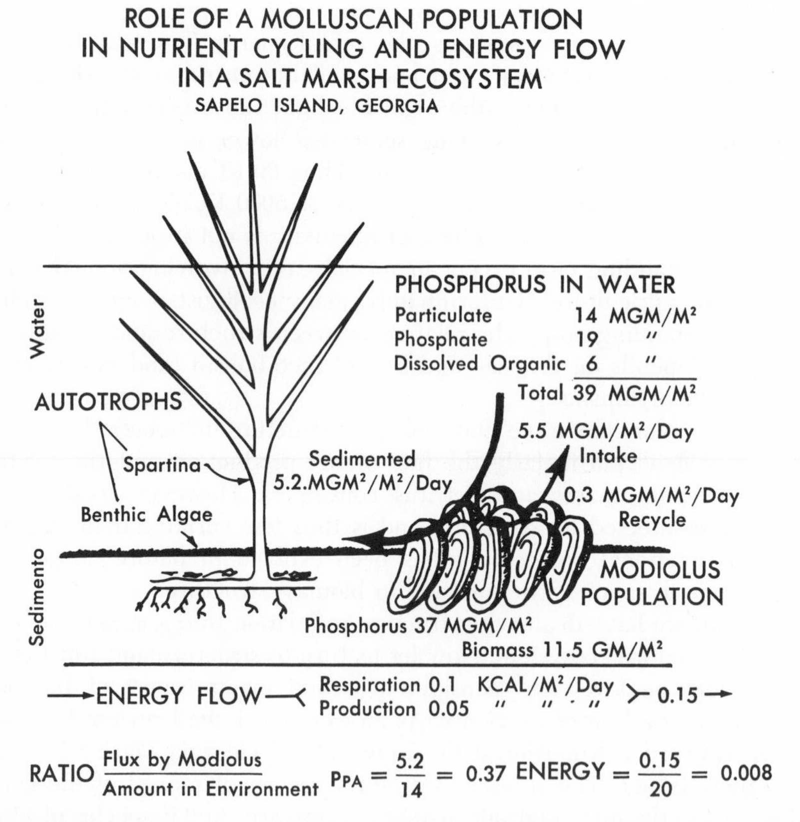

Fig. 2–2. The effect of a population of mussels (*Modiolus*) on energy flow and the cycling of phosphorus in a salt marsh ecosystem according to the study of Dr. E. J. Kuenzler at the University of Georgia Marine Institute, Sapelo Island, Georgia, U.S.A. From the standpoint of the ecosystem as a whole the population has a much greater effect on the cycling of phosphorus than on the transformation of energy. The study illustrates one often overlooked function of animals, that of nutrient regeneration. See text for details of the study.

able to microorganisms and to the autotrophs (benthic algae and marsh grass). As shown along the bottom of the diagram (Fig. 2–2) the energy flow was estimated to be about 0.15 $KCals/M^2/day$.

The most important finding of the study is summarized in the bottom line below the diagram (Fig. 2–2) which shows the ratio between flux and amount. Note that over one third of the 14 mgms of particulate phosphorus is removed from the water each day by the population, and thereby retained in the marsh. In contrast, less than one per cent of the 20 KCals of potential energy (net production estimate) available is actually utilized by the mussel population. In other words, the mussel population has a much more important effect on the community phosphorus cycle than it has on community energy flow. Or one might say that the role of the mussel in conserving nutrients in the ecosystem is more important than its role as energy transformer. In other words, the mussel population would be of comparatively little importance as food for man or animals (since population growth or production is small), but is of great importance in maintaining high primary production of the ecosystem.

To summarize, the mussel study brings out two important points: (1) It is necessary to study both energy flow and biogeochemical cycles to determine the role of a particular species in its ecosystem, (2) animals may be important in the ecosystem not only in terms of food energy, but as agents which make basic nutrient more available to autotrophs.

Finally, I think it is highly significant that the most productive ecosystems of the biosphere are those in which autotrophic and heterotrophic strata lie close together, thus insuring efficient nutrient regeneration and recycling. Estuaries, marshes, coral reefs and rice fields are examples of such productive ecosystems.

Now let us consider the third important aspect of ecological function, that is, community regulation. Ecological succession is one of the most important processes which result from the community modifying the environment. Figure 2–3 illustrates a very simple type of ecological succession which can be demonstrated in a laboratory experiment. Yet the basic pattern shown here is the same as occurs in more complex succession of natural communities. The diagram (Fig. 2–3) was suggested to me by Dr. Ramon Margelef, hence we may call it the Margelef model of succession.

At the top of the diagram (Fig. 2–3) are a series of culture flasks containing plankton communities in different stages of succession. The graph shows changes in two aspects of structure and in one aspect of function. The first flask on the left contains an old and relatively stable community; this flask represents the climax. Diversity of species is high in the climax; species of diatoms, green flagellates, dinoflagellates and rotifers are shown in the diagram to illustrate the variety of plants and animals present. Biochemical diversity is also high as indicated by the

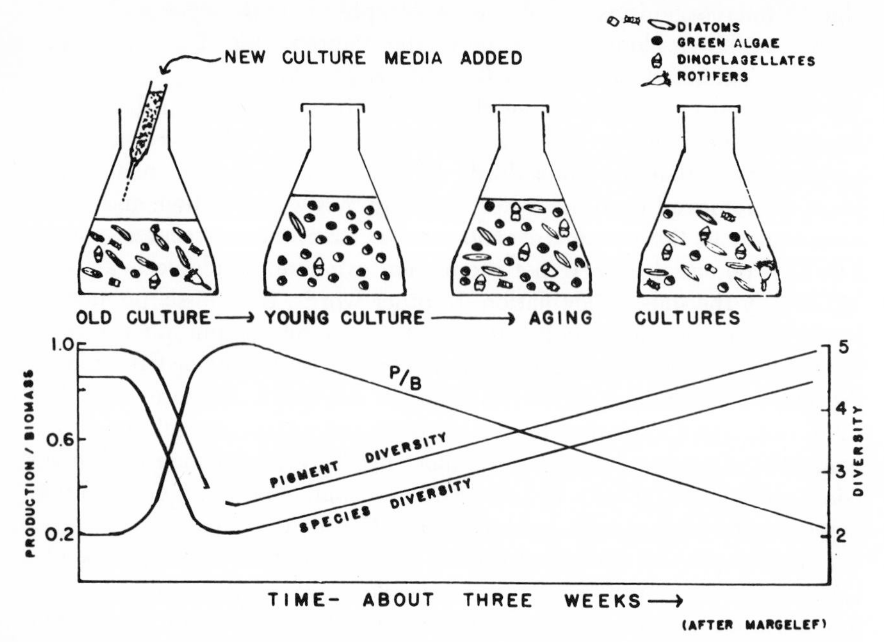

Fig. 2–3. The MARGELEF model of ecological succession showing a simple type of succession which can be demonstrated in laboratory cultures
The flasks show changes in species composition occurring when succession is set in motion by the introduction of new nutrient media into an old "climax" culture. The graph shows resultant changes in two aspects of diversity and in the relation between production and biomass (P/B). See text for details of the experiment.

ratio of yellow plant pigments (optical density at 430mμ) to chlorophyll-a (optical density at 665mμ). On the other hand the ratio of production to biomass (P/B in Fig. 2–3) is low in the old or climax culture, and gross production tends to equal community respiration. If we add fresh culture medium to the old culture, as shown in Figure 2–3, ecological succession is set in motion. An early stage in succession is shown in the second flask. Species diversity is low, with one or two species of phytoplankton dominant. Chlorophylls predominate so that the yellow/green ratio (O.D. 430/ O.D. 665) is low, indicating low biochemical diversity. On the other hand, production now exceeds respiration so that the ratio of production to biomass becomes higher. In other words, autotrophy greatly exceeds heterotrophy in the pioneer or early succession stage. The two flasks on the right side of the diagram (Fig. 2–3) show the gradual return to the climax or steady state where autotrophy tends to balance heterotrophy.

The changes which we have just described are apparently typical of all succession regardless of environment or type of ecosystem. Although much more study is needed, it appears that differences in community structure mainly affect the time required, that is, whether the horizontal scale (X-axis in Fig 2–3) is measured in weeks, months or years. Thus, in open water ecosystems, as in cultures, the community is able to modify the physical environment to only a small extent. Consequently, succession in such ecosystems is brief, lasting perhaps for only a few weeks. In a typical marine pond or shallow marine bay a brief succession from diatoms to dinoflagellates occurs each season, or perhaps several times each season. Aquatic ecosystems characterized by strong currents or other physical forces may exhibit no ecological succession at all, since the community is not able to modify the physical environment. Changes observed in such ecosystems are the direct result of physical forces, and are not the result of biological processes; consequently, such changes are not to be classed as ecological succession.

In a forest ecosystem, on the other hand, a large biomass accumulates with time, which means that the community continues to change in species composition and continues to regulate and buffer the physical environment to a greater and greater degree. Let us refer again to Figure 2–1 which compares a forest with an aquatic ecosystem. The very large biological structure of the forest enables the community to buffer the physical environment and to change the substrate and the microclimate to a greater extent than is possible in the marine community.

Recent studies on primary succession on such sites as sand dunes or recent volcanic lava flows indicate that at least 1000 years may be required for development of the climax. Secondary succession on cut-over forest land or abandoned agricultural land is more rapid, but at least 200 years may be required for development of the stable climax community. When the climate is severe as, for example, in deserts, grasslands or tundras the duration of ecological succession is short since the community can not modify the harsh physical environment to a very large extent.

To summarize, I am suggesting that the basic pattern of functional change in ecological succession is the same in all ecosystems, but that the species composition, rate of change and duration of succession is determined by the physical environment and the resultant community structure.

The principles of ecological succession are of the greatest importance to mankind. Man must have early successional stages as a source of food since he must have a large net primary production to harvest; in the climax community production is mostly consumed by respiration (plant and animal) so that net community production in an annual cycle may be zero. On the other hand, the stability of the climax and its ability to buffer and control physical forces (such as water and temperature) are desirable characteristics from the viewpoint of human population. The only way man can have both a productive and a stable environment is to insure that a good mixture of early and mature successional stages (i.e., "young nature" and "old nature") are maintained with interchanges of energy and materials. Excess food produced in young communities helps feed older stages which in return supply regenerated nutrients and help buffer the extremes of weather (storms, floods, etc.).

In the most stable and productive of natural situations we usually find such a combination of successional stages. For example, in continental shelf marine areas such as the inland sea of Japan the young communities of plankton feed the older, more stable communities on the rocks and on the bottom (i.e., benthic communities).The large biomass structure and diversity of the benthic communities provide not only habitat and shelter for life history stages of pelagic forms, but also provide regenerated nutrients necessary for continued productivity of the plankton.

A similar favorable situation exists in the Japanese terrestrial landscape where productive rice fields on the plains are intermingled with diverse forests on the hills and mountains. The rice fields, of course, are, ecologically speaking, "young nature" or early successional communities with very high rates of net community production which are maintained as such by the constant labor of the farmer and his machines. The forests represent older, more diverse and self-sustaining communities which have a lower net production, but do not require the constant attention of man. It is important that both ecosystems be considered together in proper relation. If the forests are destroyed merely for the sake of temporary gain in wood production, then the water and soil will wash down from the slopes and destroy the productivity of the plains. In my brief travels in Japan I have noted an unfortunate tendency in some areas to consider only the productive aspect of forests, and consequently to ignore their protective value. Complete deforestation of slopes may yield more wood for the time being but is ecologically a very dangerous procedure; also rebuilding the ecosystem is always more expensive than maintaining it in good condition. I believe ecologists should be more aggressive in bringing these principles to the attention of those charged with responsibility of national resources. Especially, ecologists need to provide good data which demonstrates the value of forests and other mature-type ecosystems in maintaining water and nutrient cycles. The value of forest should not be measured only in terms of net production.

My purpose in reviewing the three basic aspects of ecological function (that is, energy flow, nutrient cycles, and biological regulation) is to emphasize that we must study both structure and function if we are really to understand and control nature. Usually the study of function is more difficult than the study of structure; hence functional ecology has lagged behind the descriptive ecology. To study function we must measure the rate of change per unit of time, and not just the situation at any one time. That is, we must measure the rate of energy flow, not just the standing crop; we must measure the rate of exchange of phosphorus, not just the amount present in the ecosystem; and we must measure the degree of regulation, not just describe it. I should like to close my lecture with a brief discussion of how new techniques resulting from the peaceful uses of atomic energy may help us to make better measurements of ecological function. I refer, of course, to the use of radioactive tracers in ecological research.

We should first emphasize two points about radioactive tracers: (1) Tracer techniques do not solve any problem alone, but may be useful in conjunction with other methods. Tracers do extend our powers of observation greatly. Just as the microscope extended our powers of observation of biological structure, so tracers have extended our powers of observation of function. (2) The amount of radioactivity employed in a tracer experiment can be so small that there need be no hazard to the investigator and no effect on organisms or environment. Instruments are now so sensitive that amounts of radioactivity far less than that contained in a radium-dial watch can be easily detected.

In the intact ecosystem it is often difficult to determine the exact energy source or food being utilized by various organisms. Likewise, it is difficult to measure the rate of metabolism of free living populations. Radioactive tracers can aid in both of these important determinations in plants, animals and microorganisms.

Figure 2–4 shows the results of a simple experiment in which all individuals of one species of plant within a large quadrat in a natural grassland community were labeled with radioactive phosphorus (P^{32}). This was done by placing a small drop of P^{32} (in form of phosphate) on a leaf of each individual plant of the species in question; within a short time the labeled phosphate spreads to all parts of the plant. In the experiment shown in Figure 2–4 one of the dominant plants, *Keterotheca subaxilaris*, was labeled, while all other plants in the quadrat remained unlabeled. During the 30 days following the labeling, samples of the invertebrates were collected and the concentration of P^{32} in their tissues determined. According to this procedure, any animal which becomes radioactive must have fed upon the plant, or must have eaten an animal which had previously fed upon the plant. Such a procedure enables us to isolate a single food chain (that is, a food chain beginning with a single species of autotroph) in an intact community.

In Figure 2–4 the buildup of P^{32} radioactivity in six major populations is shown. Radioactivity in terms of disintegrations per minute per milligram of live weight is indicated on the Y-axis, and time is indicated on the X-axis. A small ant (Formicidae) and a small cricket (Orthoptera) were the first animals to reach a peak in radioactivity; these species were the most active animals as indicated by general observation. Larger plant feeders such as grasshoppers (*Melanoplus*) reached a peak at a later time, while predators such as spiders did not reach maximum levels until

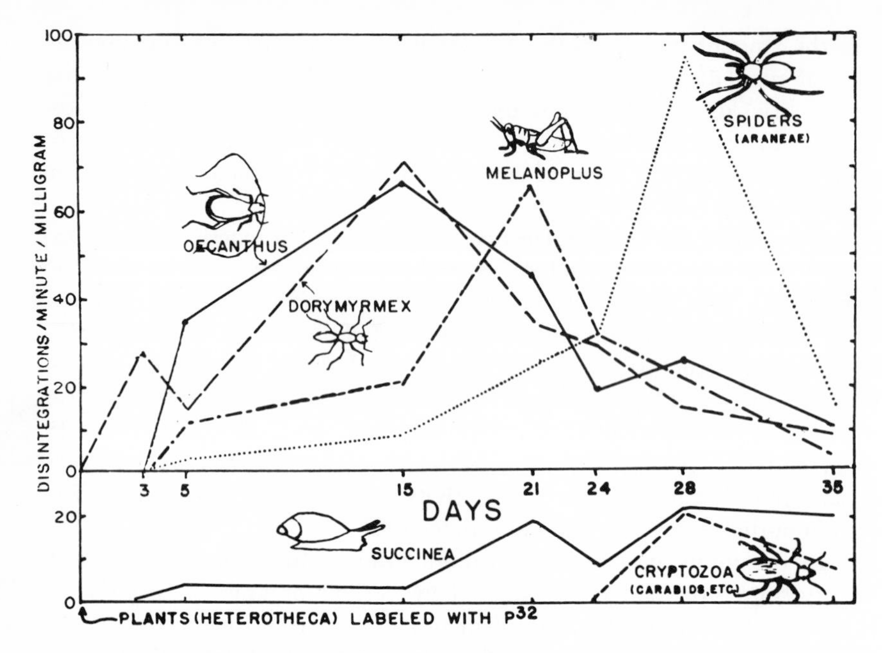

Fig. 2–4. The results of a simple experiment in which a radioactive tracer (P^{32}) was used to isolate a food chain in an intact "old-field" (i.e. early stage of succession on abandoned agricultural land) community
The buildup of P^{32} activity in the biomass of 6 major invertebrate populations following the labeling of a single species of dominant plant is shown in the diagram. This tracer technic resulted in the separation of certain trophic and habitat niches in the intact and undisturbed community. See text for details of the experiment.

about four weeks after the labeling of the plant. There was also a delay in appearance of P^{32} in animals living on the ground or in the surface litter (carabids, snails, etc.).

As the final example I would like to mention very briefly the work which Mr. Jiro Mishima did while he was studying with us in the U.S.A. During the past year our group at the University of Georgia has been interested in investigating the possibility that the excretion rate of isotopes may be used as indices of function. Theoretically, an active population should excrete a tracer more rapidly than an inactive group of organisms. If a good relation between excretion rate and activity can be demonstrated, then we would be able to estimate activity rate in wild animals provided only that we can recapture individuals or samples of the population after the labeled individuals had been released in their natural habitat. Preliminary trial experiments have now been completed in which the excretion rate of several radioactive tracers has been measured in insects, Crustacea, mollusks and fish under different environmental conditions.

As shown in Figure 2–4 plotting radioactivity in the biomass against time resulted in a good separation of certain trophic and habitat niches.

The experiment indicated that the ant, *Dorymyrmex*, was functioning as a herbivore rather than a predator since rapid buildup of radioactivity would not be expected if the species were feeding on other insects only. Since aphids could not be found on the plants it was tentatively concluded that the ants were feeding directly on plant juices at the time of the experiment (which was in the late spring of the year).

We believe that procedures involving the labeling of single energy sources in intact communities can be very useful not only in determining food intake and trophic position of free-living populations, but also perhaps in determining the rate of feeding. The more rapid the feeding the sooner the transfer of the tracer will be observed.

Mr. Mishima studied the excretion rate of Zinc-65 in the salt marsh snail, *Littorina irrorata*, in relation to air temperature and body size. The general procedure is shown in Figure 2–5. Snails freshly collected from the marsh were placed in a grass "labeling bowl" as shown at the top of

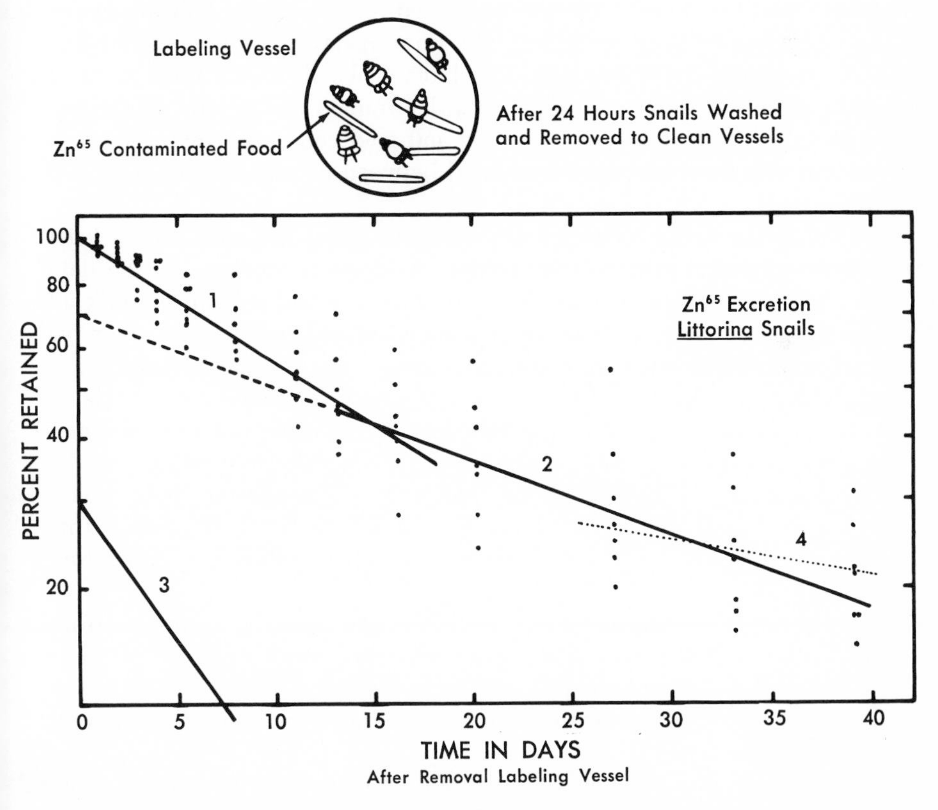

Fig. 2–5. The procedure used by Mr. Jiro Mishima in his study of the excretion (biological half-life) of the tracer, Zinc-65, by *Littorina*
The slope of regression line number 2 was found to be significantly affected by air temperature and body size in a manner parallel to expected oxygen consumption (i.e. rate of metabolism). See text for possible significance of this type of study.

the figure. A solution of $Zn^{65}Cl$ was placed on food in the bowl. After 24 hours the snails usually ingested a small amount of radioactive zinc. The snails were then washed thoroughly to remove surface contamination and placed in nonradioactive environments either under laboratory or field conditions. At intervals of one to three days marked individuals were placed in a well scintillation counter to determine how much of the tracer remained in the body. In the laboratory experiments the snails were transferred to clean bowls after each determination in order to avoid the possibility of re-ingestion of excreted material.

As shown in Figure 2–5 the Zn^{65} tracer tended to be eliminated in two phases. Excretion was rapid during the first 10 days, and less rapid for the next 20 days or so. It is the second regression line (line 2 in Fig. 2–5) which is of greatest ecological interest since it shows the rate of loss of the tracer which was more completely assimilated into the organic biomass. Regressions were calculated for each individual by IBM electronic computer from logarithms of successive counts (corrected for radiological decay), and the slope of the line converted to a biological half-life figure. By biological half-life we mean: The time in days required for one-half of the tracer to be eliminated. The biological half-life of three sizes of snails (small, medium and large) was determined under four conditions: laboratory bowls at 10°, 25° and 30° and field conditions (natural marsh habitat) where temperature fluctuated between 9° and 31° but averaged 19°C.

Since the details of this study are to be published soon only a brief mention of the results will be needed. Analysis of variance of the data showed that both air temperature and body size had a highly significant effect on the biological half-life of the assimilated tracer. The larger the snail, or the lower the temperature, the slower was the excretion rate (i.e., the longer the biological half-life). Furthermore, excretion rate as related to body size and temperature paralleled oxygen consumption rates as determined in a previous study on this species. Especially interesting was the fact that biological half-life in the field at 19°C was shorter (hence excretion more rapid) than in the laboratory at 25° or 30°C, suggesting that snails in the field were more active than when confined to laboratory bowls.

These preliminary results indicate that Zn^{65} may be useful as an index to the rate of metabolism. However, much work needs to be done before such an "atomic meter" can be perfected. Probably a mixture of tracers will eventually prove useful, each tracer measuring a different aspect of function.

If we can discover how to measure the rate or activity or metabolism in free-living individuals of populations in intact ecosystems, then we will have taken a giant step towards better understanding of the relation of function to structure at the ecological level.

3

Ecosystems, systems ecology, and systems ecologists

George M. Van Dyne

ECOSYSTEMS

Definitions

In 1935 Tansley (106) introduced the term ecosystem, which he defined as the system resulting from the integration of all the living and nonliving factors of the environment. Webster now defines the term as a complex of ecological community and environment forming a functioning whole in nature. An ecosystem is a functional unit consisting of organisms (including man) and environmental variables of a specific area (3). Macroclimate has an overriding impact on the other components, each of which is interrelated at least indirectly (Fig. 3–1). The term "eco" implies environment; the term "system" implies an interacting, interdependent complex.

Trends in Ecological Research

Although the concept of the ecosystem and many methods for studying ecosystems have been available for some time, only recently have many ecologists given more than lip service to the idea. Recently it has been suggested that the ecosystem is the rallying point for ecologists.

Research performed at Oak Ridge National Laboratory and sponsored by the U. S. Atomic Energy Commission under contract with Union Carbide Corporation. Reprinted with permission and in modified form from *Oak Ridge National Laboratory Report* 3957, pp. 1–31.

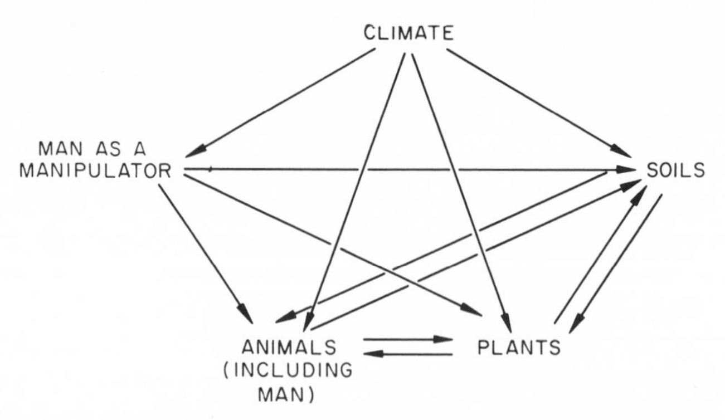

Fig. 3–1. An ecosystem is an integrated complex of living and nonliving components. Each component is influenced by the others, with the possible exception of macroclimate. And now man is on the verge of exerting meaningful influence over macroclimate.

There has been a gradual but distinct shift in emphasis in ecological studies and training from the description or inventory of ecosystems, or parts thereof, to the study of energy flow, nutrient cycles, and productivity of ecosystems. More workers are extending knowledge from the "anatomy" to the "physiology" of the environment. This requires different concepts, tools, and methods. The gradual change in emphasis from inventory to experimentation also requires more use of scientific methodology; this will be discussed below in the section, "Systems Ecology."

Ecosystem Components

The controlling factors of the ecosystems are macroclimate, available organisms, and geological materials, where the last term includes parent material, relief, and ground water. Time is considered as a dimension in which the controlling factors operate, rather than as an environmental factor. The controlling factors are partially or entirely independent of each other. Each of the controlling factors is a composite of many separate elements, and each element is variable in time or space. Operationally, we may consider each controlling factor as a multiple-dimensioned matrix. Each change in a controlling agent in the ecosystem produces in time a corresponding change in the dependent elements of the ecosystem. In space and time there is a continuum of ecosystems.

Internal properties of ecosystems, such as rate of energy flow, might be considered as dependent factors which vary through time under the influence of a series of independent controlling factors. The dependent factors of the ecosystem are soil, the primary producers (vegetation), consumer organisms (herbivores and carnivores), decomposer organisms (bacteria, fungi, etc.), and microclimate. Each of these factors is dynamically dependent on the others (Fig. 3–1), and each is a product of the controlling agents operating through time.

Producers, consumers, and decomposers are not distributed at random in the abiotic part of an ecosystem. To maintain either dynamic equilibrium or ordered change in an ecosystem requires that a tremendous number of ordered interrelations exist among its dependent elements (82). To function properly ecosystems must process and store large amounts of information concerning past events, and they must possess homeostatic controls which enable them to utilize the stored information. This information may be expressed in amino acid and nucleotide sequences in genetic codes which have developed over evolutionary time, or it may be expressed in spatial or temporal patterns (20). For example, the changing patterns of plant populations and communities in secondary succession can be considered as expressions of genetically coded information. One species, population, or community is replaced by another with greater genetic potential for utilizing the resources of the changing environment.

Dynamics

Ecosystem changes may be caused by fluctuations in internal population interactions or by fluctuations of the controlling factors. Such changes may be cyclical or directional (14). Directional change from less complex to more complex communities may be considered as progression or succession; directional change from more complex to less complex communities may be considered as regression or retrogression; both are shown in Figure 3–2.

Autogenic succession occurs when the controlling factors are stable and change is due to the effect of the system or some part of it on the microhabitat. Clements (15) formalized this process as migration, ecesis, competition, reaction, and stabilization. This type of primary succession produces changes which are usually gradual and continuous. Allogenic succession occurs when there is a change in the controlling factors. Most changes in the ecosystem are products of both allogenic and autogenic successions. Most macroecosystems can be said to be polygenetic and are the result of several climatic changes and erosion cycles. Purposeful alterations, such as disruption by man, in the controlling and controlled factors of the ecosystem may induce relatively permanent changes in the ecosystem.

Because ecosystems vary both temporally and spatially, and to prevent ambiguity, it is important to specify at least semiquantitative time and space scales. The importance of specifying a time scale is illustrated in Figure 3–2, where the time for primary succession (see T_5 for progression in Fig. 3–2) is shown as much greater than the time requirement

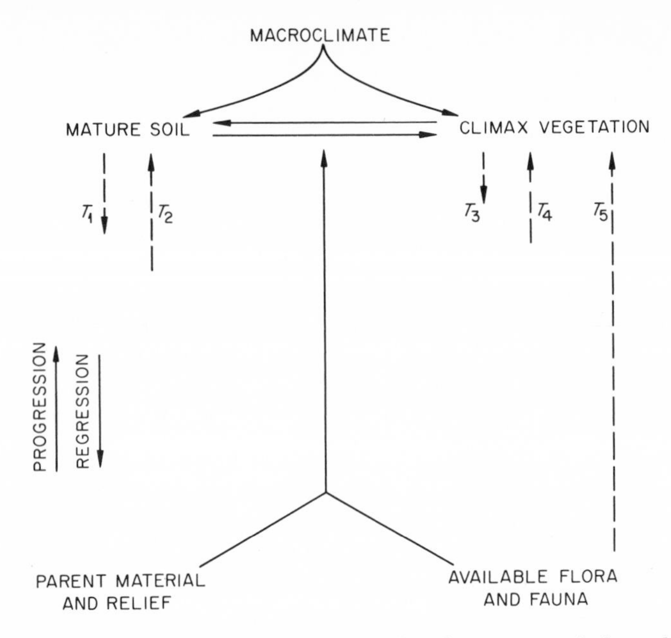

Fig. 3–2. Ecosystems develop through time, under climatic control, from the original flora and fauna under a given set of relief and parent material conditions. A final dynamic equilibrium is reached in which there exist a mature soil and climax plant and animal populations.

for man to disrupt the system and alter soil or vegetation (T_1 and T_3 in Fig. 3–2). In the process of retrogression, changes take place in the vegetation more rapidly than they do in the soil. Generally, the ecosystem will recover stable state through a progressive process called secondary succession (T_2 and T_4). Again, the rate of progressive changes of soil properties is usually lower than that for vegetation. Recovery of the vegetation to the climax state may take an amount of time similar to that required for deterioration of the soil. Change in a given ecosystem component or property may be negligible in T_3 but considerable in T_5.

During progressive succession there is usually an increase in productivity, biomass, relative stability and regularity of populations, and diversity of species and life forms within the ecosystem (74). Finally, the ecosystem reaches a steady state or equilibrium, which is characterized by dynamic fluctuation rather than by directional change. This steady state of the ecosystem is referred to as climax (119). At climax the dependent factors are in balance with the controlling factors; the climax is an open steady state (101). A diversity of species and life forms occupies every available ecological niche at climax and, because there is a maximum number of links in the food web, the stability of the system is maximized (63). A maximum amount of the entering energy is used in maintenance of life. Fosberg (34) considers "that climax communities [are those] in which there is the greatest range and degree of exploitation or utilization of the available resources in the environment." There is no net output from an ecosystem in the climax state (86). Three states of ecosystems exist with regard to energy or nutrient balance: steady state or climax, positive balance or succession occurring, and negative balance or decadence and senescence (99).

There is continual interchange of matter and energy among contiguous ecosystems. This interchange or flux is an essential property of ecosystems. The fluxes in and out of an ecosystem may be difficult to measure accurately, but there is relatively less error in measuring flux in a macroecosystem than in a microecosystem, because usually the error in measurements is inversely proportional to the magnitude of the object, rate, or processes being measured. Also, the relative amount of relevant surface or area around an ecosystem decreases as its size increases; many of the measurement errors or biases occur at such interfaces because of subjective decisions in defining boundaries. Still, we may find it convenient to study microecosystems such as a sealed bottle containing nutrients, gases, organisms, and water. Essentially, this is the type of system we need to study in preparing for interplanetary travel. But even such discrete microcosms are not adiabatic with their environment, and ultimately they are dependent upon their environment for a continuing energy input.

When flux of some element in and out of a given ecosystem is negligible for a defined period of time we consider that ecosystem to be stable with regard to that element. The equilibrium is referred to as climax only if it is reached naturally. Other equilibria, or disclimaxes, can be maintained by man's intervention. Here is the essence of renewable resource management: maintaining disclimaxes at equilibrium for the benefit of man.

Manipulation of Ecosystems

Man is a vital part of most major ecosystems, and there is an increasing human awareness of man's part in them and his influence on them (108) (Fig. 3–3). Traces of his pesticides probably can be found in living organisms throughout the world. Humans are both parts of and manipulators of ecosystems. Induced instability of ecosystems is an important cause of economic, political, and social disturbances throughout the world. In altering his environment in order to overcome its limitations to him, man learns that he often is faced with undesirable consequences of the environmental change (13, 38, 39). In manipulating his environment (e.g., felling forests, burning grasslands or protecting them from fire, and draining marshes), seldom has he foreseen the full consequences of his action (104).

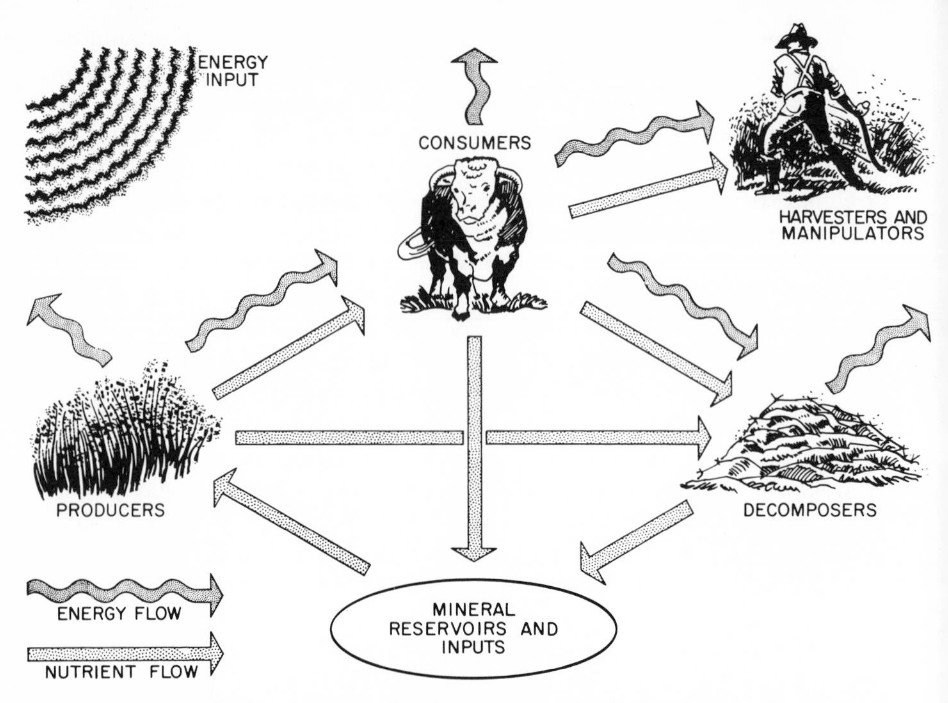

Fig. 3–3. Man is both a spectator of and a participant in the functioning of ecosystems. He has manipulated ecosystems to maximize the flow of nutrients and energy to him from the producers and primary consumers. He has attempted to minimize the respiratory losses of energy from producers, consumers, and decomposers.

Most ecosystems in our country were in climax states when civilized man began to affect them, but the economy of civilized man demanded that the ecosystems produce a removable product under his domination. In order to reach this goal he disrupted the climax ecosystems, perhaps by shortening food chains or by altering the diversity of life forms of primary producers. He has altered the rate of and amount of nutrients cycling through the system by such means as fertilization, both in aquatic and terrestrial systems (Fig. 3–3). In some instances the fertilization has been excessive and has led to undesirable side effects, such as algal blooms caused by excesses of organic wastes. In other instances man has altered the structure of ecosystems by simplifying them and diverting the flow of energy into his food products, such as in replacing a grassland and wild animals with a wheatfield. Eventually he has produced changes in some ecosystem properties, which in some instances has led to new quasi-stable levels. In other instances such changes have led to desertification, such as the result of centuries of overgrazing in the Middle East.

Man has also encountered difficulties when he attempts to return ecosystems to their native state or to preserve vegetation by the development of national parks or by control of predators (104). In several instances ungulate populations have multiplied rapidly, outstripped the natural control by predators, exceeded the carrying capacity of their ranges, and severely damaged their habitat. Examples include the classical Kaibab mule deer problem (94) and the elk problem in Yellowstone National Park (64). Man himself has had a direct and profound effect on some ecosystems he has attempted to maintain in a natural state, such as in Yosemite National Park (39).

SYSTEMS ECOLOGY

Definitions

Systems ecology can be broadly defined as the study of the development, dynamics, and disruption of ecosystems. I consider systems ecology to have two main phases—a theoretical and analytical phase and an experimental phase.

Earlier I stated that for studying function in ecology we need methods and concepts which are different from those for studying structure. Essentially, study of problems in systems ecology requires three groups of tools and processes: conceptual, mechanical, and mathematical.

STUDY OF ECOSYSTEMS

The tools and processes required for systems ecology are different from those for conventional phases of ecology because of the complexity of the total ecosystem as compared with a segment of it. When we consider the totality of interactions of populations with one another and with their physical environs—i.e., ecosystem ecology—we face a new degree of complexity (10). Other than some recent papers (e.g., ref. 41) only a few reasonably adequate functional analyses of natural ecosystems exist (80).

One of the major problems in systems ecology is that of analyzing and understanding interactions. Events in nature are seldom, if ever, caused by a single factor. They are due to multiple factors which are integrated by the organism or the ecosystem to produce an effect which we observe (45). To further complicate the matter, various combinations of factors and their interactions may be interpreted and integrated by the ecosystem to produce the same end result.

Conceptual Requirements

A first conceptual requirement in systems ecology is clearer definition of problems. It is axiomatic that ambiguous use of terminology and an ambiguous statement of the problem lead to ambiguities of thought as well (19). These statements apply to many fields, but, particularly here, clear definitions are required because of the type of people systems ecologists will be and the types of people with whom they will work (discussed further below). Furthermore, in using computers, which are essential tools for systems ecologists, it is necessary to formulate the problems precisely and to clearly delineate the factors involved.

A second conceptual requirement in systems ecology is more and better use of logic and scientific and statistical methods. Essentially, we can define scientific method as the pursuit of truth as determined by logic and experimentation. In scientific method we use the approach of systematic doubt to discover what the facts really are. Experimentation is one of several tools of scientific method used to eliminate untenable theories, that is, to test hypotheses (32). Other experiments may be conducted to determine existing conditions, or to suggest hypotheses, etc. The conclusions from experiments may be criticized because the interpretation was faulty, or the original assumptions were faulty, or the experiment was poorly designed or badly executed (88). Experimental design and statistical inference are aids in testing hypotheses.

Much past ecological research has not tested a hypothesis. There is a tendency for ecologists to pass over the primary phase of analysis. The lack of understanding of what is known already (inadequate knowledge of the literature, in part) is understandable because of the volume of material to be covered (58). Glass (40) has clearly stated this dilemma—"the vastness of the scientific literature makes the search for general comprehension and perception of new relationships and possibilities every day more arduous." But inadequate examination of facts and data and inadequate formulation of hypotheses lead to uncritical selection of experiments testing poorly formulated hypotheses, and ecologists are often at fault here (51). The experimental design is, essentially, the plan or strategy of the experiment to test clearly certain hypotheses (32). Statistical methods are especially important in experimentation with ecosystems, because not all factors influencing the system can be controlled in the experiment without altering the system (29). These uncontrollable factors lead to error or "noise" in our measurements, and inferences to be made from the results of experiments should be accompanied by probability statements (32).

Eberhardt (27) has discussed many of the problems ecologists encounter in sampling, and has stressed the importance of statistical techniques in analysis of such problems. Methods of statistical inference are also useful in suggesting improvements in our mathematical models and in suggesting alterations in the design of future experiments. Some of the work initiated and developed by the late R. A. Fisher on partial correlation and regression is invaluable to us in evaluating independent and interaction effects in complex ecosystems where experimental control is neither possible nor desirable (45).

The first two conceptual needs for systems ecology, mentioned above, lead naturally to the third, the approach of modeling (Fig. 3–4), with models which are mathematical abstractions of real world situations (17, 107). In this process some real world situation is abstracted into a mathematical model or a mathematical system. Next, we apply mathematical argument to reach mathematical conclusions. The mathematical conclusions are then interpreted into their physical counterparts. In some instances we are able to proceed from the real world situation via experimentation to reach physical conclusions. In other instances, however, we cannot experiment with a situation that does not exist but may become real; examples are such situations as thermonuclear war and wide-scale environmental pollution (50). In many cases we find it too costly to ex-

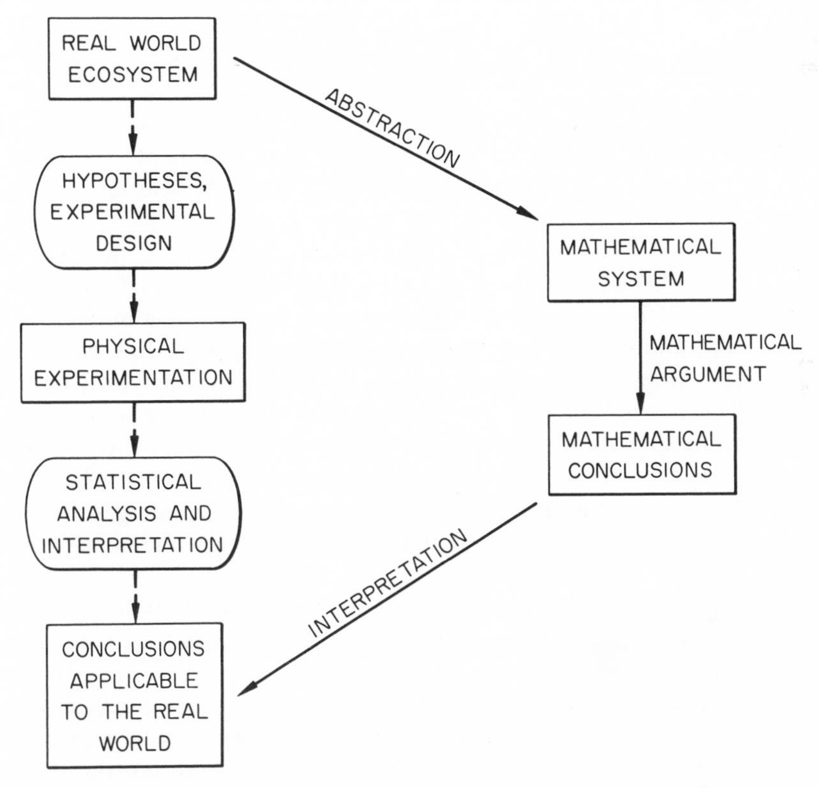

Fig. 3–4. Two ways of experimenting with ecosystems. One involves the conventional process of formulating hypotheses, designing and conducting experiments, and analysis and interpretation of results. The second involves the abstraction of the system into a model, application of mathematical argument, and interpretation of mathematical conclusions.

periment; therefore mathematical modeling or mathematical experimentation may be especially useful.

Mathematical modeling is somewhat new to many conventional ecologists and, in part, is just as much an art as a science. To ensure that the model will be valid, the mathematical axioms must be translations of valid properties of the real world system. The application of mathematical argument gives rise to theorems which we hope can be interpreted to give new insight into our real world system. However, the value of these conclusions should, where possible, be verified by experimentation. We must then accept the conclusions or reject them and start over again. This procedure of modeling, interpretation, and verification is used in many engineering and scientific disciplines. The success of the procedure, however, depends on the existence of an adequate fund of basic knowledge about the system. This knowledge permits predictive calculations. Hollister (50) outlines some of the problems to be encountered in modeling ecological phenomena.

Tools for Study of Ecosystems

The above conceptual tools should provide a framework in which to attack the complex problems of systems ecology. To implement these methods in studying ecosystems we will need both physical and mathematical tools, including digital and analog computers and electrical, mechanical, and hydraulic simulation devices, and artificial populations (44, 75, 85). The act of expressing and testing biological problems with numerical, electrical, or hydraulic analogs often reveals some unsuspected relationships and leads to new approaches in investigation. In conducting experiments in systems ecology, more refined chemical analytical equipment will be needed, such as gas chromatographs, infrared gas analyzers, and recording spectrophotometers. Physical analytical equipment required includes micro-bomb calorimeters, biotelemetric equipment, and other electronic equipment useful for rapid, nondestructive sampling and measuring of plant and animal populations and parameters under field conditions.

The importance of these chemical and physical tools is apparent when one considers the amount and variety of apparatus required to construct and maintain even the simplest aquatic ecosystems or to transplant and manipulate naturally occurring ecosystems for detailed measurements (e.g., ref. 2). A major reason for the scarcity of detailed studies of entire terrestrial ecosystems is that many ecologists are not trained to use many of the required, diversified tools. In other instances these tools may not be available to the ecologist. The systems ecologist cannot be an expert with each of these tools, but he must be aware of their applications and limitations in the study of components and processes in ecosystems. He will need to be conversant with the specialists in other disciplines who make increasing use of these modern and complex tools.

Operations Research and Systems Analysis Applications

Mathematical analysis will become increasingly important in providing advances in systems ecology. Large, fast digital computers have become available in the last 15 years and have allowed the development of special methods of analyzing and studying complex systems in industry and government. Most of these newer mathematical tools were developed in and are used primarily in two loosely defined and somewhat overlapping fields, operations research and systems analysis.

Operations research may be defined as the application of modern scientific techniques to problems involving the operation of a system looked upon as a whole (77). Included therein are any systematic, quantitative analyses aimed at improving efficiency in a situation where "efficient" is well understood (103).

Systems analysis is more difficult to define. Perhaps it can best be defined by opposites. The opposite of a systems approach is unsystematic or piecemeal consideration of problems; intuition may be taken as the opposite of analysis (46). Essentially systems analysis is any analysis to suggest a course of action arrived at by systematically examining the objectives, costs, effectiveness, and risks of alternative policies—and designing additional ones if those examined are found to be insufficient (93).

It is easily seen that operations research and systems analysis are both alike and different. They both contain elements from mathematical, statistical, and logical disciplines. In operations research, however, there usually is an unambiguous goal to be achieved, and the operations researcher is interested in optimization. The systems analyst faces a multiplicity of goals, a highly uncertain future, a frequent predominance of qualitative elements, and an exceeding low probability of building an accurate and satisfactory model for his total problem (103). Because of the methods and techniques he can use effectively, the systems engineer has much to offer in study of ecosystems but he will need considerable guidance. In systems ecology he will be facing a collection and coupling of "green, pink, and brown" boxes (plants, animals, and physical environment) rather than the black boxes with which he is familiar (56). The interconnections between these boxes may be known only imperfectly, and the functional significance of the boxes will need to be established.

Some of the mathematical tools to be employed and examined in systems ecology include scientific decision-making procedures, theory of games, mathematical programming, theory of random processes, and methods of handling problems of inventory, allocation, and transportation (77).

Linear, nonlinear, and dynamic programming, which are especially important to the operations researcher, already show promise in ecology (4, 112, 115) and in management of renewable resources (12, 60, 69). Mathematical programming has already been used widely in agro-ecological problems, such as crop or yield prediction (97), in formulation of least-cost rations for livestock (110), and in farm management decisions (5). Game theory has been applied to decisions in cultivated-crop agriculture (113) and appears to have potential in dealing with wildland resources. Queuing theory and network flow appear to offer much in looking at problems of flow rates in ecosystems (90). Margalef (74) has discussed and indicated some important applications of information theory in ecology. Cybernetics principles and techniques are also useful in studying biological systems (37). Simulation is another important tool in operations research, although not limited to it. Mathematical simulation models have been used to study important resource problems, such as salmon population biology (59, 95), and abstract systems (36).

Importance of Digital Computers

Probably most systems ecology problems will be attacked first with deterministic models as first approximations (70, 71). However, to increase their usefulness and their realism, stochastic elements will be involved in most models or an indeterministic point of view will be taken; for example, see Leslie (65), Neyman and Scott (79), and Jenkins and Halter (53). This will require extensive use of digital computers, not only in simulation but also in analysis. Most stochastic models in ecology to date have been concerned with only one or two species rather than populations or ecosystems (6). Stochastic simulation of biological models or processes has been a useful process in some problems (109, 111). Many problems of modeling and analysis will require study and examination of the underlying statistical distributions (28). In addition to the normal distribution, other distributions which will need examination and use in systems ecology problems include the Poisson, the exponential, and the log-normal. Monte Carlo methods will be especially valuable in developing, testing, and using stochastic or probabilistic models (30, 67). Computers are essential in studying and using these statistical techniques in systems ecology. Other statistical aspects are discussed by Eberhardt (29).

Compartment model methodology, implemented with both analog and digital computers, has proven its value in theoretical studies and is beginning to be put to use in analysis and extension of real data in medical (9) and ecological fields (87, 90). Thus far, however, compartmental simulation models have been restricted to relatively simple ecological and agricultural situations, because most investigators have worked with analog systems of limited capacity (1, 35), although simulation systems have been developed for and used with digital computers in the study of renewable resources (e.g., ref. 42). Most systems ecologists will find it surprisingly easy to express many problems in the pseudoalgebraic languages, such as the many dialects of FORTRAN, used to communicate with digital computers.

An ecosystem might be depicted, as in Figure 3–5, as composed of

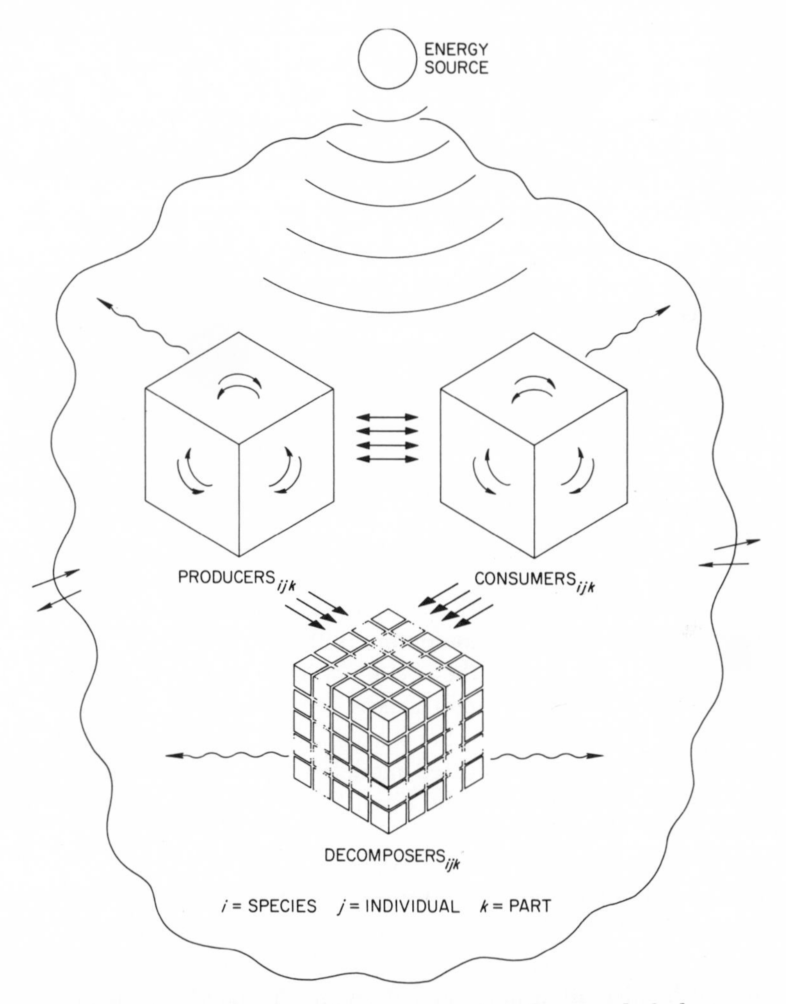

Fig. 3–5. A matrix representation of ecosystems adapted to pseudoalgebraic computer languages. Each trophic level is represented as a three-dimensional matrix. The arrows, wavy for energy and straight for matter, represent matrices of transfer functions interconnecting parts within individuals, individuals of a species to each other and to other species, etc., on up to connections between contiguous ecosystems. The transfer functions may contain probalistic components and may be probalistic functions of external variables such as macroclimate.

trophic levels represented as three-dimensional matrices. PRODUCER (I,J,K), CONSUMER (I,J,K), and DECOMPOSER (I,J,K) are matrices of species, individuals, and parts. The ranges of I, J, and K in each matrix are variable and depend upon the study. Matrices of transfer functions, depicted and simplified by the arrows in Figure 3–5, are concerned with movement of matter or energy within individuals, between individuals, or between species. The latter two types of transfers may be between or within trophic levels. Also included in the figure is the fact that flux among contiguous ecosystems may be considered in matrix representation. Some of the transfer functions themselves may contain random noise and may be functions of a driving variable, such as macroclimate, acting on the system over time. Models or functions for macroclimatic influences may be constructed from actual data or may follow some prescribed hypothetical statistical distribution.

Consider the simplified case (Fig. 3–5) with only three parts per individual, three individuals per species, three species per trophic level, three trophic levels per ecosystem, and three ecosystems per problem. This leads to 3^5 microcompartments to be accounted for in addition to the many transfer functions interrelating the compartments. Many of the transfers will be zero, but this simplified model exceeds the capacity of most analog computers even if the problems of using various random function generators with an analog computer are bypassed. This example does not indicate that analog computers will not play an important role in systems ecology, but only that they may be of limited value in many realistically complex situations. Their major role may be as teaching (and learning) tools and as components of hybrid (digital-analog) systems. The capabilities and versatility of digital computers in general are far greater than those of analog computers (62).

Maximum use of most of the above mathematical tools and others by systems ecologists depends upon access to fast digital computers with large memory capacities (115). Such access will be especially important in working with large complicated models where remote-console access to large central computers will be essential for efficient and rapid progress. Computer technology is approaching the point where the rate of debugging of programs is the limiting factor.

The role of computers in the future of systems ecology is too readily underestimated. Computers in tomorrow's technology will have larger and faster memories, remote consoles, and timesharing systems. Some may accept hand-written notes and drawings, respond to human voices, and translate written words from one language into spoken words in another (96). There will be vast networks of data stations and information banks, with information transmitted by laser channels over a global network. This network will be used not only by researchers but also by engineers, lawyers, medical men, and sociologists as well as government, industry, and the military. Computers could become tomorrow's reference library used by students in the university; they are already starting to revolutionize our present approaches to certain kinds of teaching. To utilize computers effectively in ecology we will have to state precisely what we know, what we do not know, and what we wish to know. Also, it will be necessary to assemble, analyze, identify, reduce, and store our ecological data and knowledge in a form retrievable by machine.

A Systems Approach

Systems ecology will call for an interdisciplinary team of systems ecologists, systems analysts and operations researchers (if they can be separated), conventional ecologists, mathematicians, computer technologists, and applied ecologists, including agriculturalists and natural resource managers of various disciplines. Systems ecologists studying ecosystems will devote at least as much time to delineating the problem as they will to solving it. This gives a hint as to the nature of the work of the systems ecologist.

The physical and mathematical tools to be used by this team are impressive, even though the list in the preceding sections is only partial. It serves to show that the systems ecologist will have to have more types of specialized training than did his predecessors. That different tools and methods may be needed to solve some of today's complex ecological problems is emphasized by the fact that many important contributors to advances in ecology in recent years may not be identified as ecologists (92). This will be especially true of systems ecology in the future, even though ecologists must be generalists, and systems ecologists also will, in part, have to be generalists. Still, there are probably few if any authentic generalists or truly great minds who are not firmly grounded in a specialty (18). Most systems ecologists will serve their apprenticeship in basic fields. The conventional plant, animal, and aquatic ecologists will not be acceptable as systems ecologists, because they will lack the depth required in many specialties (78).

This raises the difficult question of how to train a man to be a specialist in at least one field, to be able to converse well with specialists in several fields, and yet to have a holistic or systems viewpoint.

SYSTEMS ECOLOGISTS

Definitions

The systems ecologist of tomorrow may be defined as one type of scientist who is a specialist in generalization (100). There are few if any systems ecologists today. Of today's biologists, perhaps some scientists in applied ecology can be considered systems ecologists (Fig. 3–6). In some applied ecology fields, such as forestry, it has been noted that the field is becoming so complex that more members of the profession will likely find it to their advantage to either specialize or become exceptionally well-balanced generalists (11). This trend is well established in many professional fields. In addition to the growing need for specialists in resource management, there also is a niche for the generalist (120). Undergraduate programs have been developed to train such individuals.

The applied ecologists, shown as the second group of specialists in Figure 3–6, to some degree have in their training many elements of the training of the four groups of specialists listed below them. The applied

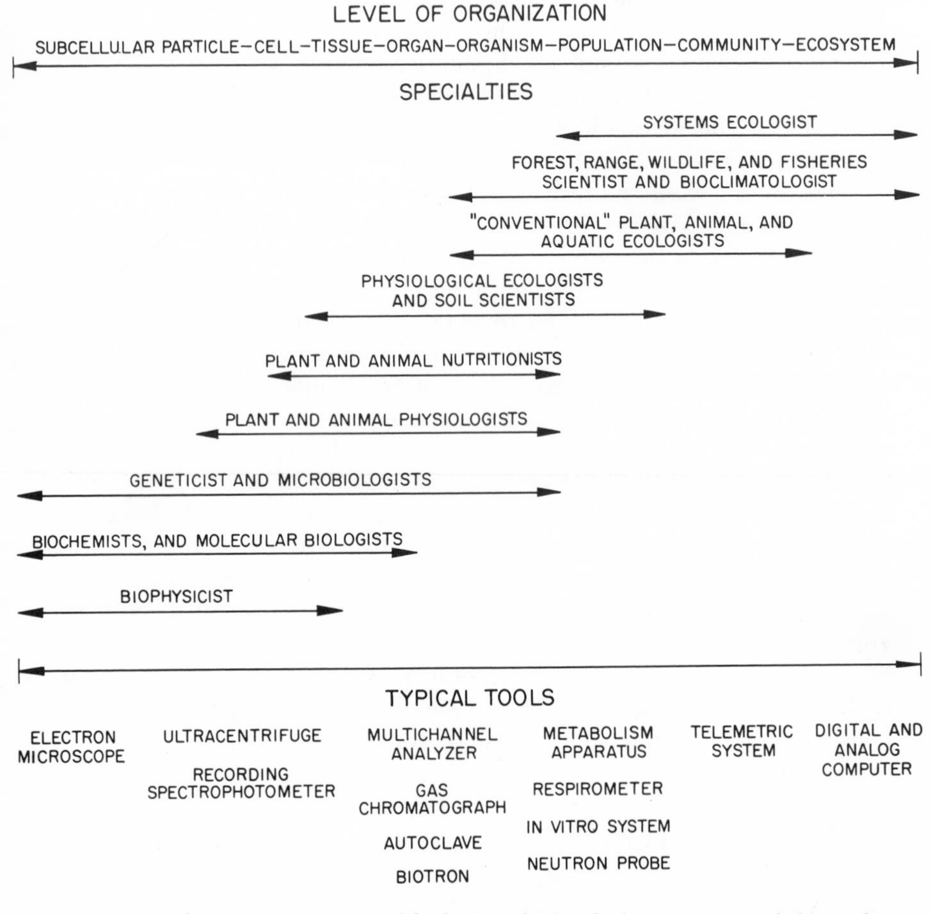

Fig. 3–6. A schematic comparison of biologists, the level of organization of the media with which they work, and the tools they use. The double-ended arrows indicate general positions of the specialties. Tools especially, and to some extent the media, overlap widely for different specialties.

ecologists are closer akin to the systems ecologists than are the conventional ecologists, because in their training and in their work they usually are more cognizant of the total ecosystem and its interrelations than are many conventional ecologists.

Consider, for example, a scientist responsible for the trout population in a Rocky Mountain forest. He realizes that the trout's well-being is inextricably related to the total environment. He must consider the impact of grazing, lumbering, mining, and road-building upon the response of the watershed to uncontrolled and fluctuating precipitation. He must consider also the inherent fertility of the watershed and its impact on populations of fish and fish foods. Superimposed upon this are other factors, such as insect control by wide-scale pesticide spraying, the problems of optimum rates and places of artificial stocking of streams, of seasons and levels of bag limits, of public relations etc., ad infinitum. In contrast, for example, few plant ecologists thoroughly appreciate aquatic problems or communicate well with aquatic ecologists; few animal ecologists understand soil problems or communicate well with soil scientists.

The systems ecologist will require better mathematical, chemical, physical, and electronic training than either the applied or the conventional ecologist. Yet he must share their holistic way of thinking or approaching problems, and he must have a broad background in ecological subject matter. A lifetime may not be sufficient for any one person to prepare adequately to perform unassisted the synthesizing function, a major effort of the systems ecologist. This function requires the cooperation of specialists, and publication in each other's journals (52). No individual will be able to direct or conduct research without consulting others to obtain a complete understanding of the processes within even most fairly simple ecosystems (80). Future leaders toward this goal must have the ability to organize concepts, things, and people.

It has long been apparent to those in physical sciences that the ecologist, in the broad sense, must be an environmental specialist. For example, Jehn (52) suggested that an ecologist must simultaneously be a meteorologist, a soils physicist, a geologist, and a geographer. But because no one man can encompass all the required specialties, he must ally himself with these specialists (61). Therefore, the greatest advances to be made in systems ecology will require the effort of an interdisciplinary team. How can this be done without losing the spontaneity and originality of the individual's personality?

Systems Ecologists: Interdisciplinarians and Multi-disciplinarians

That interdisciplinary teams are required to solve many physicobiological problems of national importance is becoming more and more apparent (43). For example, the understanding of pollution processes requires the cooperation "of [systems] ecologists, physiologists, biomathematicians, microclimatologists, geneticists, microbiologists, biochemists, chemists, morphologists, and taxonomists . . ." (108). To work effectively as a member of an interdisciplinary team, the systems ecologist will need to establish a common vocabulary, an agreed-upon ideology, a set of reasonable goals, a common context for symbols, and ways for translating ideas into action (57).

The systems ecologist is one of the types of interdisciplinary scientists who should be in great demand in the near future. It has been estimated (98)

> that about ten percent of our total national effort will be going into production, development, and research based on biophysics, biomedicine, bioengineering, and related computer projects by 1970. . . . then we must hurriedly prepare to train several thousand additional students to the Master's or Ph.D. level per year in this difficult field and we must anticipate at least a doubling of our teaching and research facilities in this interdiscipline once in each three years during this decade.

Systems ecologists can and should contribute heavily to these efforts. But the increasing importance of group effort and interdisciplinary teams in the study of major, man-created environmental problems creates new paradoxes. Large-scale, expensive research activity may decrease the flexibility and freedom which are intrinsic to research. Operation of an interdisciplinary team requires unique coordination and appreciation of contributions by different skills at many levels (23). Interdisciplinary research must be reconciled with the continuing importance of distinctve contributions by highly talented and motivated individuals. Furthermore, it is historical fact that to date many major scientific achievements have been made by specialists—by scientists wearing blinders (46).

Imagination and inventiveness, like the ability to work in an interdisciplinary team, are difficult to develop by training (47). A successful systems ecologist will be one with the imagination to perceive an important problem before others do. He must have the inventiveness to devise and weigh alternatives for its solution. This emphasizes the multidisciplinary nature of systems ecologists. In order to contribute effectively in the interdisciplinary team they must have sufficient depth in more than one specialty in order to make significant contributions to the solution of the problem. Thus, systems ecologists will need both breadth and depth of interests.

The role of the systems ecologist is complex and not well defined, nor is it easy to analyze. He may be viewed by many specialists of the interdisciplinary team as an amateur, and he may be viewed by his fellow ecologists with suspicion. Both views are justified until he proves his worth to all concerned. A major problem of the systems ecologist will be to convince an ecologist that a mathematical attack is useful and to convince the mathematician that his time and methods will be productive in ecology. Is there a natural course for this convincing to take? The interdisciplinary viewpoint does not rest solely on the biologist. A team composed of biologists with no mathematical and computer training and of engineers, mathematicians, and programmers with no biological training is doomed for failure (62).

Some Pitfalls Facing Systems Ecologists

The availability of such powerful tools and equipment as gas chromatographs, telemetric devices, and computers will not make the solution of new problems trivial, nor will it make systems ecology research routine. Some of these powerful tools themselves are raising important problems.

A special problem exists with computers. In general, the larger and faster a computer is, the more economical it is, even for small problems, if there is sufficient work available. But, of necessity, operating procedures of large computer centers are rigid in order to maintain output. The "people problems" of getting small problems into large computers grow disproportionately with computer size (98), so remote access to and time sharing on these big computers will be essential. Without direct access to the computer, such as provided by remote consoles, many problems can be completed more rapidly with a hand calculator, although they may require several hours' work. Even though they could be run in a few seconds on a computer, the long delay or "turn-around" time in using a computer without direct access leads to inefficiency. Our research output is best and most efficient when we are able to progress at full speed, regardless of time of day or day of week, rather than to take days or weeks to complete a problem. Remote-console access to the large computers which will be required for many ecological problems will allow concentrated work and will in every sense give rapid results.

As compared with his predecessors, the systems ecologist will still have to acquire empirical data by means of experimental or literature research, but he will need a better grasp of the biological and physical interactions in the system he is studying, and he will have to apply more ingenuity and invention to formulating and analyzing his problems in order to make significant advances. The easy problems have been solved.

Another pitfall facing the systems ecologist of tomorrow, who may be an undergraduate today, is that often he has been given equations and their coefficients and has been asked to produce numerical or graphic solutions. The problem he faces when he leaves the "ivy-covered halls" is first to design experiments correctly and then to conduct them effectively before he even obtains experimental results. Then his problem will be to infer and derive the form of the equations and to determine analytically the magnitude of the coefficients. Needless to say, this will be a much different and more difficult task than that which he faces as a student. Perhaps it will be desirable to develop "co-op" training programs wherein the student may intersperse practical experience in his undergraduate program. Graduate students in systems ecology, of course, will have numerous opportunities to test the effectiveness of their training, especially in their research work.

Mathematical Training of Systems Ecologists

The training of systems ecologists in mathematics and computer sciences is an especially important part of their education. Watt's review (114) of the use of mathematics in population ecology gives numerous examples of mathematical methodologies and applications. Unfortunately, many ecologists receive little mathematical, statistical, or computer training, and this is only late in graduate school. An encouraging trend is the recent development of undergraduate biomathematical courses and curricula at several universities. For example, an undergraduate biologist at Colorado State University who has had college algebra can, in 12 quarter credits, complete courses in calculus and differential equations designed for biologists. Mathematical training for attacking four types of problems has been outlined for undergraduate students in biological, management, and social sciences (24).

Consider the four combinations generated by deterministic and stochastic phenomena, each with few or with many variables. Tools required in study of organized simplicity (i.e., deterministic × few variables) include the classical analytic geometry—calculus sequence, and difference and differential equations. Disorganized simplicity (i.e., stochastic × few variables) requires probability and statistics for analysis. Organized complexity requires linear algebra and many-variable advanced calculus. Study and use of complex stochastic models are needed for analysis of phenomena characterized as disorganized complexity. Computers are especially important in these last two areas, and computing practice in numerical analysis is equally important. For those systems ecologists who wish to specialize in computers in their undergraduate or graduate training, additional courses may be recommended (25).

Recently courses have been taught to ecologists which combine many systems-oriented mathematical approaches to ecological problems with practice in the use of digital computers (117) or use of both analog and digital computers (Systems Ecology, a yearlong graduate course at the University of Tennessee, which has been taught by B. C. Patten, J. S. Olson, and the author). Systems ecologists, however, should be trained in mathematics not so much for developing their skill in performing mass computations as for having the ingenuity of escaping them (47).

Systems Ecology Research

It is considered by some that existing ecological theory often has limitations in the rapid solution of many problems (76). Also, some consider that the reliance upon analogies from physics for the solution of ecological problems has distinct limitations (102). But the field essentially is a virgin area for tomorrow's better trained and better equipped systems ecologists. The dearth of quantitative ecologists has been mentioned above, and most ecologists have been isolated and not well supported. Hopefully, we will be able to develop centers wherein "critical masses" of systems ecologists will migrate and find suitable niches.

Perhaps a routine similar to the following will be of use to the systems ecologist (7). He will often have to guess at the fundamental cause-and-effect relationships in his system and may even have to guess about the basic variables. He will then test these hypotheses by comparing the quantitative and qualitative behavior of the real world with that predicted from his model. Fortunately, he will have varied and powerful tools available for the testing of complex (and realistic) as well as simple hypotheses. But there are an infinite number of hypotheses to test about any complex system, and most of these hypotheses will be wrong. Holling (48–49) shows by example how theory and experiment can be combined in a systems analysis of predation in a way to greatly reduce the number of hypotheses to be tested. The majority of the alternatives must be excluded by means which require, not computers, but only pencils, paper, and discriminating thought. The systems ecologist will have to develop the knack of feeding on negative information and use these negative hypotheses to guide his further experimentation and theory. In instances where experimentation or measurement of parameters is impossible without disruption of the system, perturbation of the system followed by measurements may still give new insight for the definition of a model (8).

A major hurdle to overcome in systems ecology research is lack of precedent in funding detailed and integrated research on complex ecological systems. This applies both to the theoretical and analytical phase and the experimental phase. In many respects no single organization has been working in depth on complex systems ecology problems, for such work may be beyond the role, objectives, or structure of existing organizations. Universities, national laboratories, and state and federal experiment stations each have some unique resources and capabilities for studying ecosystem problems. Some advantages and disadvantages of these three types of organizations with respect to "total systems" research, based on my experience in working in these organizations, are briefly outlined as follows.

In the past, ecological research conducted at universities generally has involved one investigator, or at most a few, on a part-time basis on problems of limited extent. Extensive and intensive interdepartmental cooperation in ecological research has been the exception rather than the rule. Many sources of funds are available to these researchers, but usually in amounts insufficient to attract permanent personnel and to support long-term ecosystem research. Although the economy of graduate student use has been exploited, often there is a lack of continuity in the conduct of long-term environmental researches. By the time the student gains competence and becomes capable of independent contributions to the project, he graduates and is lost to the project. Also, universities often lack controllable research areas or have conflicting needs for them.

Applied ecologists, in state fish and game departments or in state and federal agricultural experiment stations, for instance, often have controllable research areas on which to conduct long-term ecosystem research, but continuity again is impeded because of high turnover rate of personnel. Furthermore, their research funds often are restricted to only one or a few phases of the total ecosystem problem, and their funds have become more limited in recent years. Funds from many granting agencies may not be available to them for research, special training, or foreign study.

In the past ten years considerable ecological research has been conducted at several national laboratories. Ecologists in these laboratories have available many services, tools, and consultants which the university or experiment station scientists lack. Much of the ecological work in the national laboratories, however, has been concerned with specific needs of the funding agency. Most ecological researchers in these laboratories have come directly from liberal arts departments in universities, and these laboratory staffs are divided into subject matter groups for conduct of research. A total-system, interdisciplinary approach usually has not been implemented in their research. Although these ecologists are funded comparatively well, their costs are high due to the nature of their work. Although they may have long-term control over their research areas, the number of these laboratories is limited, and some important biomes, such as grasslands, are not within the boundaries of these laboratories.

I feel that perhaps research in systems ecology could encompass the advantages held by ecological researchers of the above three categories. This would require, however, some shifts concerning funding and conduct of research, and some shifts in administrative policy of the respective agencies or institutions. The exact nature and organization of such research is uncertain, although Dubos (21) has raised some interesting questions and has made some good suggestions about similar research in environmental biology.

The long-term impact on man of fundamental, total-ecosystem research should be recognized, and the framework should be developed for extensive and intensive intercooperation of these three groups of ecologists. Analytical and experimental research on total-ecosystem complexes should be initiated as soon as possible, if man is to benefit tomorrow, because most problems of environmental magnitude require many years of study before conclusions may be reached.

The proposed International Biological Program is, in several respects, a call for systems ecology research, both experimental and theoretical. This program could provide an incentive and a means for ecologists from universities, experiment stations, and national laboratories to work cooperatively and share funds, research areas, and talent. An example follows. Other such examples of needed research on total-system problems can be found in other parts of this continent and on other continents, in both terrestrial and aquatic ecosystems.

Consider the seminatural grassland ecosystems in the Great Plains. A more complete understanding of the structure and function of these ecosystems becomes increasingly essential as these lands are called upon to provide food, water, and recreation for tomorrow's growing populations. Several state and federal agricultural experiment stations hold sizeable acreages of representative variations of grasslands, from the aspen parkland in Canada to the semidesert grasslands of Mexico. But at none of these stations is there a team equipped with suitable manpower or funds for intensive total-ecosystem research. Scientists at these experiment stations and nearby universities have accumulated considerable data and experience on and about these grassland ecosystems. There is no national laboratory in this vast area, and no university in the area can marshall many of the unique facilities, such as computing facilities and computer consultants, that are necessary for systems ecology research. Still, scientists at universities in the area can provide much necessary insight into these ecosystems and graduate students can help conduct ecosystem research in the area. With sufficient funds and planning it should be possible to combine the special skills and resources of all these groups and bring them to bear on problems of ultimate national importance at selected locations in the Great Plains.

LITERATURE CITED

1. Arcus, P. L. 1963. An introduction to the use of simulation in the study of grazing management problems. *Proc. New Zealand Soc. Animal Prod.* 23:159–168.
2. Armstrong, N. E. and H. T. Odum. 1964. Photoelectric ecosystem. *Science* 143: 256–258.
3. Bakuzis, E. V. 1959. Structural organization of forest ecosystems. *Proc. Minn. Acad. Sci.* 27:97–103.

4. Barea, D. J. 1963. Analisis de ecosistemas en biologia, mediante programacion lineal. *Archivos de Zootecnia* 12:252–263.
5. Barker, R. 1964. Use of linear programming in making farm management decisions. *New York Agr. Exp. Sta. Bull.* 993. 42 pp.
6. Bartlett, M. S. 1960. *Stochastic population models in ecology and epidemiology.* Methuen and Co., Ltd., London. 90 pp.
7. Bellman, R. 1961. *Mathematical experimentation and biological research.* Rand Corp. P-2300. 12 pp.
8. Berman, M. 1963. A postulate to aid in model building. *J. Theoret. Biol.* 4:229–236.
9. Berman, M. 1963. The formulation and testing of models. *Ann. New York Acad. Sci.* 108:182–194.
10. Blair, W. F. 1964. The case for ecology. *BioScience* 14:17–19.
11. Briegleb, P. A. 1965. The forester in a science-oriented society. *J. Forestry* 63: 421–423.
12. Broido, A., R. J. McConnen, and W. G. O'Regan. 1965. Some operations research applications in the conservation of wildland resources. *Manage. Sci.* 11:802–814.
13. Caldwell, L. K. 1963. Environment: a new focus for public policy? *Public Administration Review* 23:132–139.
14. Churchill, E. D. and H. C. Hanson. 1958. The concept of climax in arctic and alpine vegetation. *Bot. Rev.* 24:127–191.
15. Clements, F. E. 1916. Plant succession: an analysis of the development of vegetation. *Carnegie Inst. Wash. Publ.* 242. 512 pp.
17. Coombs. C. H., H. Raiffa, and R. M. Thrall. 1954. Some views on mathematical models and measurement theory. *In*: Thrall, R. M., C. H. Coombs, and R. L. Davis (eds.). *Decision Processes.* John Wiley & Sons, Inc. New York. 332 pp.
18. Dansereau, P. 1964. The future of ecology. *BioScience* 14:20–23.
19. Davis, C. C. 1963. On questions of production and productivity in ecology. *Arch. Hydrobiol.* 59:145–161.
20. Deevey, E. S. 1964. General and historical ecology. *BioScience* 14:33–35.
21. Dubos, R. 1964. Environmental biology. *BioScience* 14:11–14.
23. Duckworth, W. E. 1962. *A guide to operational research.* Methuen and Co., Ltd. London, England. 145 pp.
24. Duren, W. L., Jr. (Chr.). 1964. Tentative recommendations for the undergraduate mathematics program of students in the biological, management and social sciences. Mathematical Association of America, Committee on the Undergraduate Program in Mathematics, 32 pp.
25. Duren, W. L. (Chr.). 1964. Recommendations on the undergraduate mathematics program for work in computing. Mathematical Association of America, Committee on the Undergraduate Program in Mathematics. 29 pp.
27. Eberhardt, L. L. 1963. Problems in ecological sampling. *Northwest Sci.* 37:144–154.
28. Eberhardt, L. L. 1965. Notes on ecological aspects of the aftermath of nuclear attack. pp. 13–25 *in:* Hollister, H. and L. L. Eberhardt. *Problems in estimating the biological consequences of nuclear war.* U.S. Atomic Energy Commission TAB-R-5.
29. Eberhardt, L. L. 1965. Notes on the analysis of natural systems. pp. 27–40 *in:* Hollister, H. and L. L. Eberhardt. *Problems in estimating the biological consequences of nuclear war.* U.S. Atomic Energy Commisison TAB-R-5.
30. Elveback, L., J. P. Fox, and A. Varma. 1964. An extension of the Reed-Frost epidemic model for the study of competition between viral agents in the presence of interference. *Amer. J. Hygiene* 80:356–364.

32. Feibleman, J. K. 1960. Testing hypothesis by experiment. *Persp. Biol. and Med.* 4:91–122.
34. Fosberg, F. R. 1965. The entropy concept in ecology. pp. 157–163 *in: Symposium on Ecological Research in Humid Tropics Vegetation,* Kuching, Sarawak, July 1963.
35. Garfinkel, D., R. H. MacArthur, and R. Sack. 1964. Computer simulation and analysis of simple ecological systems. *Ann. New York Acad. Sci.* 115:943–951.
36. Garfinkel, D. and R. Sack. 1964. Digital computer simulation of an ecological system, based on a modified mass action law. *Ecology* 45:502–507.
37. George, F. H. 1965. *Cybernetics and biology.* Freeman and Co., San Francisco, California. 138 pp.
38. George, J. L. 1964. Ecological considerations in chemical control: Implications to vertebrate wildlife. *Bull. Entomol. Soc. Amer.* 10:78–83.
39. Gibbens, R. P. and H. F. Heady. 1964. The influence of modern man on the vegetation of Yosemite Valley. *Calif. Agr. Exp. Sta. Manual* 36. 44 pp.
40. Glass, B. 1964. The critical state of the critical review article. *Quart. Rev. Biol.* 39:182–185.
41. Golley, F. B. 1965. Structure and function of an old-field broomsedge community. *Ecol. Monogr.* 35:113–137.
42. Gould, E. M. and W. G. O'Regan. 1965. Simulation, a step toward better forest planning. *Harvard Forest Paper 13.* 86 pp.
43. Gross, P. M. (Chr.). 1962. Report of the committee on environmental health problems. *Public Health Service Publ. 908.* 288 pp.
44. Harris, J. E. 1960. A review of the symposium: models and analogues in biology. *Symp. Soc. Exp. Biol.* 14:250–255.
45. Hasler, A. D. 1964. Experimental limnology. *BioScience* 14:36–38.
46. Hitch, C. 1955. An appreciation of systems analysis. Rand Corp. P-699. (*Symposium on Problems and Methods in Military Operations Research*, pp. 466–481.)
47. Hoag, M. W. 1956. *An introduction to systems analysis.* Rand Corp. RM-1678. 21 pp.
48. Holling, C. S. 1963. An experimental component analysis of population processes. *Mem. Entomol. Soc.* Canada. 32:22–32.
49. Holling, C. S. 1965. The functional response of predators to prey density and its role in mimicry and population regulation. *Mem. Entomol. Soc.* Canada. 45:3–60.
50. Hollister, H. 1965. Problems in estimating the biological consequences of nuclear war. pp. 1–11 *in:* Hollister, H. and L. L. Eberhardt. (same title). U.S. Atomic Energy Commission TAB-R-5.
51. Hughes, R. D. and D. Walker. 1965. Education and training in ecology. *Vestes* 8:173–178.
52. Jehn, K. H. 1950. The plant and animal environment: a frontier. *Ecology* 31: 657–658.
53. Jenkins, K. B. and A. N. Halter. 1963. A multi-stage stochastic replacement decision model. *Ore. Agr. Exp. Sta. Tech. Bull.* 67. 31 pp.
56. Jones, R. W. 1963. System theory and physiological processes: An engineer looks at physiology. *Science* 140:461–464.
57. Kennedy, J. L. 1956. A display technique for planning. Rand Corp. P-965.
58. Kramer, P. J. 1964. Strengthening the biological foundations of resource management. *Trans. N. Amer. Wildlife and Natural Resources Conf.* 29:58–68.
59. Larkin, P. A. and A. S. Hourston. 1964. A model for simulation of the population biology of Pacific salmon. *J. Fish Res. Bd.* Canada 21:1245–1265.
60. Leak, W. B. 1964. Estimating maximum allowable timber yields by linear programming. *U.S. For. Serv. Res. Paper NE-17.* 9 pp.

61. Lebrun, J. 1964. Natural balances and scientific research. *Impact of Sci. on Society* 14:19–37.
62. Ledley, R. S. 1965. *Use of computers in biology and medicine.* McGraw-Hill Book Co., Inc. New York. 965 pp.
63. Leigh, E. G. 1965. On the relation between the productivity, biomass, diversity, and stability of a community. *Proc. Nat. Acad. Sci.* 53:777–783.
64. Leopold, A. S., S. A. Cain, C. H. Cottam, I. N. Gabrielson, and T. L. Kimball. 1963. Wildlife management in the national parks. *Amer. Forests* 69:32–35, 61–63.
65. Leslie, P. H. 1958. A stochastic model for studying the properties of certain biological systems by numerical methods. *Biometrika* 45:16–31.
67. Lloyd, M. 1962. Probability and stochastic processes in ecology. *In:* H. L. Lucas (ed.) *The Cullowhee Conf. on Training in Biomath.* Institute of Statistics, N. Car. St. U., Raleigh, N.C.
69. Loucks, D. P. 1964. The development of an optimal program for sustained-yield management. *J. Forestry* 62:485–490.
70. Lucas, H. L. 1960. Theory and mathematics in grassland problems. *Proc. Intern. Grassland Cong.* 8:732–736.
71. Lucas, H. L. 1964. Stochastic elements in biological models; their sources and significance. pp. 355–383 *in:* Gurland, J. (ed.). *Stochastic models in medicine and biology.* U. Wisc. Press, Madison, Wisc. 393 pp.
74. Margalef, R. 1957. Information theory in ecology. *Mem. Real Acad. Ciencias y Artes de Barcelona* 23:373–449.
75. Margalef, R. 1962. Modelos fiscos simplificados de poblaciones de organismos. *Mem. Real Acad. Ciencias y Artes de Barcelona* 24:83–146.
76. Margalef, R. 1963. On certain unifying principles in ecology. *The Amer. Nat.* 97:357–374.
77. Miller, I. and J. E. Freund. 1965. *Probability and statistics for engineers.* Prentice-Hall Inc., Englewood Cliffs, N.J. 432 pp.
78. Miller, R. S. 1965. Summary report of the ecology study committee with recommendations for the future of ecology and the Ecological Society of America. *Bull. Ecol. Soc. Amer.* 46:61–82.
79. Neyman, J. and E. L. Scott. 1959. Stochastic models of population dynamics. *Science* 130:303–308.
80. O'Connor, F. B. 1964. Energy flow and population metabolism. *Science Prog.* 52:406–414.
82. Odum, E. P. 1963. *Ecology.* Holt, Rinehart, and Winston. New York. 152 pp.
85. Odum, H. T. 1965. An electrical network model of the rain forest ecological system. U.S. Atomic Energy Commission PRNC 67.
86. Odum, H. T. and R. C. Pinkerton. 1955. Time's speed regulator: the optimum efficiency for maximum output in physical and biological systems. *Amer. Sci.* 43:331–343.
87. Olson, J. S. 1965. Equations for cesium transfer in a *Liriodendron* forest. Health *Physics* 11:1385–1392.
88. Ostle, B. 1963. *Statistics in research.* Iowa St. Univ. Press. 2nd Ed. 585 pp.
90. Patten, B. C. 1964. *The systems approach in radiation ecology.* Oak Ridge National Laboratory Technical Memorandum 1008. 19 pp.
92. Platt, R. B., W. D. Billings, D. M. Gates, C. E. Olmsted, R. E. Shanks, and J. R. Tester. 1964. The importance of environment to life. *BioScience* 14:25–29.
93. Quade, E. S. (ed.). 1964. *Analysis for military decisions.* Rand Corp. R-387. Rand McNally & Co., Chicago.
94. Rasmussen, D. I. 1941. Biotic communities of Kaibab Plateau, Arizona. *Ecol. Monogr.* 11:229–275.
95. Royce, W. F., D. E. Bevan, J. A. Crutchfield, G. J. Paulik, and R. L. Fletcher.

1963. Salmon gear limitation in northern Washington waters. *U. Wash. Publ. in Fisheries* (N.S.) 2:1–123.

96. Sarnoff, D. 1964. The promise and challenge of the computer. *Amer. Fed. Infor. Process. Soc. Conf. Proc.* 26:3–10.
97. Schaller, W. N. and G. W. Dean, 1965. Predicting regional crop production: an application of recursive programing. *USDA Tech. Bull.* 1329. 95 pp.
98. Schmitt, O. H. and C. A. Caceres (eds.). 1964. *Electronic and computer-assisted studies of bio-medical problems.* C. C. Thomas. Springfield, Illinois. 314 pp.
99. Schultz, A. M. 1961. Introduction to range management. U. California. Ditto notes. 116 pp.
100. Schultz, A. M. 1965. The ecosystem as a conceptual tool in the management of natural resources. *In:* Parsons, J. J. (ed.). Symposium on quality and quantity in natural resource management (in press), manuscript 33 pp.
101. Sears, P. B. 1963. The validity of ecological models. pp. 35–42. *In: XVI* International Congress of Zoology. Vol. 7. Science and Man Symposium–Nature, Man and Pesticides.
102. Slobodkin, L. B. 1965. On the present incompleteness of mathematical ecology. *Amer. Scientist* 53:347–357.
103. Specht, R. D. 1964. Systems analysis for the postattack environment: some reflections and suggestions. Rand Corp. RM-4030. 34 pp.
104. Stone, E. C. 1965. Preserving vegetation in parks and wilderness. *Science* 150: 1261–1267.

106 Tansley, A. G. 1935. The use and abuse of vegetational concepts and terms. *Ecology* 16:284–307.

107. Thrall, R. M. 1964. Notes on mathematical models. U. Mich. Engin. Summer Conf.—Foundations and Tools for Operations Research and Management Sciences. Multilith 97 pp.
108. Tukey, J. W. *et al.* (Environmental Pollution Panel, President's Science Advisory Committee). 1965. Restoring the quality of our environment. Superintendent of Documents, U.S. Govt. Printing Office, Washington, D.C. 317 pp.
109. Turner, F. B. 1965. Uptake of fallout radionuclides by mammals and a stochastic simulation of the process. pp. 800–820 *in:* Klement, A. W. (ed.). Radioactive fallout from nuclear weapons test. *U.S. Atomic Energy Commission Symp. Ser.* 5. 953 pp.
110. van de Panne, C. and W. Popp. 1963. Minimum-cost cattle feed under probabilistic protein constraints. *Manage. Sci.* 9:405–430.
111. Van Dyne, G. M. 1965. Probabilistic estimates of range forage intake. *Proc. West. Sect. Amer. Soc. Animal Sci.* 16(LXXVII):1–6.
112. Van Dyne, G. M. 1965. Application of some operations research techniques to food chain analysis problems. *Health Physics* 11:1511–1519.
113. Walker, O. L., E. O. Heady, and J. T. Pesek. 1964. Application of game theoretic models to agricultural decision making. *Agronomy J.* 56:170–173.
114. Watt, K. E. F. 1962. Use of mathematics in population ecology. *Ann. Rev. Entom.* 7:243–260.
115. Watt, K. E. F. 1964. The use of mathematics and computers to determine optimal strategy and tactics for a given insect pest control problem. *Canad. Entomol.* 96:202–220.
117. Watt, K. E. F. 1965. An experimental graduate training program in biomathematics. *BioScience* 15:777–780.
119. Whittaker, R. H. 1953. A consideration of climax theory: the climax as a population and pattern. *Ecol. Monogr.* 23:41–78.
120. Yamber, P. A. 1964. Is there a niche for the generalist? *Trans. N. Amer. Wildlife and Natural Resources Conf.* 29:352–372.

II
Problems related to ecosystem energy flow

MANAGEMENT OF WILDLIFE POPULATIONS

4

Adaptability of animals to habitat change

A. Starker Leopold

All organisms possess in some measure the ability to adapt or adjust to changing environmental conditions. But the degree to which different species are capable of adjusting varies enormously. This chapter concerns the nature and extent of adaptability and demonstrates the truism that in a world undergoing constant and massive modification by man, the animals with the highest capacity for adjustment are those that persist in abundance. Specialized animals with narrow limits of adjustment are those that have become scarce or in some instances extinct.

RELATION OF UNGULATE POPULATIONS TO PLANT SUCCESSIONS

In any given ecosystem, there are animals that thrive best in the climax stages of plant succession and others that do better when the climax has been destroyed in some way and the vegetation is undergoing seral or subclimax stages of succession, working back toward restoration of the climax. This can be interpreted to mean that the climax animals are more specialized in their environmental needs, while the seral or successional species are more adaptable and able to take advantage of transitory and unstable situations. The principle can be well illustrated by considering the status of various native ungulates in North America.

The mass of data accumulated in studies of North American deer permits us to draw some general deductions about population dynamics in these animals, particularly in relation to food supplies. The following remarks apply equally to the white-tailed deer (*Odocoileus virginianus*) and the mule deer or blacktail (*O. hemionus*). The quality and quantity of forage available to a deer population during the most critical season of the year has proven repeatedly to be the basic regulator of population level. Usually this means winter forage, but not always. In desert areas or regions of Mediterranean climate, like coastal California, summer may be the critical season. In any event, the nutritive intake of the individual deer during the critical season determines both productivity of the herd (Cheatum and Severinghaus, 1950; Taber, 1953) and mortality in the herd, whether death be caused by starvation, disease, parasites, or even to some extent by predation or accidents (Longhurst et al., 1952, and others). Average population level is a dynamic function of these two opposing variables—rate of productivity and rate of mortality. Hunting is a source of mortality artificially interposed in the formula, and although it is intercompensatory with other forms of loss (that is, hunting kill will reduce starvation losses, etc.) it is not regulated by nutrition, but by legislative fiat. However, since hunting is generally controlled in North America to remove no more than annual increment, and usually less, it cannot be construed as a primary determinant of population level in most areas. Putting all this in much simpler form, good forage ranges generally have many deer; poor ranges have few. All other influences are secondary.

Good deer ranges characteristically include stands of nutritious and palatable browse which as a rule are produced in secondary stages of plant succession (Leopold, 1950). Burned or cutover forest lands support most of the deer in the continent; some brush-invaded former grasslands are of local importance. In a few special cases, as for example that of the burro deer (*O. h. eremicus*) on the desert, sparse populations live in climax floras. But on the whole the association between deer and secondary brushlands (the connecting link being nutritional) is so general as to permit classification of deer as seral or successional species.

Assuming that range relationships are equally dominant in determining populations of other North American ungulates, and much evidence indicates that this is so, a general characterization can be made of each species, permitting classification along lines of range affinities, as had been done in Table 4–1.

This rather subjective classification requires some explanation.

Climax species

The northern caribou is a classic example of an animal that depends heavily in winter on undisturbed climax vegetation of the subarctic zone.

TABLE 4–1. General association of North American ungulates with climax or subclimax successional stages

	Biotic zone		
	Boreal	*Temperate*	*Tropical*
1. Associated primarily with climax forage types:			
Caribou (*Rangifer arcticus*)	x		
Bighorn (*Ovis canadensis* and allied species)	x		
Mountain goat (*Oreamnos americanus*)	x		
Musk ox (*Ovibos moschatus*)	x		
Bison (*Bison bison*)		x	
Collared peccary (*Pecari tajacu*)		x	x
White-lipped peccary (*Tayassu pecari*)			x
Tapir (*Tapirella bairdii*)			x
Brocket (*Mazama americana*)			x
2. Associated primarily with subclimax forage types:			
Moose (*Alces americana*)	x	x	
Elk (*Cervus canadensis*)		x	
White-tailed deer (*Odocoileus virginianus*)		x	
Mule deer (*Odocoileus hemionus*)		x	
Pronghorn antelope (*Antilocapra americana*)		x	

The lichens which supply much of the caribou's winter food grow either as an understory to the spruce forest or suspended from the spruce limbs. Any disturbance such as fire or grazing that depletes this particular vegetative complex lowers the carrying capacity for caribou.

Similarly bighorn sheep and mountain goats in their alpine retreats, bison on the great prairies, and musk ox on the arctic plains are adapted to feed on climax species of forbs, sedges, grasses, and a few shrubs.

In the southern reaches of the continent the two species of peccaries are generalized in their food habits, like other pigs, but the mast of oak and of many tropical fruit trees contributes heavily to their diet. Besides mast, the bulbous roots, palmettos, cacti, forbs, and grasses on which these pigs feed are on the whole characteristic of climax associations. The tapir and brocket are even more typical of climax rain forest (Leopold, 1959).

It is notable that of nine species of North American ungulates associated with climax vegetation, four are of boreal or arctic affinities, four are tropical, and only the bison and the collared peccary in part of its range occur in temperate latitudes.

Subclimax species

Nearly all of the ungulates that thrive best on kinds of weeds and brush that characterize disturbed vegetative situations are native in the temperate zone. This includes the two common deer, elk, and the moose, which extends northward through the boreal zones as well. The pronghorn antelope is predominantly a weedeater (Buechner, 1950), although it may consume much sage and other browse at times. On the Great Plains, the weeds and forbs that supported antelope originally may have resulted from local overgrazing by the native bison. On the deserts of Mexico, however, the antelope almost certainly depended on climax vegetation, but this is the fringe of its continental range.

Whereas most boreal and tropical ungulates have climax affinities, the temperate-zone species thrive largely on successional vegetation. In an evolutionary sense this would suggest that these adaptive species, all highly successful today, developed in an environment subject to frequent disturbance, presumably fire. Even the bison, here classed as a climax species, would fall in this category if one accepts the prairie as a subclimax, maintained by fire (Sauer, 1950).

Recognition of successional affinities of big-game species is basic to determination of sound management policy. The subclimax species (two deer, antelope, elk, moose) fit nicely into multiple-use land programs, including logging, grazing, and controlled burning. The climax species do not. Preservation of wilderness areas, without competing or disturbing uses, is particularly important in sustaining remnants of the climax forms designated in Table 4–1.

PLANT SUCCESSIONS AND OTHER WILDLIFE

The principle illustrated above with ungulates applies generally to wild animals. In areas heavily modified by human action, the abundant species are those adapted to take advantage of disturbed ecologic situations. Over much of the United States, the upland game species that supply most of today's recreational hunting are the bobwhite quail, cottontail rabbit, ruffed grouse, mourning dove, and the introduced ring-necked pheasant—all typical subclimax or successional species.

Game species once abundant on the continent, but now localized and scarce because of shrinkage of particular climax vegetational types on which they depended, are the prairie chicken, sharp-tailed grouse, sage hen, upland plover, and wild turkey. Extinct are the heath hen and passenger pigeon.

The case of the passenger pigeon illustrates particularly well the dilemma facing an unadaptive species. The fabled legions that "darkened the sky" were supported in large part by mast crops produced in climax stands of mature timber, especially oak, beech, and chestnut. The flocks were highly mobile and searched the eastern half of the continent for favorable feeding grounds. When a good food supply was found, millions of birds would congregate to establish one of the massive colonial nestings so well documented by Schorger (1955). With the settlement of the country, two things happened concurrently that contributed to the swift collapse of the pigeon population: uncounted millions of the birds were slaughtered in the nesting areas, and the mature timber stands that produced the mast crops were felled to make way for farms. The demise of the pigeon is traditionally blamed on the market hunters; but had there been no hunting, it is doubtful that the pigeon would have survived the depletion of its food supply. So specialized was this bird that it seemingly had no capacity to adjust to the modest, scattered food source that certainly continued after the main hardwood forests were felled. When the big pigeon flocks were reduced, the survivors simply perished without a single pair exhibiting the ability to feed and reproduce under changed conditions. Its close relative, the mourning dove, on the other hand, adjusted very well indeed to the conversion from forest to farm, and today is undoubtedly much more numerous than in primitive times, despite heavy and persistent shooting.

THE NATURE OF ADAPTATION

Precisely what is this character of "adaptability" that some animals have and others do not? What are the mechanisms by which animals adapt?

The paleontological record tells us that over the eons of time there have been enormous changes in climate and hence in habitat. With each major shift many animal species became extinct; these presumably were the unadaptable ones. Other animals persisted but evolved and were modified to meet the new conditions. One component of adaptability, therefore, is the capacity for genetic change.

At the same time, current experience offers many examples of individual animals learning new tricks of survival that contribute to longevity and hence to persistence of the population. A coyote can learn to be trap-shy; a raccoon learns to search for eggs in wood-duck boxes; mallards learn the precise hour when legal shooting ceases, which signals the exodus from a refuge to go in search of food. Some species are quick to pick up new behaviorisms; others are not. Adaptability, therefore, may include the capacity to learn.

Genetic and learned adaptations will be discussed in that order.

MORPHOLOGIC ADAPTATION

One manifestation of genetic adaptation to local environment is the demonstrable evidence of subspeciation in animals. Many widely distributed species are segmented into local populations that show marked and persistent differences in morphology. Some of the characters that vary and are easily observed and measured are body size, proportion of body parts, and color of plumage or pelage. The bobwhite quail (*Colinus virginianus*) is an example of a resident (nonmigratory) game bird that varies greatly from place to place. This bird occurs throughout the eastern half of North America, from New England and South Dakota south to Chiapas in southern Mexico. Within this range, twenty-one well-differentiated races or subspecies are recognized (Aldrich, 1955). In size, the bobwhite decreases from over 200 grams in the north to slightly over 100 grams in Chiapas. Likewise there is a general north-south gradient in plumage color, the palest birds occurring in the open or arid ranges, such as the Great Plains, the darkest forms being found in the wet tropics or subtropics of southern Florida, the coast of Veracruz, and the interior valleys of Chiapas. It is presumed that each population is particularly adapted to the local habitat in which it exists. The capacity to be molded genetically by local environment doubtless underlies the bobwhite's success in occupying such an extensive range in North America.

Commenting on this general question of genetic plasticity, Grant (1963, p. 434) states:

> The great role of natural selection in the formation of races [sub-species] can be inferred from the observation that racial characteristics are often adaptive. The adaptiveness of the racial characters in many plants and animals is demonstrated by two sets of correlations. First, the different races of a species have morphological and physiological characters that are related to the distinctive features of the environment in their respective areas. Second, the same general patterns of racial variation frequently recur in a parallel form in separate species inhabiting the same range of environment.

He goes on to comment on some of the generally accepted "rules" of morphologic adaptation that have been summarized and analyzed by many other authors, including Mayr (1942, p. 90). These are,

1. *Bergmann's rule:* The smaller races of a species are found in the warmer parts of a species range, the larger races in cooler parts.
2. *Allen's rule:* Protruding body parts, such as ears, tails, bills, and other extremities, are relatively shorter in the cooler parts of the range of a species than in the warmer parts.
3. *Golger's rule:* Dark pigments (eumelanins) increase in the warm and humid parts of a range, paler phaeomelanins prevail in arid climates.

The bobwhite illustrates all of these rules of local adaptation. The same may be said of white-tailed and mule deer, the raccoon, the bobcat, hares of the genus *Lepus*, cottontails of the genus *Sylvilagus*, and many other widely distributed birds and mammals. In the case of the white-tailed deer, the size gradient is extreme: in Wisconsin an adult buck weighs well over 200 pounds, in parts of Mexico scarcely seventy pounds.

The larger size and smaller ears of northern animals presumably give an advantageous ratio of body mass to exposed surface, for heat conservation. The opposite is true in warmer climates.

PHYSIOLOGIC ADAPTATION

More difficult to measure, but perhaps even more important in fitting local populations to their environments, are the physiologic adaptations. To be successful, a population must breed at the right time of year, produce only as many young as can be cared for, be able to digest and assimilate the foods locally available, and otherwise adjust its life processes to the local scene. Migratory birds lay on fat (fuel) for their travels and require elaborate navigational machinery. Research to date has scarcely scratched the surface of this enormously complicated area of animal adaptation.

A species that has been studied in some detail and that well illustrates several facets of physiologic adaptation is the common white-crowned sparrow of the Pacific coast (*Zonotrichia leucophrys*). There are two races of this bird, very similar in appearance, that winter together in central California; but in spring one race migrates to British Columbia to breed while the other breeds locally, on the winter range. Blanchard (1941) showed a number of differences in the life cycles of these two populations. Though living together all winter, the migrants laid on fat in spring and departed for the north; the residents did not accumulate fat but went leisurely about the business of nesting. The migrants, having less time on the breeding grounds, compressed the reproductive cycle into approximately two thirds of the time used by resident birds. Subsequent investigation by a number of workers has demonstrated that the mechanism triggering these events is changing length of day in spring, but the important differences in response reflect inherent, physiologic adaptations peculiar to the two populations.

Differences in timing of breeding are demonstrable in many other species. Black-tailed deer along the California coast fawn in May, mule deer in the Sierra Nevada in July, whitetails in northwestern Mexico in August. In each case fawning corresponds to the period of optimum plant growth—spring in California, summer rains in Mexico. Time of mating (seven months before fawning) is presumably timed by changing day length—in this case by shortening days, since the breeding occurs in fall or early winter. Ian McTaggart Cowan has kept a number of races of blacktailed deer in pens in Vancouver, and notes that the southern Alaskan and British Columbian stocks breed at almost the same time, whereas the Californian stocks have retained a response that induces antler growth, shedding of velvet, breeding, antler drop, and pelage molt a month or more in advance of the northern races kept in the same pens.

The number of young produced by a breeding population is regulated by physiologic controls. Lack (1954) presents examples of clutch size in birds varying apparently with food availability. He cites the work of Swanberg on the thick-billed nutcrackers, in which it was shown that in years when the autumn crop of nuts was below average, the birds laid only three eggs. In years of good or excellent nut crops, clutches of four eggs were normal. When the experimenter supplied nuts in winter for certain wild nutcrackers, those particular individuals had clutches of four eggs, even in years of poor mast crop. The change in number of eggs was therefore apparently a physiologic adjustment to the amount of food available. But in all cases the birds laid no fewer than three eggs, nor more than five, the limits presumably set by hereditary factors.

Clutch or litter size likewise may be a function of predator populations and the likelihood of losses of eggs or young due to predation. The mallard of continental North America lays eight to twelve eggs and predictable losses are high. The closely related Laysan duck, on isolated Laysan Island where there are no predators, lays only three to four eggs.

Certain deep-seated physiologic differences have been detected between wild and domestic turkeys which shed some light on how the wild birds are adapted to live successfully in the woods (Leopold, 1944). In the Missouri Ozarks the native turkey persists even under highly adverse circumstances and populations respond readily to protection and management. Domestic turkeys cannot exist away from farmyards. Hybrids between the two barely hold their own in refuges under intensive management. Differential reproductive success seemed to underlie the disparities in population behavior. Time of breeding is earlier in domestic and hybrid turkeys, leading to loss of eggs and chicks in late-spring storms. Behavioral differences between hens and chicks suggested other reasons why wild birds raised more young. These differential reactions were related to size of brain and relative development of some of the endocrine organs that control behavior, suggesting a few of the components that may be involved in "local adaptations."

DANGER OF TRANSPLANTING LOCAL RACES

If indeed some kinds of animals are delicately attuned to life in specific local environments, one may question the advisability of trapping and shifting these populations about in an effort to restock underpopulated ranges.

During the era 1920–40 there was a very large trade in Mexican bobwhites, imported into various midwestern states for release to augment local populations. Actual measurements of the results of this endeavor are lacking, but there seemed to be a consensus among observers that such releases never led to sustained increase in bobwhite numbers, and in fact some thought that in years following a liberation, local populations were depressed. This may well have been the case, since birds from the tropical coast of Tamaulipas (the main source of stock), and their progeny if crossed with northern birds, would not likely have been winter-hardy. In any event this program was abandoned, attesting to its failure.

Dahlbeck (1951) reported a similar failure when gray partridges from southern Europe were imported to Sweden and mixed with the hardy northern populations. A catastrophic drop in number followed. He also relates a case of shipping in Carpathian red deer stags to "improve" the stock on an island off the Scandinavian coast. The resulting hybrids apparently were unable to stand the rigors of the northern climate, and the population on the island fell to near extinction. Following these experiences Sweden adopted regulations to prohibit import of game birds and mammals from outside the country.

INDIVIDUAL ADAPTABILITY

Certain adaptive responses to a changing environment appear to be nongenetic. Some animals seem capable of internal physiological and behavioral adjustments and as a consequence can tolerate wide fluctuations in weather and other environmental factors. A classic example would be the mourning dove (*Zenaidura macroura*).

There is no more widely distributed or successful game bird in the North American continent than the dove. Its breeding range extends from the Atlantic to the Pacific and from the prairie provinces of southern Canada to Oaxaca in southern Mexico. Two weakly differentiated subspecies are recognized—an eastern and a western race. But each of these races successfully occupies a great variety of habitats. The western mourning dove, for example, breeds in the pine zone of the mountains, in the bleakest southwestern deserts, and along the tropical Mexican coast with equal success. If there are local physiologic adaptations, no one has detected them. In our present state of knowledge we must assume that the individual birds are capable of this range of adjustment.

The same can be said for some migratory birds like the mallard, which breeds from the arctic tundra to northern Baja California and from coast to coast. There are no detectable morphologic differences among North American mallards, nor is there any hint of local physiologic variation. Not only is the mallard adaptable in the sense of occupying a variety of breeding situations, but it has shown a remarkable capacity to adjust to the changes wrought in its wintering habitat. In primitive North

America the mallard wintered in the natural marshes, sloughs, and backwaters and ate aquatic foods along with other ducks. Today most of these waterways are drained or otherwise made unattractive, and during the autumn much of the remaining habitat bristles with the guns of eager duck hunters. The mallard copes successfully with this situation by several adjustments in its habits. First, it feeds at night, spending the day in safety of a waterfowl refuge or on some open bay or sandbar. Secondly, it has learned to feed on the waste grain of stubble fields—wheat and corn in the midwest, rice and kafir in Texas and California. Each day with cessation of legal shooting the birds rise in great masses and fly to the stubbles for the evening repast. For a period in the 1940s shooting closed at 4 P.M. and the flight began at 4:15. When the law was changed to permit shooting till sunset, the birds adjusted their exodus to fifteen minutes after sunset, attesting to their capacity for quick reaction to circumstances. As a result, the mallard today is by far the most abundant duck in North America.

Some other species of waterfowl have learned the same tricks. The pintail and widgeon in the west, and various geese, feed on crop residues and avoid guns during the day by flocking in safe refuges. But many of the ducks have not adjusted and are steadily decreasing in number. The redhead, canvasback, wood duck, and shoveler continue to feed in the marshes and along shorelines where they are exposed to heavy shooting. These nonadaptive species require special protection and their situation will not likely improve in the future.

Another example of an adaptable species is the coyote. Originally it occurred in modest numbers through western North America, something of a hanger-on in the range of the wolf, scavenging scraps left by this lordly predator and catching such rodents as were available. In the remaining climax forests of the Mexican highlands, where wolves still occur, coyotes are scarce or absent (Leopold, 1959). But over most of the continent where the virgin flora and fauna (including wolves) have been eliminated, conditions for the coyote have been vastly improved. The scourge of rodents that came with agriculture and with over-grazing of the western ranges, plus the carcasses of domestic stock, offered a food supply much superior to that originally available. As a result, the coyote has thrived and extended its range far to the north and east. It invaded Alaska in the 1930s and currently has moved as far east as New York State. The coyote, in other words, is an example of a successional or subclimax predator that has profited from alteration in the climax biota, as much so as the deer. Because it occasionally preys on sheep and poultry, it has been the object of intensive control efforts, more so perhaps than any other carnivore in the world. Yet so adept is the coyote at learning the tricks of avoiding guns, traps, poisons, dogs, and even airplanes (from which it sometimes is shot) that it persists over nearly all of its original and adopted range, at least in modest numbers. The coyote will be among the surviving wild species long into the future.

EVOLUTION OF BEHAVIOR

When species like the mallard and the coyote show adaptive behavior, as described above, it is difficult to say what part of this adaptation may be genetic and what part is learned. Many mallards are shot and many coyotes are trapped or poisoned. Are these the slow-witted ones? Is man, acting as a predator himself, applying a strong selective force to hunted species that may be bringing genetic changes in the survivors? If so, nothing is known of this force, but there is room for speculation.

Consider first the mallard. Much of a duck's behavior we know to be learned. The quick adjustment of the birds to a change in legal shooting hours could hardly be based on anything but experience. This quickly could become a tradition, transmitted from older experienced birds to young ones as some migratory habits are transmitted (Hochbaum, 1955). Yet over the years many individual mallards depart from this tradition and decoy into small ponds during shooting hours. They are among the missing when the breeders migrate northward in spring. Shooting, then, may be creating a new strain of mallard that tends to conform to mass behavior patterns and is less prone to make mistaken individual judgments.

In the southeastern United States, where the bobwhite has been heavily hunted for a century or more, it is generally reported that the birds have changed their habits. Old hunters claim to remember the day when a covey, flushed before a dog, would fly 100 to 200 yards and scatter in the broom-sedge or weed fields where they could be taken easily over points. Today covies tend to fly 300 to 400 yards and to seek shelter not in open fields but in dense oak thickets. Often such coveys put a "hook" on the end of their flight, turning to the right or left after entering the woods, thus being much harder to relocate. Is this change in behavior, if true, strictly learned and transmitted from adults to young? Or is there a genetic change involved as well, favoring the birds that fly far, seek woody cover, and change directions after entering the cover?

Much of the coyote's skill in avoiding peril is clearly learned. Individuals known to have escaped from a trap or to have survived a dose of strychnine become wary and are much more difficult to capture than young, inexperienced individuals. But the innate capacity for wariness may be strengthened and bolstered over the years by constant removal of the least wary individuals.

Thus, it may be that the hunted animals are evolving under a new selective force not affecting those animals that are permitted to live without persecution. In this sense, the adaptability which may be expressed as a genetic trait—or put in other words, as the ability to learn—is not a biological constant but a shifting attribute of a species.

SUMMARY

There are notable differences in the response of wild animals to the sweeping changes in environment brought on by man. Some species are clearly associated with and dependent upon undisturbed climax situations, and these suffer the most from environmental change. They are here designated as nonadaptive species. The list includes all the rare or endangered species and some that have become extinct.

On the other hand, other animals adjust very well to changes in vegetation and in land use, and these on the whole persist or may even increase in abundance. Included in this group are many of the common game birds and mammals that supply the bulk of the recreational hunting in North America today. There appears to be a direct correlation between the affinity with seral or subclimax biotas and adaptability in the sense of the capacity to adjust to change.

The ability to adapt seems to involve two distinct components: (1) genetic plasticity, or the capacity for segments of a population to evolve rapidly to fit local conditions; and (2) the capacity for individuals to learn new habits of survival under altered circumstances. These cannot readily be separated, since the capacity to learn is itself a genetic trait.

BIBLIOGRAPHY

Aldrich, J. W. 1955. Distribution of American Gallinaceous Game Birds. U.S. Fish and Wildl. Serv., Wash., D.C. Circ. 34.

Blanchard, B. D. 1941. The White-crowned Sparrows (*Zonotrichia leucophrys*) of the Pacific Seaboard: Environment and Annual Cycle. Univ. Calif. Pub. Zool. 46: 1–178.

Buechner, H. K. 1950. Life History, Ecology, and Range Use of the Pronghorn Antelope in Trans-Pecos Texas. Amer. Midl. Nat. 43:257–354.

Cheatum, E. L., and C. W. Severinghaus, 1950. Variations in Fertility of White-tailed Deer Related to Range Conditions. Trans. N. Amer. Wildl. Conf. 15:170–90.

Dahlbeck, N. 1951. [Commentary During U.N. Conf., Fish and Wildl. Res.] Proc. U.N. Sci. Conf. on Conserv. and Utiliz. of Res. Lake Success, N.Y. Aug. 17–Sept. 6, 1949. 7:210.

Grant, V. 1963. The Origin of Adaptations. Columbia Univ. Press, New York and London.

Hochbaum, H. A. 1955. Travels and Traditions of Waterfowl. Univ. Minn. Press, Minneapolis.

Lack, D. 1954. The Natural Regulation of Animal Numbers. Oxford Univ. Press.

Leopold, A. S. 1944. The Nature of Heritable Wildness in Turkeys. Condor 46:133–97.

——. 1950. Deer in Relation to Plant Succession. Trans. N. Amer. Wildl. Conf. 15: 571–80.

——. 1959. Wildlife of Mexico: the Game Birds and Mammals. Univ. Calif. Press, Berkeley.

Longhurst, W. M., A. S. Leopold, and R. F. Dasmann. 1952. A Survey of California Deer Herds, Their Ranges and Management Problems. Calif. Fish and Game, Game Bul. 6.

Mayr, E. 1942. Systematics and the Origin of Species. Columbia Univ. Press, New York.

Sauer, C. O. 1950. Grassland Climax, Fire, and Man. J. Range Mgt. 3:16–21.

Schorger, A. W. 1955. The Passenger Pigeon: Its Natural History and Extinction. Univ. Wisc. Press, Madison.

Taber, R. D. 1953. Studies of Black-tailed Deer Reproduction on Three Chaparral Cover Types. Calif. Fish and Game Bul. 39(2):177–86.

5

The dynamics of three natural populations of the deer **Odocoileus hemionus columbianus**

Richard D. Taber
Raymond F. Dasmann

INTRODUCTION

In the course of an investigation of the ecology of the Columbian black-tailed deer [*Odocoileus hemionus columbianus* (Richardson)] in relation to chaparral management in Lake County, California, information on the dynamics of three natural populations has been obtained. These populations occupy different habitats. The original plant community and environment were common to all, but secondary modification has created three distinct range types.

The climax plant cover consists on south slopes of *chamise chaparral* and on the north slopes of *broad sclerophyll forest* as described by Cooper (1922). These associations are dominated by fire-tolerant shrubs which either sprout from the root-crown when burned or have seeds which germinate readily following heating.

On the southerly exposures the most abundant plant is chamise (*Adenostema fasciculatum*). Other species include yerba santa (*Eriodictyon californicum*), wedgeleaf ceanothus (*Ceanothus cuneatus*), and toyon (*Photinia arbutifolia*). Occasional burning apparently took place on these slopes during prehistoric times.

Contribution from Federal Aid in Wildlife Restoration Project California W-31-R and the Museum of Vertebrate Zoology, University of California. Reprinted with permission from *Ecology*, 38:233–246 (1957).

The northerly exposures were largely covered, before white settlement, with a broad-sclerophyll forest, in which the dominant trees were interior live-oak (*Quercus wislizenii*), canyon oak (*Q. chrysolepis*), California laurel (*Umbellularia californica*) and madrone (*Arbutus menziesii*). Since that time (1855–65) fires have become more frequent and at present much of the north-exposure vegetation must be called mixed or mesic chaparral, consisting of broad-sclerophyll species which have been reduced to shrub form by burning. The principal constituent of this association is interior live-oak *(Quercus wislizenii).* A large assemblage of other woody plants including scrub oak (*Q. dumosa*), Eastwood manzanita (*Arctostaphylos glandulosa*), and deerbrush (*Ceanothus integerrimus*) is also found on north exposures.

At present the general cycle of events is that every five to twenty years or more a fire sweeps the region bare, except for isolated patches of brush and the bare, charred trunks of the burned shrubs. Within five to ten years the sprouts and seedlings have grown up to cover the ground with a dense thicket of shrubs. Two of the range-types that were compared in this study are the two extremes of this burn-and-recover pattern: the newly burned area on which crown-sprouts and seedlings are abundant—here called the "wildfire burn"; and the area which has not been burned for at least 10 years—here called the "chaparral." The wildfire burn is admittedly a transient stage, but while it lasts it constitutes a special type of deer range, being rich in food and poor in cover.

Another modification of chaparral is possible. Certain portions may be burned deliberately in either the early spring or late summer, and the burned areas seeded to herbaceous species. After the rains, when the seedlings emerge, those of the herbaceous species compete successfully with those of the woody species. The brush-sprouts are kept hedged by deer use or reduced by re-burning. The net result is a scattering of shrubs with the intervening area occupied by herbaceous plants. When an area is managed in this way portions are left in heavy brush for cover. This range type is now called "shrubland."

The technique of managing chaparral to create shrubland is still in the experimental stage so the areas of shrubland are limited in extent. The area most intensively studied covered about 400 map acres (horizontal projection) and was followed from before burning, in 1949, until 1955. Areas which had been longer established were also studied.

Some discussion of the ecology of these chaparral cover types has been included in a previous publication (Biswell *et al.* 1952), and a fuller description of the plant cover, with special reference to use by deer, will form the body of a later report.

The three range-types under consideration may occur in close juxtaposition, and it was in such an area that the present studies were made. The area lies about five miles southwest of Lakeport, California, between the elevations of 1500 and 2500 feet. The substrate consists of Pliocene sandstones elevated in Pleistocene times (Manning and Ogle 1950). Erosion has been rapid and the soils are correspondingly thin. They are classified principally as *Maymen*, with lesser areas of *Los Gatos*. Small pockets of still other residual soils occur on the uplands, and alluvial soils are found in the stream-bottoms. The topography is moderately steep (average slope = 22 degrees), and consists of irregular drainages and ridges with rounded tops. Rainfall averages about 28 inches annually. Winter snows usually melt soon after they fall, especially on the warmer south exposures where the deer spend most of their time in winter.

The adjustment of the vegetation to the Mediterranean climate has a profound effect upon the food-regime of the deer. Most of the shrubs are evergreen, but do not grow in the winter, when it is moist enough but too cold, or in summer, when it is warm enough but too dry. Ordinarily there is some growth in crown-sprouts in late fall, but little or none in mature shrubs. The principal growth period, then, is spring. In April and May, the deer, feeding on the new shoots, gain weight rapidly. Most fawns are born in the second and third weeks of May. The last rains usually fall in May. Increasing drought dries the herbaceous plants and the shrubs become dormant in July. The moisture and crude protein levels in the browse fall steadily all summer, and the condition of the deer, especially lactating does, falls accordingly. Occasionally there is a heavy set of acorns and the deer begin to browse these directly from the low oaks in August and September, months when otherwise the forage is of low quality. In the fall come cool weather and the first rains. The timing of the first substantial rain is especially important in the shrubland, where large quantities of annual plants appear if about one inch of rain falls. Sufficient rain for germination usually falls in October, but occasionally the first heavy precipitation may be as early as August or as late at November.

The present study began in the fall of 1948 and continued through the fall of 1955. Many people aided the investigation: of the field forces of the California Department of Fish and Game, Norman Alstot, Gordon Ashcraft, Bonar Blong, Herbert Hagen and Manley Inlay; of the State Fish and Game Laboratory, John Azevedo, Art Bischoff, Oscar Brunetti and Merton Rosen. Help in the field was also provided by Richard Genelly and Gerald Geraldson of the University of California. Aid, advice, and encouragement were constantly available from H. H. Biswell, Project Leader, A. M. Schultz, A. S. Leopold and W. M. Longhurst, of the University of California, and William Dasmann and Robert Lassen of the Department of Fish and Game. To these, and to Glen Keithley and Harold Manley on whose land much of the work was done, and to others too numerous to list, we are indebted for help.

In addition we wish to acknowledge the critical reading of the manuscript, with suggestions for improvement, by W. Leslie Robinette, J. J. Hickey, Robert F. Scott and staff members of the Museum of Vertebrate Zoology.

METHODS AND RESULTS

The basic data from which the life tables were derived consist of information on individual movement, population density, reproduction according to age-class, population structure, and mortality according to age and sex. These data were gathered seasonally for the three populations under study; notes on the methods involved are given below.

Movement

By studying the movements of marked deer daily, seasonally and annually, it has been found that the deer in the study region are nonmigratory and that most of them occupy home ranges with diameters of about one-half mile (does) to three-quarters of a mile (bucks). Populations occupying neighboring ranges may, therefore, be considered separately (Dasmann 1953).

Population density

Censuses were made at least twice a year, first by the pellet-group-count method and later by the sample-area-count method, both of which were found to be accurate when checked against populations of known density. These methods are described elsewhere (Dasmann and Taber 1955).

Table 5–1 shows population densities observed at various times on the three types of range under study. Counts in the chaparral gave a summer density of about 30 per square mile. The wildfire burn, at the same season, showed 120 the first year after the fire. The summer density dropped to 106 the second year, 52 the third and 44 the fourth. The shrubland went from 98 the year following burning to 131 the second year and then down to about 84 the fifth and sixth years, at which level the population presumably stabilized.

Natality

Data on ovulation rates were obtained by collecting pregnant and post-partum does. Forty-eight does over 17 months of age were taken on the three range types. Younger does were also collected and were found not to breed under our conditions (Taber 1952).

TABLE 5–1. Deer density (individuals per map-square-mile) on three range types

A. Chaparral

	April-June	July-October	November-December
1949	28	..	30
1950	13	30	26
1951	..	..	30
Average.......	20	..	29

B. Wildfire burn

Growing season from time of burning	Early May	Mid-July	Early December
1	..	120	86
2	75	106	56
3	48	52	50
4	32	44	32

C. Shrubland

	Early May	Mid-July	Early December
1951	88	131	88
1952	69	112	99
1953	69	103	..
1954	53	85*	..
1955	55	82*	..

* mid-June

The presence of *corpora lutea of pregnancy* is an indicator of the successful shedding, fertilization and implantation of ova, but not every *corpus luteum* represents a developing fetus. The ratio was found to be 94 fetuses per 100 corpora lutea, based on all pregnant does collected. No evidence of abortion or resorption was noted, so this six percent loss of ova must occur before fertilization or in very early pregnancy.

Does often breed first, in this region, at the age of 17 months, although on the poorest ranges some may not breed until the age of 29 or even 41 months. No yearling does (17–24 months) were collected in the poorest range type, the chaparral. Two from the shrubland had one fetus apiece and two out of three from the new wildfire burn had one fetus each. Two yearling does were taken from an older wildfire burn and neither was pregnant. These samples are so small that little confidence can be placed in them. However, some values for the contribution of yearling does to the annual fawn-increment must be assumed in the calculations which follow, so these figures for the shrubland and new wildfire burn will tentatively be accepted. In addition, since no data are available for reproduction in yearling does in the chaparral, it will be assumed that the rate of fawn production by them is 0.5 per doe.

Among adult does the samples are larger, ranging for the three range types from 10 to 16. Fawn production in the shrubland is significantly higher than that in the chaparral, the values being 1.65 and 0.77 fawns per doe respectively. The does on the new wildfire burn show an intermediate average of 1.32 fawns apiece. Fawn production is summarized in Table 5–2.

TABLE 5–2. Fawn production by yearling and adult does on three range types

Range type	Yearling Does			Adult Does		
	Number examined	Mean number of corpora lutea	Fawns produced per doe	Number examined	†Mean number of corpora lutea	Fawns produced per doe
Chaparral	0	....	*0.50	11	0.82 (0.42 - 1.22)	0.77
Wildfire burn after one growing season	3	0.66	0.62	10	1.40 (1.03 - 1.77)	1.32
Wildfire burn after three growing seasons	2	0.0	0.0	4	0.75 (0.00 - 1.54)	0.71
Shrubland	2	1.0	0.94	16	1.75 (1.51 - 1.99)	1.65

* Assumed (see text).
† Values in parentheses indicate the range with a confidence limit = .05.

Population structure

Population structure was determined by observing and classifying undisturbed deer at ranges consistent with accuracy, at times during which all age and sex classes were equally visible. The most favorable seasons were late July, when fawns were at heel and before hot weather caused the bucks to seek heavy cover, and early December, when the rut had subsided and the bucks had returned to their normal level of mobility but before the antlers were shed (Dasmann and Taber 1956). In shrubland, where a more intensive study was carried out, determinations were made throughout the year. The values in Table 5–3 are expressed in terms of ratios, where the number of adult does is always taken as 100.

The low density of deer and the high density of cover made herd composition counts difficult to obtain in the chaparral. Therefore use was made of the fact that when an area of chaparral is burned, and sprouts appear, as happened in 1949 and 1950, the deer whose home ranges impinge upon the burned area congregate on it to feed and are then (December) easily observed. This population may be taken to represent the

TABLE 5–3. Population structure according to herd composition counts on three range types

Month	Number of deer classified	Adult ♂ ♂	Yearling ♂ ♂	Adult ♀ ♀	Yearling ♀ ♀	Fawns
A. Chaparral						
Dec. 1949	124	72	23	100	38	85
Dec. 1950	47	20	5	100	25	60
Average	85	46	14	100	31	72
B. Wildfire burn (December counts)						
During growing season						
1 (1948 burn) counted in 1949	90	63	30	100	33	107
4 (1948 burn) counted in 1952	37	50	14	100	21	79
1 (1950 burn)	54	37	21	100	39	89
2 "	45	66	20	100	33	80
3 "	90	48	7	100	27	34
C. Shrubland						
July, 1951	82	70	27	100	46	127
Dec., 1951	55	30	30	100	45	70
May, 1952	43	38	31	100	44	56
July, 1952	70	48	9	100	30	117
Dec., 1952	62	39	9	100	30	92
May, 1953	43	32	9	100	14	41
June, 1953	67	36	16	100	20	96
May, 1954	33	39	29	100	14	57
June, 1954	53	56	25	100	25	125
May, 1955	34	46	31	100	8	77
June, 1955	51	71	36	100	36	121

population in the chaparral, so far as structure is concerned, if one correction is made. Since the home ranges of bucks are larger than those of does, a burned area attracts proportionally more bucks than does, if only those animals whose home ranges impinge upon the burned area come to it. The counts in Table 5–3(A) are corrected for this; i.e., adult buck counts were reduced one-third, because average buck home-range diameter is 3⁄2 average doe home-range diameter.

Mortality

Every carcass encountered was classified, if possible, according to age, sex, season of death, cause of death and range type. Occasionally special systematic searches were made for carcasses, especially along the beds of steep-walled canyons, where dead deer were most likely to accumulate. Altogether 222 carcasses were tallied; these are listed in Table 5–4. Aging was by tooth eruption and wear (Severinghaus 1949; Moreland 1952).

A carcass-count of this sort is not a true reflection of mortality in all

TABLE 5–4. The classification of deer found dead on three range types

Sex	0 - 3 months	4 - 6 months	7 - 9 months	10 - 12 months	13 - 24 months	25 - 36 months	37 months and over	Total
A. Chaparral								
♂	0	7	6	0	8	3	9	33
♀	1	6	4	0	1	1	23	36
?	16	0	0	0	0	0	0	16
					Sub-total........		85	
B. Wildfire burn—one to four years after burning								
♂	7	1	2	0	1	6	12	29
♀	1	4	1	0	1	1	11	19
?	16	0	0	0	0	0	0	16
					Sub-total........		64	
C. Shrubland								
♂	3	7	1	2	5	4	4	26
♀	1	5	0	0	0	4	18	28
?	19	0	0	0	0	0	0	19
					Sub-total........		73	
					Grand total...........			222

classes; the very young deer are under-represented, because their fragile carcasses soon disintegrate. However, if this is taken into account, the carcass tally aids the study of mortality.

The principal cause of mortality was starvation, not caused by a quantitative lack of food, but rather by a seasonal drop in the quality of available forage (Taber 1956). The effects of starvation were often augmented by exposure to unfavorable weather and occasionally by disease or parasitism. Mortality from this cause was heaviest among fawns, but adults of both sexes were also affected. Principal starvation losses occurred in fall and winter. Next in importance as a mortality factor was hunting. Almost all adult buck mortality, except for old deer, was caused by bullet-wounds. A few deer of other classes were also affected. The loss of protected classes to hunting varies widely. It was found in the study region to be quite low. A few deer were killed in accidents or by predators, but these are relatively unimportant mortality factors.

The hunting season extends from early August to mid-September and bucks usually become legal game during their third year and remain so for the rest of their lives. The kill at this time has a profound effect on the population structure. The ratios existing between two-year, three-year and older bucks in the kill, which are given in Table 5–5, are affected by three factors: the age distribution of bucks in the population; the relative vulnerability of the various age classes to hunting; and the deliberate selection of the larger (hence older) bucks by the hunter.

Where hunting pressure is only moderate and escape cover is adequate, bucks over four years old are much less vulnerable than two-year-

TABLE 5–5. Distribution by age (in per cent) of bucks killed and taken home from three range types

Range type	Two years old	Three years old	Four years old and older	Number in sample
Chaparral	40	22	38	194
Wildfire burn after first growing season	36	22	*42	59
After second growing season	63	21	15	52
After third growing season	37	37	25	83
After fourth growing season	50	22	28	32
Shrubland	49	19	32	43

* Old bucks are unusually vulnerable during the first hunting season after a large wildfire has removed most of the escape cover.

olds, with three-year-olds being intermediate. For example: in an intensively studied area of shrubland the combined buck populations for the seasons of 1951 and 1952 were 14 two-year-olds, of which 10 were killed, and 13 older bucks, of which two were killed. This difference in vulnerability is probably above the average because little drive-hunting with dogs took place. This type of hunting results in a proportionally higher kill of old bucks.

Selective hunting, involving the deliberate attempt of the hunter to bag a trophy buck, is generally not practiced. Most hunters attempt to take the first legal target that they see. If the antlers are small and inconspicuous the hunter may not recognize the deer as legal game, or if two bucks appear together, the hunter will select the larger, but these factors do not appear to affect the kill appreciably. These remarks apply to hunters on public land, and private land with public access. All the areas studied except part of the shrubland could be so classed. However, there are certain lightly hunted areas of private land where there is a definite selection of larger (hence older) bucks by the hunter. Part of the shrubland of the present study was in this category, and for that reason the percent of old bucks in the kill (Table 5–5) is believed to be higher than would ordinarily be found.

In addition to the bucks which are killed and taken home by the hunters, there are those which are shot but not found. In the study region hunters often shoot across canyons at running deer at long range. This fact, and the heavy cover and the hard ground, which makes tracking difficult, lead to the loss of many wounded deer. Few of these recover. Intensive studies of small known populations have shown this loss to equal about 40 per cent, or slightly more, of the take-home kill.

RECONSTRUCTION OF POPULATION DYNAMICS

The information presented above has been used to deduce the detailed changes taking place within each population in the course of a year or, in the wildfire burn population, four years. In order that values per square mile may be readily derived, the tables of population dynamics given in this section are constructed to represent the deer population occupying 100 square miles of the range-type in question. In a later section, where life tables are presented, values in per cent may be more readily apprehended.

Annual population cycle: chaparral

From data presented above it may be seen that the deer population in the chaparral is one of low density, with low reproductive rate. Since the population is stable, this low reproductive rate is matched by a relatively low mortality. Does probably do not breed as early as they do on better ranges, and they certainly do not bear as many fawns in maturity. It is possible that some individuals bred only in alternate years. This partly relieves the population of the heaviest drain on adult vitality—gestation and especially lactation, where lactation takes place at a season of a falling nutritive plane. Mortality among fawns, though high, is not as high proportionately as it is on also fully stocked but better ranges, like shrubland. Presumably this is due in some measure to the fact that most fawns are born singly. Mortality among bucks during the hunting season is lighter than in more open range because the abundance of dense escape cover in the chaparral makes hunting difficult. The general pattern is of a population with a rather low rate of replacement and a correspondingly high life expectancy from adulthood. An annual cycle is shown in Table 5–6.

TABLE 5–6. The population dynamics, through one year, of a deer population inhabiting 100 square miles of chaparral

Season	4+ yr. ♂♂	3 yr. ♂♂	2 yr. ♂♂	1 yr. ♂♂	3+ yr. ♀♀	2 yr. ♀♀	1 yr. ♀♀	♂ Fawns	♀ Fawns	Total
Late May (fawn drop)	343	121	190	239	1063	262	280	504	446	3448
Early summer loss	...	...	...	...	...	...	...	180	114	294
July herd composition count	343	121	190	239	1063	262	280	324	332	3154
Hunting season bag	39	27	49	...	...	...	...	...	...	115
Crippling loss	16	11	20	...	...	...	...	...	...	47
Late summer and fall loss	...	...	...	43	152	17	8	31	19	270
Dec. herd composition count	288	83	121	196	911	245	272	293	313	2722
Late winter loss	28	...	...	6	80	13	10	54	33	224
Early May population	260	83	121	190	831	232	262	239	280	2498
Late May (fawn drop)	343	121	190	239	1063	262	280	504	446	3448

Annual population cycle: wildfire burn

The general trend of population density characteristic of wildfire burns is from a high point the spring following burning to lower and lower levels thereafter. Movement is the most important element in these population changes.

When the area burns the deer move ahead of the fire and are seldom directly injured. Lack of food, however, keeps them from re-occupying their home ranges, unless there happens to have been a heavy acorn crop. The brush on the warmer slopes sprouts from the root-crown about October or November, in areas burned in late summer. The wildfire burn thereupon becomes an attractive feeding area, and large numbers of deer appear on it. The deer densities observed on burned areas can be adequately explained by supposing that the deer feeding on the burned area are those whose home ranges included part of that area before it was burned. If this is true, deer are drawn to a burned area from a peripheral zone one-half mile wide for does and three-quarters of a mile wide for bucks; *i.e.*—the peripheral zone equals one home-range diameter in width. Thus if burns are small the deer density feeding upon them is high, whereas if burns are large the density is lower, because in the former case the peripheral zone is larger in proportion to the burned area and in the latter it is smaller.

Actual observations of increase in deer density following burning correspond, for the areas representative of wildfire burns, to that to be expected if a strip one-half mile in width were burned through the country. Therefore in the reconstruction of population dynamics (Table 5–7) it has been assumed that the burn consists of a strip one-half mile in width extending indefinitely across the country, and the population shown is that inhabiting 100 square miles of this burn.

In order to follow the dynamics of a deer population through four years, as has been attempted in Table 5–7, it is necessary to use some interpolated values for productivity. It has been seen (Table 5–2) that the average fawn production per doe in the new wildfire burn is about 0.62 for yearlings and 1.32 for adults. As the wildfire burn grows older the quality of the forage produced on it drops, and this is reflected in the reproductive rate. By the fourth year after burning reproduction appears, according to evidence from limited collecting and herd composition counts, to be about the same as that in the heavy brush, namely 0.5 fawns per yearling doe and 0.77 fawns per adult doe. It remains to interpolate intermediate values for productivity during the second and third years following burning. We will assume that yearling does each produce an average of 0.58 and 0.54 fawns and adult does 1.14 and 0.95 during the second and third years following burning, respectively.

Table 5–7 traces the dynamics of a deer population inhabiting a

TABLE 5–7. Dynamics of a deer population inhabiting 100 square miles of wildfire burn for the first four years following burning

Season	4+ yr. ♂♂	3 yr. ♂♂	2 yr. ♂♂	1 yr. ♂♂	3+ yr. ♀♀	2 yr. ♀♀	1 yr. ♀♀	♂ Fawns	♀ Fawns	Total
Original population (Dec.)	406	83	99	196	740	344	362	276	395	2901
Influx from periphery	406	83	225	444	1247	580	610	695	1003	5293
Early May population	812	166	324	640	1987	924	972	971	1398	8194
Fawn-drop	978	324	640	971	2911	972	1398	2098	1856	12148
Early summer loss	...	...	...	...	...	...	...	92	56	148
July herd composition count	978	324	640	971	2911	972	1398	2006	1800	12000
Hunting bag	126	66	108	...	...	...	...	...	...	300
Crippling loss	50	26	43	...	...	...	...	...	...	119
Late summer and fall loss	...	...	...	...	...	...	...	100	61	161
Exodus	145	42	107	175	799	267	377	524	478	2914
Dec. herd composition count	657	190	382	796	2112	705	1021	1382	1261	8506
Exodus	46	11	29	48	117	39	57	77	70	494
Late winter loss	...	...	...	...	7	...	...	473	106	586
Early May population	611	179	353	748	1988	666	964	832	1085	7426
Fawn-drop	790	353	748	832	2654	964	1085	2176	1925	11527
Early summer loss	...	...	...	...	...	...	...	589	437	1026
July herd composition count	790	353	748	832	2654	964	1085	1587	1488	10501
Hunting bag	66	92	277	...	...	...	...	...	...	435
Crippling loss	26	37	111	...	...	...	...	...	...	174
Late summer and fall loss	...	...	...	...	...	...	...	279	133	412
Exodus	259	69	55	386	1062	386	434	635	595	3881
Dec. herd composition count	439	155	305	446	1592	578	651	673	760	5599
Exodus	114	...	...	196	400	150	169	175	198	1402
Late winter loss	35	...	...	...	152	...	...	218	42	447
Early May population	290	155	305	250	1040	428	482	280	520	3750
Fawn-drop	445	305	250	280	1468	482	520	1067	943	5760
Early summer loss	...	...	...	...	...	...	...	380	230	610
July herd composition count	445	305	250	280	1468	482	520	687	713	5150
Hunting bag	64	93	93	...	...	...	...	...	...	250
Crippling loss	26	37	37	...	...	...	...	...	...	100
Late summer and fall loss	...	...	...	...	18	...	...	165	101	284
Dec. herd composition count	355	175	120	280	1450	482	520	522	612	4516
Late winter loss	94	...	...	...	646	...	...	336	204	1280
Early May population	261	175	120	280	804	482	520	186	408	3236
Fawn-drop	436	120	280	186	1286	520	408	822	727	4785
Early summer loss	...	...	...	...	...	...	...	261	174	435
July herd composition count	436	120	280	186	1286	520	408	561	553	4350
Hunting bag	44	34	104	...	...	...	...	...	...	182
Crippling loss	18	14	42	...	...	...	...	...	...	74
Late summer and fall loss	...	...	...	33	307	10	10	210	140	710
Dec. herd composition count	374	72	134	153	979	510	398	351	413	3384

wildfire burn from the December following burning for four full years. The population, if there had been no fire, would have been about 29 per square mile. The influx of deer whose home ranges abutted on the burned area added another 53 per square mile. These deer, which bred in early November, showed a reproductive rate somewhat greater than that found in the chaparral. Presumably this was due either to a high-quality diet of new sprouts for a few weeks prior to breeding, or to the psychic stimulation of crowding in the periphery of the burn during the rut. The former seems more probable.

With the onset of cold weather there was apparently an exodus of a portion of the population from the burned area, where there was little heavy cover, to the periphery, and these did not return. There was apparently a similar movement the following winter. In both cases the population dropped but a lack of carcasses indicated that movement rather than mortality was the cause.

Mortality on new wildfire burns was found to be very low, as might be expected from the high quality of the feed there. However, as forage quality declined, mortality increased.

By the end of the fourth year the deer population in the wildfire burn had declined to about the levels in the chaparral both in density and reproduction.

Annual population cycle: shrubland

The management of chaparral to create shrubland resulted in the attraction of a heavy deer population, in much the same manner as has been described for the wildfire burn. There were, however, several important points of difference between the shrubland and the wildfire burn. In the shrubland there were areas of cover closely adjacent to feeding grounds; the wildfire burn had little cover. In the shrubland there was an abundance of herbaceous forage, which the deer eat in quantity from December through March; in the wildfire burn there was little herbaceous forage. In the shrubland the browsing-pressure of the deer was sufficient to keep many shrubs hedged and within reach; the browsing pressure on the wildfire burn was not usually sufficient for this and shrubs tended to grow beyond reach. Because of these, and perhaps other factors, the shrubland continued to support a high deer density throughout the study. This density was very high during the first years due to influx and a high survival. By the fifth and sixth year following burning, however, it had stabilized at a lower, but still substantial density. It appears that this level, about 84 per square mile in July, can be maintained for some time, so it has been taken as the basis for the annual cycle of population dynamics reconstructed in Table 5–8.

The shrubland range is fully stocked and population gain is matched by loss. The reproductive rate is high. Most yearling does produce a fawn

TABLE 5–8. Dynamics of a deer population inhabiting 100 square miles of shrubland

Season	4+ yr. ♂♂	3 yr. ♂♂	2 yr. ♂♂	1 yr. ♂♂	3+ yr. ♀♀	2 yr. ♀♀	1 yr. ♀♀	Fawns ♂♂	Fawns ♀♀	Total
Late May (fawn-drop)	434	182	482	556	1935	706	858	2042	1810	9005
Early summer loss	...	...	...	...	29	...	...	350	224	603
July herd composition count	434	182	482	556	1906	706	858	1692	1586	8402
Hunting season bag	60	46	214	...	...	...	...	...	...	320
Crippling loss	24	19	86	...	...	...	...	...	...	129
Late summer and fall loss	...	...	...	20	115	37	20	856	547	1595
Dec. herd composition count	350	117	182	536	1791	669	838	836	1039	6358
Late winter loss	33	...	...	54	414	111	132	280	181	1205
Early May population	317	117	182	482	1377	558	706	556	858	5154
Late May (fawn-drop)	434	182	482	556	1935	706	858	2042	1810	9005

and adult does produce an average of 1.65 fawns apiece. The mortality in these fawns is high; during the first year of life 73 percent of the males and 53 percent of the females die. Among adult does mortality is also high—about 25 percent per year. Adult bucks, because of heavy hunting and the prevalence of open country, lost 62 percent of two-year-olds, 35 percent of three-year-olds and about 20 percent of older bucks to hunting, on the average.

Among yearling (12- to 24-month-old) deer not all the loss to the population was due to mortality. It is usual for a dispersal movement of yearlings to occur about fawning-time, when the adult does become antagonistic toward their previous offspring. In a large, homogeneous area, of uniform deer density, this movement of yearlings from a local area would be balanced by others moving into it. A small area of dense population, however, would tend to lose more than it gained; this is the situation on the 400-acre shrubland study area. Here the average loss due to emigration during the four year period 1951–55 amounted to about 6 percent of the yearling class of bucks and 38 percent of the yearling class of does. It seems probable that this higher emigration rate in yearling females is connected with the tendency toward spacing and mutual antagonism among breeding does (Dasmann and Taber 1956).

Since this measurement of emigration was made on a small area of dense population, it is large. In Table 5–8, which is an attempt to describe shrubland population dynamics for a larger area, a yearling female emigration of 15 percent is assumed.

Life tables for wild ungulates

It has been apparent, in preceding sections, that in any of the populations under study the dynamics of the male portion of the population differed from that of the female portion. This was due partly to a differential mortality in early life and partly to the fact that when adulthood was reached entirely different decimating factors were at work—physiological reproductive drain, coupled with reduced nutrient quality, for females, and predation (in the form of hunting) for males. In other populations it is probable that still other factors might be operative. As an analytical aid, then, it seems proper to consider the male and female components of each population separately.

The life table, as so ably described and used by Deevey (1947) and Hickey (1952), is a convenient vehicle for describing a population so that certain characteristics can be readily ascertained. The data, or values derived from them, are cast under the headings: x—the age-interval, in our case one year; l_x—the number alive at the beginning of each year; d_x—the number dying within the year; q_x—the rate of mortality during the year ($100\ d_x/1_x$) and e_x—the mean length of life remaining to each individual alive at the beginning of the year—the life expectation.

Deevey (*op. cit.*) points out that sources of ecological data for the construction of life tables are of three kinds: (1) knowledge of age at death for a random and adequate sample of the population; (2) knowledge of the fate of individuals of a single cohort at frequent intervals; and (3) knowledge of the age-structure among the living. He goes on to remark that information of the first and third types can be used only if one is prepared to assume that the population is stable in time.

For the black-tailed deer we have accumulated data of all three sorts. The deer populations in the chaparral and shrubland we have considered, for the purposes of this analysis, as relatively stable. Those in the wildfire burn are not stable. Life tables, then, may be constructed for the deer populations of the chaparral and the shrubland.

It is of interest to compare the elements of population dynamics displayed by these populations with those of other wild ungulates. Adequate information on which to base sex-separate life tables is available for relatively stable populations of the roe-deer (*Capreolus capreolus*) from Andersen (1953), the red-deer (*Cervus elaphus*) from Evans (1891), and

In constructing life tables from the data previously presented, it has been necessary to extend Table 5–4 (the carcass-tally) so as to break

TABLE 5–9. Life tables for the black-tailed deer of the chaparral

A. Males

x (years)	l_x	d_x	q_x	$\mathring{e}_x$
0 - 1	1000	526	526	2.3
1 - 2	474	97	204	3.3
2 - 3	377	137	363	3.0
3 - 4	240	75	313	3.4
4 - 5	165	20	121	3.7
5 - 6	145	14	97	3.2
6 - 7	131	14	107	2.5
7 - 8	117	30	256	1.7
8 - 9	87	35	402	1.1
9 - 10	52	52	1000	0.5

B. Females

x (years)	l_x	d_x	q_x	$\mathring{e}_x$
0 - 1	1000	372	372	4.2
1 - 2	628	41	65	5.3
2 - 3	587	66	112	4.6
3 - 4	521	68	131	4.2
4 - 5	453	67	148	3.7
5 - 6	386	54	140	3.1
6 - 7	332	54	163	2.6
7 - 8	278	54	194	2.0
8 - 9	224	33	147	1.4
9 - 10	191	191	1000	0.5

TABLE 5–10. Life tables for the black-tailed deer of the shrubland

A. Males

x (years)	l_x	d_x	q_x	$\mathring{e}_x$
0 - 1	1000	728	728	1.3
1 - 2	272	36	132	2.5
2 - 3	236	147	623	1.8
3 - 4	89	31	348	2.9
4 - 5	58	12	207	3.2
5 - 6	46	9	196	2.9
6 - 7	37	7	189	2.5
7 - 8	30	6	200	1.9
8 - 9	24	6	250	1.3
9 - 10	18	18	1000	0.5

B. Females

x (years)	l_x	d_x	q_x	$\mathring{e}_x$
0 - 1	1000	526	526	2.4
1 - 2	474	83	175	3.6
2 - 3	391	82	210	3.2
3 - 4	309	77	249	3.0
4 - 5	232	58	250	2.8
5 - 6	174	44	253	2.5
6 - 7	130	32	246	2.2
7 - 8	98	25	255	1.8
8 - 9	73	18	247	1.3
9 - 10	55	55	1000	0.5

the Dall sheep (*Ovis dalli*) from Murie (1944), providing that certain reasonable assumptions (stated in each case) are allowed.

The black-tailed deer

down the 37 months-and-older category into annual age-groups from three to ten. The number of individuals in each age-group was small, but the general pattern seemed to be one of fairly constant mortality rates from three to nine years for does and four to eight years for bucks. This has been tentatively assumed. In extreme old age (arbitrarily considered by us to be about 9–10 years) there is an accelerated mortality, as the oldest deer all die. When techniques of aging become more accurate it may be found that senescence is reached at 12–15 years rather than 9–10, and that mortality is roughly constant all through adulthood. In other words, our assumption that no deer live beyond 10 years may create an artificially abrupt drop-off at the end of the survivorship curve.

It may be seen (in Tables 5–9 and 5–10) that in both range types there is a higher proportional mortality in male fawns; this has been attributed to the higher metabolic rate and nutrient requirements among males (Taber and Dasmann 1954). There is a markedly unbalanced sex ratio by the time two years is reached. At this point the males become vulnerable to predation (hunting) and the females (individuals of which breed for the first time at either 17 or 29 months) to the strains of reproduction. The weight of hunting falls most heavily on the two-year-old class of males, while the loss of females is roughly constant from two to nine years. Both sexes have a higher mortality in the shrubland than in the chaparral. In the adult males this is known to be due to the easier hunting conditions of the shrubland; the deer are easier to kill. In the adult females the higher mortality found in the shrubland is probably related to the higher reproductive rate there. The loss in yearlings from the shrubland range is due, in part, to emigration rather than mortality. This makes an interesting parallel to the situation in roe-deer, described below.

The roe-deer

In 1950 a relatively stable roe-deer population inhabiting some woods on the Danish Game Research Farm at Kalø was exterminated by shooting. Forty-six males, 76 females and 91 fawns were taken. Andersen (1953) has reported on the age-distribution and given data on the occurrence of corpora lutea of pregnancy in the females.

Assuming that the annual reproductive rate was relatively constant; that fawns born amounted to 90 percent of the corpora lutea of pregnancy and that the sex ratio at birth was 100 ♂ ♂ : 100 ♀ ♀, we may construct life tables (Table 5–11) for the two sexes.

It will be noted that there is a high loss not only in the first year of life, but also in the second and especially in the third, in both sexes. Andersen (1953 and personal communication) points out that the hunting kill at Kalø has averaged only 4 ♂ ♂, 4 ♀ ♀ and 6 fawns per year, from 1943 to 1949. He accounts for the loss in young adults in terms of dispersal, citing as evidence the fact that each year roe-deer appear in small outlying woods which are normally uninhabited by roe-deer, and that as many as possible of these are killed. Thus the principal factor limiting this population is apparently not food, as it is in both examples of the black-tailed deer, but emigration. The woods of Kalø seem to have a saturation point for roe-deer, presumably based on intraspecific stress. Darling (1937) has commented on the intolerance of roe-deer for their young of previous years, which he attributes to the patriarchal nature of their social organization.

The red deer

From 1879 onward Henry Evans leased the red-deer shooting on the island of Jura, in Scotland. He conducted an annual census of males (in

TABLE 5–11. Life tables for the roe-deer of Kalø (after Andersen)

A. Males

x (years)	l_x	d_x	q_x	$\mathring{e}_x$
0 - 1	1000	740	740	1.2
1 - 2	260	60	230	2.2
2 - 3	200	110	550	1.7
3 - 4	90	20	222	2.1
4 - 5	70	20	286	1.5
5 - 6	50	30	600	0.9
6 - 7	20	20	1000	0.5

B. Females

x (years)	l_x	d_x	q_x	$\mathring{e}_x$
0 - 1	1000	630	630	2.6
1 - 2	370	50	135	4.2
2 - 3	320	150	469	3.0
3 - 4	170	50	294	3.3
4 - 5	120	40	333	2.7
5 - 6	80	30	375	2.2
6 - 7	50	20	400	1.7
7 - 8	30	10	333	1.2
8 - 9	20	20	1000	0.5

July) and antlerless deer (in February) and classified adult males, yearling (1–2 year) males and adult females and deer under one year of age. He kept a record of deer found dead and, having examined many pregnant females, estimated the production per female. During the first six years he did not burn the heather, and his deer population was roughly stable, the adult females increasing slightly. In the next five years, he was active in heather-burning and his deer population had increased markedly and was still increasing at the time he drew up his report. For present purposes, the information from the first six years only has been used. The basic data are (averages for 1879–1884) inclusive):

Stags (2 years and above)
- Alive in July– 460
- Shot– 60
- Dying from other causes– 31

Hinds (one year old and above)
- Alive in Feb.– 612
- Shot– 5
- Dying from other causes– 47

Hinds first breed at 2½ years, 100 hinds (one year old and above) bring 37 calves to February and 30 to one year of age. Of these 30, 13 are ♂ ♂–these have little loss to 3 years, and 17 are ♀ ♀–these become 15 by second birthday and are still 15 at third birthday. The sex ratio at birth is 100 ♂♂ : 100 ♀♀ (Evans 1891).

A maximum life span is assumed as 15 years (by the present authors). The life tables have been derived from these data (Table 5–12).

As in the black-tailed deer described above, these red deer died principally of starvation and exposure, except for a light harvest of adult stags. An effort was made by Evans to kill only mature stags, so there is not the heavy mortality of young adult males noted for the black-tailed deer.

The Dall sheep

From 1939 to 1941 Adolph Murie, working in Dall sheep habitat near Mt. McKinley, where wolves were common, recorded the age, sex

TABLE 5–12. Life tables for the red deer of Jura (after Evans)

A. Males

x (years)	l_x	d_x	q_x	$\mathring{e}_x$
0 - 1	1000	610	610	2.9
1 - 2	390	...	...	5.6
2 - 3	390	5	128	4.6
3 - 4	385	26	675	3.6
4 - 5	359	65	181	2.9
5 - 6	294	86	293	2.4
6 - 7	208	61	293	2.2
7 - 8	147	60	408	1.7
8 - 9	87	35	402	1.8
9 - 10	52	22	423	1.7
10 - 11	30	13	433	1.5
11 - 12	17	8	471	1.3
12 - 13	9	5	556	0.9
13 - 14	4	4	1000	0.5

B. Females

x (years)	l_x	d_x	q_x	$\mathring{e}_x$
0 - 1	1000	485	485	3.1
1 - 2	515	60	117	4.5
2 - 3	455	...	...	4.0
3 - 4	455	126	277	3.0
4 - 5	329	91	276	3.0
5 - 6	238	65	273	3.0
6 - 7	173	48	278	2.9
7 - 8	125	35	280	2.8
8 - 9	90	25	278	2.7
9 - 10	65	18	277	2.6
10 - 11	47	13	277	2.4
11 - 12	34	10	294	2.1
12 - 13	24	7	293	1.7
13 - 14	17	5	294	1.2
14 - 15	12	12	1000	0.5

and probable year of death of every sheep carcass he could find (Murie 1944). Taking only those which died from 1937 to 1941, when the population was relatively stable, it is possible to construct life tables (Table 5–13) for the two sexes if the following surmises are accepted.

Ewes first breed at 2.5 years.
Average production per adult ewe is 1.0 lamb.
The sex ratio at birth is 100 ♂♂ : 100 ♀♀.
The loss of yearlings is not more than 10 percent.

The sheep which could not be aged exactly because the horns were missing, but which were known to be at least nine years old, had the same age-distribution as the 9-year-plus-older sheep which could be aged from the horns.

These data show for both sexes a low rate of loss between the first and ninth year. The animals were not subjected to hunting, but they were

TABLE 5–13. Life tables for the Dall sheep of Mt. McKinley (after Murie)

A. Males

x (years)	l_x	d_x	q_x	$\mathring{e}_x$
0 - 1	1000	718	718	3.3
1 - 2	282	20	71	9.3
2 - 3	262	3	11	9.0
3 - 4	259	...	...	8.1
4 - 5	259	3	11	7.0
5- 6	256	5	20	6.1
6 - 7	251	...	...	5.2
7 - 8	251	14	56	4.2
8 - 9	237	6	25	3.5
9 - 10	231	28	120	2.5
10 - 11	203	40	200	1.8
11 - 12	163	70	429	1.1
12 - 13	93	87	935	0.6
13 - 14	6	...	...	1.5
14 - 15	6	6	1000	0.5

B. Females

x (years)	l_x	d_x	q_x	$\mathring{e}_x$
0 - 1	1000	718	718	2.8
1 - 2	282	28	99	7.6
2 - 3	254	6	24	7.4
3 - 4	248	3	12	6.6
4 - 5	245	3	12	5.6
5 - 6	242	8	33	4.7
6 - 7	234	17	73	3.8
7 - 8	217	9	41	3.1
8 - 9	208	11	53	2.2
9 - 10	197	93	472	1.3
10 - 11	104	59	567	1.0
11 - 12	45	36	800	0.7
12 - 13	9	9	1000	0.5

constantly under heavy pressure from wolves, which took mainly the lambs and very old adults. This apparently left the prime animals, according to Murie's range observations, with ample food supplies and a high life expectancy.

Population characteristics

When survivorship curves, based on the l_x columns in the life tables, are plotted, as in Figures 5–1 and 5–2, various segments may be compared. Both sexes of all the species are alike in having a very steep initial slope, indicating a high mortality during the first year of life.

Among the males the loss during the second year of life is small in all species except the roe-deer, where there is emigration, and the black-tailed deer of the chaparrel where some yearlings are lost, partly through being mistaken for older deer by hunters. During the third year of life there is a heavy loss among roe-deer and black-tailed deer from both range types. The roe-deer loss is due to emigration. That in the black-tailed deer is due to hunting. In the Dall sheep and the red deer there is little loss during the third year. From the fourth year onward to old age

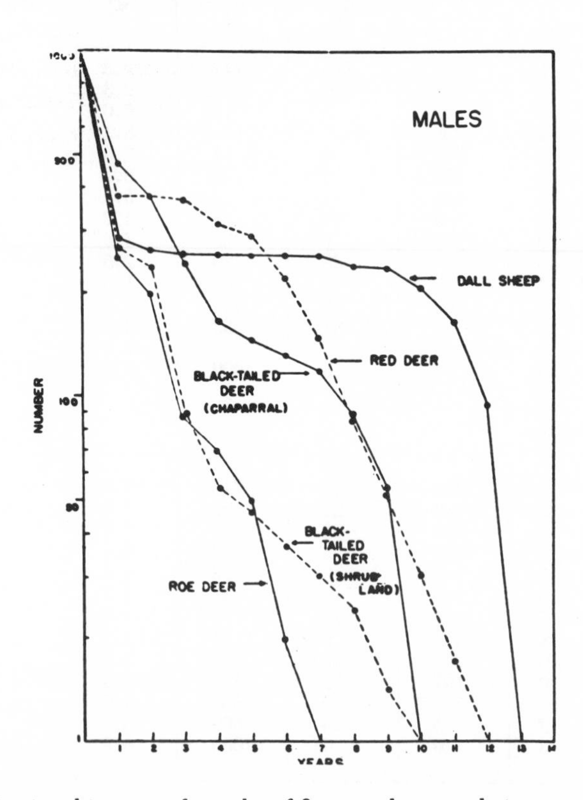

Fig. 5–1. Survivorship curves for males of five ungulate populations.

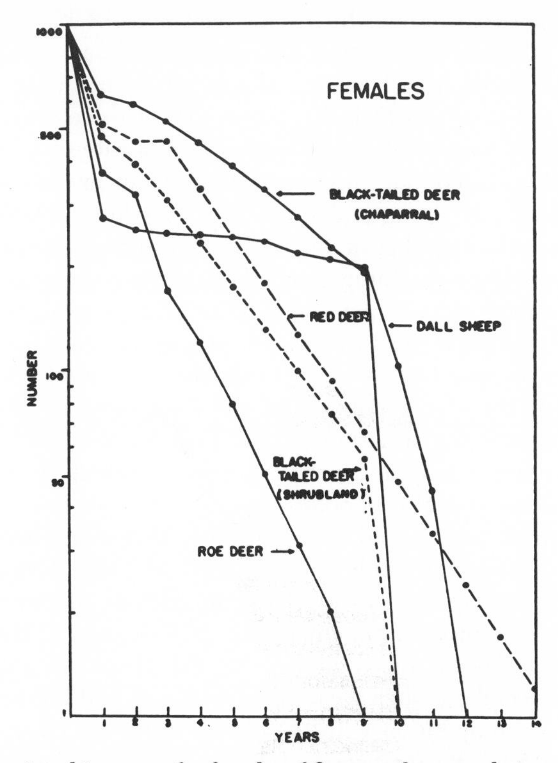

Fig. 5–2. Survivorship curves for females of five ungulate populations.

the hunted populations (roe-deer, black-tailed deer and red deer) show fairly steep losses. The rate of loss tends to become less in full adulthood in the black-tailed deer, because learning and behavior make these individuals less vulnerable to hunting. In the red deer the rate of loss becomes heavier in full adulthood because of the selection of prime stags by the hunter. The Dall sheep, which is not hunted, shows very little loss from adulthood to old age. If it were not for hunting, the other male populations would exhibit survivorship curves more nearly like that of the Dall sheep, but it is doubtful if they could ever attain as high a survivorship as long as their ranges were fully stocked and starvation was a common cause of death. The effects of competition are discussed below. The stress imposed by the annual growth and shedding of antlers, peculiar to the Cervidae, is another factor, of unknown magnitude.

In old age there tends to be a steepening of the survivorship curve—the accelerated loss due to senescence—especially in the Dall sheep. The red deer do not display this, because the high kill by hunting permits few individuals to grow old.

Among the females there is a small loss during the second year of life in all examples. In none of these populations do the females regularly bear young at one year of age. In the roe-deer and the black-tailed deer the usual time of first bearing is on the second birthday, although some black-tailed deer, especially from the chaparral, do not bear until three years old. The red deer and Dall sheep bear first at three years.

After the third year of life all examples have a rate of loss similar to that maintained until old age. In general, there is an inverse relationship between production of young and survival of breeding females. Presumably this is due to the fact that on any fully-stocked herbivore range there tends to be competition for forage, and the more successful is the reproduction, the keener is the competition. The survival of adult females may be lowered still further if they are hunted, as was true to a limited extent for the red deer of Jura. On the other hand, the survival of adult females may be raised if there is a removal of the competing members of the population. This was the situation for the Dall sheep of Mt. McKinley, where both the very young and very old were removed by wolves. The roe-deer population was also periodically lowered, reducing competition; the reduction there was caused by emigration. However, there was also a hunting loss among the adult females, and this tended to cancel the advantage conferred upon them by the emigration of the young.

The most clear-cut example among the present cases is the comparison of survival between the adult deer of the chaparral and those of shrubland. In December, when weaning is long past and the young of the year are in direct competition with the old for food, the chaparral population contains 0.53 fawn for every adult doe. The shrubland population contains 0.76 fawn for every adult doe (Tables 5–6 and 5–8). The values for life expectation of prime (3- to 6-year-old) does ranged in chaparral from 4.2 to 2.6 years and in shrubland from 3.0 to 2.2 years. Fawn production to December in chaparral is about 70 percent of that of shrubland; life expectation in shrubland is about 78 percent of that in chaparral. The comparison could be expanded to include the whole population of both sexes—taking the fawns plus yearlings on the one hand as the production, and all older deer together as the producers. The result would be much the same; the greater the proportional production, the shorter the life expectancy of the producers.

It seems that most of the differences between these various populations are not inherent in the species, but rather are imposed by environmental conditions. The limitation of the roe-deer population by emigration, for example, would be effective only if the population occupied an island of habitat, surrounded by uninhabited areas into which the emigrants could disappear and from which none would come to replace them. The situation in the shrubland is similar, but the island-like nature of the habitat is less marked, and an immigration from the surrounding low-density population may partially balance emigration. In the chaparral, which is merely an artificially designated area in a large region of similar habitat, emigration and immigration should balance.

If a roe-deer population is fenced, Andersen points out (*in litt.*) that winter mortality due to starvation is high. In such a case it would be expected that the population dynamics would be different from those found at Kalø, since the older deer would be competing with younger, physiologically more efficient animals.

Similarly, if a Dall sheep herd were not culled by predators, the population would presumably be limited by food supplies, and the mortality among prime adults would increase accordingly.

These considerations point up the danger of considering the population dynamics of a given animal under given circumstances as typical of that species in general.

SUMMARY

Population dynamics have been reconstructed for three natural populations of the Columbian black-tailed deer (*Odocoileus hemionus columbianus* (Richardson)) which differed from each other in habitat. The three habitats, chaparral, shrubland and wildfire burn, were all modifications by man of the chaparral association of the North Coast Range of California. Chaparral is a densely-growing association of shrubs; shrubland is a shrub-herb interspersion; wildfire burn is chaparral recently burned, where the shrubs are sprouting from the root-crown. These populations were compared for stability, density, structure, reproduction, mortality and movement. The principal causes of mortality were hunting (adult bucks) and starvation (other classes). The details of population gain and loss through one typical year (four years for wildfire burn) were reconstructed from these data. Some characteristics of the three populations are shown in Table 5–14.

TABLE 5–14.

	Chaparral	Shrubland	Wildfire burn (First year following burning)
Population density (Deer per square mile)			
July	33	84	112
December	27	64	86
Reproduction (fawns per 100 adult ♀♀)	0.77	1.65	1.32
Importance of movement in population gain or loss	none	little (yearling does)	great (all classes)

Populations in chaparral and shrubland were relatively stable, so life tables were prepared for them, the sexes being treated separately. The population on the wildfire burn was unstable, high at first due to influx and then dropping rapidly due to egress, increased mortality and decreased reproduction.

Sex-specific life tables for the deer of chaparral and shrubland were compared with those for: a hunted red-deer population on a fully-stocked island; a hunted roe-deer population from which there was heavy emigration; and an unhunted Dall sheep population under heavy predation by wolves.

Some characteristics of these populations are shown in Table 5–15.

TABLE 5–15.

	Average annual population loss (per cent)		Loss during first year of life (per cent)		Life expectancy from birth (years)	
	♂♂	♀♀	♂♂	♀♀	♂♂	♀♀
Black-tailed deer						
Chaparral	36	22	53	37	2.3	4.2
Shrubland	55	34	73	53	1.3	2.4
Red deer	30	28	61	49	2.9	3.1
Roe deer	59	45	74	63	1.2	2.6
Dall sheep	42	30	72	72	3.3	2.8

On fully-stocked ranges there is an inverse relationship between production of young and survival of breeding adults, due presumably to competition for food between young and adults. When young (or senescent) are removed, by predation or emigration, the chances for survival of breeding adults appear to be improved.

The differences in dynamics between these various populations of different species, are apparently due more to environment than to heredity.

REFERENCES

Andersen, J. 1953. Analysis of a Danish roe-deer population. Danish Rev. Game Biol., 2:127–155.

Biswell, H. H., R. D. Taber, D. W. Hedrick, and A. M. Schultz. 1952. Management of chamise brushlands for game in the north coast region of California. Calif. Fish. and Game, 39:453–484.

Cooper, W. S. 1922. The broad-sclerophyll vegetation of California. Carnegie Inst. of Wash., Pub. 319. 14 pp.

Darling, F. F. 1937. A herd of red deer. London: Oxford Univ. Press, 210 pp.

Dasmann, R. F. 1953. Factors influencing movement of nonmigratory deer. Western Assn. State Game and Fish Comm., 33:112–116.

Dasmann, R. F., and R. D. Taber. 1955. A comparison of four deer census methods. Calif. Fish. and Game, 41:225–228.

———. 1956. Determining structure in Columbian black-tailed deer populations. Jour. Wildl. Mangt., 20:78–83.

Deevey, E. S. 1947. Life tables for natural populations of animals. Quart. Rev. Biol., 22:283–314.

Evans, H. 1891. Some account of Jura red deer. Privately printed by Francis Carter, of Derby. 38 pp.

Hickey, J. J. 1952. Survival studies of banded birds. U.S. Fish & Wildl. Serv., Spec. Sci. Rept. (Wildlife) No. 15, Washington. [Processed.]

Manning, G. A., and B. A. Ogle. 1950. The geology of the Blue Lake quadrangle. Calif. Dept. Nat. Res., Div. Mines, Bull. 148. 36 pp.

Moreland, R. 1952. A technique for determining age in black-tailed deer. Western Assn. State Game and Fish. Comm., 32:214–219.

Murie, A. 1940. The wolves of Mount McKinley. U.S. Dept. Interior, Nat. Park Serv., Fauna of Nat. Parks of the U.S., Series 5, 238 pp.

Severinghaus, C. W. 1949. Tooth development and wear as criteria of age in white-tailed deer. Jour. Wildl. Mangt., 13:195–216.

Taber, R. D. 1952. Studies of black-tailed deer reproduction on three chaparral cover types. Calif. Fish and Game, 39:177–186.

———. 1956. Deer nutrition and population dynamics in the North Coast Range of California. North Amer. Wildl. Conf., Trans., 21:159–172.

———, and R. F. Dasmann. 1954. A sex difference in mortality in young Columbian black-tailed deer. Jour. Wildl. Mangt., 18:309–315.

6

History, food habits and range requirements of the woodland caribou of continental North America

Alexander Thom Cringan

The eastern and western woodland caribou, *Rangifer tarandus caribou* (Gmelin) and *Rangifer tarandus sylvestris* (Richardson) (Cringan, 1956) are two of the generally accepted thirteen native races of North American caribou recorded within recent times. Their populations have declined greatly since the settlement of North America by Europeans, as have the populations of all other native caribou. The purpose of this paper is to review the history of these two forms and their present status, and to present the results of a food habits and range study of the woodland caribou of the Slate Islands, Lake Superior, Ontario, which was done in 1949.

Woodland caribou formerly ranged from Prince Edward Island and Nova Scotia to western Alberta or British Columbia, south into New York, New Hampshire, Vermont and Maine in the east, and Minnesota, Wisconsin and Michigan in the Great Lakes Region, north to southern Ungava in the east and the North West Territories in the west. The morphological distinctions between these two supposed races are unclear, as is the line separating their ranges. Indeed, neither the structural differences nor the geographic ranges of races of North American caribou in general

Reprinted with permission from *Trans. 22nd North American Wildlife Conference*, pp. 485–501 (1957). Field investigations were under the sponsorship of the Research Council of Ontario and the Carling Conservation Club. The author is currently affiliated with the Department of Zoology, University of Guelph, Guelph, Ontario.

are well worked out or commonly agreed upon. For this reason, I will not further distinguish between the eastern and western woodland caribou.

HISTORY

Eastern Range The woodland caribou formerly occurred throughout a large area south of the St. Lawrence River, where its range included the Gaspé Peninsula of Quebec, New Brunswick, Nova Scotia, Prince Edward Island, extreme northern New Hampshire and Vermont, and extreme northeastern New York.

Caribou permanently disappeared from Prince Edward Island between 1672 and 1873 (Adams, 1873), and from New York prior to 1800 (De Kay, 1842). They occasionally appeared in Vermont until 1840 and in New Hampshire until about 1865 (Seton, 1953). The caribou population in Maine fluctuated considerably (G. M. Allen, 1942), but decreased noticeably before 1900 (Palmer, 1938), and was exterminated about 1916 (Seton, 1953). The only caribou reliably reported in Maine since then was in 1946 (Palmer, 1949).

The woodland caribou began to decline in Nova Scotia before 1900, and disappeared from the mainland of this Province by 1915 and from Cape Breton Island by 1924 (Anderson, 1946). The population in New Brunswick was decreasing noticeably by 1915 (Anderson, 1939), but a few animals remained until about 1927.

There has been a permanent population of caribou on the Gaspé Peninsula. It was much more numerous between 1900 and 1915 than at present (Moisan, 1956a). Despite this decline, there are still about 700 woodland caribou occupying the remaining 400 square miles of winter range, suggesting that the population is still high in relation to the amount of habitat available.

In Quebec to the north of the St. Lawrence River, woodland caribou declined first in the northern part of their range, east of James Bay, late in the 19th century, and a little later in the Laurentians to the south (Low, 1896; Riis, 1938).

Reports of eastern woodland caribou in Labrador are questionable owing to possible confusion with the Ungava caribou *Rangifer tarandus caboti* (G. M. Allen.)

Central Range In the central part of its range, the woodland caribou originally occupied the northern half of Minnesota, extreme northern Wisconsin, the Upper Peninsula of Michigan and all of Ontario north of the French and Mattawa Rivers. Its status in these areas appears to have changed little until after 1800. By 1850, it had disappeared from Wisconsin (Hoy, 1882), and decreased in Minnesota, Michigan and the southeastern part of its range in Ontario.

Caribou were absent from much of their former range in Minnesota and Michigan in 1900, but still occupied most of their orginal range in Ontario, although in reduced numbers in the eastern half of the Province. Woodland caribou generally diminished after 1900 in this section, with the possible exception of in the Patricia Portion of Ontario for which specific information was lacking until 1920 or later.

The last caribou reported from the Upper Peninsula of Michigan was in 1906, and from Michigan (Isle Royale) in 1931 (Riis, 1938). Caribou remained in the Superior National Forest, Minnesota, until 1925 (Adams, 1926) and in northwestern Minnesota until 1942 (Nelson, 1947). A caribou was seen near Manitou Rapids, Minnesota, in the winter of 1954–55 (L. Magnus, personal communication, 1955).

In Ontario, woodland caribou disappeared from 50,000 square miles of range to the east of Lake Superior and 30,000 square miles to the northwest of Lake Superior between 1900 and 1950. About 250,000 square miles of the original range of woodland caribou in Ontario remains occupied, although in many places the distribution is spotty.

Western Range Woodland caribou formerly occurred in all of Manitoba but the extreme southern and northern parts, the northern half of Saskatchewan, northern Alberta, the southwestern part of the North West Territories, and possibly in northeastern British Columbia. It decreased markedly in Manitoba but did not experience such range reductions as in Ontario. After reaching a low in this Province between 1930 and 1950, it now seems to be increasing (G. W. Malaher, personal communication, 1957). The history of woodland caribou in the rest of the western range is poorly recorded, but apparently the species has declined generally since settlement.

PRESENT STATUS OF WOODLAND CARIBOU

The eastern and western woodland caribou are now confined for practical purposes to the Provinces of Quebec, Ontario, Manitoba, Saskatchewan and Alberta and the North West Territories.

In Quebec, there are about 700 woodland caribou on the Shickshock Mountains in the Gaspé Peninsula, and possibly between 1,500 and 2,000 to the north of the St. Lawrence River between Anticosti Island and the Saguenay River (G. Moisan, personal communication, January 22, 1957). An additional number occur in the vicinity of James Bay and further north (Banfield, 1949). Woodland caribou are thought to be increasing in Quebec.

Woodland caribou currently occur right across northern Ontario, in the west, south to a line from Minaki to Savant Lake to the Black Bay Peninsula of Lake Superior, and in the east, south to a line from Pukaskwa on Lake Superior to Swastika. Along the southern edge of this range scattered isolated herds of caribou occur. Distribution is more continuous to the north, and along Hudson Bay, scattered voids occur. Distribution varies slightly from year to year. The total of estimated caribou populations of forest districts in Ontario was 7,200 in 1953–54 (Cringan, 1956). There are few places in Ontario where estimates exceed 1 caribou per 20 square miles over large areas (of several thousand square miles), yet there are some small areas such as the 15-square mile Slate Islands with estimated densities of between 2 and 3 woodland caribou per square mile. The woodland caribou in Ontario has undoubtedly increased since 1950, on the basis of the best information available at that time (de Vos and Peterson, 1950).

The woodland caribou is currently increasing in Manitoba, according to G. W. Malaher, Director of Game and Fisheries (personal communication, 1957), who thinks there now may be approximately 4,000 or more woodland caribou in that Province. I have no information on the current status of woodland caribou in Saskatchewan.

J. G. Stelfox (personal communication, January 11, 1957) estimated that there are between 700 and 1,000 woodland caribou in Alberta and remarked that the population was reasonably stable.

There are a few recent records of woodland caribou in the North West Territories (Banfield, personal communication, February 1, 1957), but no estimates of populations in this area are at hand. Reports of woodland caribou in British Columbia (Banfield, 1949) are subject to some doubt, as the form is not considered to occur in that Province (J. Hatter, personal communication, January 10, 1957).

The total of these estimates is between 14,000 and 15,000 woodland caribou, to which estimates for parts of Quebec, Saskatchewan and the North West Territories must be added. It seems probable that complete estimates for Canada would total between 20,000 and 25,000 animals. Between two-thirds and three-quarters of these are in Manitoba, Quebec and Ontario, in which provinces a phase of population increase is being experienced.

FACTORS INFLUENCING POPULATIONS OF WOODLAND CARIBOU

Emigration has been postulated as a cause of woodland caribou declines in the Maritimes (Moisan, 1956a), New England (Allen, 1942), Quebec (T. Fortin in personal communication to L. A. Richard, 1940) and Ontario (Millais, 1907). Since confirming reports of immigration into neighboring areas are lacking, and increasing moose and deer populations were usually associated with such declines, it seems likely that in these cases caribou decreased because of changes in ecological conditions rather than as a result of emigration.

Shooting has been frequently listed as a cause of woodland caribou declines. The gregarious instincts, nomadism and low reproductive potential of the form render it very vulnerable to overexploitation. Kills of 13 caribou in 3 days on Caribou Island, Ontario (Quaife, 1921), 120 caribou in a day in Nova Scotia (Anderson, 1939) and 2,400 in a winter on Manitoulin Island, Ontario (Blair, 1911), prove that populations were sometimes overexploited. As there was no real check on caribou hunting through law enforcement until late in the 19th century, it seems likely that shooting was formerly an important factor. Yet since then, without exception, law enforcement itself has failed to prevent the ultimate extinction of woodland caribou in areas in which the range has been drastically modified by man, such as Nova Scotia, New Brunswick, New England, Michigan, Minnesota and certain parts of other Canadian Provinces.

There is no adequate evidence that predation, disease or parasitism could account for general declines in woodland caribou populations. An epizootic may have accompanied one decline in the Gaspé Peninsula (Moisan, 1956a), and an infestation of warble flies (probably *Oedemagena tarandi* Linnaeus) may have been associated with a decline in central Quebec (Robertson, personal communication to L. A. Richard, 1939). Similarly, there is no adequate evidence establishing the effect of weather and climate on woodland caribou populations. Still, none of these factors should be disregarded in future investigations.

Lasting diminution and eradication of woodland caribou populations have been almost always associated with important changes in the environment. The woodland caribou seems to require extensive areas of mature coniferous or mixed-wood forest. In Nova Scotia, the final decline coincided with reduction of the areas of virgin forest to less than 2 percent of the total forested area (Fernow, 1912). The remaining woodland caribou range in the Gaspé Peninsula is being reduced through encroachment by logging, mining and forest fires (Moison, 1956a). In Ontario, there is a relationship between presence of caribou and remaining amount of mature forest as shown in Table 6–1.

Intergeneric competition with moose and deer has also been suggested as a factor in the decline of caribou. Since moose and deer are mammals of the early successional stages, caribou of the climax, this competition is probably unimportant. Intrageneric competition, such as between the woodland and barren-ground caribou on the latter's winter range (Clarke, 1940) logically would have a much greater effect on woodland caribou populations.

Range, and the factors affecting it, forest fires, logging, settlement and the caribou themselves, seemed likely to hold the answer to the great

TABLE 6–1. Occurrence of woodland caribou in relation to forest cover in Ontario

Forest District or Wildlife Management District	Per Cent of Productive Forest in Mature Coniferous or Mixedwood and Muskeg	Status of Woodland Caribou	Square Miles Per Estimated Woodland Caribou
Patricia West	67% or more	uncommon	20-25
Patricia Central	67% or more	uncommon	20-25
Patricia East	67% or more	scarce to uncommon	?
Kapuskasing	63.2%	scarce	40
Cochrane	54.5%	very scarce	300
Geraldton	53.2%	scarce	50
Gogama	48.6%	absent	....
Chapleau	47.6%	absent	....
White River	41.1%	scarce	60-70
Sioux Lookout	39.8%	uncommon	20
Kenora	31.4%	very scarce	?
Port Arthur	31.3%	very scarce	300
Sault Ste. Marie	25.2%	absent	....
Fort Frances	24.3%	absent	....
North Bay	24.0%	absent	....
Sudbury	20.1%	absent	....
Timiskaming	18.4%	very scarce (1 in district)	

Basis: Reports of the Ontario Department of Lands and Forests.

woodland caribou decline of 1750–1950. Therefore I selected a study of woodland caribou range and food habits as the subject of graduate research.

THE STUDY AREA AND METHODS

Field work was carried out in 1949 on the Slate Islands, Lake Superior, because the only ungulate present was the woodland caribou, the population was sufficiently isolated to be relatively unaffected by egress and ingress, and the population was sufficiently dense (between 2 and 3 caribou per square mile) for study purposes.

The Slate Islands are a group of 8 islands of 30 acres or more totaling some 15 square miles in area, situated in Lake Superior 8 miles south of the village of Jackfish. They are made up of Prekeeweenawan and Keeweenawan rocks (Parsons, 1918). The topography is rugged with high hills and steep cliffs being prominent. They are within the Superior Section of Halliday's (1937) Boreal Forest Region.

White birch, *Betula papyifera* Marsh., and balsam fir, *Abies balsamea* (L.) Mill. are the commonest trees of the islands. The flora of the islands is interesting as it includes certain relict species like *Dryas integrifolia* Vahl and *D. Drummondii* Richards.

The vegetation of the Slate Islands has been influenced by an extensive fire between 50 and 70 years ago, two periods of logging activity, the first coinciding with the fire and the second about 1930 (A. L. Parsons, personal communication, 1950) and in recent years by the actions of woodland caribou.

The only species of mammals recorded on the Slate Islands in 1949 were woodland caribou; varying hare, *Lepus americanus*; muskrat, *Ondatra zibethica*; red-backed vole, *Clethrionomys gapperi*; meadow vole, *Microtus pennsylvanicus*; colored fox, *Vulpes fulva*; short-tailed weasel, *Mustela eriminea;* and little brown bat, *Myotis lucifugus*. Evidence of a former population of beaver, *Castor canadensis*, was noted. The history of woodland caribou on the islands is poorly recorded. Parsons (1918; personal communication, 1950) did not see caribou there in 1918. Possibly the current population is of fairly recent origin.

The principal technique used to study range and food habits of the woodland caribou of the Slate Islands was a composite range study technique. Among information recorded was the following.

General Forest Description

Subjective appraisals of forest type, height of trees, crown density, forest age and site were made on the basis of classes set forth by Seely (1949). Tree species present, aspect, slope and moisture were also recorded. Six forest types, based on two moisture regimes and three compositions, were subsequently recognized.

Browse Analysis

Winter browsing by woodland caribou was sampled by using the Aldous system to measure availability and utilization of woody browse on 495 plots each of 1/100th of an acre in area (Aldous, 1944; Aldous and Krefting, 1946). A stratum between two and ten feet from the ground was considered to have browse available to caribou in the winter. A weighted average degree of browsing was used in analyzing the results. Individual densities were multiplied by their respective browsing values, the results totalled and divided by the sum of densities, rather than the browsing values being totalled and divided by the number of occurrences, to compute the unweighted average degree of browsing used by Aldous and Krefting.

Tree Lichens

Species of lichens growing on branches of trees such as old man's beard, *Usnea* spp. and oakmoss, *Evernia prunastri*, were grouped together and their abundance on 1/100th acre plots and utilization subjectively estimated, using the same coefficients as in the Aldous browse analysis system. Abundance was appraised on the basis of amounts of tree lichens just beyond reach of the caribou in relation to the densest stands encountered. Utilization was judged on the basis of the difference in amounts of tree lichens within reach of caribou and beyond their reach that appeared explainable due to grazing.

Ground Lichens

Lichens growing on the ground were tallied in three groups, as reindeer mosses (*Cladonia rangiferina, C. alpestris* and similar forms), foliose lichens (*e.g., Umbillicaria* spp., *Peltigera* spp.), and other ground lichens. Occurrence and areal density of these forms on 1/100th acre quadrats were recorded. Where the density of reindeer mosses exceeded 5 percent, degree of grazing was also estimated, again using the Aldous coefficients for utilization.

Additional information on caribou food habits was gained through observation of feeding animals and through detection of grazing of herbs and leaf-browsing of shrubs.

WINTER FOOD HABITS OF THE SLATE ISLANDS CARIBOU

Results of the Aldous browse survey are given in Table 6–2. No species of woody forage was more than lightly utilized. As the population

TABLE 6–2. Woodland caribou winter browse analysis Slate Islands—495 plots, 1949

Species	Frequency Per Cent of Plots Present	Mean Density	Weighted Average Degree Browsing	Utilization Factor	Per Cent of Woody Food Eaten	Per Cent of Browse Available
Mountain maple	38	6.5	4.5	27.9	41.5	11.7
Mountain ash	35	2.1	4.2	8.7	13.1	3.7
Willows	22	2.1	3.1	6.6	9.9	3.8
Red-osier dogwood	26	3.7	1.5	5.5	8.2	6.6
Highbush cranberry	20	1.0	3.9	4.0	6.0	1.8
Red-berried elder	24	1.3	3.0	3.9	5.8	2.3
Ground hemlock	14	3.0	1.1	3.2	4.8	5.3
Red cherry	7.5	0.73	2.4	1.8	2.6	1.3
White birch	52	3.7	0.47	1.7	2.6	6.5
Juneberry	10	0.52	2.2	1.1	1.7	0.9
Balsam fir	86	18.1	0.04	0.72	1.1	32.3
Trembling aspen	12	0.69	0.81	0.56	0.8	1.2
Salmonberry	16	2.5	0.16	0.41	0.6	4.5
Balsam poplar	2.0	0.20	1.5	0.30	0.4	0.4
Bush honeysuckle	7.3	0.46	0.22	0.10	0.2	0.8
Ribes spp.	11	0.66	0.15	0.10	0.2	1.2
Raspberry	13	0.62	0.08	0.05	0.1	1.1
White cedar	3.4	0.32	0.16	0.05	0.1	0.6
Buffaloberry	1.0	0.10	0.50	0.05	0.1	0.2
Blueberry	0.2	0.01	5.0	0.05	0.1	tr.
Rose	2.8	0.14	0.36	0.05	0.1	0.3
Black spruce	18	1.6				2.9
Speckled alder	12	4.0				7.1
Mountain alder	1.2	0.16				0.3
White spruce	15	1.2				2.2
Labrador tea	1.3	0.28				0.5
Common juniper	0.6	0.08				0.1
Tamarac	0.4	0.02				tr.
Sweet Gale	0.4	0.28				0.5
Andromeda	0.2	0.01				tr.
Leatherleaf	0.2	0.06		...		0.1

density of caribou was high at the time, it appears that browse is relatively unimportant in the winter diet of the Slate Islands caribou.

Mountain maple, *Acer spicatum* Lam. was the principal local woody winter food of Slate Islands caribou at the time of the study, and was followed by mountain ash, *Pyrus americana* (Marsh.) DC, willows, *Salix* spp., red-osier dogwood, *Cornus stolonifera* Michx. and highbush cranberry, *Viburnum rafinesquiana* Shultes. These five species together contributed about 75 percent of the winter browse consumed. At least 16 other species of woody plants were also eaten in the winter.

Reindeer mosses were heavily grazed. The mean degree of grazing of reindeer mosses within dry series types, which occupied 35 percent of the island and supported 90 percent of the reindeer mosses, was 50 percent. This suggests a lower-than-actual degree of utilization, for 70 percent was the maximum recordable grazing for any one plot, yet many

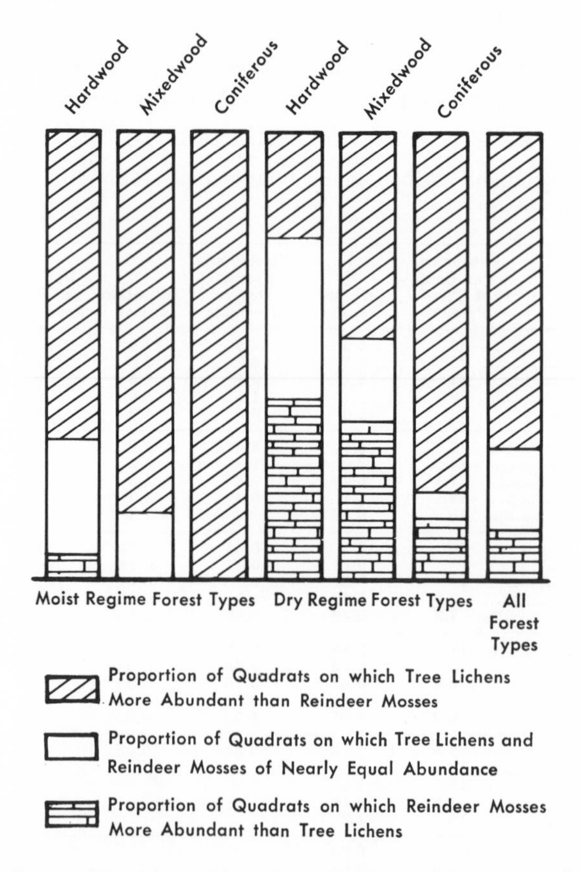

Figure 6–1. Relative abundance of tree lichens and reindeer mosses within principal forest types on the Slate Islands, Lake Superior, Ontario.

stands were reduced to 10 percent and less of their potential volume of reindeer mosses through the action of many years' grazing and accompanying trampling.

Utilization of tree lichens was also heavy, averaging about 41 percent out of a possible maximum of 70 percent. Utilization of tree lichens was heavy in all mixed-wood and coniferous types.

The utilization of woody foods by Slate Islands caribou was so low contrasted to that of arboreal and terrestrial lichens, that clearly the supply of lichens is critically important, while that of browse is relatively unimportant. No measure contrasting the volumes or weights of browse and lichens consumed in winter can be made from the results of this study.

It is more difficult to assess the relative importance of reindeer mosses and tree lichens. Both were heavily utilized, but as reindeer mosses grow in a nearly two-dimensional environment, and tree lichens in one which is three-dimensional, a direct contrast of abundance and utilization is not possible. To solve this problem, I examined all plot tally sheets to determine whether tree lichens or reindeer mosses were most abundant on the individual plot. If any doubt existed, they were classified as plots where the two groups were of nearly equal abundance. Figure 6–1 shows the result of this procedure. Tree lichens were more abundant than reindeer mosses in 70 percent of all plots, while reindeer mosses exceeded tree lichens on only about 10 percent of all plots! The only forest type in which reindeer mosses were generally more abundant than tree lichens was the dry hardwood type. Because of availability, it appears that tree lichens were in 1949 much more important than reindeer mosses in the diet of woodland caribou of the Slate Islands.

SPRING AND SUMMER FOOD HABITS OF THE SLATE ISLANDS CARIBOU

In the spring, caribou on the Slate Islands paw for roots and shoots of herbs, and for mosses, lichens and fungi. In the summer, they eat leaves of deciduous shrubs, herbs, lichens and some aquatic plants, possibly including algae.

Foods most frequently taken in the spring included the large-leafed aster, *Aster ciliolatus* Lindl.; bunchberry, *Cornus canadensis* L.; mosses and lichens. Herbs most frequently grazed in the summer included large-leafed aster; sarsaparilla, *Aralia nudicaulis* L.; Ferns; and fireweed, *Epilobium angustifolium* L. The shrubs most frequently leaf-browsed in summer were highbush cranberry, mountain ash, mountain maple and bush honeysuckle, *Diervilla Lonicera* Mill.

ENVIRONMENTAL FACTORS GOVERNING PRODUCTION OF LICHENS

On the Slate Islands, reindeer mosses grow best in dry, open places. They occur more than twice as frequently in dry regime forest types than in moist regime forest types of the same composition and are more than ten times as abundant in the dry types. These relationships are shown in Table 6–3. Relationships between reindeer moss and tree crown density

TABLE 6–3. Lichen characteristics of principal Slate Islands forest types

	Forest Type					
	Medium Moisture Regime			Dry Moisture Regime		
Reindeer Mosses	Hardwood	Mixedwood	Coniferous	Coniferous	Mixedwood	Hardwood
Per cent of plots with reindeer mosses	31	38	27	67	79	74
Per cent of plots where reindeer moss density 5 per cent or more	1.5	0.5	3.3	35	41	38
Mean density of reindeer mosses	0.4	0.4	0.4	4.9	10	4.4
Mean degree of grazing of reindeer mosses	?	?	?	54	59	26
Per cent of islands' area occupied by type	14.4	44.4	6.5	7.3	18.3	9.1
Per cent of all reindeer mosses grown within type	2.1	6.7	0.9	12.4	64.2	13.7
Tree Lichens						
Mean density of tree lichens—Per cent	6.0	25	47	34	20	4.9
Mean degree of browsing of tree lichens—Per cent	20	52	46	42	40	8
Per cent of all tree lichens grown within type—Per cent	4	52	14	11	17	2

in dry regime forest types are shown in Figure 6–2. Mean density of reindeer mosses exceeded 10 percent in dry regime stands with crown densities of less than 45 percent, but was only a fifth as abundant where the tree crown density exceeded 65 percent. A vegetation survey of tundra winter range of woodland caribou in the Gaspé Peninsula showed that ground lichens occupied about 12 percent of the area (Moisan, 1956b), a condition not unlike that in the more open dry regime forest types of the Slate Islands. The influence of the caribou themselves on the abundance of reindeer mosses could not be measured, although it was obviously great.

Abundance of tree lichens depends on forest composition, moisture conditions, tree crown density and age of the forest. Table 6–3 shows that the ratio of tree lichen densities in coniferous, mixed-wood and hardwood stands is about 7.5 : 4 : 1, and that the densities are about one-quarter greater in medium regime types than in dry regime types of the same composition. Abundance of tree lichens varies directly with age of the forest and tree crown density, the latter being shown for the six prin-

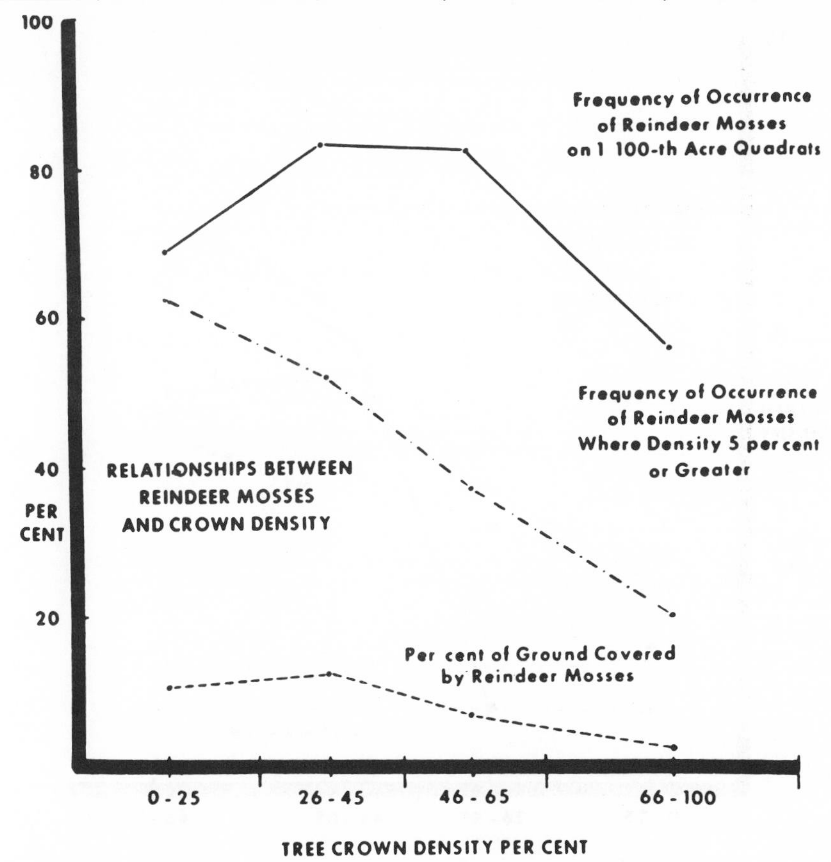

Figure 6–2. Relationship between frequency and areal density of reindeer mosses and tree crown density within dry regime forest types of the Slate Islands, Lake Superior, Ontario.

cipal forest types in Figure 6–3. A further factor influencing supplies of tree lichens available to caribou is the past browsing by caribou. For this reason, availability may be very low even if excellent supplies of tree lichens are to be found in the tree tops. Both reindeer mosses and tree lichens require long periods of time to attain high densities, and both grow slowly, even when abundant.

RANGE REQUIREMENTS OF THE WOODLAND CARIBOU

The results of the Slate Islands investigation suggest that a supply of lichens as winter food is of critical importance to woodland caribou and that tree lichens may be more important than reindeer mosses. In

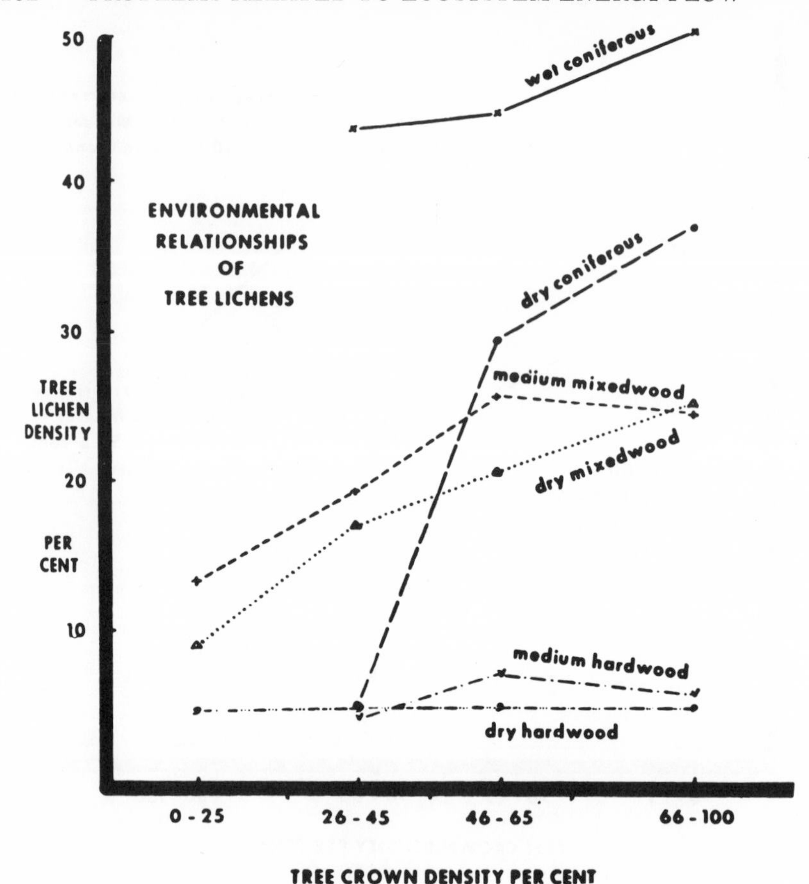

Figure 6–3. Relationship between mean density of tree lichens, and tree crown density within six principal forest types of the Slate Islands, Lake Superior, Ontario.

addition, findings point to some of the major environmental factors controlling production of lichens. The results therefore contribute to an understanding of woodland caribou range requirements.

The distribution of woodland caribou is closely related to that of the northern coniferous forest. It is generally, although not always, associated with climax stands. Climax stands of that forest are typically uneven-aged and usually coniferous or mixed in composition, with spruces, balsam fir and white birch being the principal trees.

There are some communities in this forest which have dense stands of reindeer mosses. If such communities become inhabited by woodland caribou, there is nothing to inhibit the development of a predator-prey oscillation between the caribou and its principal lichen food, the reindeer moss. *If the supply of reindeer moss is adequate to support a healthy population of woodland caribou, and does so, that supply must inevitably become reduced because of the caribou, if for no other reason, simply since the reproductive potential of caribou is much greater than the rate of increase of stands of reindeer moss.* Reindeer mosses might also be subjected to trampling by other ungulates and grazing by rodents, which would act to the detriment of woodland caribou.

There are other communities in the northern coniferous forest, which, although deficient in reindeer mosses, have fair stands of tree lichens. Because they occur as uneven-aged stands, they are fairly stable when considered over large areas. Each year, a certain number of trees mature, die, and a few years later, fall to the ground. With that, tree lichens which were previously inaccessible to woodland caribou come within their reach. Thus a mechanism for a sustained supply of essential woodland caribou foods is set up. Under these conditions, caribou would have but little effect on the supply of the principal lichens.

In full consideration of the instability of reindeer moss stands and the stability of tree lichen yields, it may be theorized that:

Although dense stands of reindeer moss often support populations of woodland caribou, these are apt to be irruptive; the continued existence of steady populations of this animal in many parts of the northern coniferous forest is due to the peculiar tree lichen-producing characteristics of the climax stand of that forest.

If this is the basic essential of caribou range, then any factors affecting the northern coniferous forest so as to reduce total production of tree lichens or impair their sustained production is bound to be detrimental to the woodland caribou. Forest fire, logging and settlement are obviously among factors which could do so. It is reasonable to conclude that these have been important in the decline of woodland caribou. It seems unlikely that other factors which have been mentioned in connection with declines, emigration, shooting, predation, disease, parasitism and weather either could or did lead to extirpation over large areas in the absence of any basic range-altering factor.

FUTURE WOODLAND CARIBOU MANAGEMENT

The woodland caribou requires specific conditions, as provided by the typical mature northern coniferous forest. Disturbances affecting this forest influence caribou through their effects on lichens produced by the forest. Extensive forest fires create even-aged stands; centuries may pass before the uneven-aged climax stand essential to sustained caribou populations is re-established. Logging results in the cropping of trees before they have acquired dense stands of tree lichens; few lichens are produced by forests managed on short rotations for production of pulpwood. Each disturbance affects either abundance or production of tree lichens; each influences woodland caribou. These facts are of basic importance to woodland caribou management.

Forest management, forest fire protection, and harvesting of surplus animals are among the most important tools of manipulative woodland caribou management. The future of woodland caribou within the merchantably forested area of Canada is limited. As logging increases, this interesting wilderness mammal will retreat farther, as it has already done from the United States and from 80,000 square miles of former range in Ontario. However, there is a huge area of non-merchantable forest farther north in Canada which should always have sufficient mature stands to support woodland caribou.

Forest fire protection in this nonmerchantably forested area, for the primary purpose of caribou management, will probably become necessary in the future.

Even though an adequate supply of woodland caribou habitat may be assured as a result of forest management and fire protection, that habitat will be subject to damage by caribou. It is therefore necessary to prepare a harvesting plan, so that herds can be kept within the carrying capacity of their ranges.

Only a few years ago the woodland caribou was on the list of most endangered mammals, yet now it seems to be thriving, and perhaps is too numerous in certain places. If there are between 20,000 and 25,000 woodland caribou in Canada as estimated, an annual harvest of 2,500 or more animals per year would be necessary to stabilize this population. If the experience of Canadian Provinces in attempting to achieve desired harvests of moose from semi-wilderness areas serves as a guide, it will be difficult to harvest the required number of woodland caribou from wilderness areas. Failing such control, the woodland caribou will likely continue to fluctuate markedly in the wilderness portion of its range, just as it has always done.

LITERATURE CITED

Adams, Andrew L. 1873. Field and forest rambles, with notes and observations on natural history of eastern Canada, H. S. King, London.

Adams, Charles C. 1926. The economic and social importance of animals in forestry. Roosevelt Wildlife Bull. 3, Syracuse, N.Y.

Aldous, Shaler, E. 1944. A deer browse survey method. Journ. Mamm., 25(2):130–136.

——, and Laurits W. Krefting. 1946. The present status of moose on Isle Royale. Trans. N. Am. Wildlife Conf., 11:296–308.

Allen, Glover M. 1942. Extinct and vanishing mammals of the western hemisphere. Amer. Com. Internatl. Wildlife Protect., special pub. no. 11, pp. 1–620.

Anderson, Rudolph Martin. 1939 (1938). The present status and distribution of the big game mammals of Canada. Trans. N. Am. Wildlife Conf., 3:390–406.

——. 1946 (1947). Catalogue of Canadian recent mammals. Natl. Mus. Can., Bull. No. 102, Biol. Ser. No. 31, pp. 1–238.

Banfield, A. W. F. 1949. The present status of North American caribou. Trans. N. Am. Wildlife Conf., 14:477–491.

Blair, Emma Helen. 1911. The Indian tribes of the upper Mississippi Valley and region of the Great Lakes as described by Nicholas Perrot. Clark, Cleveland, Vol. 1, pp. 1–372.

Clarke, C. H. D. 1940. A biological investigation of the Thelon Game Sanctuary. Natl. Mus. Can., Bull. No. 96, Biol. Ser. No. 25, pp. 1–133.

Cringan, Alexander Thom. 1956. Some aspects of the biology of caribou and a study of the woodland caribou range of the Slate Islands, Lake Superior, Ontario. Unpublished. M. A. Thesis, University of Toronto, pp. 1–10 and 1–300.

De Kay, J. E. 1842. Natural history of New York. Zoology, pt. 1, Mammalia. D. Appleton and Co. and Wiley and Putnam, New York. pp. 1–146.

de Vos, Antoon and Randolph L. Peterson. 1951. A review of the status of woodland caribou (*Rangifer caribou*) in Ontario. Journ. Mamm., 32(3):329–337.

Fernow, B. E. 1912. Forest conditions of Nova Scotia. Canada Commission of Conservation. pp. 1–93.

Halliday, W. E. D. 1937. A forest classification for Canada. Forest Serv. Bull. 89, Can. Dept. Mines and Resources, pp. 1–50.

Hoy. 1882. Trans. Wisc. Acad. Sci., 5:256.

Low, A. P. 1896. Report on explorations in the Labrador Peninsula along the Nain, Koksoak, Hamilton, Manicuagan and portions of other rivers. Geol. Sur. Can., 8:70L. 86L and 318–320L.

Millais, J. G. 1907. Newfoundland and its untrodden ways. Longmans, Green & Co., New York. pp. 1–340.

Moisan, Gaston. 1956a. Le caribou de Gaspé I. Histoire et distribution. Le Naturaliste Canadien, 83(10):225–234.

———. 1956b. Le caribou de Gaspé II. Le Naturaliste Canadien, 83(11–12):262–274.

Nelson, E. C. 1947. The woodland caribou in Minnesota. Journ. Wildlife Mgt., 11(3): 283–284.

Palmer, Ralph S. 1938. Late records of caribou in Maine. Journ. Mamm., 19(1):37–43.

———. 1949. *Rangifer caribou* in Maine in 1946. Journ. Mamm., 30(4):437–438.

Parsons, A. L. 1918. Slate Islands, Lake Superior. Rept. Ont. Bureau Mines, 27(1).

Quaife, Milo Milton (ed.). 1921. Alexander Henry's travels and adventures in the years 1760–1776. Lakeside Press, Chicago. pp. 1–340.

Riis, Paul B. 1938. Woodland caribou and—time. Parks and recreation. 21(10–12).

Seely, H. E. 1949. Meeting of the committee on surveys research, C.S.F.E., held in Ottawa, Feb. 4th, 1949. Forestry Chronicle. 25:62–65.

Seton, Ernest Thompson. 1953. Lives of game animals. Charles T. Branford Co., Boston, 4 vols.

7

Wolf predation and ungulate populations

Douglas H. Pimlott

Studies of wolf *(Canis lupus)* ecology have become fairly common during the past two decades. In many instances the programs were stimulated by "... apprehension concerning the welfare of the big game herds," as Adolph Murie phrased it in the foreword to his classical study, "The Wolves of Mt. McKinley" (1944).

Murie's was the first of the studies that dealt intensively with the interaction of wolves and their prey. Since he completed his field work, Cowan (1947), Thompson (1952), Stenlund (1955), Mech (1966), and Shelton (1966) have also reported on the wolves of North America. Pulliainen (1965) has presented an account of the species in Finland. In addition to these published accounts other work has been in progress, some of which will be reported for the first time at this meeting.

Although virtually all the studies which have been mentioned have dealt, at least in part, with the effects of predation by wolves on the population levels of the animals on which they prey, quantitative data in many cases have been sparse and their lack has precluded a very detailed consideration of the subject. Murie (1944) obtained considerable data on predation on Dall sheep *(Ovis dalli)* and on caribou *(Rangifer rangifer)*. Mech (1966) and Shelton (1966) published data on predation on moose *(Alces alces)*, and studies in Ontario (Pimlott, *et al.*, 1967) presented data on predation on white tailed deer.

In most cases a considerable element of the problem has been, and

Reprinted with permission from the *American Zoologist*, 7:267–278 (1967). The author is currently affiliated with the Department of Zoology, University of Toronto, Toronto, Canada.

continues to be, the difficulty that is encountered in obtaining sufficiently detailed data on the population levels of both the wolf and its principal prey species. The studies on Isle Royale, which are directed by Durward Allen and which have been conducted by Mech (1966), Shelton (1966), and Jordan (unpublished), have come the closest to laboratory studies of any big-game species which have been undertaken. They, and similar future studies, will undoubtedly provide a much firmer quantitative basis from which the principles of wolf predation will be developed.

Although we have not yet reached the stage where a broad definitive statement can be made on the role of wolf predation in controlling the populations of species on which they prey, the studies that I have mentioned have added a great deal of fresh insight on the question.

My objective in this paper is to review the state of our knowledge on wolf predation, to attempt to clarify some of the areas where thinking on the subject has not been clear and, finally, to present my preliminary thoughts on the interaction of wolves and their prey.

VARIABLES AND COMPONENTS OF PREDATION

The literature on predation and its influence on prey populations is extensive. The great majority of the detailed studies have, however, been conducted on situations where both predator and prey were insects (Thompson, 1939) or where a vertebrate predator was preying on insect prey (Tinbergen, 1955, 1960; Holling, 1959, 1961; Morris, *et al.*, 1958; Kendeigh, 1947).

Studies of predation that have been reported make it apparent that many variable factors can influence, moderating or intensifying, the effect of predation. Leopold (1933) classified the factors into five groups: (1) the density of the prey population, (2) the density of the predator population, (3) the characteristics of the prey, *e.g.*, reactions to predators, (4) the density and the quality of alternate foods available to the predator, (5) the characteristics of the predator, *e.g.*, food preferences, efficiency of attack, and other characteristics.

Holling (1959, 1961) has developed a comprehensive theory of predation, based on his studies of small-mammal predation on the European pine sawfly; following the scheme proposed by Leopold (1933) he classified the factors into basic and subsidiary variables. The variables that are always present, predator and prey density, he referred to as universal variables. Since they are part of every predator-prey situation he considered that "... the basic components of predation will arise from these universal variables" (Holling, 1961, p. 164). The remaining variable factors (environmental characteristics, prey characteristics, and predator characteristics) are either constant or absent, so he called them subsidiary variables and the components represented by their effects as subsidiary components.

To describe the dual nature of predation he adopted terminology proposed by Soloman (1949) and used the terms "functional response" to indicate the numbers of animals consumed per predator and "numerical response" to indicate the change in the population level of the predators. The basic components of predation that he described are the functional response to prey density, the functional response to predator density, and the numerical response, which arose from the functional response and from other population processes.

Holling (1961) considers that there are two types of functional response to prey density. In one type more prey, or hosts, are attacked as host density increases. The relationship is curvilinear and the slope of the curve decreases until the curve becomes level. In the second type, predators attack more prey as prey density increases; however, the rising phase of the curve has an S-shape. Holling documented the form of the curve by his studies, both in the laboratory and in the field, of small mammals preying on the cocoons of the European pine sawfly. However, he also stated (Holling, 1961) that the curve for functional response of vertebrate predators to the density of their prey seemed in general to be of this type. He pointed out that Leopold (1933) predicted this type of response when he suggested that vertebrates attack scarce prey by chance but develop the ability to find a greater proportion when the prey becomes abundant.

The curves of both types ultimately level off because of satiation of the predator or, if for no other reason, simply because of the time expended in finding, attacking, and killing prey. In reviewing the components of the equations which describe the two types of curve Holling (1961, p. 170) stated,

> The two types of functional response to prey density therefore can be explained by combinations of the five components: time predator and prey are exposed, searching time, handling time (including identification, capture, and consumption), hunger, and stimulation of predator by each prey discovered. The first three are universally present and hence basic and, by themselves or in conjunction with the effects of hunger, can explain those response curves that rise with a continually decreasing slope to a plateau. If stimulation by prey discovery is added to those four components, an S-shaped response results.

The third type of response curve which Holling (1961) stated might be expected in response to prey density is a domed type which may result from a predator attacking fewer prey, when the prey are very abundant. In some cases at least this may result from the "confusion effect" described by Allee (1951) as a result of studies, by J. C. Welty, of goldfish feeding on *Daphnia.*

Holling (1961) pointed out that, in the past, studies of predation have concentrated on direct numerical responses. He cited the works of Lack (1934) and Andrewartha and Birch (1954) in which survival, fecundity, and dispersal are related to consumption of food. Holling (1961) pointed out that studies of vertebrates preying on insects (*e.g.*, Kendeigh, 1947; Morris, *et al.*, 1958; Holling, 1959) have demonstrated direct and inverse responses, as well as no response, to increasing density of prey species.

I suggest that this scheme, or structure, of predation that has been proposed by Holling (1959, 1961) is worthy of detailed consideration by students of vertebrate predation. It does a great deal to clarify this area of population dynamics that has long been a rather nebulous one. It could be valuable in guiding our thought as we seek to understand the background principles of predation by wolves on the large ungulates.

WOLF POPULATIONS

Obtaining accurate data on the two basic variables, predator and prey densities, has proven to be the principal stumbling block to understanding the influence that wolves have on prey populations. It is mandatory, if we are to gain an understanding of the processes involved, that we continue our efforts to develop census methods that will provide accurate data at costs that are economically feasible.

The early estimates of numbers of wolves were based to a considerable extent on impressions that the individual investigators obtained as a result of their observations on the occurrence of wolves and as a result of packs reported to them by other individuals. When areas of moderate size were involved the estimates were probably quite close to the actual population. When very large areas were involved too many unknown factors entered the picture and the "estimates" could hardly warrant being called anything but guesses.

Cowan (1947) worked for three years on wolves and ungulates of the Rocky Mountain National Parks of Canada. He had the close cooperation of the wardens and made estimates of the wolf populations of Banff and Jasper Parks. In the latter, the wardens regularly patrolled the principal wolf ranges and provided him with details of their observations. The area of the park is 4200 square miles; the minimum and maximum estimates of the wolf population made by Cowan were 33 and 55 wolves (Table 7–1). Based on summer range he estimated the population density at between one wolf per 87 and one wolf per 111 square miles. He stated (Cowan, 1947, p. 150), "At the time of maximum winter compression, however, this population is present on an area that averages approximately 10 square miles per wolf."

An estimate of wolf numbers that has been quite widely quoted is the one made by Clarke (1940) for the range of the barren-ground caribou, which he estimated at 600,000 square miles. He considered that there probably was a pack (6 animals) for every 100 square miles and on this basis estimated the wolf populations at 36,000 animals. It has been suggested by both Banfield (1954) and Kelsall (1957) that the estimate was too high. Kelsall (1957) suggested, on the basis of observations made in the course of 43,624 miles of transit flying on caribou surveys, that a population of 8,000, or one wolf per 60 square miles of caribou range, would be more realistic (Table 7–1). Kelsall also pointed out that the kill of

TABLE 7–1. Estimated densities of wolf population in North America

Location	Author	Area (sq. miles)	Population	Density: General range	Density: Winter range
N.W. Territories (Canada)	Clarke (1940)	600,000	36,000	16+	
N.W. Territories	Kelsall (1957)	480,000	8,000	60	
Mt. McKinley (Alaska)	Murie (1944)	2,000	40-60	50±	
Jasper Natl. Park (Canada)	Cowan (1947)	4,200	33-55	87-111	10
Superior Natl. For. (Minnesota)	Stenlund (1955)	4,100	240	16+	
Isle Royale (Michigan)	Mech (1966) Shelton (1966)	220	20-22	10±	
Algonquin Park (Ontario)	Pimlott, *et al.* (1967)	1,000	90-110	10±	

wolves (2000 to 3000 annually) made during the height of the control program would not have had any influence on a population of 36,000 but did appear to have considerable influence on the population that was present.

The relationship of the area occupied by a species to its population density can be most accurately appraised where the area occupied is essentially the same at all periods of the year. Three of the more recent studies conducted in white-tailed deer and moose ranges of eastern North America were in areas where this situation applied. In Minnesota (Stenlund, 1955), Isle Royale (Mech, 1966; Shelton, 1966), and Algonquin Park (Pimlott, *et al.*, 1967), the density of the wolf population was determined, primarily, by the use of aerial surveys during the winter (Table 7–1). Isle Royale has proven to be a particularly excellent area and the study there has provided accurate data on the density of the wolf population. On the island the population has remained at a level of approximately one wolf per 10 square miles. During this period the wolves were completely protected, a moose population of high density (approximately three per square mile) was present and, as far as could be determined, there was no movement of wolves from the island.

In Algonquin Park, the boundaries of the study area were fairly well delineated but those of the wolf ranges rarely coincided with them. Because of this, it was not possible to state with as high a degree of certainty what the relationship was between the number of wolves and the size of the study area (Pimlott, *et al.*, 1967). The work, however, was quite intensive, and extended over several years, so that the estimate of the density of the wolf population at between one wolf per 9 and one wolf per 11 square miles was considered to be very close to the actual size of the population. During the greater part of the study the wolves were protected as was the principal prey species, white-tailed deer, and the secondary prey species, moose and beaver.

Another study of a wolf population in Ontario indicated that this high density of wolves does not occur generally throughout the province. Aerial surveys in an area of 10,000 square miles of moose range, in conjunction with an experimental wolf control program, suggested a population density of between one wolf per 100 and one wolf per 200 square miles (Pimlott, *et al.*, 1961; Shannon, *et al.*, 1964).

In summary, data on wolf populations in North America indicate that densities of one wolf per 10 square miles are high, and they show that populations of a much lower density are common over very large areas.

FOOD HABITS OF WOLVES AND SELECTION OF PREY

The evidence from the studies of the food habits of wolves in Alaska (Murie, 1944), western Canada (Cowan, 1947), Wisconsin (Thompson, 1952), Minnesota (Stenlund, 1955), Isle Royale (Mech, 1966; Shelton, 1966) and from work in Algonquin Park (Pimlott, *et al.*, 1967) shows clearly that wolves are dependent to a very marked degree on large mammals for their food.

Summer Food

It has been fairly generally accepted that large mammals serve as prey in winter; however, it is often stated that in summer wolves utilize small animals to a considerable degree. For example, Olson (1938, p. 329), writing about the wolf in the Superior National Forest in Minnesota, stated that, "The major portion of the food of the wolf during the summer is grouse, woodmice, meadow voles, fish, marmots, snakes, insects, and some vegetation. In fact anything that crawls, swims, or flies may be included in their diet." His conclusions have not been borne out by quantitative studies.

The greatest degree of uncertainty about the food habits of wolves in summer is for tundra areas, the range of the barren-ground caribou. Banfield (1951) stated that the observations of Farley Mowat, made near Nueltin Lake in Keewatin, N. W. T., "indicated a drastic change in diet between the denning period and the nomadic period."

He stated that there were no caribou in the vicinity of the wolves between June 17 and August 20, and during that period the wolves were observed hunting for small mammals and eating dead fish and a dead gull. Unfortunately, an intensive study of their food habits was not undertaken, and only 61 scats were examined (Kelsall, 1957). In spite of the apparent absence of caribou, 42 of these contained caribou remains, while the remains of small mammals occurred in 17. Kelsall (1957) pointed out that almost all wolf scats collected in caribou country contain caribou hair; however, he suggested that much of this may be the result of scavenging activity by the wolves.

One aspect of the uncertainty about the importance of caribou in the summer diet of wolves is caused by the comparative behavior of the two species. Barren-ground caribou are highly migratory and, it would seem, must often leave the wolves behind during the period when the pups are young and relatively immobile. Possibly under such circumstances the wolves are much more dependent on small animals, or if there are such breaks in the contact with the primary prey, they may constitute an important limiting factor to populations of tundra wolves.

It is also possible that Murie's work (1944) may provide an answer to the question that was not apparent to Mowat and other investigators. In the area of Mt. McKinley Murie found that even after the main movement of the caribou through the zone, there were usually stragglers left behind. The wolves were able to locate these animals and thus subsist on caribou long after the main herds had disappeared. It is possible that such a situation exists in caribou range much more frequently than has been realized.

In a study of the food habits of wolves in Algonquin Park during the snow-free period of the year, white-tailed deer comprised 80 percent of the food items that occurred in 1435 scats; moose comprised 8 percent and beaver 7 percent. The remaining 5 percent included snowshoe hare *(Lepus americana)*, muskrat *(Ondatra zibethica)*, marmot *(Marmota monox)*, porcupine *(Erethizon dorsatum)*, raccoon *(Procyon lotor)*, and three species of mice (Pimlott, *et al.*, 1967).

In addition to Algonquin Park, studies were conducted in a number of areas in other parts of Ontario. In two of these which lie west and north of Algonquin Park (the Pakesley area of Parry Sound Forest District and the Marten River area of the North Bay Forest District) and where the same species of prey occurred, the collections were compared with the data from Algonquin Park. For the Pakesley area (206 scats) the frequency of occurrence of the three most important species was beaver, 59 percent, deer, 27 percent, and marmot, 7 percent. For the Marten River area (226 scats) the frequency of occurrence of the three most important species was deer, 42 percent, beaver, 37 percent, and moose, 17 percent. The deer population had undergone a marked decline in the Marten River area as a result of losses during the severe winters of 1958-59 and 1959-60 and was very low the year when the study was conducted. The relatively high occurrence of deer hair in wolf scats suggested that predation on deer may have been disproportionate to their abundance in the area.

The data from Pakesley are the only ones, of which I am aware, that indicate that ungulates have comprised less than 50 percent of the summer food of wolves. However, even in this area beaver cannot be considered to be a primary food for they are unavailable three to five months of the year; wolves could not persist in the area during this period if deer and moose were not available to them.

Selection of Prey in Summer

In Ontario the percentage of wolf scats in summer (July 1 to September 30) that contained fawn hair and calf moose hair was high. Fawn hair comprised 71 percent of the occurrences in the scats that contained deer hair, and calves 88 percent in the scats that contained moose hair (Pimlott, *et al.*, 1967). It has not been shown to what extent the frequency of occurrence of juvenile and adult remains reflects the proportion of animals in the age classes that are killed by wolves. However, Pimlott, *et al.* (1967) considered that the best assumption is that the proportion of remains in scats approaches the actual proportion in the kill; in this respect we disagreed with the conclusion of Mech (1966) that juveniles in the kill are over-represented by the occurrence of their remains in scats.

Selection of Prey in Winter

The food habits of wolves in winter on Isle Royale and in Algonquin Park were known primarily from the remains of animals found during the aerial searches. In the latter the remains of 676 deer that were believed to have been killed by wolves were located and the mandibles of 331 (47 percent) were collected.

The age distribution of the deer killed by wolves was not a normal one. Animals under five years of age included 42 percent of the specimens while those five years of age and older comprised 58 percent. The comparable percentages for a sample of 275 deer that were killed by cars or collected for research purposes were 87 percent and 13 percent. The comparative percentages of fawns, the age class most likely to be under-represented in the collection from wolf kills, was 17 percent and 20 percent, respectively.

The only other data on the age classes of deer killed by wolves were reported by Stenlund (1955) for the Superior National Forest. The collection (33 deer) did not show the preponderance of animals in the older age class, but the sample was too small for statistical comparisons to be valid.

The data on the kill of moose by wolves on Isle Royale show a somewhat similar trend to those from Algonquin Park. Of 80 animals examined, 50 by Mech (1966) and 30 by Shelton (1966), 22 were calves, one was a yearling, and 57 were 6 years of age (Age Class VI, Passmore, *et al.*, 1955) or older.

In Alaska, Burkholder (1959) tracked a pack of wolves from the air and reported on the ages of eight caribou and eight moose. Six of the moose that were killed were calves, one was a yearling, and one an adult of unknown age. The caribou were all adults, three of unknown age, four were between two and six, and one was over 10 years of age.

Fuller (1962) stated that the evidence from stomach samples of wolves (95) and an analysis of 63 scats, collected in Wood Buffalo Park, indicated that bison *(Bison bison)* form a staple food of wolves in both summer and winter. He found the remains of eight animals that had been killed by wolves and he observed wolves attacking bison on three occasions. Five of the bison killed were very old animals, three were calves, and three were in middle-age classes. All three of the latter animals were injured or diseased. He could not determine whether the leg of one had been broken before or during the attack by the wolves.

Food Requirements of Wolves

The studies on Isle Royale (Mech, 1966; Shelton, 1966) have permitted an estimation to be made of the food requirements of wolves. Mech (1966) obtained data on 48 moose that were killed over a total period of 110 days. He estimated the average daily consumption at 12.3 pounds per wolf (Mech, 1966). Scrutiny of his data suggested that he overestimated the size of moose and underestimated the amount of wastage. Pimlott, *et al.* (1967) recalculated Mech's data and concluded that 10 pounds per day would be a better estimate. The suggested daily rate of consumption was 0.14 pound of food per pound of body weight (winter conditions). In their calculations they arbitrarily lowered the per diem rate to 0.12 pounds per pound for the summer (June to September) period.

THE DYNAMICS OF WOLF PREDATION

The most intensive studies of vertebrate predation were conducted by the late Paul Errington (1934, 1943, 1963; Errington, *et al.*, 1940). A fundamental aspect of his theory of threshold phenomena is that vertebrate predators take a high toll of prey only when the prey are living in insecure situations, in marginal or submarginal habitats. However, he frequently referred to predation by the genus *Canis* on the ungulates and indicated that he considered that there were at least some occasions when it might be of a non-compensating nature. In his major review of the topic he stated (Errington, 1946, p. 158):

> Intercompensations in rates of gain and loss are evidently less complete in the life equations of the ungulates, however, than in the muskrats. There is vastly more reason that I can see for believing that predation can have a truly significant influence on population levels of at least some wild ungulates.
> Without losing sight of the fact that much more than predation or lack thereof may be involved in the great changes recorded for American deer populations of recent decades . . . , we may detect pretty strong indications of the depressive influence of predation upon the numbers of the deer.

He summed up:

> Most examples of predation upon wild ungulates showing a reasonably clear evidence of population effect have one thing in common: the predators involved had special abilities as killers—indeed were usually *Canis* spp., members of a subhuman group inferior as mammals only to man in adaptiveness and potential destructiveness to conspicuous, relatively slow-breeding forms.

The environments in which wolf predation occurs in North America are extremely variable. The rigorous arctic environment of Ellesmere Island, where the wolves prey on musk oxen and caribou, contrasts sharply with the mixed forests of the Superior National Forest in Minnesota, where the wolves prey on white-tailed deer, and with Isle Royale National Park, where the wolves prey on moose. The nature of the universal variables of predation, predator, and prey density, and the nature of the subsidiary variables are very different in the various environments. The studies that have been or are being conducted suggest that we are likely to find that the interaction of the variables of predation produce such complexities that few generalizations are possible on the influence of predation by wolves on populations of prey.

Influence of Wolves on Ungulates in North America

In the case of the caribou, Murie (1944) suggested that predation was apparently an important limiting factor on a population in Alaska through predation on fawns. However, Banfield (1954) suggested that the mortality caused by wolves in the western Canadian Arctic did not exceed 5 percent of the population. Kelsall (1957) stated that an annual kill of four caribou was a likely average kill of a tundra wolf, although he estimated that it would take 14 caribou to sustain a wolf for a year. Banfield's and Kelsall's estimates, which are of approximately the same magnitude, will be subject to upward revision if future studies indicate that caribou, and particularly fawns, are primary prey of tundra wolves in summer in northern Canada, as they are in Alaska. However, if future studies confirm that small mammals are a primary component of the summer diet of wolves, then the estimates of Banfield and Kelsall may be quite realistic. Dall sheep are prey of wolves in Mt. McKinley National Park in Alaska (Murie, 1944). Although the data from scat analyses suggested that they were not as important as caribou, Murie considered that the wolves were controlling the population. The control appeared to be exercised through periodic, heavy predation on yearlings. When Murie (1961) returned to the Park in 1945, after an absence of four years, he found that the population of sheep had declined to 500 from a minimum of 1000 to 1500 in 1941. He considered that poor survival of the young, combined with the loss of old sheep that had been predominant in 1941, had caused the decline. The wolf population had also delined but, unfortunately, there was no knowledge of their role in the decline of the sheep population.

The sheep quadrupled in numbers by 1959, but the wolf population did not show any parallel increase. Murie (1961) believed that predator control operations, conducted outside the Park, were the factor that prevented an increase in wolf numbers.

Cowan (1947) gave rough estimates of the population of elk *(Cervus canadensis)* and mule deer *(Odocoileus hemionus)* in Jasper National Park. His data suggest that the ratio of wolves to the combined populations of elk and mule deer was of the order of 1:100. In addition to these two species, moose, bighorn sheep *(Ovis canadensis)*, mountain goat *(Oreamnos americanus)*, and caribou occurred in the area and were utilized to a lesser degree by the wolves. Their numbers would appreciably increase the ratio of predator to prey. Cowan (1947) discussed the overpopulations of ungulates that existed in the park and pointed out that, in addition to not removing the net increment of the populations, the wolves were not even removing the diseased and injured animals, which he referred to as the "cull group," from the population.

The work of Thompson (1952) in Wisconsin, though covering the ranges of only two packs of wolves, provides informative data. He showed that the two areas in which he studied wolves developed the same symptoms of an overpopulation of white-tailed deer as did areas where there were no wolves. Data on the deer population indicated that their density increased very rapidly in the late 1930's, from 10 to 30 per square mile, following extensive changes in habitat that resulted from fire and logging. The density of the wolf population was of the order of one per 35 square miles so that the ratio of wolves to deer would have been greater than 1:300.

Mech (1966) estimated the population of moose in Isle Royale, in late winter, at approximately 600 animals, and, as mentioned previously, the wolf population at 20 to 22. The ratio of wolves to moose in this case was approximately 1:30. Mech (1966) and Shelton (1966) concluded that the wolves were controlling the moose population. They estimated that the control was being accomplished by the kill of between 142 (Mech, 1966) and 150 (Shelton, 1966) moose, or approximately 25 percent of the late winter population.

The data on the deer population in Algonquin Park suggest a density of 10 to 15 per square mile, or a ratio of wolf to deer of between 1:100 and 1:150. The deer are primary prey of the wolves and predation may have been important in preventing major irruptions such as those that have occurred in many deer ranges where wolves are absent (Leopold, *et al.*, 1947). The population of deer has not been in perfect balance with the environment, however, for there have been periodic reductions caused by starvation during severe winters (Pimlott, *et al.*, 1967). The interpretation of the influence of wolves on the deer population in Algonquin Park is made difficult by the fact that wolves in the Park were subject to control by Park personnel for many years prior to the inauguration of the research program. The deer, however, were protected from hunting.

Calculations based on the data on rates of food consumption by wolves, and on the data obtained from studies of food habits of wolves in Algonquin Park, suggest that a population of a wolf per 10 square miles would require and would utilize 3.7 deer per square mile per year. This would require a deer population of a minimum density of 10 per square mile and a productivity rate of approximately 37 percent to support the wolf population (Table 7–2) (Pimlott, *et al.*, 1967).

TABLE 7–2. Calculation of number of deer required to support a wolf population of one per 10 square miles

Basic Assumptions			
Size of area			100 sq. miles
Wolf population			10
Gross food consumption by wolves (avg. wt. 60 lbs.)			
Oct.-May			8.4 lbs./day
June-Sept.			7.2 " "
Wastage			20%
Species other than deer—winter			10%
summer			20%
Age-composition and weight of deer killed			
winter— Fawns	30%	80 lbs.	
Adults	70%	150 lbs.	
summer—Fawns	80%	40 lbs.	
Adults	20%	150 lbs.	
Total kill of deer—winter	177		
summer	190		
	367 deer		
Density of 10 deer/sq. mile, productivity of 37% is required to support 1 wolf/10 sq. miles.			

DISCUSSION

Since Errington's review (1946) of vertebrate predation, there has been a great increase in knowledge of the population dynamics of both wolves and the ungulates. The marked variation in reproductive performance that has been shown to exist among the ungulates permits considerable compensation for adverse or favorable environmental factors (*e.g.*, Cheatum and Severinghaus, 1948; Pimlott, 1959). It is conceivable that predation is a factor in triggering an increase in the reproductive rate, and, if so, it could be considered to be of a compensatory nature.

In the discussion of the selection of prey by wolves it has been shown that predation tends to be concentrated on the very young and the very old. When the old animals in a population are eliminated it probably has very little influence on the population level of the prey species, for they, like animals in submarginal habitats, would soon have died of other causes anyway. Predation then on the old animals in the population also appears to be of a compensatory nature.

A great weakness that exists in the study of wolves in summer is that there does not appear to be any way of making concrete determinations about the condition of the young that are eaten by wolves. A number of studies show conclusively (Thompson, 1952; Murie, 1944; Mech, 1966; Shelton, 1966; Pimlott *et al.*, 1967) that wolves feed heavily on the young of the year—but, what percentage of these animals was actually killed by wolves? To what extent is the feeding on young animals a scavenging activity? What percentage of those killed by wolves would have survived in the absence of wolves? Studies in a number of areas where wolves have been extirpated indicate that a significant mortality of young ungulates occurs between spring and fall. To the extent that predation by wolves removes young that would have died anyway, as in the case of adults, it is of a compensatory nature.

Although Murie's (1944) work indicated that predation fell heavily on young animals, I do not think that the full import of this fact has been realized. If a considerable portion of this predation is noncompensatory, a population of wolves of high density would exercise a considerable influence on ungulate populations. Allee, Emerson, Park, Park and Schmidt (1949) listed a series of principles that arose from their review of predation. The third is of particular interest to this discussion "predation is frequently directed against the immature stages of the prey and as such may constitute an effective limiting factor." (p. 374).

The question of whether or not wolves constitute an effective limiting factor on ungulates, and particularly on deer, moose, and caribou, is one that has only been partially answered. In considering the population dynamics of some big-game species, deer and moose in particular, the question arises as to why intrinsic mechanisms of population control have not evolved to prevent them from increasing beyond the sustaining level of their food supply. It seems reasonable to postulate that it may be because they have had very efficient predators, and the forces of selection have kept them busy evolving ways and means not of limiting their own numbers but of keeping abreast of mortality factors.

Contemporary biologists often have a distorted viewpoint about the interrelationships of ungulates and their predators. We live in an age when there is a great imbalance in the environments inhabited by many of the ungulates. In the case of deer and moose the environmental changes, or disturbances, have been favorable and populations are probably higher than they have ever been. Under such circumstances it is not much wonder that we have been inclined to argue that predators do not act as important limiting factors on deer and moose populations. I doubt, however, that it was a very common condition prior to intensive human impact on the environment. In other words, I consider that adaptations between many of the ungulates, particularly those of the forest, and their predators probably evolved in relatively stable environments that could not support prey populations of high density.

The history of wolves and moose on Isle Royale is an interesting example. There, as I have mentioned, in the presence of abundant food and complete protection, the wolf population stabilized at a level of one wolf per 10 square miles. In Algonquin Park the estimates indicated a population of the same magnitude; there was no significant difference between 1959 and 1964, although during most of this period the wolves were protected (Pimlott, *et al.*, 1967). These examples suggest that a wolf per 10 square miles is close to the maximum density that can be attained by a population.

The data from Isle Royale suggest that a state of equilibrium has been reached between the wolves and the moose at a ratio of approximately one wolf per 30 moose. A similar calculation, based on the data from Algonquin Park, suggests that a ratio of one wolf per 100 deer may be close to an equilibrium. On the basis of these data, and on the basis of the previous discussion of the evolution of wolf-prey population mechanisms, I suggest that wolves may not be capable of exercising absolute control of white-tailed deer at ratios that exceed 1:100. I also suggest that predation by wolves may cease to be an important limiting factor when densities of deer exceed 20 per square mile.

The fact that no animal smaller than the beaver has been shown to be the predominant food of wolves for any significant period is not surprising. Their size, and the complex social organization of the packs, are such that it would rarely be efficient for them to live on small animals. The organization of the pack is undoubtedly an adaptation which has developed because wolves prey on animals larger and often fleeter than themselves. Such an organization would be unlikely to persist if small animals became their primary source of food. Energy relationships are undoubtedly also involved. An adult wolf may weigh between 50 and 150 pounds and it would rarely be efficient to obtain the energy to maintain this biomass by the utilization of animals that weigh a few ounces or even a few pounds, especially when these are often difficult to capture.

I suggest that energy demands alone make it very unlikely that tundra wolves regularly subsist on small animals during the summer. When the question is studied intensively it is likely that the successful rearing of a litter will, in the great majority of cases, be found to be dependent on the availability of caribou or of other large ungulates as food for the wolves.

A study of the interaction of wolves and their prey indicates that there are a number of characteristic aspects of predation that are worthy of review. They serve to sum up this discussion of the dynamics of wolf predation; a knowledge of their existence may also contribute to the further development of understanding of the underlying principles of vertebrate predation.

1. In all but one instance, intensive studies of the food habits of wolves indicate that the large ungulates are the primary prey of wolves both in summer and in winter. It remains to be demonstrated that wolves can live and raise young in areas where they must subsist on small animals.

2. The process of wolf predation does not come about simply as a result of random contacts between predator and prey but is complicated by a process in which the ability of the prey to escape is tested. The dynamic aspects of the process have been observed in a number of areas (Murie, 1944; Crisler, 1956) and have been particularly well documented on Isle Royale (Mech, 1966; Shelton, 1966).

3. Among the ungulates, wolves prey primarily on the young-of-the-year and on animals in older age-classes. Predation is most heavy on the young during the summer but is less intensive during the winter, when old animals are vulnerable.

4. Intensive utilization of prey animals that are captured is a characteristic of wolf predation. One study (Pimlott, *et al.*, 1967) has demonstrated, however, that utilization was less complete during a winter when severe snow conditions prevailed.

REFERENCES

Allee, W. C. 1951. Cooperation among animals. Henry Schuman, New York.

Allee, W. C., H. E. Emerson, O. Park, T. Park, and K. P. Schmidt. 1949. Principles of animal ecology. W. B. Saunders Co., Philadelphia, Pa.

Andrewartha, H. G., and L. C. Birch. 1954. The distribution and abundance of animals. The Univ. of Chicago Press, Chicago.

Banfield, A. W. F. 1951. The barren-ground caribou. Canad. Wildl. Serv., Dept. Res. and Dev.

Banfield, A. W. F. 1954. Preliminary investigation of the barren-ground caribou. Canad. Wildl. Serv., Wildl. Mgmt. Bull. Ser. 1, No. 10B.

Burkholder, B. L. 1959. Movements and behavior of a wolf pack in Alaska. J. Wildl. Mgmt. 23:1–11.

Cheatum, E. L., and C. W. Severinghaus. 1950. Variations in the fertility of the white-tailed deer related to range conditions. Trans. N. Am. Wildl. Conf. 15:170–190.

Clarke, C. H. D. 1940. A biological investigation of the Thelon Game Sanctuary. Natl. Mus. Canad., Bull. No. 96, Biol. Ser. No. 25.

Cowan, I. McT. 1947. The timber wolf in the Rocky Mountain National Parks of Canada. Canad. J. Res. 25:139–174.

Crisler, L. 1956. Observations of wolves hunting caribou. J. Mammal. 37:337–346.

Errington, P. L. 1934. Vulnerability of bob-white populations to predation. Ecol. 15: 110–127.

Errington, P. L. 1943. An analysis of mink predation upon muskrats in north-central United States. Agric. Expt. Sta., Iowa State Coll. Res. Bull. 320:797–924.

Errington, P. L. 1946. Predation and vertebrate populations. Quart. Rev. Biol. 21:144–177, 221–245.

Errington, P. L. 1963. Muskrat populations. Iowa State Univ. Press, Ames.

Errington, P. L., F. Hammerstrom, and F. N. Hammerstrom, Jr. 1940. The great horned owl and its prey in north-central United States. Agric. Expt. Sta., Iowa State Coll., Res. Bull. 277:757–850.

Fuller, W. A. 1962. The biology and management of the bison of Wood Buffalo National Park. Canad. Wildl. Serv., Wildl. Mgmt. Bull., Ser. 1, No. 16.

Holling, C. S. 1959. The components of predation as revealed by a study of small mammal predation of the European pine sawfly. Canad. Entomol. 91:293–320.

Holling, C. S. 1961. Principles of insect predation. Ann. Rev. Entomol. 6:163–182.

Kelsall, J. P. 1957. Continued barren-ground caribou studies. Canad. Wildl. Serv., Wildl. Mgmt. Bull., Ser. 1, No. 12.

Kendeigh, S. C. 1947. Bird population studies in the coniferous forest biome during a spruce budworm outbreak. Div. of Res., Ont. Dept. of Lands and Forest. Biol., Bull. No. 1.

Lack, D. 1954. The natural regulation of animal numbers. Oxford Univ. Press, London.

Leopold, A. 1933. Game management. Charles Scribner's Sons, New York.

Leopold, A., L. K. Sowls, and D. L. Spencer. 1947. A survey of over-populated deer ranges in the United States. J. Wildl. Mgmt. 11:162–177.

Mech, L. D. 1966. The wolves of Isle Royale. U.S. Natl. Park Serv., Fauna Ser. No. 7.

Morris, R. F., W. F. Cheshire, C. A. Miller, and D. G. Mott. 1958. Numerical response of avian and mammalian predators during a gradation of the spruce budworm. Ecol. 39:487–494.

Murie, A. 1944. The wolves of Mt. McKinley. U.S. Dept. Interior, U.S. Natl. Park Serv., Fauna Ser. No. 5.

Murie, A. 1961. A naturalist in Alaska. Devin Adair Co., Ltd., New York.

Olson, S. F. 1938. A study in predatory relationships with particular reference to the wolf. Sci. Monthly 46:323–336.

Passmore, R. C., R. L. Peterson, and A. T. Cringan. 1955. A study of mandibular tooth wear as an index to age of moose, pp. 223–246. *In:* R. L. Peterson, North American moose. Univ. Toronto Press, Toronto, Canada.

Pimlott, D. H. 1959. Reproduction and productivity of Newfoundland moose. J. Wildl. Mgmt. 23:381–401.

Pimlott, D. H., J. A. Shannon, and G. B. Kolenosky. 1967. The interrelationships of wolves and deer in Algonquin Park. Trans. Northeast Wildl. Conf. 16 pp. mimeo.

Pimlott, D. H., J. A. Shannon, W. T. McKeown, and D. Sayers. 1961. Experimental timber wolf poisoning program in northwestern Ontario. Res. Branch, Dept. Lands and Forests, Ontario. Unpubl. 11 pp. (typed).

Pulliainen, E. 1965. Studies on the wolf in Finland. Ann. Zool. Fenn. 2:215–259.

Shannon, J. A., J. L. Lessard, and W. McKeown. 1964. Experimental timber wolf poisoning program in northwestern Ontario, 1963–64. Res. Branch, Dept. of Lands and Forests, Ontario. Unpubl. 4 pp. (typed).

Shelton, P. S. 1966. Ecological studies of beavers, wolves, and moose in Isle Royale National Park, Michigan. Ph.D. thesis. Purdue Univ., Lafayette, Ind. 308 pp.

Soloman, M. E. 1949. The natural control of animal populations. J. Animal Ecol. 18: 1–35.

Stenlund, M. H. 1955. A field study of the timber wolf (*Canis lupus*) on the Superior National Forest, Minnesota. Minn. Dept. Conserv., Tech Bull. No. 4.

Thompson, D. Q. 1952. Travel, range, and food habits of timber wolves in Wisconsin. J. Mammal. 33:429–442.

Tinbergen, L. 1955. The effect of predators on the numbers of their hosts. Vakblad voor Biologen. 28:217–228.

Tinbergen, L. 1960. The natural control of insects in pinewoods. I. Factors influencing the intensity of predation by song birds. Arch. Neerl. Zool. 13:265–343.

8

Primary productivity and fish harvest in a small desert impoundment

William J. McConnell

INTRODUCTION

One of the most important goals of fishery investigation is an understanding of the trophic structure upon which fish production is based. Only through such an understanding can we estimate the future potential of fisheries which currently are not under intensive management or harvest. Limnologists and fishery biologists have made many attempts to classify inland waters with regard to basic productivity or the ability to synthesize new organic matter. The most commonly used criterion of productivity has been the instantaneous standing crop of forage animals or plants. Most investigators have been content to base their comparisons directly on the number or weight of forage organisms measured, while others (Lindemann, 1942; Dineen, 1953; Juday, 1940) have applied vital population statistics to standing crops and thereby arrived at an annual production. It cannot be doubted that some sort of a relationship must exist between productivity of an environment and instantaneous biomass; however, there are no theoretical reasons for expecting a close relationship. The alternative of attempting to convert standing crops into true productivity by applying vital population statistics for each organism present is clearly too laborious for any extensive application. Potential fish productivity may be computed in some situations from fishery statistics (Ricker, 1958), and is sometimes a valid criterion of aquatic productiveness. In the present context, however, fish productivity is the dependent variable.

Reprinted with permission from *Trans. Amer. Fish. Soc.*, 92:1–12 (1963).

Within recent years the problem of aquatic productivity has been approached from viewpoints other than that of standing crop. Most important and basic of these has been the primary-productivity concept. Ryther (1956) and Odum (1959) review the development and present status of this approach; both discuss the superiority of the productivity concept over the standing-crop concept. Ideally the most desirable measurement of primary productivity would be that of net primary productivity (or actual total growth of the aquatic plants present in a habitat). For reasons discussed by both Ryther (1956) and Odum (1959), this is virtually impossible to measure in most environments. In contrast, gross primary productivity, or gross photosynthesis, is the criterion of productivity which can be measured most readily in the majority of fisheries. With the exception of net primary productivity it is theoretically the most useful single datum in the array of measurable productivity criteria.

Previous to the present paper the only reported effort to relate fish productivity to gross photosynthesis was that of Odum (1957). Odum calculated fish productivity as the product of estimated standing crop and turnover coefficient. This is obviously a rather precarious method, but the only one available in most situations. Direct measurement of fish harvest rate, where possible and meaningful, appears to have advantages over calculated fish productivity. Harvest rate must approach potential fish productivity, however, before it may be legitimately compared to indices of basic habitat productive capacity. It is the purpose of this paper to present data on primary productivity and fish harvest in a small desert lake wherein this condition is approximated as closely as might be expected in an actual fishery.

DESCRIPTION OF THE STUDY AREA

Pena Blanca Lake was constructed by the Arizona Game and Fish Department as Federal Aid to Fisheries Project FW-5-D. It is located in the Pajarito Mountains about 14 airline miles west-northwest of Nogales on the international border shared with Mexico. Annual precipitation averages about 18 inches near the lake site, and annual solar energy income is about 200,000 gram calories per cm^2. Altitude of the lake site is 3,800 feet; latitude, 33°30′ N.

The surface area of Pena Blanca Lake is 49 acres at spillway level. Average depth is 20 feet; maximum depth, 57 feet; and shoreline development factor, 2.7. It is a warm monomictic lake (Hutchinson, 1957) and is usually stratified from March to November. During the first 3 years of impoundment extreme surface water temperatures were 85° and 42° F. Turnover usually occurred in November, at which time the entire water mass was close to 60° F. Total alkalinity remained at about 80 mg/liter calcium carbonate equivalent, while pH varied from 6.8 below the thermocline to 8.3 in protected coves.

Water was first impounded in the spring of 1958 and immediately stocked with 1,000 largemouth bass, *Micropterus salmoides* (Lacépède), brood stock; 30,000 channel catfish, *Ictalurus punctatus* (Rafinesque), fingerlings; and 38 adult black crappie, *Pomoxis nigromaculatus* (LeSueur). An unknown number of threadfin shad, *Dorsoma pentenense* (Gunther), were also introduced at this time. The entire drainage contributing to Pena Blanca Lake is intermittent; therefore no native species of fish were present.

METHODS

Creel census

Fish harvest was estimated by creel census of the sport fishery. Rate of success and creel composition were sampled by examination of creels of departing anglers. Fishing pressure in terms of angler-days was estimated from trafficounter tallies and road block interviews. During road blocks, information was collected concerning the proportion of angler cars, the number of anglers per car, the number of days fished, ratio of shore to boat fishermen, and counter accuracy. The foregoing data were used to convert total vehicle entries into fishing pressure estimates.

Averages and mean variances were computed for strata during which angling pressure and success were homogeneous relative to other periods of the year. Although no important gains in precision resulted, disproportions in sampling effort were minimized by stratification. Means and variances were weighted and combined by procedures outlined by Snedecor (1946, Chap. 17). The magnitude of the inherent variability in both the fish harvest estimate components and in the productivity measurements are discussed more completely later in the paper.

Photosynthesis

Productivity theory and terminology used in this study are summarized in literature previously cited (Ryther, 1956; Odum, 1959). A detailed exposition of productivity methods is not intended here.

Plankton photosynthesis was estimated from chlorophyll content of the phytoplankton in measured amounts of photic-zone water. Conventional light- and dark-bottle techniques were abandoned early in the study because of the frequency of disturbance of bottle sets by trolling fishermen. Phytoplankton samples were taken by a sampling technique described by Lund (1949) in which a weighted polyvinyl tube was slowly lowered to the lower limit of the photic zone, at which time the part of the tubing just entering the water was collapsed with a clamp. The weighted end was then returned to the surface by an attached line. The contents of the tubing were next discharged into a brown glass bottle by raising the upper end of the tube and releasing the clamp. This was done at four to five locations over the pelagial area of the lake to obtain a representative pooled sample. During all but the winter months, samples were refrigerated within 30 minutes after collection. Photic depth was obtained with a Schueller submarine photometer and a multirange milliammeter. The lower limit of the photic zone was considered to coincide with the depth where 1 percent of light incident to the surface at midday remained. Chlorophyll extraction was based on a technique described by Richards with Thompson (1952). Optical densities of the acetone extract were measured with a Bausch and Lomb "Spectronic 20" as described by Odum *et al.* (1958). The relation between chlorophyll content and capacity for photosynthesis (oxygen/chlorophyll ratio) was established experimentally by concurrent measurements of chlorophyll and oxygen production in the open water of the pelagial zone. The technique used in measuring the oxygen production of pelagial phytoplankton was essentially the same as that used to determine littoral oxygen production which is described later in this section.

In the present study reference is made to average hourly rates of photosynthesis in the photic zone, not to rates at optimum light intensity or at one time of the day. The variability and practicality of chlorophyll-based productivity measurement are discussed in a later section.

No attempt was made to compute average plankton photosynthesis from less than 20 individual chlorophyll measurements because of the variability of individual oxygen/chlorophyll ratios (Table 8–1). Total kilogram-hours of chlorophyll for each day sampled were estimated as the following product: (day length) (chlorophyll concentration) (total photic volume). An average daily value was computed for each month, and these averages were used to compute total kilogram-hours of chlorophyll for each year. The annual chlorophyll values were then multiplied by the average hourly oxygen/chlorophyll ratio of 2.04 (Table 8–1).

Average chlorophyll concentration per unit volume of the littoral region was not significantly different from that of the pelagial region (Table 8–1). Because of the bottom slope of the lake and consequent gradual reduction in volume of the photic zone from the bottom up, however, there was a simultaneous decrease in plankton chlorophyll per unit area and an increase in average light intensity within the remaining photic volume. Using a curve of relative photosynthesis based on light penetration as calculated by Ryther (1956), it was possible to estimate average plankton photosynthesis per unit area of the littoral. Average littoral

TABLE 8–1. Reliability of estimates of fish harvest, gross photosynthesis, the most variable components of their computation, and derived ratios and products

Estimate	Year	Sample size	Mean or total	Relative standard error[1]
Fishing pressure (angler-days)	1959	740 cars	15,200	5.5
	1960	1,265 "	18,600	11.0
	1961	613 "	25,300	10.5
Fishing success as fish per angler-day (all warm-water species)	1959	585 parties	1.350	8.9
	1960	588 "	0.306	24.2
	1961	290 "	1.020	13.0
Fish harvest (numbers of all warmwater species)	1959	derived	20,500	11.5
	1960	"	5,700	26.5
	1961	"	24,800	16.9
	All years	"	51,000	11.0
2-year centrarchid harvest as pounds live weight per acre	1960–61	derived	277	14.0
Oxygen/chlorophyll ratio as g/g/hr	All years	15	2.04	10.7
Replicate chlorophyll samples on same day (mg/m^3)	All years	20 pairs	25	2.4
Chlorophyll (mg/m^2)	1959	23 days	134	14.3
	1960	28 "	85	13.8
	1961	20 "	100	5.7
	All years	71 "	105	7.4
Littoral *vs.* pelagial plankton chlorophyll on same day	All years	22 pairs	0.91[2] (mean difference)	
Littoral gross production measured directly (g O_2/m^2/day)	All years	31 days	7.0	13.4
Pelagial gross production computed from chlorophyll (g O_2/m^2/day)	All years	derived	2.66	13.9
Pelagial gross production measured directly (g O_2/m^2/day)	All years	15 days	2.40	16.7
Total gross production (g O_2/m^2/day)	All years	derived	4.0	10.0
Ratio of fish harvest to gross production (live weight of centrarchids)	1960–61	derived	0.0098	16.0

[1] (Standard error/mean) (100).
[2] Differences between pairs not significant at 0.05 confidence level.

plankton photosynthesis per unit area was thereby calculated to equal 75 percent of pelagial photosynthesis. The estimate of littoral plankton photosynthesis was subtracted from total littoral photosynthesis to estimate total benthic photosynthesis, and added to the pelagial estimate to get total plankton photosynthesis.

Estimation of combined photosynthesis of planktonic plus benthic plants in the littoral portions of the lake was accomplished by measuring the diel changes in oxygen concentration of the open water at several stations. An average change was computed from these. Water samples were taken to the lower limit of the photic zone or to the lake bottom in a way similar to that described for chlorophyll sampling. Instead of the tubing being drained into a bottle, atmospheric contact with the sample was prevented by withdrawing the photic-zone water into a glass syringe of 120 ml capacity. A 3-volume overflow occurred when the sample was transferred from the syringe to a 30-ml sample bottle. Winkler reagents were added from polyethylene dropper bottles. In order to detect oxygen concentration changes caused by handling, several samples of anaerobic water were taken from the hypolimnion of the lake by the method described. Presumably it would be impossible to take a sample of oxygen-free water if any significant gas exchange with the atmosphere occurred.

Diffusion error

Titration was accomplished with N/100 sodium thiosulfate solution delivered from a microburette.

When making measurements of oxygen change in unconfined water it is necessary to consider the magnitude and direction of gas exchange at the air-water interface. Odum and Hoskins (1958) outline a technique for correcting for diffusion error in a detailed discussion of productivity measurements of unconfined aquatic communities. One of the prime assumptions of their technique for correcting for diffusion error is that the rate of eddy diffusion, expressed by them as K, be fairly constant through the 24-hour measurement period. In lakes situated in canyons of the Southwest, a constant degree of turbulence rarely occurs. On most days windy afternoons are followed by calm evenings. The diffusion correction used by Odum and Hoskins must be computed for the period between sunset and midnight, and does not apply where diffusion conditions during the evening period are not representative of those that occurred during the day. Because it was not possible to correct for water turbulence at Pena Blanca Lake, all measurements of diurnal oxygen change were taken during periods when turbulence was minimal.

When diffusion correction is not planned it is only necessary to measure oxygen concentration at three cardinal times of the day rather than at more frequent intervals. Under calm water conditions and with small diel water-temperature changes, diffusion error is serious only when the highest diel oxygen concentration exceeds the lowest by several milligrams per liter, and the saturation value occurs midway between the two extremes. This situation never occurred on days of measurement at Pena Blanca Lake, and probably is rare in all unpolluted lakes. For further discussion of the 3-measurement technique and the effects of diffusion on photosynthesis measurements, see McConnell (1962).

As might be expected with the diel wind velocity changes at Pena Blanca Lake, some measurements were begun in calm weather but never completed because of strong winds arising later in the measurement period. Often the respiration rate for the dark part of the diel period was obtained, but excessive water movement during the following daylight period invalidated the measurement of daytime increase in oxygen concentration. If measurements of plankton photosynthesis in the pelagial region were interrupted by wind, the data were discarded. In the littoral region, however, it was possible to estimate gross photosynthesis from night community respiration. Figure 8–1 indicates the strong tendency for community respiration and gross photosynthesis to be equal. Data in Figure 8–1 are the 15 measurements of pelagial photosynthesis and respiration made in unconfined pelagial water which are referred to in Table 8–1. Similar measurements in microcosms maintained by the present

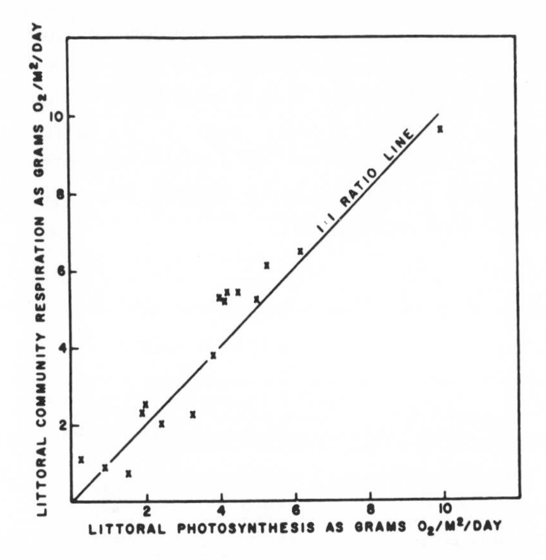

Figure 8–1. Relation of littoral gross photosynthesis to littoral respiration.

author have invariably yielded gross photosynthesis values which were very close to 24-hour respiration not corrected for diffusion (McConnell, 1962). Using diel CO_2 changes of unconfined lake water, Verduin (1956) reported a similar relationship between respiration and photosynthesis. About half of the littoral photosynthesis values in the present study were computed from night respiration only.

RESULTS AND DISCUSSION

Variety and precision of estimates

The estimates of precision of fish harvest, gross photosynthesis, the ratio between them, and the factors used in their computation are presented in Table 8–1. They are primarily of interest in making comparisons between this and similar studies. Information on variability encountered in the several aspects of measuring gross photosynthesis, because of its present rarity in the literature, is also valuable in judging the practicality of this approach in routine management surveys. There are already many references available on the sampling hazards involved in creel census. It will suffice to briefly present corroborating evidence for the accuracy of the fishing-pressure estimate.

Boat-rental records were available for several months during the study, and afforded an independent check on the accuracy of the fishing-pressure estimate (Table 8–2). Information concerning number of boaters who did not fish and ratio of shore anglers to boat anglers was gathered at road blocks. From these data it was possible to convert number of boaters to an estimate of total angling pressure. No correction was made for the small but unknown number of anglers who brought their own boats, which probably accounts for the slightly lower pressure estimate calculated from the boat rental records.

TABLE 8–2. Comparison of fishing pressure estimates derived by independent procedures

Period and year	Number of angler-days from boat rentals	Number of angler-days from trafficounter
March–May 1960	6,210	6,240
June–Aug. 1960	5,100	5,490
Sept.–Nov. 1960	3,300	4,394
March–May 1961	9,850	10,175
Total of periods	24,460	26,874

The high natural variability encountered in using chlorophyll to estimate plankton photosynthesis casts considerable doubt on the efficiency of the technique where alternate methods are possible. Odum *et al.* (1958) review this problem and tabulate oxygen/chlorophyll ratios from investigations of their own and from the literature. Wright (1959) also discusses the cause of variability of oxygen/chlorophyll ratios in a new reservoir. To halve the standard error of the mean oxygen/chlorophyll ratio for the present study, it can be computed that 60 rather than 15 oxygen/chlorophyll ratio determinations would be necessary. With the 60 direct measurements of oxygen production that would be necessary one could make a satisfactory estimate of average photosynthesis without resorting to chlorophyll measurements. While oxygen production values might be somewhat more variable than chlorophyll measurements (Table 8–1) one estimate (chlorophyll) would thereby be eliminated. If relative standard errors for chlorophyll and oxygen/chlorophyll ratio happened to be equal, the relative standard error of their product would be half again as great as that for mean oxygen production alone (Snedecor, 1946, Chap. 17). The relative standard error is the absolute standard error as a percent of the mean. Total littoral photosynthesis in Pena Blanca Lake was estimated directly from oxygen production, and, with only 31 measurements during the 3-year period, the relative standard error for the mean littoral rate was about equal to that for pelagial photosynthesis based on 71 chlorophyll measurements.

Relative precision of the mean amount of chlorophyll per square meter showed a steady increase in successive years of the study (Table 8–1). There is a suggestion that as the lake aged the phytoplankton population became more stable, and the variability encountered in 1961 might be more normal for an older established environment.

The temporal variations in the basic data used in the computation of gross photosynthesis are shown in Figure 8–2. No predictable pattern is evident which might be used in a plan of sample stratification. The apparent relation between photic depth and plankton chlorophyll concentration is expected because phytoplankton were usually the chief cause of light attenuation. Even when storm runoff caused inorganic turbidity, phytoplankton growth was stimulated by the influx of dissolved nutrients that accompanied the suspended matter. When the photic depth was reduced by phytoplankton, benthic photosynthesis was suppressed. The reciprocity between benthic and planktonic photosynthesis is probably why no evident correlation exists between intensity of total littoral photosynthesis and photic depth.

Benthic vs. planktonic photosynthesis

Fish harvest in this study is compared directly to total photosynthesis of Pena Blanca Lake. It should be pointed out, however, that this is done in the absence of data on the relative forage value of equal incre-

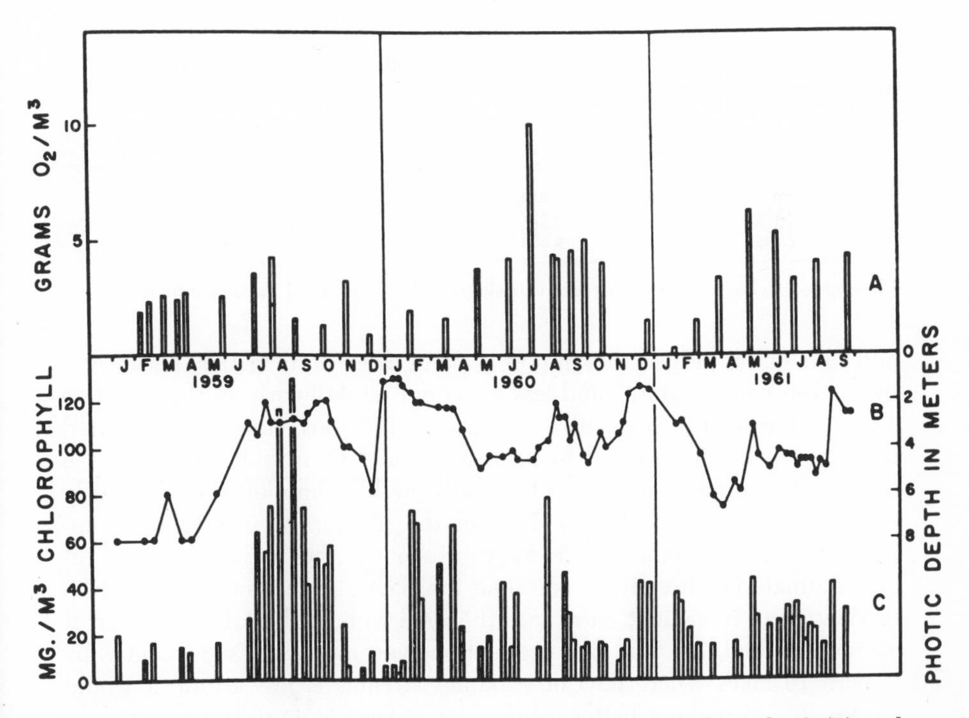

Figure 8–2. Season variations in littoral gross photosynthesis (A); photic depth (B); and plankton chlorophyll (C).

ments of benthic planktonic photosynthesis. It is a common supposition that photosynthesis by microscopic forms of algae in phytoplankton (or *Aufwuchs*) provides more fish forage than that by coarse filamentous algae or vascular aquatics. While this is probably so, critical evidence is lacking. In Pena Blanca Lake approximately 50 percent of total photosynthesis was due to benthic plants. In a larger lake or in one with a less intricate shoreline, plankton photosynthesis would be relatively more important. It is conceivable that a higher rate of fish yield per unit of photosynthesis might prevail in lakes where phytoplankters are the dominant producers.

Organic import

Fish production in a lake might be enhanced by organic import short of pollution. This undoubtedly occurred at Pena Blanca Lake where import exceeded export by a considerable margin. The exact magnitude of import during the entire study was impossible to estimate; however, a crude estimate for the summer of 1959 is available. Surface runoff caused by the usual cloudburst type of summer rainfall carried in about 385 tons of leaves, grass, bark, and manure of range cattle. This quantity is equal to total gross photosynthesis for that year (Table 8–3). In 1960

TABLE 8–3. Average annual rates of gross photosynthesis in Pena Blanca Lake as kilograms of oxygen per square meter

Year	Total littoral	Littoral[1] plankton	Littoral benthos	Pelagial plankton	Total plankton	Lake total	Lake total as pounds per acre
1959	2.610	0.920	1.690	1.230	1.120	1.915	16,900
1960	2.990	0.570	2.420	0.760	0.705	1.455	12,800
1961	2.490	0.600	1.890	0.800	0.711	1.430	12,600
Average	2.695	0.697	2.000	0.930	0.843	1.600	14,100

[1] Estimated to have been 75 percent of pelagial rate on an areal basis (see methods).

and 1961 it was judged that import was not as great as in 1959 because of more even winter rainfall and less torrential rainfall early in the monsoon season. It probably was still in excess of 100 tons, however. Imported material usually floated out to deep water and then sank. Presumably a large part of organic import sinking into the hypolimnion was effectively removed from the food web. Few of the fish-forage invertebrates important at Pena Blanca Lake were present in the hypolimnion to eat organic material directly or consume the bacteria decomposing it. Even in view of an inefficient entry into the food web of the lake, however, it cannot be doubted that fish production was increased significantly by import. In lakes where flow of tributary streams is less erratic it would be possible to make a better estimate of organic import and export.

Areal vs. volumetric presentation

For purposes of comparison, productivities are most conveniently quoted on an areal basis. Ruttner (1953), however, discusses the greater range of values to be encountered when plankton productivity is considered on a unit volume basis. One might suspect that the products of photosynthesis would be more efficiently used where concentrated in a shallow photic zone than in a clear lake, which, by virtue of a deep photic zone, has an equal rate of photosynthesis per unit area. In the present study fish harvest is compared to gross photosynthesis per unit area, but it is acknowledged that the intensity of photosynthesis per unit volume should somehow be used to qualify the areal estimate. A few purely theoretical approaches suggest themselves, but it is probably wiser to await development of more experimental data before corrections for differences in intensity of photosynthesis per unit volume are considered.

Fish harvest

The ideal expression of potential fish productivity is the annual rate of weight increase of a total population under the maximum degree of exploitation which is consistent with a high sustained yield. Fish harvest may approach potential productivity in special fisheries where the following general conditions prevail:

1. Efficient harvest is the chief cause of mortality.
2. The forage producing capacity of the habitat is saturated by harvestable species of fish.

A commercial or experimental food-fish-pond operation would be the kind of fishery most likely to fulfill these conditions. Unfortunately, however, no one has attempted to measure primary productivity as related to fish harvest in such pond operations.

Fish harvest in sport fisheries probably does not come as close to potential productivity as that in pond fisheries where harvesting techniques more effective than hook and line are used. With regard to sport fishery applications, however, the relation of sport fish harvest to environmental factors may be more realistic. Inefficiency in harvest and in exploitation of the habitat are probably inherent factors which will always be related to sport fishing gear and the species of fish involved. The largemouth bass and possibly black crappie harvest for Pena Blanca Lake (Table 8–4) probably approach maximum potential yield to the creel even though short of potential production. Reasons for believing this are discussed in the following paragraphs.

Angling pressure was heavy and well distributed in space and time at Pena Blanca Lake. The average annual fishing effort was in excess of

400 angler-days per acre. All parts of the lake were easily reached in a rowboat so that pressure was exerted on all segments of the fish population. Water temperatures were sufficiently high (over 60° F.) to cause warmwater game fish to feed actively during 7 months of the year. Stocking of catchable trout during the remaining 5 months of the year encouraged heavy fishing pressure and further increased the harvest of warmwater species. Because of the remoteness of other public fisheries in southeastern Arizona, anglers were willing to fish for and keep yearling game fish which had attained a total length of 7 or 8 inches. The preceding factors imply that harvest of game species should have been maximal.

Further evidence concerning the actual degree of exploitation of each species exists in the length frequencies and inferred age compositions of the populations. Length frequencies of largemouth bass samples taken by electrofishing[1] indicate that few largemouth bass reached the third year of life (Fig. 8–3). Such early harvest of a year class undoubtedly reduces natural mortality over that which occurs when harvest is deferred. Total lengths of largemouth bass taken by fishermen within a month of the electrofishing dates were distributed in about the same fashion as electrofishing samples except that young-of-the-year were excluded from harvest. This means that the great majority of largemouth bass were harvested in their second year of life, a condition which attests to the effectiveness of harvest of this species at Pena Blanca Lake.

Length-frequency data for the black crappie (Fig. 8–4) do not yield much information regarding survival because there was no significant year class spawned before 1960.

The ability of channel catfish to evade anglers is apparently much greater than that of largemouth bass. The initial plant of 30,000 fingerlings in 1958 did not become important in the fishery until 1961 (Table 8–4), although they had reached an acceptable size the first year (Fig. 8–4). It is suspected that the fishermen of the region took this long to learn how to fish effectively for channel catfish. At any rate, we may not assume that the harvest of this species in Pena Blanca Lake was representative of the lake's actual production because of the delay in harvest and the nonfishing mortality associated with it.

The change in rate of fishing success as the largemouth bass population was reduced is also informative with regard to intensity of harvest. Rate of harvest of largemouth bass in 1959 was 1.29 per angler-day, while that in 1960 was 0.27 per angler-day. If rates of harvest are proportional to the population present, it may be assumed that the 80 percent decline in harvest rate was caused by a like reduction in the largemouth bass population between midseason 1959 and midseason 1960.

[1] These data were kindly provided by Mr. Roger Gruenwald, Regional Fisheries Biologist for the Arizona Game and Fish Department.

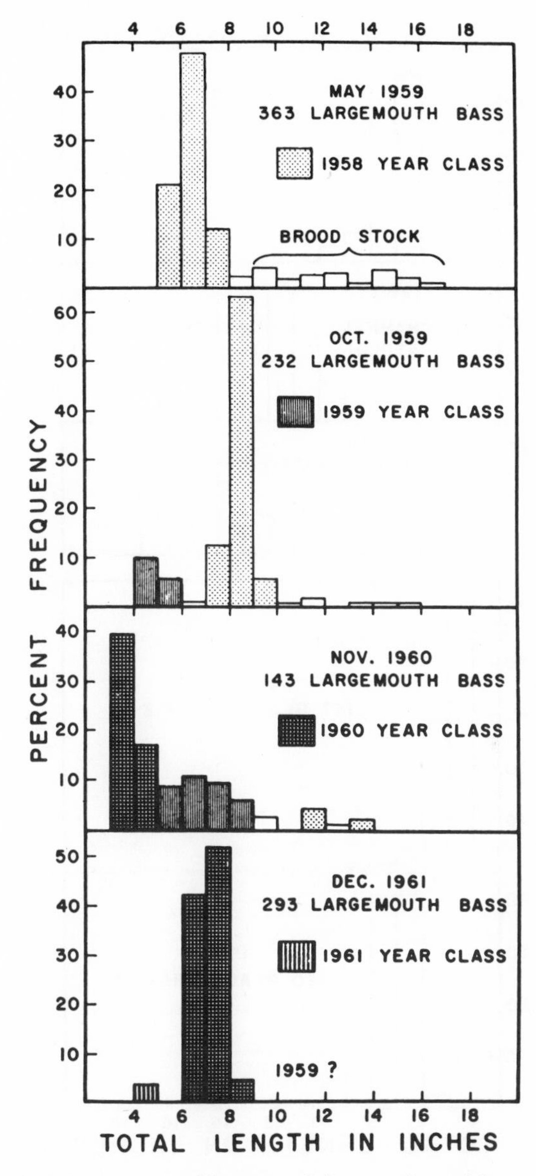

Figure 8–3. Length frequencies of largemouth bass in electrofishing samples.

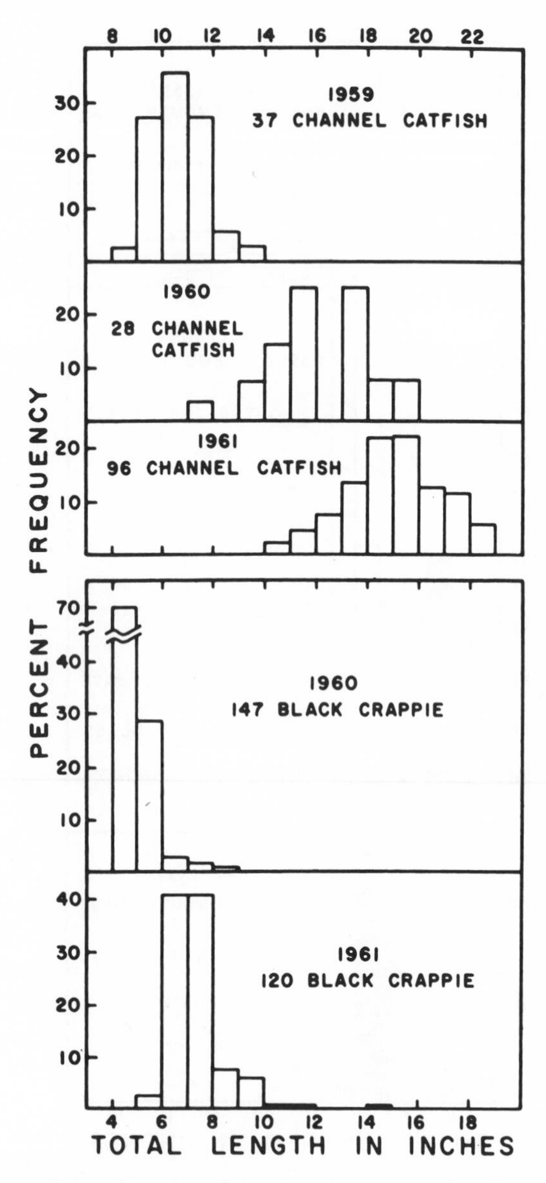

Figure 8–4. Length frequencies of channel catfish and black crappie in creel samples.

TABLE 8–4. Estimated harvest of game fish from Pena Blanca Lake

[All values except dry weight expressed on a per-acre basis]

Year	Black crappie		Channel catfish		Largemouth bass		All species	
	Number	Live weight (pounds)	Number	Live weight (pounds)	Number	Live weight (pounds)	Live weight (pounds)	Dry weight[1] (g/m²)
1959	—	—	16	12	404	178	190	3.24
1960	—	—	14	20	102	71	91	1.55
1961	80	18	65	160	360	188	366	6.25
						Average:	216	3.67

[1] Computed as 15 percent of live weight.

Habitat exploitation

A factor other than food supply could limit the reproduction or survival of a consumer to such a degree that its production would not reflect the potential that might be achieved if more intensive exploitation of the food supply occurred. The length-frequency data in Figure 3 suggest the degree to which the habitat was exploited by the fish population of Pena Blanca Lake. The presence of large numbers of yearling largemouth bass in the spring of 1959 reduced survival of the black crappie and largemouth bass hatched that spring. For this reason the population of largemouth bass during 1960 was probably not great enough to saturate the potential of the environment. The small harvest of largemouth bass and rarity of black crappie in 1960 is evidence for this interpretation. It appears, then, that the most efficient use of the forage available for largemouth bass occurs on alternate years under the conditions existing at Pena Blanca Lake. Although only 3 years of fishing have occurred, the length distribution of the largemouth bass population in December 1961 (Fig. 8–3) indicated an even more pronounced shortage of young-of-the-year than did the 1959 distribution. The failure of black crappie to produce a significant year class in 1961 (Fig. 8–4) suggests that their numbers will fluctuate with those of the largemouth bass and probably for the same reasons. Unpublished data on food habits of black crappie and largemouth bass in Pena Blanca Lake show an almost identical diet composed mostly of cladocerans and insects. For purposes of this study both species are considered as trophic equivalents.

Gross photosynthesis—fish harvest relations

A consideration which complicates interpretation of short-term gross photosynthesis estimates is the "use lag." Odum and Smalley (1959) have shown how use of plant forage by an aquatic consumer can be deferred for many months after the growth of the forage occurred. Pelagial photosynthesis probably results in the production of plankton forage which is used almost immediately or which sinks into the hypolimnion where it is in effect exported from the fishery. By contrast, littoral plant material may conceivably contribute more to the growth of fish-forage organisms in later years than in the year in which the photosynthesis that produced it was recorded. It is therefore impossible to ascribe with certainty any single year of fish biomass increase to a single year of primary production. Furthermore, a fish harvested in a given year has gained only part of its biomass that year. By virtue of deferred harvest of fish and deferred conversion of the products of photosynthesis into fish-forage organisms, the effect of a year of unusual gross photosynthesis on fish harvest can be spread over 2 or 3 years at least.

The foregoing discussions of deferred entry into the food chain by plant material, deferred fish harvest, and cyclical habitat exploitation indicate the risks involved when harvest for one year is compared to gross photosynthesis for that year. Using data from two or more years of measurement would largely compensate for the first two of the above considerations. If the fish population under harvest has recognizable cycles of abundance, the duration of the measurement period should be a multiple of the length of the cycle. In Pena Blanca Lake the cycle was 2 years. The most informative and reliable ratio, then, is the one derived by dividing mean centrarchid harvest for 1960 and 1961 by mean gross photosynthesis for 1959, 1960, and 1961. This ratio is 0.0098 when computed on a live-weight basis, or 0.00142 on a dry-weight basis. One may conclude from this that about 1 pound of largemouth bass, or their trophic equivalent, may be harvested per 100 pounds of gross photosynthesis per year in a fishery similar to that in Pena Blanca Lake.

The ratio of channel catfish harvest (live weight) to gross photosynthesis for the three years of the study is 0.0045, a figure which is probably 3 to 4 times lower than one based on potential yield. There is no evidence concerning the degree of competition between centrarchids and channel catfish. If significant competition did exist, the centrarchid harvest presented here is somewhat short of the harvest that might be expected in the absence of channel catfish. Considering the usual difference in the range of foods accepted by both groups of fish, the increment of centrarchid harvest missing because of competition from channel catfish could not exceed a minor part of the realized average channel catfish harvest of 64 pounds per acre per year. If channel catfish did saturate their trophic niche to the degree that intraspecific competition was obvious, centrarchid production would probably be significantly less than that realized in this study.

Comparison of the centrarchid production ratios of the present study with those estimated by Odum (1957) for Silver Springs, Florida, is interesting, but complicated by the more diverse fish population of Silver Springs. We may, however, compare the Pena Blanca Lake ratio with the ratio for the approximate trophic equivalents in Silver Springs. Odum calculated that the combined annual production (dry weight) of obligatorily predaceous fish in Silver Springs was equal to 0.29 percent of gross photosynthesis, while in Pena Blanca Lake the comparable value was 0.14 percent. Top minnows, centrarchids, and gars were placed in this predaceous fish category to allow the comparison to be made. It is logical to expect that Odum's estimate should be higher because it is based on the total calculated biomass increase and not on actual harvest. The principal producers in Silver Springs were benthic phanarogams (or *Aufwuchs* growing on them), which might also have an effect on the efficiency of production of predatory fish. The precision of estimate of ratios for both Silver Springs and Pena Blanca Lake, however, is probably too low to allow acceptance of the difference in ratios as significant.

In a more typical fishery than Pena Blanca Lake, where predatory species did not become important in the fishery until a larger size was attained, trophic efficiency would drop because of the decreasing importance of small arthropods in the diet and the tendency to feed on fish and larger arthropods. This factor plus increased natural mortality, lower population turnover, and competition from unharvested species would probably lead to a considerably lower harvest per unit of gross photosynthesis than was experienced in Pena Blanca Lake. However, in view of predicted future increases in fishing pressure, it will be necessary to know what potential exists for each fishery under more intensive fishing pressure and management. Gross photosynthesis appears to be the most correct of the practical indices on which to base judgments of fishing potential. Furthermore, the economics of changes in basic fertility of fisheries, decreases as well as increases, are probably best understood through interpretation of gross photosynthesis. Before gross photosynthesis may be interpreted with confidence, however, more data on its relation to potential fish harvest in a variety of circumstances will be necessary.

ACKNOWLEDGMENTS

This study is a contribution of the Arizona Cooperative Wildlife Research Unit. Financial support was provided by the Arizona Game and Fish Department through Federal Aid to Fisheries Project F-8-R, and by the University of Arizona.

LITERATURE CITED

Dineen, C. F. 1953. An ecological study of a Minnesota pond. Am. Midland Nat., 50: 349–376.

Hutchinson, G. E. 1957. A treatise on limnology. Vol. 1. John Wiley and Sons, New York. xiv + 1,015 pp.

Juday, Chauncey. 1940. The annual energy budget of an inland lake. Ecology, 21: 438–450.

Lindemann, R. L. 1942. The trophic-dynamic aspect of ecology. Ecology, 23:399–418.

Lund, J. W. 1949. Studies on *Asterionella formosa.* I. The origin and nature of the cells producing seasonal maxima. Jour. Ecol. 37:389–419.

McConnell, Wm. J. 1962. Productivity relations in carboy microcosms. Limnol. and Oceanogr., 7(3):335–343.

Odum, E. P. 1959. Fundamentals of ecology. W. B. Saunders Co., Philadelphia and London. vii + 546 pp.

Odum, E. P., and Alfred E. Smalley. 1959. Comparison of population energy flow of a herbivorous and a deposit-feeding invertebrate in a salt marsh ecosystem. Proc. Nat. Acad. Sciences, 45:617–622.

Odum, H. T. 1957. Trophic structure and productivity of Silver Springs, Florida. Ecological Monogr., 27:55–112.

Odum, H. T., and C. M. Hoskins. 1958. Comparative studies on the metabolism of marine water. Publ. Inst. Marine Sci., Univ. Texas, 5:15–46.

Odum, H. T., W. McConnell, and W. Abbott. 1958. The chlorophyll "A" of communities. Publ. Inst. Marine Sci., Univ. Texas, 5:65–96.

Richards, F. R., with T. G. Thompson. 1952. The estimation and characterization of plankton populations by pigment analysis. II. A spectrophotometric method for the estimation of plankton pigments. Jour. Marine Res., 11:156–172.

Ricker, W. E. 1958. Handbook of computations for biological statistics of fish populations. Fish. Res. Bd. Canada, Bull. 119, 300 pp.

Ruttner, F. 1953. Fundamentals of limnology. Translated by D. G. Frey and F. E. Fry. Univ. of Toronto Press. xi + 242 pp.

Ryther, John H. 1956. The measurement of primary production. Limnol. and Oceanogr., 1:72–84.

Snedecor, G. W. 1946. Statistical methods, 5th ed., Iowa State College Press, Ames, Iowa. xiii + 485 pp.

Verduin, Jacob. 1956. Energy fixation and utilization by natural communities in western Lake Erie. Ecology, 37:40–50.

Wright, John C. 1959. Limnology of Canyon Ferry Reservoir. II. Phytoplankton standing crop and primary production. Limnol. and Oceanogr., 4:235–245.

CONTROL OF UNDESIRABLE SPECIES

9

An ecological approach to pest control

P. W. Geier
L. R. Clark

In common usage, the term "pest control" designates empirical attempts at curbing intolerably dense populations of noxious organisms.

The need for pest control, i.e., for relief from undesirable species, has increased to such an extent over the past half-century that a world-wide situation of unending struggle now exists. Since modern control began, we have become confronted by an increasing number of pest problems, the fundamental nature of which continues to be ignored in our attempts at solving them. An endless sequence of *ad hoc* programmes of destruction has kept the available trained personnel constantly mobilized in practical pursuits, with little time and energy for thinking about fundamentals.

This largely explains why the concepts which generally inspire pest control practices are so much out of step with modern ecological ideas. It must be conceded that the current approach to pest control has not progressed far beyond the original idea of striking down noxious populations by a series of blows delivered almost blindly but with a maximum of violence. This may be described as the "fly-swat approach."

We shall endeavor to indicate how ecological knowledge, and more particularly an understanding of population dynamics, ought to be utilized in pest control if lasting solutions are to be achieved. The basic ideas and descriptive terms employed in the discussion are those of Nicholson

---

Reprinted with permission from *Symposium of the 8th Technical Meeting of the International Union for Conservation of Nature and Natural Resources, Warsaw, 1960,* E. J. Brill, Leiden, pp. 10–18.

(1933, 1954, 1957, 1958). The essentials of his theory provide the most realistic basis available for a critical appreciation of the nature and problems of pest control.

The acceptance of such ideas leads us immediately to question the soundness of the term "pest control." A "pest" is a living organism which sometimes occurs in numbers inconvenient to man. The emphasis is on *population density*, i.e., on intensity of infestation, since it alone determines pest status, exclusively of any inherent property of the organism. In the field of crop production, one may usefully consider that pest status is attained only when a potentially injurious species reaches a threshold density at which it begins to cause a significant reduction in crop yield. However, for practical purposes, it is better to consider another more critical threshold density which may be defined as the level at which the loss caused by a pest just exceeds in value the cost of the control measures available. Here we have a relative criterion of injuriousness whose yardstick is fixed according to the general level of the economy at any time and place. In other words, an undesirable species is not an economic pest if it does not pay to combat it.

Thus, the rather ambiguous concept "pest" may be taken in two connotations—the first absolute—the second relative and economic. This distinction may be illustrated as follows. Assuming conditions of uniform infestation, kangaroos may reduce the carrying capacity of many Australian grazing properties for sheep, but it does not follow that kangaroo control would always be profitable. The only properties which would benefit by control are those where the increase in carrying capacity could be fully utilized to make up for the cost incurred. No immediate benefit would be derived from kangaroo control on other properties because a corresponding increase in financial return could not ensue. The difference between the two sorts of exploitation is one of economic intensity and technical development. In the first instance, the values of the absolute and economic threshold densities of the pest population are not very different. This is indicative of a high level of economic development. In the second instance, there is a large difference between the two threshold values—the density of the pest population would have to be substantially higher for control to be economical. Large differences between the absolute and economic thresholds occur at low levels of economic development. Since the value of the economic threshold density of a pest population will finally determine if and how protection is feasible commercially, no ecological assessment of a pest situation is complete without consideration of the economic and technical factors involved.

Since relative scarcity rather than great abundance is characteristic of most phytophagous organisms, one may wonder how species reach pest status. In most cases, if not all, they reach pest status as a consequence of man's activities. These activities are pest-inducing whenever they upset long established patterns of intra- and interspecific relationships, which have evolved in the course of natural selection. The disturbing influences of intensive land use, the breeding and distribution of man-improved organisms, the practice of mono-culture, the introduction of exotic animals and plants, all contribute to the production of pest species, which, however, are surprisingly few in number.

Man's attempts to deal with his unintentional creation of pests has not only been inadequate; it has also been conducive to the production of further problems. In general, he has developed no better solution to pest problems than that of making his fly-swat bigger and more destructive. There is no essential difference between the effect of a fly-swat and that of, say, a modern insecticide applied indiscriminately. Each is a frontal attack on inconveniently dense populations delivered with the intention of knocking down numbers to a more acceptable level. Since eradication of the noxious species is not usually achieved by such means, their success is necessarily of short duration and, sooner or later, repeated application becomes indispensable. The fly-swat approach bears within itself the cause of its limited success—it can only reduce symptoms. It leaves unaffected the circumstances which induce pest-level densities.

This statement suggests that the term "control" must be divested of the meaning it assumes in the expression "pest control." There can in fact be no entirely man-made "control" of living organisms, since control is implicit in the existence of every persistent population. Every such population is kept under control as a matter of course by the interplay of its innate tendency to multiply and the inherent environmental opposition to population growth. This condition of continuing and universal control determines the state of balance in the absence of which populations would multiply indeterminately wherever individuals are able to reproduce at more than the replacement rate. Control, in its true meaning, bears no necessary relation to the anthropocentric concept of "pest." In general, pest populations are under natural control like any others, however inconvenient their numbers may be (Milne, 1959).

Since the density of a population is determined primarily by the intrinsic favourability of its environment, it follows that a population's vitality is not necessarily reduced in proportion to the numbers of individuals destroyed by an insecticide treatment or similar "control" measure. Thus, it is wrong to assume *a priori* that the greater the relative kill due to a treatment, the greater will be the long-term relief obtained. Populations of living organisms, as Nicholson has demonstrated by laboratory experimentation, are well-adapted to withstand high accidental mortalities. It appears that their resilience is a natural outcome of the governing mechanisms which implement natural control. Such mechanisms effectively oppose numerical increase when the population density is relatively high and facilitate it when numbers are relatively low. Any drastic reduction of population density due, for example, to the application of an insecticide, will induce the density governing mechanisms to relax the severity of their action to compensate for the losses incurred by the population. The relaxation induced may allow pest numbers to bounce back to peak population densities in as little as one generation. Compensation reactions probably explain the tendency for sprays to cause a sharp increase in the magnitude of the fluctuations normally displayed by pest populations.

In general, modification or elimination of the influence of density governing mechanisms probably accounts for the necessity of repeating treatments to keep numbers down in a population which has been reduced artificially. However, more may be involved than this: populations are genetically heterogeneous and often include small numbers of individuals "resistant" to the particular chemical agent used to limit population density. Relieved by insecticidal treatment from the competitive pressure of nonresistant forms, the resistant individuals may gradually form a new population against which the current means of protection are no longer effective. Moreover, many commonly-used pesticides are nonselective. Besides killing individuals of the pest population, they often affect numerous other species, harmless or even beneficial, with consequences similar to those described above. The void due to the treatment will soon be filled under peculiar selective conditions which favour species with high reproductive capacities. Repeated treatment may result in a drastic change of faunal composition characterized by an overall reduction in the number of species present, and by a sharp increase in the abundance of surviving populations.

In brief, the fly-swat approach is unsatisfactory because it is applied without regard to the existing process of natural regulation to which all persistent populations of living organisms are subjected—however inadequate this limitation may appear to be for practical purposes. In blindly forcing alien mortality factors into balanced systems of natural control, current pest control practices not only fail to take advantage of the regulatory mechanisms available in nature, but even tend to interrupt or destroy the automatic operation of destiny governing agents.

Assuming these views to be correct, we obviously cannot solve the pest problem by relying only on natural control for protection against intolerable infestations. Having started to manage nature to our own ends, we can only overcome the unexpected difficulties encountered by the further development and extension of management. In other words, we must cease thinking in terms of traditional "pest control" and begin to think about pests in terms of *protective population management.* Considered in this light, protection against undesirable species appears as the mirror-image of species conservation, a field in which much lasting success has been achieved.

We must return to population theory to foresee how the protective management of noxious species can be achieved. The regulation of population density may be regarded as a dual process. Primarily, limits to numerical abundance are set "legislatively" by density factors (properties of essential requisites and properties of the species itself) which determine the intrinsic favourability of a given environment and the ability of a population to persist in it. The intrinsic favourability of a given environment may vary considerably from time to time. For an environment to be favourable to a species, it must at least generate a certain minimum supply of each requisite and provide a certain minimum level of security from attack or interference by inimical organisms. Otherwise the population could not persist.

In a favourable environment, the numbers of a species tend to increase progressively, eventually inducing an increase in environmental resistance by depleting one or more requisites, e.g., food, or by inducing an increase in the intensity of attack by natural enemies. The intrinsic environmental *rate of supply* of such a requisite, or the *level of security* from inimical organisms which the environment will permit, plays a very important part in numerical regulation. However, these components of the favourable environment, being unaffected by population density, cannot adjust numbers to the prevailing conditions.

Numerical adjustment, i.e., density governance, which constitutes the second part of the dual process of regulation, is effected by the interaction of population density and the requisite brought into short supply by numerical increase. The governing agent is commonly a modification induced by population density in the availability of the requisite concerned, or in the intensity of attack by natural enemies: the mechanism of adjustment is usually some form of intraspecific competition. Where intensity of attack by a natural enemy is the environmental component involved in numerical adjustment, competition between individuals of the inimical population provides the mechanism. Density-induced processes hold populations about a constant or variable equilibrium density related to the properties of the species and to the rate of supply of the limiting requisite(s), both of which are modified by other components of the favourable environment, e.g., weather.

When faced with the problem of "managing" a population which has reached the status of an economic pest, an ecologist might decide either that the given environment favours, permanently or at irregular intervals, a general equilibrium level of population density which is intolerably high, or that the outbreaks of a pest are essentially the peaks of violent oscillations about an economically acceptable level. In the first case, management calls for a reduction in the general equilibrium level of the pest population, or stabilization at the low level which persists between outbreaks. In the second, management requires damping of the oscillations. In both cases, the approach would be either to induce a systematic reduction in the intrinsic favourability of the environment, or to modify adversely some specific properties of the noxious organism. In the present state of our knowledge, the former method seems the more promising generally. It could be implemented, for example, by reducing the supply of a requisite (e.g., shelter), or by increasing the overall intensity of attack by natural enemies.

The higher man's economic development of the environment, the greater are the possibilities for protective management. Where land use is primitive and agriculture extensive, protective management is necessarily reduced to its simplest form by lack of financial resources and limitations of technique. It can go no further, possibly, than the implantation of competitive or antagonistic organisms. A case in point is that of prickly pear and *Cactoblastis* (Nicholson, 1958, p. 109). As the level of economic development rises, so do the potentialities of protective management. The more man has altered the environment, the greater are the chances that some new or old environmental factor could be modified critically to the disadvantage of a pest without adversely affecting crop production. A perfect illustration is provided by the method of defence adopted against *Phylloxera vastatrix* in European vineyards, where the economy of the wine industry permits the use of a range of vine rootstocks on which the insect cannot complete its life cycle.

Protective management does not imply the adoption or rejection of any mode of defence against undesirable populations, for it is not concerned primarily with means but with the way in which means are employed. For instance, there is certainly no incompatibility between protective management and the use of chemical pesticides. In fact, chemical pesticides represent an invaluable asset in protective management, initially as working tools and later as safety factors.

Where chemical pesticides have become indispensable, treatments obviously cannot be discontinued abruptly just because of their ecological inadequacies. The establishment of protective management must be gradual and may require a transitory period of considerable length before any significant departure from common routine has been achieved. Thus, in most cases, protective management must begin as an attempt to integrate current pest treatment with the recognized processes of natural control.

The first step in protective management is taken when treatments are no longer applied blindly, but made to conform objectively with local circumstances. This implies a systematic limitation of the disruptive interference with regulatory processes which a chemical treatment must entail. It calls for deliberate timing and careful selection of active ingredient, formulation, concentration, and method of application. Judicious timing requires accurate forecasting of pest population densities and a knowledge of the most vulnerable stage in the life cycle of an undesirable species. Moreover, timing should also restrict the effects of insecticidal treatments on harmless or beneficial populations. It is desirable that pesticides be not only effective but also selective in their action. The more specific a pesticide, the less likely it is to induct adverse side-effects. If eradication cannot be achieved, treatments should be designed to reduce noxious populations no more than is commercially desirable.

This first stage in protective management was reached fully—and profitably—several years ago in the apple orchards of Nova Scotia thanks to Pickett and his co-workers (Pickett *et al.*, 1946, 1953, 1958). Although crop protection is still achieved there largely by chemical means, the methods adopted have stabilized a pest situation previously as fluid and unpredictable as that faced by orchardists elsewhere.

In general, progress in the first stage will lead automatically to the second stage of protective management. This requires a working knowledge of the natural numerical regulation of the noxious species concerned, and, particularly, the identification of limiting density factors. It is effected directly by the deliberate modification of limiting density factors found to be susceptible to economical manipulation by man. Chemical destruction is by no means excluded at this stage, but treatments need no longer be as frequent or as heavy as before. Conversely, their efficacy is enhanced by diminution of the population pressure against which they have to measure. This stage, for example, has been attained successfully in the case of the cockchafer *Melolontha melolontha* in western Switzerland by Murbach, Bourqui & Geier (1952).

The third possible stage in protective management is that of pest suppression as the ultimate consequence of comprehensive husbandry. This complete mastery is illustrated by the case of *Phylloxera vastatrix*. While this stage may sometimes be reached without detailed ecological study, such investigations will greatly enhance the chances of achieving it. The process of natural regulation is frequently both complex and subtle, involving critical density factors not readily identified by *ad hoc* methods.

The rate at which these successive stages may be reached depends firstly on the complexity of the initial situation, particularly on the number of potential pests to be dealt with simultaneously, and secondly on the intensity of the research effort devoted to the subject.

Nova Scotian experience suggests that the careful control of spraying is sufficient alone to allow the re-establishment of a measure of natural regulation which will maintain some undesirable populations below the economics threshold density (Lord, 1947, 1949). These populations are of species which attained economic pest status because of indiscriminate applications of pesticides. Their number may not be inconsiderable, and it is essential to distinguish them from the true problem species whose intolerable abundance is induced by the environment proper. Whereas the former need no longer be taken into account after Stage I, the latter must be dealt with by Stage II methods.

Whether Stage III can ever be reached for a majority of current pests is in fact not so much a matter of scientific possibility as of future economic expediency. It is economically expedient *now* to begin replacing pest control by some sort of protective management. Although the changeover will require much creative effort from all concerned, it should prove to be less expensive in the long run than the continued use of present day methods.

Obviously, one cannot expect a proposal of this order to be implemented forthwith. However, we share with others the hope that the need for protective management will be realized before much more time and money have been wasted in the pursuit of remedies which ignore the ecological realities of pest control.

REFERENCES

Lord, F. T. 1947. The influence of spray programs on the fauna of apple orchards in Nova Scotia. II: Oystershell scale. Canad. Ent., 79:196–209.

———. 1949. The influence of spray programs on the fauna of apple orchards in Nova Scotia. III: Mites and their predators. Canad. Ent., 81:217–230.

Milne, A. 1959. Weather, enemies and natural control of insect populations. J. Econ. Ent., 52:532–533.

Murbach, R., P. Bourquoi, and P. Geier. 1952. Le hanneton et le ver blanc. Nouveaux enseignements, nouveaux procédés de lutte. Rev. Rom. Agric., 8:73–84.

Nicholson, A. J. 1933. The Balance of Animal Populations. J. Anim. Ecol., 2:132–178.

———. 1954. An outline of the dynamics of animal populations. Australian J. Zool., 2:9–65.

———. 1957. The self adjustment of populations to change. Cold Spr. Harb. Symp. Quant. Biol., 22:153–173.

———. 1958. Dynamics of insect populations. Ann. Rev. Ent., 3:107–136.

Pickett, A. D., N. A. Patterson, H. T. Stultz, and Lord, F. T. 1946. The influence of spray programs on the fauna of apple orchards in Nova Scotia, I: An appraisal of the problem and a method of approach. Sci. Agr. 26:590–600.

———, and N. A. Patterson. 1953. The influence of spray programs on the fauna of apple orchards in Nova Scotia. IV: A review. Canad. Ent., 85:472–478.

Putnam, W. L., and E. J. Leroux. 1958. Progress in harmonizing biological and chemical control of orchard pests in Eastern Canada. 10th Inter. Congr. Ent., 3: 169–174.

10

Species diversity and insect population outbreaks

David Pimentel

The structure and dynamics of the community system are based on the successful evolution of plants and animals together in a habitat. Diversity of species and complexity of association among species are considered vital to the stability and balance of the community system. Animal population outbreaks occur most frequently in cultivated areas where man has restricted the vegetation to a single species of plant and in the Arctic where simplified communities naturally exist. Ample evidence is available in the literature to support this concept. Marchal (1908) wrote that, "Man in planting over a vast extent of country certain plants to the exclusion of others, offers to the insects which live at the expense of these plants conditions eminently favorable to these excessive multiplication." Later, Graham (1915) wrote, "that pines (white) growing in mixture with other species such as Norway pine (*P. resinosa*) are somewhat less subject to attack" from the white-pine weevil (*Pissodes strobi* Peck). Further support of this observation has come from the works of Peirson (1922), Graham (1926), and MacAloney (1930) who noted that the white-pine weevil was excessively abundant in pure stands of white pine; but when the pine was in with mixed hardwoods, infestations were insignificant. According to von Hassel (1925), insect outbreaks were common in the monocultured forests of Germany but seldom occurred in the virgin forests of South America. Trägardh (1925), comparing outbreaks in the

Reprinted from *Ann. Ent. Soc. Amer.*, 54:76–86 (1961), by permission of the copyright owner, The Entomological Society of America.

forests of Germany and Sweden, concluded that fewer outbreaks took place in Sweden because about two-thirds of their forests are mixed pine-spruce-birch. Rohrl (1928) and Friederichs (1928) observed that catastrophes occur in forests where man has changed the environment by monoculture, and they suggested that lack of diversity characteristic of mixed stands was responsible for the outbreaks noted. Damage from two moths was found to be negligible to spruce trees in mixed woods, but in pure stands severe damage often resulted (Barbey 1931). A study of the mango beetle (*Cryptorhynchus gravis* F.) in Java showed that while it causes little damage to the mango fruit when the trees are growing in natural forests, it severely attacks the fruit where the trees are grown by themselves or are otherwise cultured near man's habitation (Voute 1935). In Sumatra, Schneider (1939) reported a lepidopteran (*Oreta carnea Saalm*) in great numbers on cultivated gambir, but in the virgin forests where gambir grows naturally no outbreaks of the pest had ever been observed. Working in Turkey, Schimitschek (1940) reported that the beetle (*Ips sexdentatus* Börn.) is found primarily in pure woods planted by man. Eidmann (1942, 1943) observed that the African virgin forests are relatively free of insect pests in contrast to areas in which man has altered the environment.

The evidence supporting the proposition that more outbreaks occur in single-species stands under man's cultivation than in natural-mixed-species communities, has led some workers to propose that in virgin forests outbreaks never occur (Schimitschek 1937, in Voute 1946). Schneider (1939) wrote further that the virgin forests of the tropics are unsuited for insect plagues because of species diversity. Podhorsky (1933) reported that forests not affected by man seldom suffer from insect pests. In contrast, however, both Schenk (1924) and Eidmann (1942) have pointed out that catastrophes are not restricted to cultivated areas, but that damage occurs occasionally in virgin forests. They emphasize "occasionally" because, as mentioned earlier, both recognize the fact that outbreaks most frequently take place in cultivated areas.

Population outbreaks occur frequently in the simple natural communities of the Arctic (Elton 1927). Kalabukhov (1947) concludes that population fluctuations of greater amplitude occur in mammal populations in the Arctic because the community contains few species. Further south in the Russian forestlands and steppes, the animal communities are more complex and these fluctuations are less severe. Support of this analysis is provided by Odum (1953) and Hutchinson (1959).

Insecticides, because they are general poisons, tend to sterilize habitat fauna and in so doing reduce diversity. This encourages population outbreaks, as documented in Ripper's (1956) review and the more recent research of Stern and van den Bosch (1959) and Smith and Hagen (1959). This information has lead the University of California ecologists to propose the new concept, "integrated insect control" (Stern et al. 1959).

Thus, considerable evidence in the literature suggests that the lack of species diversity, characteristic of communities modified by cultivation and of natural communities in the Arctic, may be responsible for the population outbreaks which are so typical of these simplified communities. The above information is based on field observation, but in this investigation the relationship between species diversity and population outbreaks was explored by experimentation. Here the quantitative changes in structure of the animal community associated with *Brassica oleracea* L., family Cruciferae, as a single-species stand and then as a member of a more complex plant community were studied and evaluated.

The investigator wishes to acknowledge the conscientious and able assistance of Messers F. A. Streams, E. F. Menhinick, and W. L. Freeman in establishing and maintaining the experimental plots and in aiding in some of the population counting. I wish to especially thank Professors L. C. Cole and B. V. Travis for their helpful suggestions during the preparation of this manuscript.

METHODS

The design of these experiments was influenced by the following hypotheses. First, from evidence available in the literature, more animal species were expected to be present in mixed-species planting than in single-species planting. Second, the density of the specific populations would in general be less in a mixed than in the single-species planting. Third, a greater degree of population stability would be characteristic of the community associated with the plant in a mixed-species planting (complex community) than in the single-species planting (simple community).

The test plots were relatively large, to minimize variability due to either immigrants or emigrants. The plots were separated by distances great enough to reduce variability due to immigration and emigration of animals. Sampling thoroughly and frequently over a long period reduced variability due to changes in plants, temperature, and moisture. Recording the changes found in density in more than one species added support to the general trends. The investigator personally did most of the sampling to reduce variability due to individual differences. Replicating the experiments during two summer seasons reduced the chance of any misleading influence because of weather. The general design, explained in detail below, provided the most satisfactory method for obtaining the information desired using the facilities available to the investigator.

The animal community associated with *B. oleracea* was selected for the study because members of this community are rich in numbers both of species and individuals (Pimentel, unpublished data). In addition, all animals occur on both natural and cultivated Cruciferae in the Ithaca region. Preliminary observations further showed that the invertebrate herbivores are more attracted to and prefer the cultivated *B. oleracea* species for feeding and oviposition over most wild Cruciferae. Of importance was the fact that the biennial growth characteristics of the plant provides a certain amount of stability to the plant factor in the test environment.

The varieties of *B. oleracea* used in 1957 were cabbage, collards, broccoli, Brussels sprouts, and kale. These differed little in their attractiveness to animals (Pimentel, unpublished data), but collards and broccoli blossomed readily and provided nectar for those individuals (such as adult Lepidoptera and Hymenoptera) that required it, and the broad, flat leaves of collards facilitated insect counting. Hence, collards alone were used in 1958.

At the start of each experiment plants were from 4 to 8 weeks of age. Generally, measurements were made on younger plants in 1958, and this is indicated in the size measurements of the plants.

In the mixed-species planting *B. oleracea* plants were set out in a 15-year fallow field (500 x 600 feet) in which approximately 300 species of plants grew (Fig. 10–1). The mixed-species planting was large so that a greater number of *B. oleracea* plants could be planted than would be possible in a 75 x 100 feet plot. The *B. oleracea* plants were planted in rows at 9-foot intervals both in the row and between rows so as to leave as much as possible of the natural vegetation still growing in the mixed plot. To set out the plants, the soil in a 1-square-foot area was well worked and fertilized and limed. Fertilizer (5-10-5) and lime were applied at a rate of 2,000 pounds per acre to an area of 2 square feet which included the worked area of 1 square foot. A piece of black plastic, 1 foot square, was laid over the worked soil. Through a hole in the center of the plastic, a plant was set in the soil below. The plastic offered the *B. oleracea* plants some protection from competition of the surrounding vegetation.

In addition to the planted *B. oleracea*, five wild species of Cruciferae were present: *Brassica nigra* (L.), *Barbarea vulgaris* R. Br., *Armoracia rusticana* Gaerth., *Cardamine pennsylvanica* Muhl. and *Lepidium campestre* (L.) R. Br. Ample plant food was available for the herbivorous animals associated with *B. oleracea* because most of these species can utilize the wild Cruciferae (Hering 1932). *Barbarea vulgaris* was the most abundant species in the area and from the standpoint of available leaf surface area totaled more than three times that of *B. oleracea* introduced into the mixed-species planting.

The single-species planting was located about 2,000 feet east of the mixed-species planting and measured 75 x 100 feet. This was large enough to prevent any noticeable effect on the populations due to immigrants.

Fig. 10–1. A view of the mixed-plant-species habitat.

In 1957 and 1958 the same varieties of *B. oleracea* were planted there as were planted in the mixed-species planting.

In 1958, another spacing of the single species was added in order to have a plot with plants spaced similarly to those in the mixed-species planting, or 9 x 9 feet from each other. This plot measuring 300 x 900 feet was located about 800 feet east of the single-species planting. Hereafter, this sparse planting of *B. oleracea* will be referred to as the "sparse-single-species planting" in contrast to "single-species planting."

Early in the 1957 season, slugs and field mice destroyed many of the young *B. oleracea* plants. Since this was a behavior phenomenon, not a population phenomenon, and was not present in the single-species planting, 10 percent metaldehyde dust was sprinkled on the plastic below the plants for slugs, and phosphorus-treated corn was distributed to control the mice. These poisons were specific for the slugs and mice. Similar practices were followed in 1958. Any plant destroyed during the experimental period for any reason was immediately replaced with another of similar size obtained from replacements maintained in the greenhouse.

Weekly, 100 plants were measured and animal populations sampled in each plot. Starting the last week in May, the plants were selected systematically to insure a representative sample of the plot. To avoid bias in the choice of individual plants to be sampled, one plant out of every so many, depending upon the total number of plants, was designated for inspection. The outer rows were not included in any sample because of the chance of having edge-effects. An estimate of the total surface area was made by placing a sheet of plastic, which had been etched in 1-inch squares, against the plant and estimating the number of square inches of surface area on the plant. Only one side of a leaf was included in the measurement. The area estimated in this way was demonstrated to have an error of less than 10 percent.

The animal species which make up the community associated with *B. oleracea* are varied, but consist principally of insects. Previous to the initiation of this project, a thorough study was made of the relationship and interactions of the animals of the test community. This information was available to the investigator for reference (Pimentel, unpublished data).

In estimating the number of individuals of any taxon, only a single stage in the life history of each one was used. The stages selected were those which would give the best estimate of the density of the species and parasitism rates in the species. The following is a list of the types found and the methods by which their densities were determined.

Lepidoptera All in the last larval instar were counted. As the larvae were counted, they were collected and placed into separately labeled vials and taken to the laboratory and held to determine rates of parasitism and disease. The pupal stages were not included in the count, but they were collected and held to reveal any parasites which were not common to the larvae.

Homoptera All were counted in the adult stage. With the Aphididae, those parasitized and attached to the plant were counted as parasitized. Those individuals from which parasites had emerged were not counted. When more than 100 were present a sample of these was taken; otherwise all parasitized aphids were taken to the laboratory for holding to determine the parasites and their ratios.

Hemiptera All were counted in the adult stage. Parasitism was recorded when observed.

Coleoptera All were counted in the adult stage. Last-stage larvae of these were collected and held for parasite emergence.

Diptera *Liriomyza*, *Scaptomyza*, and *Aphiodoletes* groups were counted in the third-larval instar. Pupae were held for parasite emergence. The Syrphidae were counted in the pupal stage and held to determine parasitism rates. All the remaining Diptera types were counted in the adult stage, and the larvae when seen were collected to determine parasitism rates.

Hymenoptera All except those species belonging to the family Formicidae were counted as so many parasitic attacks upon their hosts. Their hosts were collected in the field and held for parasite emergence. All other types were counted in the adult stage, and parasitism rates were determined where possible.

Neuroptera All were counted in the last larval stage, and they were held in the laboratory to determine parasitism rates.

Arachnida All stages were counted and parasitism rates determined where possible.

Other invertebrates Any in this category were counted in the adult stage and parasitism was determined where possible.

Microorganisms All were counted as so many individual host infections.

In field-sampling the populations, it was difficult to distinguish the differences between some species, genera, and in some cases families. Such organisms were classed into taxa which could be accurately determined in the field. The taxa in all the test communities were handled similarly. A list of the taxa and a discussion of the ecological role of each will be given in a separate paper. The methods of grouping species for counting reduced the categories of taxa in this study to one-fourth of the total noted in the initial survey.

RESULTS

The taxa associated with *B. oleracea* differed in a single-species planting compared to those in a mixed-species planting. In 1957, a total of 27 taxa were found associated with *B. oleracea* grown in the mixed-species planting, 50 taxa in the single-species planting (Table 10–1). In 1958, the differences between the mixed-species and single-species plantings, although significant (39 and 50 taxa, respectively), were not as marked as in the 1957 season (Table 10–1). A listing of the taxa observed in each of the experimental plots and the number of individuals counted each week for the 15-week period is filed in the Comstock Memorial Library of Cornell University, and copies of these data may be obtained on interlibrary loan. These taxa were only the ones associated with the *B. oleracea* in the mixed-species planting. Considering the estimated number of 300 plant species in that planting, I would estimate at least 3,000 species of heterotrophs for the mixed-species habitat.

During 1958, the 54 taxa which were reported for the sparse-single-species planting agrees favorably with 50 taxa reported in the single-

TABLE 10–1. The number of taxa which were the same or different in the respective mixed- and single-species stands during the 1957 and 1958 seasons

Classes	Mixed			Single		
	In 1957 only	Common to both years	In 1958 only	In 1957 only	Common to both years	In 1958 only
Herbivores	4	16	9	1	22	5
Parasites	2	2	4	3	10	1
Predators	0	3	5	5	9	3
Totals	6	21	18	9	41	9

species planting. Since the spacing of *B. oleracea* plants did not alter the number of taxa appreciably, the differences in number of taxa found in the single and mixed planting cannot be attributed to differences in spacing of the plants.

In the mixed-species planting the taxa number increased from 27 in 1957 to 39 in 1958 (Table 10–1). Of the 27 taxa found in 1957, 21 were common to both seasons. In the single-species planting the number of taxa remained 50 for both seasons (Table 10–1). There was a change in the composition and a total of 41 taxa were common to both seasons. Although a change was made from a few collards plus other varieties of *B. oleracea* planted in 1957 to all collards in 1958, these changes, especially in the mixed-species planting, are not considered responsible for the changes noted in the number and kinds of taxa. Variation in weather is a more valid explanation of the changes. Appreciably lower temperatures in 1958 accompanied by a 42 percent increase in rainfall would be expected to cause significant changes in the general makeup of the community (Table 10–2).

Seasonal change in taxa was found to be greater in the mixed planting than in the single planting because the taxa in the mixed planting existed at low densities which made them difficult to measure by the

TABLE 10–2. Mean monthly temperatures and rainfall for the experimental periods in 1957 and 1958

Year	May	June	July	Aug.	Sept.
		Monthly temperatures (° F.)			
1957	53.9	67.4	67.4	63.6	60.0
1958	52.4	59.2	68.1	66.6	59.2
		Monthly rainfall (inches)			
1957	3.30	3.09	5.55	1.43	3.15
1958	3.37	5.94	5.23	4.83	4.17

sampling technique. Those taxa which were sparse were the ones most susceptible to variation in detection by sampling. Although a large sample was taken in each plot, not all the plants in any one plot were sampled. A species with few individuals distributed on two or three plants could be missed. Any reduction in number of those taxa which were already marginal for detection would greatly reduce their chances of detection, and, on the contrary, a slight increase in number would increase chances of detection. Thus, because densities of individuals in the mixed planting were marginal for detection by the sampling technique used, a greater change in taxa took place in this plot compared with the single-species planting. This, however, does not alter the fact that there were significantly more taxa per unit plant area in the single compared with the mixed.

In 1957, 20 taxa were common to both the mixed-and single-species planting, and, in 1958, 34 taxa were common to both (Table 10–3). The

TABLE 10–3. The number of taxa found only in the mixed- and single-species planting compared with those common to both plantings during 1957 and 1958 seasons

Classes	Taxa common to both	Mixed only	Single only
	1957		
Herbivores	14	6	7
Parasites	3	1	11
Predators	3	0	12
Totals	20	7	30
	1958		
Herbivores	21	4	6
Parasites	6	0	5
Predators	7	1	5
Totals	34	5	16

additional taxa present in the mixed planting were principally herbivores, while in the single-species planting the majority of the additional taxa were parasitic and predaceous. Because 74 percent of the taxa in the mixed-species planting consisted of herbivores compared with 46 percent in the single-species planting, a major change in taxa between years in the mixed planting occurred in the herbivore taxa. Many general feeding herbivore taxa in the mixed moved onto the *B. oleracea* plants from the various other plant species in the habitat. However, in the single-species planting herbivores existed at high density and parasites and predators flourished. Thus the greatest change in taxa between years occurred in parasite and predator taxa in the single-species planting.

The total number of taxa reported for the sparse-single-species planting was 54, which was significantly greater than the 39 reported for the mixed-species planting. A total of 33 taxa were similar in the mixed-species planting and sparse-single-species planting (Table 10–4). The

TABLE 10–4. The number of taxa found only in the mixed- and sparse-single-species planting compared with those common to both plantings during the 1958 season

Classes	Taxa common to both	Mixed only	Single only
Herbivores	20	5	9
Parasites	6	0	7
Predators	7	1	5
Totals	33	6	21

additional taxa found in the mixed were principally herbivores, while most of the other taxa found in the sparse-single-species planting were predominantly parasites and predators. This is similar to findings in mixed-species versus the single-species planting.

Significant changes in the density of the taxa occurred in all the experimental plantings. Variation in plant size and relative abundance of some taxa complicate the presentation of the results. The *B. oleracea* plants in the mixed planting were about one-seventh the size of the plants in the single-species planting at the close of the experimental period in 1957, and about one-fifth in 1958. The mature plants in the sparse-single-species planting were about twice as large as the plants in the single-species planting in 1958. With such differences in size, certain qualitative differences in nutritional content might be expected. No allowances could be made for such qualitative changes, but the quantitative differences in size of plants were considered. To do this the measurements of the plant surface area recorded for each plant were used in adjusting animal densities. All animal densities were made equivalent per unit plant area by adjusting densities to a constant of 20,000 square inches.

Density trends of various groups of taxa are discussed instead of treating each one individually. In 1957 the aphids, especially *Brevicoryne brassicae* (L.), reached outbreak levels in the single-species planting but were scarce in the mixed-species planting (Fig. 10–2). The aphids were not as abundant in 1958 as 1957 in the single-species planting, but they were still more abundant in the single-species planting than in the mixed-species planting (Fig. 10–3). The difference in aphid density between the single-species and sparse-single-species plots was insignificant.

The other herbivores to reach outbreak levels in the single-species planting were the flea beetles, *Phyllotreta cruciferae* Goeze, *P. vittata*

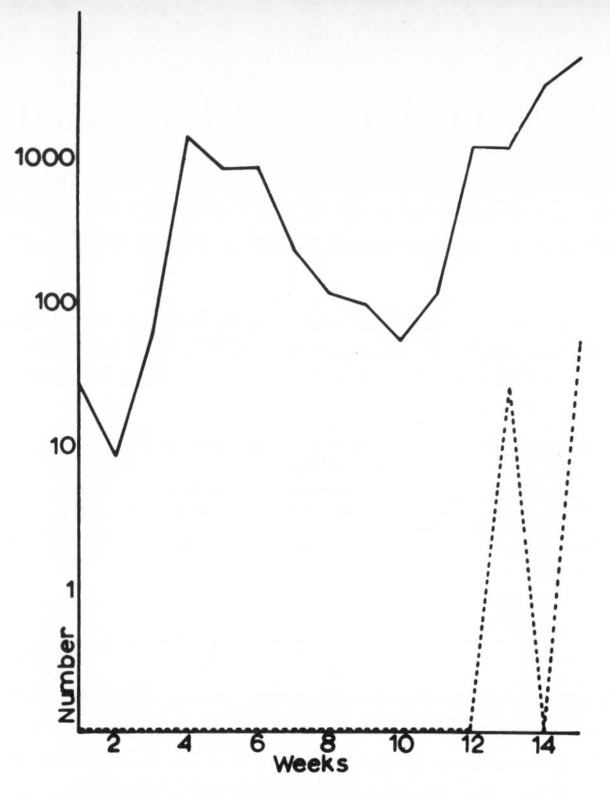

Fig. 10–2. The log number of aphids per unit plant area in the mixed- (- - -) and single-species (———) stands during the 1957 season.

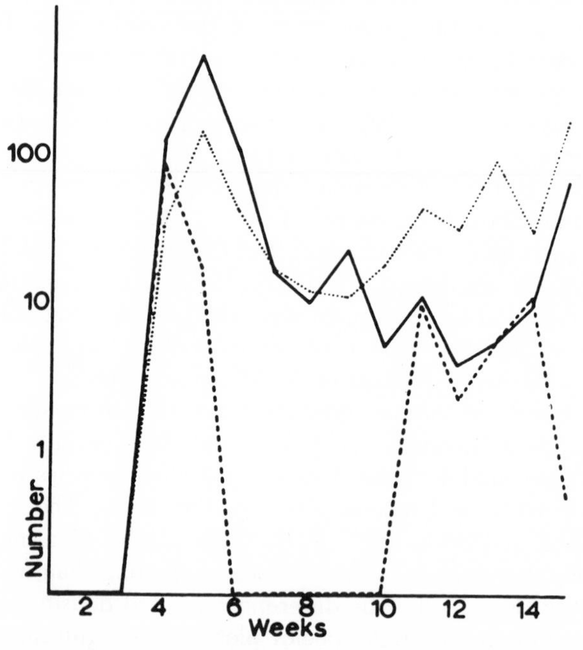

Fig. 10–3. The log number of aphids per unit plant area in the mixed- (- - -), single- (———), and sparse-single-species (........) stands during the 1958 season.

(F.), and *Psylliodes punctulata* Melsh (Figs. 10–4 and 10–5). The flea beetles were most abundant in 1958. The spacing of the plants in the sparse-single-species planting had little affect from the view of diversity on the outbreaks of the flea beetles (Fig. 10–5). As for the Lepidoptera and all other herbivores not mentioned specifically, little seasonal difference was found (Figs. 10–6, 10–7, 10–8, 10–9).

In 1957, the significantly greater number of parasites present in the single-species planting as compared with the number found in the mixed-species planting was directly related to the high *B. brassicae* population in the latter (Fig. 10–10). Because the parasite population had also increased on the aphid population, the ratio of parasites to herbivores was greater in the single-species plot than in the mixed, 18.9 percent to 8.1 percent, respectively. These parasite trends were not apparent in 1958, and no difference was found in the number of parasites in either the single-species or mixed-species or sparse-single-species planting (Fig. 10–11). In 1958, however, the parasite-herbivore ratio changed to 12.0

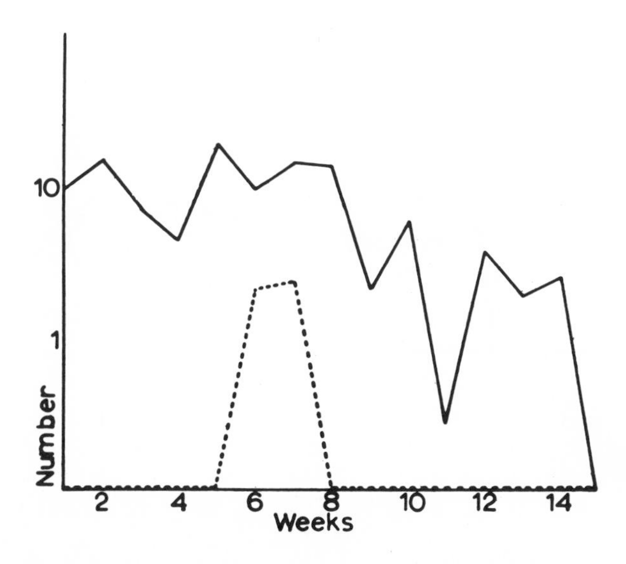

Fig. 10–4. The log number of flea beetles per unit plant area in the mixed- (- - -), single- (———), and sparse-single-species (........) stands during 1957.

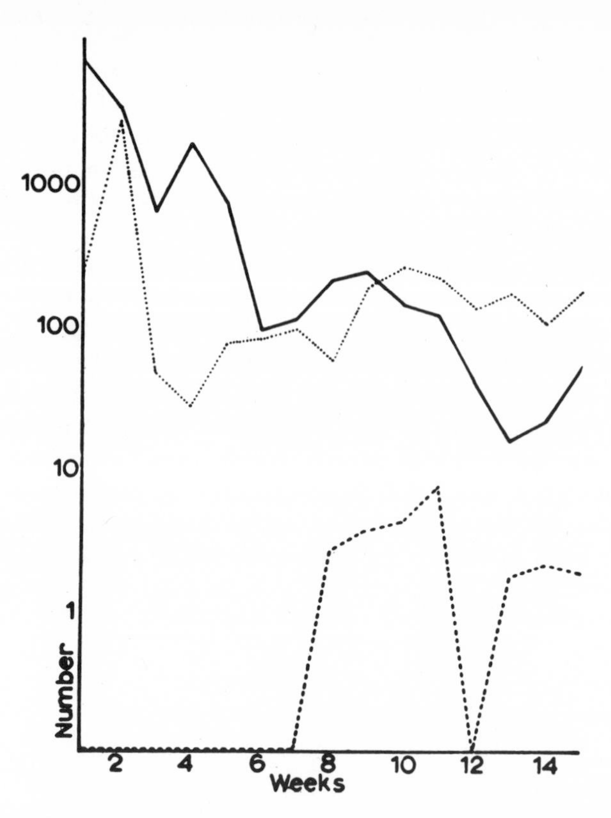

Fig. 10–5. The log number of flea beetles per unit plant area in the mixed- (- - -), single- (———), and sparse-single-species (........) stands during 1958.

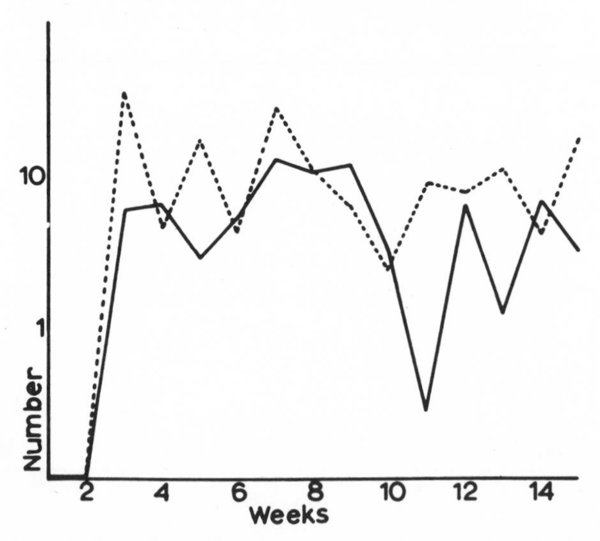

Fig. 10–6. The log number of Lepidoptera per unit plant area in the mixed- (- - -) and single-species (———) stands during 1957.

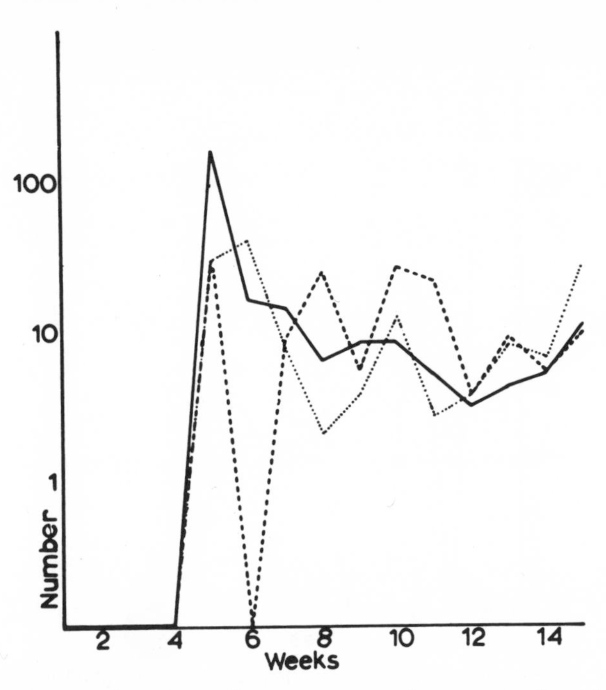

Fig. 10–7. The log number of Lepidoptera per unit plant area in the mixed- (- - -), single (———), and sparse-single-species (........) stands during 1958.

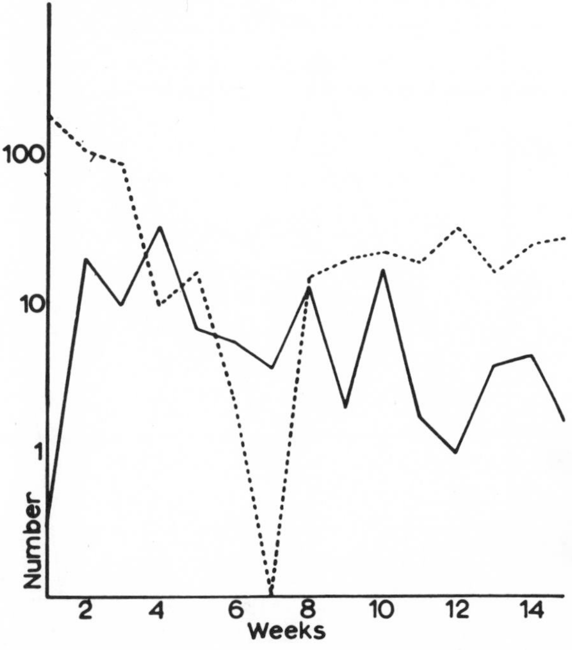

Fig. 10–8. The log number of herbivores, other than those graphed in figs. 2, 4, and 6, per unit plant area in the mixed- (- - -) and single-species (———) stands during 1957.

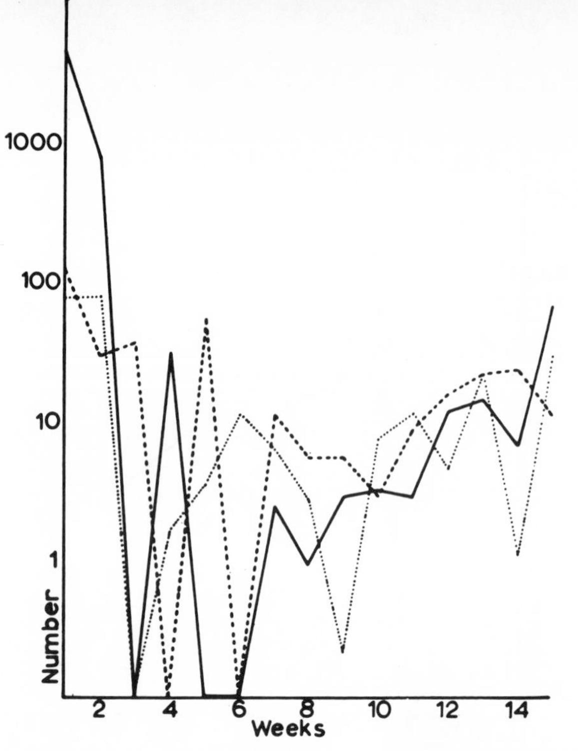

Fig. 10–9. The log number of herbivores, other than those graphed in figs. 3, 5, and 7, per unit plant area in the mixed- (- - -), single- (———), and sparse-single-species (........) stands during 1958.

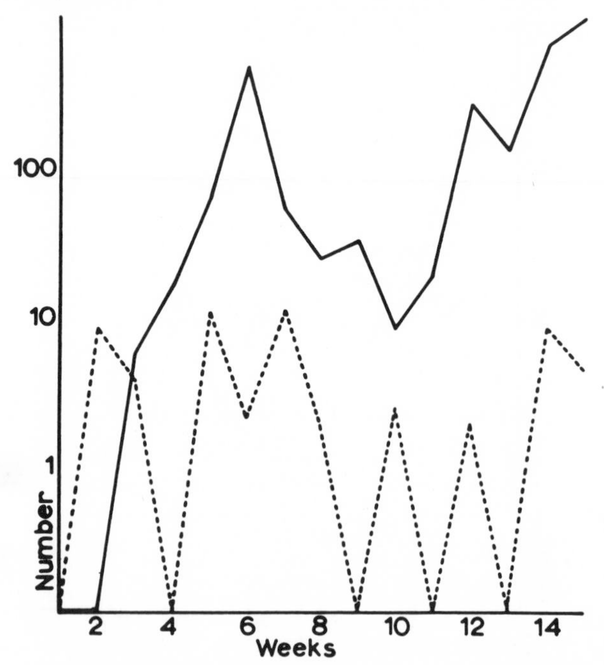

Fig. 10–10. The log number of parasites per unit plant area in the mixed- (- - -) and single-species (———) stands during 1957.

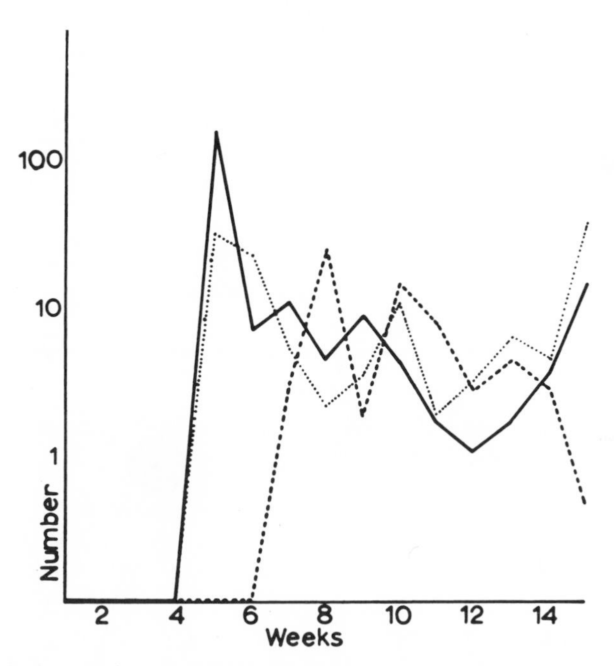

Fig. 10–11. The log number of parasites per unit plant area in the mixed- (- - -), single- (——) and sparse-single-species (........) stands during 1958.

percent in the mixed, followed by ratios of 4.6 percent in the single-species, and 4.4 percent in the sparse-single-species planting. This reversal was mostly due to the change in number of the aphid *B. brassicae* from outbreak level to a markedly lower level in 1958.

Predators which were especially abundant in the single-species plantings in 1957 (Fig. 10–12) were less dense the next season (Fig. 10–13) and paralleled the density of the aphid population. This predator population consisted principally of coccinellids, syrphids, and *Aphidoletes* which were feeding on the tremendous number of aphids present there in 1957. Although in 1957 the total number of predators was greater in the single-species planting, the predator-herbivore ratio was 9.5 percent in the mixed and 1.1 percent in the single-species planting. In 1958 a similar trend was evident: a 102.5 percent ratio in the mixed and 4.5 percent in the single. As indicated in Figure 10–11, little difference in predator density was noted between the sparse-single and the other two plantings. The ratio of predators to herbivores in the sparse-single was 6.8 percent,

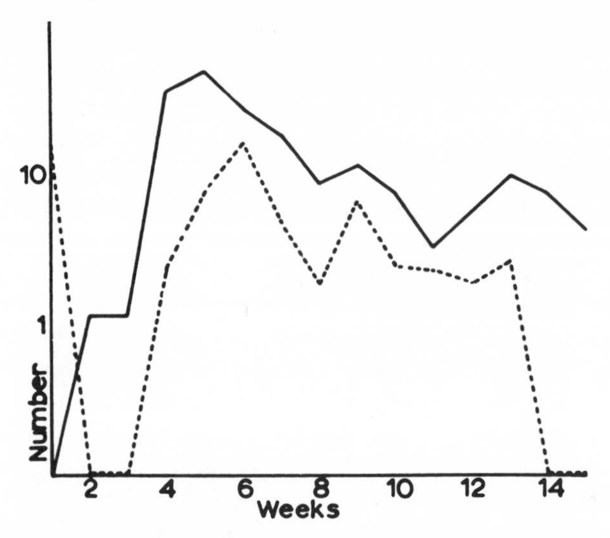

Fig. 10–12. The log number of predators per unit plant area in the mixed- (- - -) and single-species (——) stands during 1957.

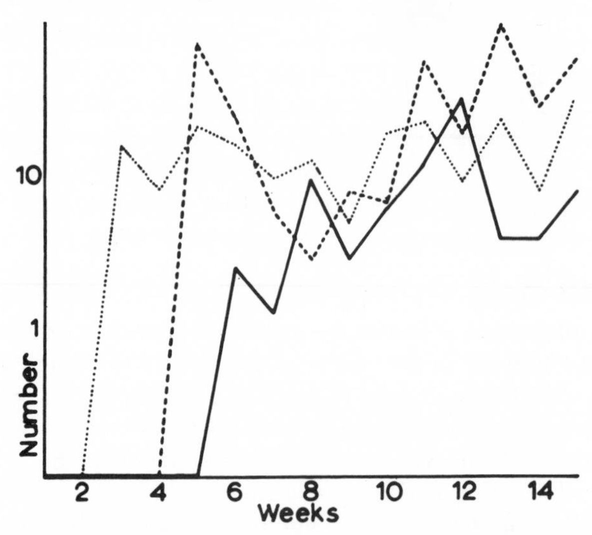

Fig. 10–13. The log number of predators per unit plant area in the mixed- (- - -), single- (——), and sparse-single-species (........) stands during 1958.

which is significantly lower than the 102.5 percent recorded in the mixed. Excluding the single-species planting in 1957, spiders made up the largest proportion of the predaceous fauna in all plantings. One reason for their large number was that all stages were counted, as often nymphal stadia could not be distinguished from adults in the field. However, this trend would still have been evident if only adults had been counted.

DISCUSSION AND CONCLUSION

The significance of this study lies in evaluating the changes in species composition, the density of species, and stability of the community of organisms associated with *Brassica oleracea* when the plant was grown in the single-species stand (simple community) and when grown in mixed-species stand (complex community). These results and the evidence cited in the literature will be considered in the light of some ecological principles.

The ecological principle that species diversity or community complexity functions to stabilize ecologic assemblages was recognized by workers in the early 1900's. More recently the principle has been restated by Odum (1953), MacArthur (1955), Elton (1958), and Hutchinson (1959). Workers in the field of economics have long relied on the principle of diversity to reduce financial risks. In 1857 the philosopher Herbert Spencer proposed that the trend toward diversity which results in stability has universal application and is the drift of all evolving systems.

Diversity is related in at least three ways to community stability. First, diversity of host and prey species provides alternate food for the parasites and predators and this provides greater stability in these population systems (Nicholson 1933). Each host or prey species reacts differently to the same environmental conditions. One host population may decline as another host population increases. This tends to dampen the oscillations of the interacting host and parasite populations and provides greater stability to the system as a whole. Significantly more host and prey species were present in the mixed-species planting than in the single-species planting, especially when all the host and prey species which were present on the other species of vegetation in the mixed-species planting are included in the total. In this way diversity contributed to the relative population stability observed in the mixed-species planting.

Second, diversity in types of parasitic and predaceous species feeding on one species of herbivore may result in greater stability in these population interactions. With most carnivores there exists a quarry-control range in which the ratio of the number of carnivores to quarry at some quarry density is such that over the range a carnivore species is capable of controlling the quarry population (Holling 1959). At ratios below this range, the quarry population tends to increase until it reaches the effective control range of the carnivore, provided the carnivore population is maintained on supplementary food previous to quarry population increase. If the quarry population can by some chance attain a density greater than the control range, its population then escapes carnivore control. Because the quarry-control range for each species of carnivore is specific for each species, increasing the number of carnivore species generally increases the effectiveness of the control range. Hence, increases in the diversity of species of parasites and predators tend to improve the stability of the natural community. The diversity of parasites and predators in the mixed-species planting probably contributed to the stability of the taxa association with *B. oleracea.*

Third, increased diversity of feeding habits of the species members of a community results in more stability of the organization. Polyphagous predators usually contribute more stability than parasites which are specific in their feeding habits. Predominating in the single-species planting were coccinellids and syrphids which attack mainly aphids. In the mixed-species planting spiders existed, and these are relatively nonspecific in their feeding habits and tend to attack whatever is abundant (Bristowe 1958). Consequently, spiders have more prey species subject to their predation, and this contributed to the greater stability noted in the mixed planting.

The lack of diversity in both the sparse-single-species planting and single-species planting allowed outbreaks to occur. Two other factors may also have been involved in the observed outbreaks. The first concerns a time-lag in the establishment of both carnivores and herbivores in the single-species planting. All taxa had to move into this planting because vegetation was completely destroyed previous to the establishment of the *B. oleracea* plants. In this instance, the investigator doubts that the time-lag factor played a major role in the outbreaks, because wild Cruciferae were flourishing adjacent to all plots and provided ample sources of taxa for invasion of the single-species plots.

The second factor operating in the outbreaks of the single species stand might be termed lack of "community organization." The simple community in the single-species planting is comparable to a subunit of a more complex community unit. Since the simple community is not a functional whole, as the complex community is thought to be, it is considered less stable. Little is known concerning the exact nature of interspecific integration in communities, but as Emerson (in Allee et al. 1949) has pointed out, it is considered an important factor in community stability. Species members have evolved together to form an integrated and functional system in the community. The parts of this system are balanced as a unit analogous to the functional organization of the parts within an organism (Pimentel 1959). The removal of a subunit will in both instances result in an unbalanced and unstable unit. While this mechanism cannot be singled out as an only cause of the herbivore outbreaks in the simple communities of the single-species planting, it undoubtedly influenced the outbreaks.

The principle that species diversity is important in the control of population outbreaks has application in agricultural crop protection. In fact, the principle has already been applied in silviculture. Graham (1959) recommends diversification as a means of controlling forest pests. Belyea

(1923) made a similar recommendation for the control of the white-pine weevil (*Pissodes strobi*) based on his findings that weevil infections were reduced from 290 to 85 per 1,000 trees when the white pines were grown in mixed stands. Although the principle does not have application to all crops, there are some in addition to silviculture where diversification would prove of value for insect control.

In the light of the present findings concerning the value of mixed plantings, the universal recommendation for "clean culture" (Peairs 1947) needs re-evaluation. The rule is sound as far as it applies to the removal of crop remains, but the extension of "clean culture" to include the destruction of weeds and hedge rows surrounding the crop area can be seriously questioned. Hedge rows, including weeds and plants of many kinds, may provide diversity with additional species of animals which when added to the community existing in the sterile-crop fields increase its stability. Ullyett (1947) has shown that "where weeds occur around headlands and in hedges, they should be left" for the purpose of supporting parasites and predators important in the natural control of the diamond-back moth (*Plutella maculipennis* Curt.). These hedge rows "form a reservoir for enemies and parasites of insects and mite pests of crops" (Elton 1958).

In conclusion, species diversity is an important factor in preventing population outbreaks in communities. Where possible, diversification of both faunal and floral varieties should be encouraged. The use of specific pesticides and biological control agents which attack only the pest without sterilizing the other fauna should prove to be a sound agricultural practice.

REFERENCES CITED

Allee, W. C., A. E. Emerson, O. Park, T. Park, and K. P. Schmidt. 1949. Principles of Animal Ecology. Philadelphia: W. B. Saunders Co. 837 pp.

Barbey, A. 1931. Une invasion de la tordeuse du sapin dans les forets d'Alsace. Rev. Euax et Forets. 69:132–6.

Belyea, H. C. 1923. The control of the white-pine weevil (*Pissodes strobi*) by mixed planting. Jour. Forest. 21:384–90.

Bristowe, W. S. 1958. The World of Spiders. London: Collins. 304 pp.

Eidmann, H. 1942. Der tropische Regenwald als Lebensraum. Kolonialforst. Mitteil. 5:91–147.

1943. Zur Oekologie der Tierwald des Afrikanischen Regenwaldes. Beitr. Kolonialforsch. 2:25–45.

Elton, C. 1927. Animal Ecology. London: Sidgwick and Jackson, Ltd. 209 pp.

Elton, C. S. 1958. The Ecology of Invasions by Animals and Plants. London: Methuen & Co., Ltd. 159 pp.

Friederichs, K. 1928. Waldkatastrophen in biozönotischer Betrachtung. Anz. Schädlingsk. 4:139–42.

Graham, S. A. 1915. The biology and control of the white-pine weevil (*Pissodes strobi*). M.F. thesis, Cornell University, Ithaca, N.Y.

1926. Biology and control of the white-pine weevil. *Pissodes strobi* Peck. Cornell Agric. Expt. Sta. Bull. 449:3–32.

1959. Control of insects through silvicultural practices. Jour. Forestry. 57:281–3.

Hering, P. E. 1932. The insect fauna of the mustard family: Cruciferae. Ph.D. thesis, Cornell University, Ithaca, N.Y.

Holling, C. S. 1959. The components of predation as revealed by a study of small-mammal predation of the European pine sawfly. Canadian Ent. 91:293–320.

Hutchinson, G. E. 1959. Homage to Santa Rosalia or why are there so many kinds of animals? Amer. Nat. 93:145–59.

Kalabukhov, N. T. 1947. Dynamics in numbers of terrestrial vertebrates. Zool. Zhur. 26:503–20. (In Russian; translation by R. S. Hoffman.)

MacAloney, H. J. 1930. The white-pine weevil. Its biology and control. Bull. New York State Coll. Forestry (Tech. Publ. 28) 3(1):87 pp.

MacArthur, R. H. 1955. Fluctuations of animal populations and a measure of community stability. Ecology 35:533–6.

Marchal, P. 1908. The utilization of auxiliary entomophagous insects in the struggle against insects injurious to agriculture. Pop. Sci. Monthly 72:352–419.

Nicholson, A. J. 1933. The balance of animal populations. Jour. Anim. Ecol. 2:132–78.

Odum, E. P. 1953. Fundamentals of Ecology. Philadelphia: W. B. Saunders Co. 387 pp.

Peairs, L. M. 1947. Insect Pests of Farm, Garden, and Orchard. New York: John Wiley & Sons, Inc. 549 pp.

Pierson, H. B. 1922. Control of the white-pine weevil by forest management. Harvard Forest Bull. 5:1–42.

Pimentel, D. 1959. Nature's Community. Cornell Plan. 15:35–8.

Podhorsky, J. 1933. Kann ein vollständig sich selbst überlassener Hochgebirgswald durch Überhandnehmen von Forstinsekten zu einer Gefahrenherd werden? Allg. Forst- u. Jagdzeitung 109:151–6.

Ripper, W. E. 1956. Effect of pesticides on balance of arthropod populations. Ann. Rev. Ent. 1:403–38.

Rohrl, Dr. 1928. Waldkatastrophen. Forstwissenschaftl. Zentralbl. 73:293–315.

Schenk, C. A. 1924. Der Waldbau des Urwalds. Allg. Forst- u. Jagdzeitung 100:377–88.

Schimitschek, E. 1940. Die Massenvermehrung der *Ips sexdentatus* Börner im Gebiete der orientalischen Fichte. Zeitschr. angew. Ent. 26:545–88.

Schneider, F. 1939. Ein Vergleich von Urwald und Monokultur in Bezug auf ihre Gefährdung durch phytophage Insekten, auf Grund einiger Beobachtungen an der Ostkuste von Sumatra. Schweiz. Zeitchr. Forstw. 90:41–55, 82–89.

Smith, R. F., and K. S. Hagen. 1959. Impact of commercial insecticide treatments. Hilgardia 29: 131–54.

Spencer, H. 1857. First Principles. New York: Dewitt Revolving Fund, Inc. 599 pp.

Stern, V. M., and R. van den Bosch. 1959. Field experiments on the effects of insecticides. Hilgardia 29:103–30.

Stern, V. M., R. F. Smith, R. van den Bosch, and K. S. Hagen. 1959. The integrated control concept. Hilgardia 29:81–129.

Trägårdh, I. 1925. On some methods of research in forest entomology. Trans. III. Internatl. Ent. Kongr. Zurich 2:577–92.

Ullyett, G. C. 1947. Mortality factors in populations of *Plutella maculipennis* Curtis (Tineidae: Lep.) and their relation to the problem of control. Union S. Africa. Dept. Agric. and Forest. Ent. Mem. 2(6):77–202.

von Hassel, G. 1925. Das Geheimniss des Urwaldes. Anz. Schädlingsk. 1:42–43.
Voute, A. D. 1935. *Cryptorhynchus gravis* F. und die Ursachen seiner Massenvermehrung in Java. Arch. Néerland. Zool. 2:112–43.
———. 1946. Regulations of the density of the insect-populations in virgin-forests and cultivated woods. Arch. Néerland. Zool. 7:435–70.

11

Experimental studies on predation: predation and cyclamen-mite populations on strawberries in California

C. B. Huffaker
C. E. Kennett

The work here reported is an experimental study of predation under field conditions. This study is the result of an attempt to develop a practical program for control of the cyclamen mite, *Tarsonemus pallidus* Banks, in strawberry fields in this state, by use of a native predatory mite, *Typhlodromus cucumeris* Oudemans, or *Typhlodromus reticulatus* Oudemans, or both.[1] The exploratory greenhouse experiments were conducted at Albany, the field experiments in the Santa Clara Valley.

THE NATURE OF THE PROBLEM

The cyclamen mite is the most important pest of strawberries in California, a crop with an annual farm value of about $50,000,000. No

Reprinted with permission from *Hilgardia*, 26(4):191–222 (1956). This reprinted version omits the section *Ecological Significance and Theories*, pp. 208–218, and related material in the introduction, summary, and bibliography sections.

[1] In these experiments the form present in any given field was not commonly determined. The mite is hereinafter referred to as the "predator" of the cyclamen mite or occasionally—to distinguish it from still other species of the genus—as *Typhylodromus cucumeris*; but this usage does not imply absence of *T. reticulatus*.

suitable chemical means for the control of the pest on this crop are known. Heat treatment of nursery plants, although effective at the time (Smith and Goldsmith, 1936), is no longer practiced since nursery stock is not now the source of most infestations. First-year plants are, during their early growth, sparse and open to the sun and wind—a condition not favorable to this pest. However, a survey in late 1953 in the Santa Clara Valley showed that nearly all fields had small foci of infestation by late fall after the first season's growth. From that period onward, the mite is a serious threat to berry production during the three crops years, the second, third, and fourth seasons, except as it is kept naturally under control by native *Typhlodromus.*

The present problem, even when clean planting stock is used, arises almost invariably

from a complex of circumstances, involving a closer association of fields and of grower activities conducive to rapid spread of the mite from field to field; a shift from a rather resistant variety, the Marshall, to more susceptible ones; and the *destructive* effects of recently developed chemicals on natural enemies of the mite,

as stated by Huffaker and Kennett (1953).

The problem is reciprocally related to infestations of other pests, two-spotted mite, *Tetranychus telarius* (Linn.), and the strawberry aphid, *Capitophorus fragaefolii* Ckll., in particular. The former commonly causes damage only during the spring period in the major strawberry area and if not controlled by acaricides can cause damage just as severe as that caused by the cyclamen mite during midsummer and fall. If plants have suffered severe injury from two-spotted mite in the spring, vigor and nutritional qualities are impaired, summer growth is sparse, and the plants remain open to sun and wind. The microenvironment and food material are thus less favorable to subsequent cyclamen-mite populations. Furthermore, *Typhlodromus occidentalis* Nesbitt and *Orius* sp. may develop high populations in fields with high populations of two-spotted mite. These two predators, although more characteristically predators of the two-spotted mite, also prey on cyclamen mite, particularly if the latter occurs at high densities. Both, however, *may* prey on *T. cucumeris*, the effective predator of cyclamen mite. Presence of less specialized predators resulting from two-spotted mite infestations may, therefore, affect subsequent cyclamen-mite densities in either a positive or a negative way.

In its direct injury, the strawberry aphid is not commonly a serious pest of strawberry. Its seasonal abundance is largely during the periods in spring and late fall when the berries are not in high production, and the direct injury appears to be short in duration. High populations of aphids appear to compete very little with cyclamen mites. However, this aphid has been considered (Thomas and Marcus, 1953) as a vector of a virus-disease complex of strawberries, the etiology of which is not fully understood. Virus-weakened plants probably have a decreased physiological potential for development of high cyclamen-mite densities. In addition, the cyclamen-mite predator is known to feed on aphid honeydew, which is considered an important source of both nutrients and water during periods when prey densities are low. An alternate prey species, *Tarsonemus setifer* Ewing, which is harmless to strawberries and lives on moldy, dead leaves is also at times important in the economy of this predator.

The necessity for chemical treatments for two-spotted mites and, to a lesser degree, aphids complicates the improvement of predatory control of cyclamen mite. This phase of the work has generally been conducted in conjunction with W. W. Allen of the Department of Entomology and Parasitology of the University of California. The objective has been to use chemicals which would give satisfactory control of these other pests, but not destroy too many of the predators of cyclamen mite, and to stock new fields with predators in order to reduce the costly time lag in appearance of effective predation.

Aphids are generally secondary to cyclamen mites and two-spotted mites. When chemical control is required, treatment with TEPP prior to stocking a field with predators often gives adequate control. After a field has come under biological control by predators, a single treatment with TEPP either during the spring or late fall periods of trouble, would not seriously interfere with cyclamen mite/predator balance because of the timing and limited toxicity of the material to the predators.

SOME BIOLOGICAL ATTRIBUTES

Habits and Requirements of the Mite

A female cyclamen mite deposits about three eggs per day for a period of 4 to 5 days. The males constitute only about 5 percent of the adult population, reproduction being possible by parthenogenesis for successive generations (Smith and Goldsmith, 1936). The species is reproductively inactive during the colder winter months. Relatively small numbers of adult females, hidden in crevices and folds deep in the crown of the plant, constitute the overwintering population during normal years. The strawberry plant is unusually favorable for the mite's microenvironmental requirements, particularly in the coastal valleys of California, where extreme summer temperatures combined with excessively low humidities are uncommon. The many folds, crevices, and hairs characteristic of the young leaf shoots and blossoms prior to their opening and maturity furnish an ideal microenvironment for this mite: abundant food, maximum humidity and shelter, and protection from any general predators which might otherwise thrive on them.

Through its feeding, the mite prevents or retards elongation of leaf petioles and growth of leaflets, and may cause malformed leaflets which never fully open. Thus, to a degree, it is capable of prolonging the favorability of its own habitat. Nevertheless, the opening and maturing of new growth eventually forces the mites to seek new growth emerging at the crown position. Therefore, even in the absence of predation there is definite limitation to numbers. The continuing sequence of movement from old growth to new, accompanied by high mortality, imposes a marked restriction, so that the mite's high potential for reproduction is not realized even prior to conditioned unfavorableness of the plants, such as occurs after prolonged infestation.

Predator-Prey Relations

The predator is admirably adapted to hold the densities of its prey at levels noninjurious to the strawberry plant. When cyclamen mites are abundant, its rate of reproduction is close to that of its prey—two or three eggs per day per female for a period of 8 to 10 days or more. Reproductive *rate*, however, is only a small asset, almost meaningless taken alone. Well-adapted capacities for searching, synchrony of environment and seasonal activity, and behavioral patterns and adaptations that are highly specialized to reduce prey densities and to survive at low prey densities play a far more vital role.

There are notable checks on the cyclamen mite (such as food and shelter limitations) other than predation at small foci on the same plant, which may prevent achievement of its reproductive potential at those points, while the predator may be relatively unchecked. Thus, a favorable ratio of predator to prey may be achieved fairly quickly. Once such a balance is achieved, overpopulation is not normally again attained even on the small foci because the predator is more capable of moving from leaf to leaf, and thus it cancels the reproductive power of its prey while its own is temporarily unhampered.

The predators may often be found secluded in the petiole sheaths at the base of the plant and are thus in position to readily capture surviving mites which must enter that area when moving to new leaf shoots. This predator has not exhibited cannibalism, a means of survival of many predators at very low densities of their prey species. Its utilization of honeydew of aphids or white flies, or possibly plant exudates, enables its survival at very low densities of its prey. So far as known, it is limited on cultivated strawberries to cyclamen mite and related tarsonemids for reproductive nutrients. It utilizes other materials only as supplemental food and for survival, and does not produce eggs when not feeding on regular prey. Hence, its density is directly and reciprocally geared to the density of its prey, but it is cushioned, without resort to cannibalism, against extermination through overconsumption of its prey. This enables it to recover quickly in numbers and to a point of adequate frequency of occurrence, so that it does not undergo oscillations of very wide amplitude in both space and time, such as those discussed by Nicholson (1954).

In this connection, physical barriers, representing a security threshold that results from the great heterogeneity in the microenvironment, seem to preclude actual extermination of the prey on any unit as large as an entire plant, or at least a group of adjacent plants. Hence, equilibrium is reached and a rough, although disturbed, balance at very low densities is characteristic.

The predators do not become established in very new fields, but delay their appearance until after the mite has become established. The foci of mite infestation, which appear in the late first year, "blossom" into serious infestations during the second year at about the time of greatest population ascent in June and July. Predators naturally invade the fields, a few being found even in the late first-year fields, but there is a definite lag which occasions various degrees of economic loss, especially during the second-year crop. Waves of cyclamen-mite damage from many or several spots, followed by "clean-up" by predators, are characteristic. Loss during the important second year may vary from relatively little to nearly complete. Very occasionally in Santa Clara Valley, a warmer area, and rather commonly in the cooler regions near Watsonville, California, where the industry is still new and sources of infestation by cyclamen mites are not so plentiful, infestations do not reach a troublesome stage until the third year. Action of predators consequently is delayed. Once "clean-up" by predators has been achieved, no case has been found where this pest has again become a serious problem during later years of this four-year crop, except where use of chemicals destroyed the predators.

EXPERIMENTAL APPROACH AND POPULATION ESTIMATES

The approach in these experiments has been through altering a single environmental factor—manipulation of the predator population by artificially introducing the pest species and the predator species—and by using acaricides to eliminate or exclude the naturally occurring predators in control plots so as to encourage earlier predator-prey equilibrium at noneconomic levels of the pest.

Except for the green house experiment on the nature of predator/prey oscillations, the experiments were carried out in commercial fields where the coöperating growers followed a suggested program for the control of two-spotted mite; hence, they constitute a test of the compatibility of biological and chemical methods of control of these two strawberry pests in this area.

Aphid control did not prove necessary except in the greenhouse experiment; there spraying with nicotine was a compatible solution.

The specific data upon which this paper is based were obtained in 1953 and 1954. However, the basis of analytical discussion comprises experimental work done over a five-year period, partly covered in a preliminary way by Huffaker and Spitzer (1951) and Huffaker and Kennett (1953).

Estimating Cyclamen-Mite Populations

The method of partial census or sampling used in this study has varied somewhat according to the specific nature of certain experiments. The procedure was determined in a statistical manner. Some general aspects were stated by Huffaker and Spitzer (1951):

> The population estimates must be reproducible within reasonable confidence limits. In this study, this problem has been complicated by many factors. The estimates can be made only by removing a portion of the population. The most characteristic plant part for use as the sample item is the unopened, rolled leaf-shoot, and such leaves in the proper stage for sampling are always few in number and at certain seasonal periods are very scarce. The logical plant unit for sampling would otherwise be the individual strawberry plant, but since it was learned that only one or two leaflet units can be taken with any hope of taking repeated samples from the same plant at the required intervals, groups of plants had to be used as the plant unit of replication. Fortunately, it was found that the variation in mite numbers was low between the three leaflets of a leaf, and therefore greater economy in labor and less disturbance of the populations was experienced by taking only one of the three leaflets of a given leaf. It was also found that in a fairly late stage of an infestation the variation between leaflets was of about the same statistical order whether five leaflets were taken from five different but adjacent plants or whether one leaflet was taken from each of five different leaves of the same plant.

Hence, a group of fifteen adjacent plants became the plant unit for replication.

In arriving at sample size, a total of thirty leaflets was taken, one each from thirty plants. By randomized selection of the leaflet results from groups of variously sized lots, it was found, by drawing twelve successive randomized combinations of each size group, that only when twenty-four leaflets were used was the standard error of the means sufficiently low so that reproducibility was achieved. However, the sample size was increased to thirty leaflets per lot or position.

The method of estimating populations has been more satisfactory for the cyclamen mite than for the predator. Cyclamen-mite densities were relegated to size classes, and after some practice, and by use of occasional counts when densities appeared to be borderline, estimates of prey populations were found to be reproducible and adequate. Estimates consistently run from 30 to 40 percent lower than actual counts. The relegation to size classes for the population of each leaflet has been on the following basis: 0; 1 to 5, or a nil density; 6 to 20, or low density; 21 to 50, or medium density; 51 to 100, or high density; and 101 or over, as very high density. Only in 1954 was a category above 100 utilized. It has been found that the estimate of 75 for averaging purposes (51 to 100) was far too low for the highest densities encountered. For averaging purposes, the figures 0, 3, 13, 35, 75, and 150 were used to represent these range classes. The requirements in the practical operation of this program did not call for or permit actual counting of the prey species.

Estimating Predator Populations

With regard to the predator populations, it was found by trial analysis that frequency of occurrence—that is, enumeration of the number of leaflets, among the thirty taken, having one or more predators—is a more reproducible estimate than is obtained from the numbers counted in the sample. In an area where predators were in abundance (a mean of means of 27.8 per thirty leaflets, based on twelve separately randomized lots of thirty leaflets each), it was found that the mean of means for frequency, although lower, at 15.6, had a range of only 6 and a rule-of-thumb standard deviation of 2, while the range for the actual count was 20 with a rule-of-thumb standard deviation of 7. Particularly at low populations, that is, those at equilibrium, the concept of frequency distribution was considered more important in this study. Moreover, since the quantitative order of changes at high populations is easier to follow and to interpret than are those at the low levels, the frequencies of predator occurrence have been used. For example, the occurrence of large numbers of predators on a given leaf or strawberry plant is not indicative of predator action throughout the sample unit (the group of thirty leaflets) and is not as closely related to capacity to increase intensity of action quickly with increase in prey densities. The predators are very small and have limited powers of movement even though they are much more active than their prey.

When populations are low, the sample size has been too small with respect to the predators. In the absence of knowledge in advance of population densities of the plots to be sampled, the sample size would have to be such as to be adequate for the lowest densities. However, since the low densities were normally so very low as compared with the high densities, and consistently so from one sample date to another, a standard sample size of thirty leaflets was adopted.

Validity of Methods and Procedures

The enormous contrast between compared plots as to prey densities, the frequency of sampling (each sample being a curvilinear check on the previous and subsequent samples) and the fact that differences between estimates made in this manner and actual counts on the same sample were uniform and dwarfed by the differences between the experimentally compared plots (that is, predator-present vs. predator-free plots), seem to be ample evidence for the placing of general reliance on recorded estimates of prey densities. The over-all conclusions in any case seem inevitable.

The method of evaluating predation has consisted basically of employing an acaricidal check method for removing predators from one plot of two paired field plots, and either allowing natural predators to enter the other unaided, or artificially stocking it initially with predators.

As a significant evaluation, a check was also made on the "check method" by mechanical hand-removal of predators from one group of plants, so as to eliminate the possible alternative explanation that the chemical used in destroying the predators in the field studies may have "stimulated" the cyclamen-mite, either directly or indirectly, to greater abundance. The latter technique was reported by Huffaker and Kennett (1953). They concluded, "The results from tests on hand-removal of the predators removed any objection to the interpretations [of effective control] on the basis that parathion might itself favor increase of cyclamen-mite."

Samples were taken on one occasion at night for comparison with samples from the same plots taken during daytime because of the known diurnal pattern of movement of many species of mites, both predatory and phytophagous. The object was to learn if significant predation occurs by species not encountered by daytime sampling. Such was not the case, the counts of all faunal elements being very similar.

The experimental design and objectives of each phase of experiment, as well as any basic departures from the methods as herein outlined, will be discussed with the respective experimental data.

SUMMARY OF EARLIER TESTS

Data obtained from 1950 to 1952, inclusive, published by Huffaker and Spitzer (1951) and Huffaker and Kennett (1953), form an integral part of the experimental evidence, and their results may be reviewed briefly. Work was first done in 1950 in third- and fourth-year fields because of grower experience of much less cyclamen-mite severity in older berry fields as compared with that in second-year fields. It was desired to learn if the predators in those fields accounted for the condition. In the former paper, results on four pairs of plots in third-year fields, one plot of each pair with natural predator populations, the other predator-free, were reported. The authors concluded, "The results, though not excluding the possibility of a chemical or residual 'stimulative' influence, were sufficiently logical as to suggest strongly a definite control of its host [prey] by the predator."

During 1951 and 1952, results from twenty pairs of field plots, in which one plot of each pair was kept predator-free and one had predators, were obtained (Huffaker and Kennett, 1953). These included second-, third-, and fourth-year berries of six different varieties in plantings of many different operators throughout the Santa Clara County strawberry district. The field population studies conclusively showed that the predators

> exert very effective control of the cyclamen mite in the third- and fourth-year fields in this area when their activities are not inhibited by detrimental chemical treatments used for control of other pests. Control in second-year fields is somewhat erratic due, it is felt, to the lag in appearance of predators and to the more vigorous (chemical) treatment practices which are applied to the second-year crop.

Either hand-removal or chemical removal of predators resulted in striking increase in cyclamen mites, as compared to the predator-present controls. Increase in berry yield due to the action of the predators in the predator-present control was in one test seven-fold. The predator maintained marked control of the prey population both in old weakened fields and when seasonal and berry-culture conditions were not conducive to the highest densities of cyclamen mite, such as would be necessary for demonstration of the most striking effects.

GREENHOUSE TEST

Methods

On April 8, 1952, seventy-two potted plants were arranged in two randomized groups on greenhouse benches. Each group had been treated in January in order to destroy all predators. By April the prey species had developed light infestations. The predator had recovered to low density levels. Parathion was again applied on April 8 to control the predators on one group after the first samples were taken. This group is hereinafter referred to as the "predator-free" group, the term meaning relative but not complete absence of predators at all times. Three additional predators were added to each plant of the predator-stocked group on that date. The directly compared predator-present/predator-free paired groups were sampled at approximately 11-day intervals until April 14, 1953, at which time the plants in the predator-free group had become badly weakened and some were dying. The experiment consequently was terminated.

The strawberry plants were cultured in a greenhouse maintained at temperatures of 65° F at night; daytime temperatures occasionally rose during summer, as a result of insolation and inadequate cooling controls, to 80° to 85°. Generally, however, temperatures were fairly even at a daily mean of about 70°. Relative humidity was not controlled, but the humidities of the special leaf-shoot habitats of the cyclamen mite are naturally high beneath the lush growth of the strawberries, which were irrigated frequently.

The few predators shown in Figure 11-1C in the predator-free group at the end of June and early July were taken from only three of the thirty-six

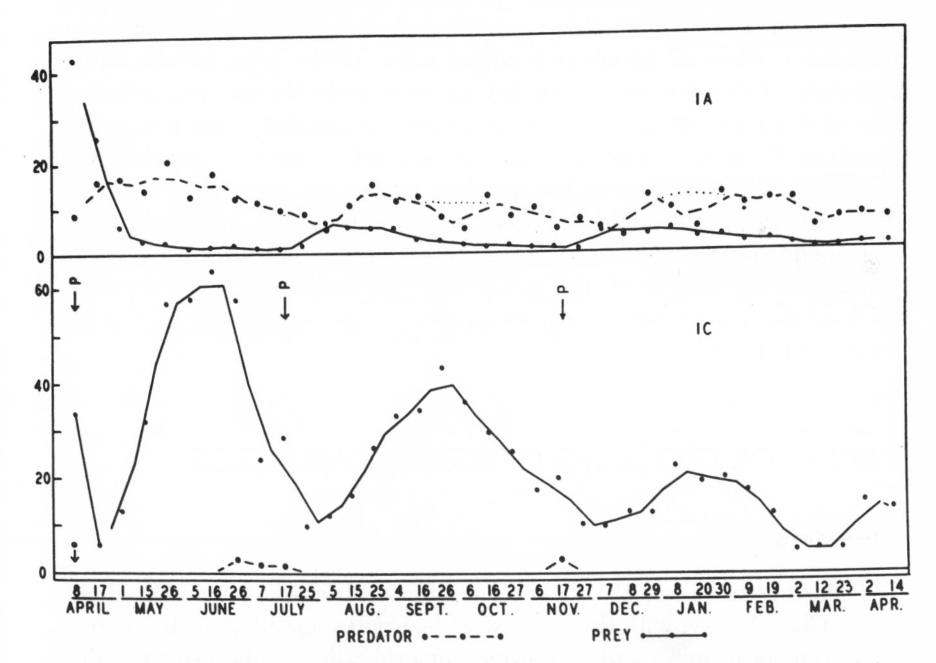

Fig. 11–1. 1952–1953. Greenhouse plots: Changes in densities of *Tarsonemus pallidus* (= prey) in predator-present and predator-free plots, and *Typhlodromus* (= predator) frequencies. One pair of plots; one plot (1A) with predators, the other kept predator-free (plot 1C). The *T. pallidus* in the two plots are plotted on a per-leaflet basis. The predator frequencies express the number of leaflets among 36 which had one or more *Typhlodromus*. "p"s by arrows indicate dates of parathion treatment.

plants. Retreatment on July 18 with parathion became necessary, therefore, in order to reëstablish a truly predator-free group. Retreatment was again necessary on November 18, but not thereafter because no predators were encountered after that date. Genite, a material which does not harm either the predator or prey, was used to control two-spotted mites on both groups of plants.

Control with Oscillations

The preference of Allee *et al.* (1949) for the term "fluctuation" is well taken. However, the changes in populations of prey species and predator species as shown in Figure 11–1 are such that a reciprocal relation of an oscillatory nature is indicated. These data furnish the best example of long-term, undisturbed interaction. The field data in no case appear to controvert these remarks.

At the initiation of the experiment, cyclamen-mite and predator densities were subequal in the groups shown in Figures 11–1A and 11–1C. Both prey populations at that time were in a crash phase as a result of predator action. The diminished predator effect at that time, resulting from predator destruction by parathion, was offset by the temporary depressive effect of parathion's killing some of the prey species. Some influence of the previous predation on eggs probably was registered as part of this crash effect, since the eggs were not counted in the population densities. From this point of simultaneous crash in the two populations—a holdover from equivalent history prior to April 8—there is never again any similarity whatever in the prey densities of the two groups. Instead, the unequivocal contrast persisting thereafter was always a measure of the regulatory action of the predatory *Typhlodromus*. For reference, contrast the magnitudes of cyclamen-mite, or prey, densities in Figures 11–1A and 11–1C.

PREDATOR STOCKING OF FIRST-YEAR FIELDS

Methods

During the fall of 1953, plots were established in San Martin in the Santa Clara Valley, with the purpose of learning something of the pattern of movement of mites and predators, but primarily to establish the value of stocking early mite infestations with predators, so as to achieve early equilibrium and, consequently, control of the prey species. Six foci of infestation were located in several test varieties of berries. These were considered the "zero" distance, designated "A" plots. Two of these, 1A

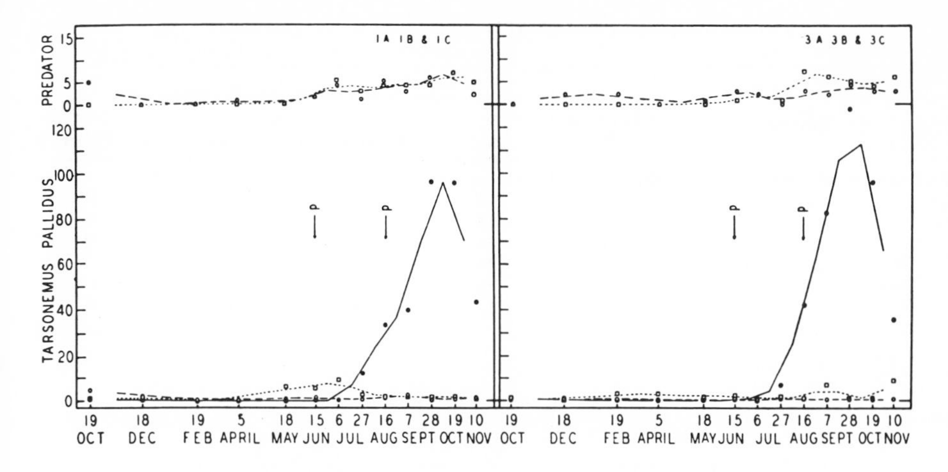

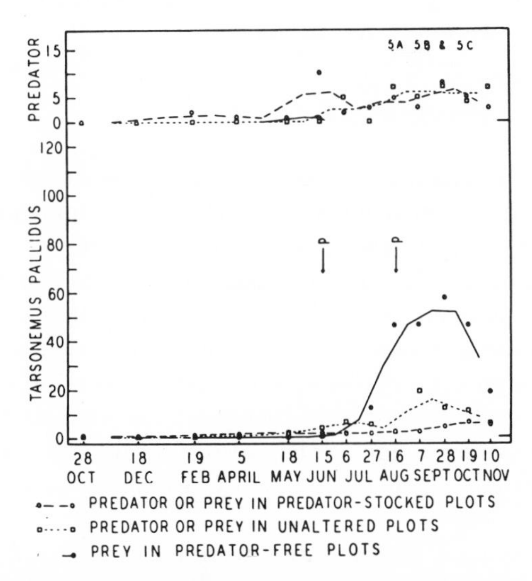

Fig. 11–2. 1953–1954. First-year plantings: 1A, 1B, and 1C; 3A, 3B, and 3C; 5A, 5B, and 5C; experimental varieties at San Martin. Changes in densities of *Tarsonemus pallidus* (= prey) in predator-stocked, predator-free and unaltered plots, and *Typhlodromus* (= predator) frequencies. Three triplicated series; one plot of each series predator-stocked, one unaltered and one kept predator-free. The *T. pallidus* in the three plots of a series are plotted on a per-leaflet basis. The predator frequencies in the predator plots express the number of leaflets among thirty which had one or more *Typhlodromus*. "p"s by arrows indicate dates of parathion treatment in the C plots.

and 3A, of Figure 11–2, were stocked with predators on September 30; and one, 5A, Figure 11–2, on October 28. The other three A plots, 2A, 4A, and 6A, had no predators added at any time. Six B plots were set up 25 feet down row from the corresponding A plots. Six C plots were set up three rows lateral to the six A plots and six D plots were set up 75 feet

TABLE 11–1. Arrangement of first-year field plots

Text figure	Replicate no.	Replicate no., position and fig. designation			
		Zero distance	3 rows lateral to zero dist.	25 feet down row	75 feet down row
2	1	1A*	1C	1B	1D
2	3	3A*	3C	3B	3D
2	5	5A*	5C	5B	5D
None	2	2A	2C	2B	2D
None	4	4A	4C	4B	4D
None	6	6A	6C	6B	6D

* Predators added only to 1A, 3A, and 5A above.

down row from the A plots. The arrangement is shown in Table 11–1.

The B, C, and D positions were otherwise random as to known occurrence of cyclamen-mite infestations.

Occurrence of Predators and Prey

The degree to which first-year fields may have developed infestations by the end of the first growing season (October) is indicated by the fact that 5, 4, and 5, respectively, of each of the six plots at the lateral (C) positions, the 25-foot (B) positions and the 75-foot (D) positions, were infested at the beginning of the experiment, although only very lightly so in most instances. Hence there was no test of the relation between proximity to known centers of cyclamen-mite infestation and earliness of infestation. For that reason the 75-foot positions were later dropped from the experiment. Other evidence has also indicated that incipient cyclamen-mite infestation is general in Santa Clara Valley in a majority of late season first-year plantings.

On the other hand, the predators were not nearly so regular in occurrence in the field at the initiation of the experiment. There was a fairly rapid establishment of predators in the three plots, 1A, 3A, and 5A, where they were added, for they were taken in those plots at the first or second samplings subsequent to stocking. None of the plots which were not stocked showed any definitely or persistently established colony of predators prior to May of the following year, although in plots 3D, 6C, and 6B a single predator was taken in each case, either in December or October. Subsequent samples showed none for several successive sample periods.

There was no appreciable movement of predators from the stocked positions to the plots three rows to the side or either 25 feet or 75 feet down row; that is, proximity to the stocked plots did not add significantly to the earliness of predator establishment in the plots at those positions. Natural dispersion of predators throughout the field occurred in adequate numbers in May, June, and July of the following year to minimize the importance of any predators received naturally from the stocked positions.

Since plots 2A, 2B, 2C; 4A, 4B, 4C; 6A, 6B, 6C; and 1D, 2D, 3D, 4D, 5D, and 6D are not directly of use in the present discussion, having been established in order to obtain information on movement, the data from those plots are not presented graphically in this paper.

Resulting Control in Second-Year Berries

Stocking of predators during the fall of the first growth year resulted in control of cyclamen mites throughout the economically important second-year season. This was achieved by early establishment of predator/prey equilibrium before the season of population surge of the prey species. Once equilibrium was achieved, and in the absence of detrimental chemical treatments, no degree of conduciveness to prey-species fecundity seemed enough to change seriously the level of equilibrium. In spite of changes in predator population densities, sometimes hard to explain in the absence of abundant specific prey, the cyclamen mite was held at levels entirely noneconomic in plots 1A, 3A, and 5A. Maximal densities of cyclamen mites per leaflet during the summer of 1954 were 3.1, 0.9, and 6.5, respectively. On the other hand, the corresponding predator-free plots 1C, 3C, and 5C developed prey densities highly injurious to the crop, with peak populations of 93.5, 128.6, and 57.4, respectively.

A comparison between plots 1A, 3A, and 5A and plots 1B, 3B, and 5B (Fig. 11–2) (the latter were permitted natural immigration of predators but had none added) revealed a definite difference in prey densities in relation to the previous predator history for the respective plots. Natural build-up of predator populations was adequate shortly after the initial spurts in prey populations to curtail the development of very high densities, such as characterized the predator-free, or C plots, 1C, 3C, and 5C of Figure 11–2.

It may be seen that in plots 1B, 3B, and 5B the prey population levels attained were intermediate between those of the comparable predator-free and predator-stocked plots, although much nearer the condition in the latter. This is because the natural predator establishment in the area during that season was unusually early, but a delay of one or two months longer might well have been enough to have allowed cyclamen-mite densities to surge to maximal levels and injury before natural predator populations would have sufficed to control the infestations.

Consideration of certain segments of ascendance in population in the absence of either predators or chemical-treatment history, in the various plots which had delayed entry of predators, lends further support to results obtained by Huffaker and Kennett (1953), by use of hand-removal of predators, to the effect that the parathion used as a chemical "check method" of evaluation did not itself cause the marked increases in cyclamen-mite densities observed when that chemical was used. For example, plots 1B and 5B (Fig. 11–2) and plots 2A and 4A (not presented) had delayed entry of predators and rates of population increase in April, before the entry of predators, which were just as pronounced as that shown in the initial periods of increase subsequent to chemical treatment in June for the C plots, that is, 1C, 3C, and 5C.

These data verify those from the greenhouse experiment in every general respect, although correlative aspects of predator/prey populations in plots having predators are not as close chronologically. The longevity characteristic of the predator population, even in trough positions when prey densities are at their lowest, is evident. Upon the approach of fall and winter under these field conditions, the predator species does not show the marked decline that is characteristic of the prey species. The latter declines sharply in numbers encountered on young shoots during early winter, partly because of a movement downward to points of concealment from sampling. On the other hand, the predator population increases as the reproductive season progresses, offset only partially and belatedly by declines in prey densities. The population passes the winter in the adult state, and mortality during the late winter and spring is heavy, despite utilization of alternate foods. Even in old infestations, cyclamen mites as a rule are very low in density from December until April, and in some places until May or June. However, relatively earlier increases in mites have been observed during years of exceptionally mild winters or unseasonably early summers.

Actual counts of populations of cyclamen mites, or prey species, on the sample leaflets in the predator-stocked, unaltered plots, and predator-free plots—that is, comparison of 1A with 1B and 1C; 3A with 3B and 3C; and 5A with 5B and 5C—were made on September 7 so as to permit a critical statistical analysis of differences at a season that was representative of high infestations. Upon pairing the data appropriately for September 7, the results were 92, 21, and 57, respectively, for the predator-stocked plots, 1A, 3A, and 5A; and 1,132, 2,480, and 1,392, respectively, for the predator-free plots, 1C, 3C, and 5C. The differences were significant. Viewing these data in the light of extensive chronological population estimates throughout several seasons, with the data lining up in similar manner in all cases and with significance greatly exceeding the 1 percent level, there is no room for doubt concerning significance of differences between the two experimental conditions.

PREDATOR STOCKING OF SECOND-YEAR FIELDS

Methods

In April of 1953, two or three predators per plant were added to plots in ten new fields or sections of fields in the Santa Clara Valley that were just starting their second year of growth. Check plots were arranged on each side of the plots. Predators were not added nor removed in one of the check plots, although normal grower field practices were continued, including primarily the grower's use of Aramite or TEPP or both for control of two-spotted mite during the spring months. The objective was to gain knowledge of the pattern of natural predator invasion and the resultant control under such treatments. The other check plot was the acaricidal check plot in which predators were kept out by use of parathion. Hence, there was again the direct comparison between predator-free and predator-stocked plots and an intermediate condition characteristic of normal grower practices. That year predators entered the second-year fields very early, the fields being very close to older ones, which were thus near at hand as a source for predator invasion. In some years the natural predator invasion was more delayed and more erratic, and the control exhibited in second-year fields in the absence of predator stocking was very much less effective than was the case in these plots (Huffaker and Kennett, 1953).

Resulting Control in Second-Year Fields

The conclusions from the data, plots 3A, 3B, and 3C; 4A, 4B, and 4C; 5A, 5B, and 5C; 6A, 6B, and 6C; 7A, 7B, and 7C; 8A, 8B, and 8C; 9A, 9B, and 9C; and 10A, 10B, and 10C, of Figures 11–3 and 11–4, are in complete agreement with the results from stocking first-year fields, previously discussed.

It appears obvious that predator-stocking can be done during the spring of the second year, a time when it is more practical from an operational viewpoint. In each of these cases there was complete economic control of the prey species. In plot 5B, the natural predator population was very low on June 15 and July 14; hence, a subsequent increase in cyclamen mites occurred in August in that plot—a demonstration of a more definite intermediate condition between the predator-stocked and predator-free plots than existed in many of the triplicated series. Some of the plants had high prey populations, the predators having been erratically distributed until August, when, as a result of abundant food, they increased to a higher general level than in the predator-stocked plot; they then reduced the rather mild infestation to a condition of no

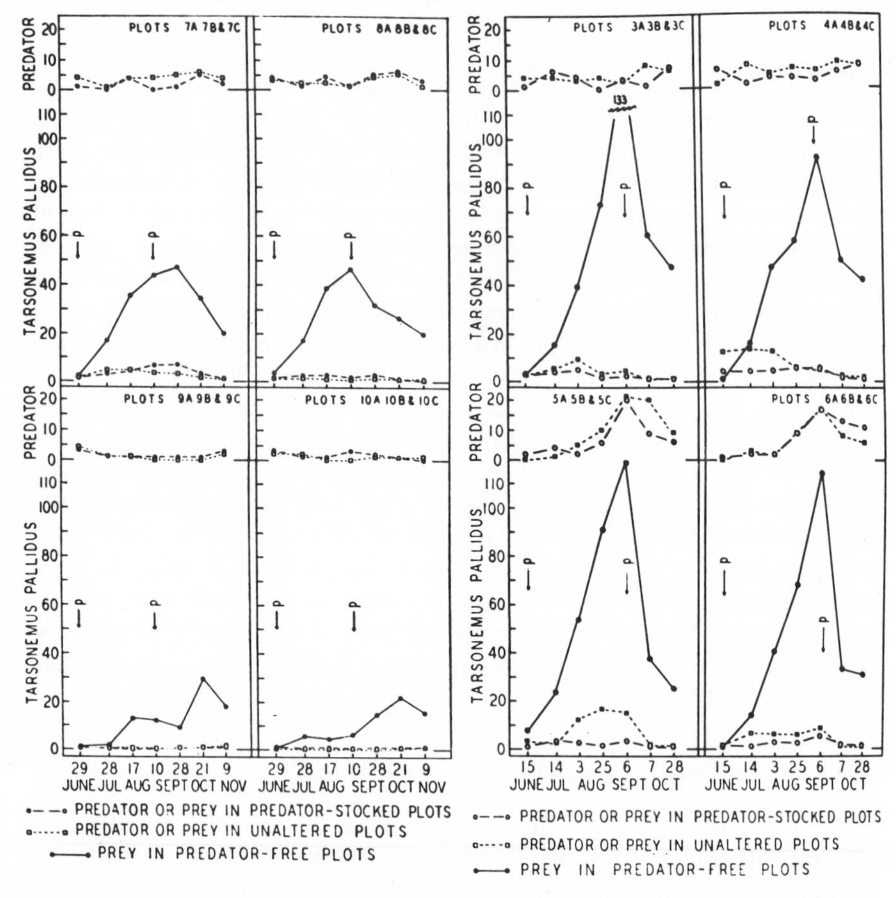

Fig. 11–3. 1953. Second-year plantings: 7A, 7B, and 7C; 8A, 8B, and 8C, Lassen at Madrone; 9A, 9B, and 9C; 10A, 10B, and 10C, J-7 at Madrone.

Fig. 11–4. 1953. Second-year plantings: 3A, 3B, and 3C; 4A, 4B, and 4C, Shasta at Morgan Hill; 5A, 5B, and 5C; 6A, 6B, and 6C, Shasta at Edenvale.

Changes in densities of *Tarsonemus pallidus* (= prey) in predator-stocked, predator-free and unaltered plots, and *Typhlodromus* (= predator) frequencies. Eight triplicated series: one plot of each series predator-stocked, one unaltered, and one kept predator-free. The *T. pallidus* in the three plots of a series are plotted on a per-leaflet basis. The predator frequencies in the predator plots express the number of leaflets among thirty which had one or more *Typhlodromus*. "p"s by arrows indicate dates of parathion treatment in the C plots.

economic importance. In this connection, although the predator densities in the unaltered plots were often as high as those in the predator-stocked plots at a given time, the more even pattern of distribution and the persistence at low densities during the *early* summer, as a result of early spring stocking, seemed, in every case, to account for more consistent

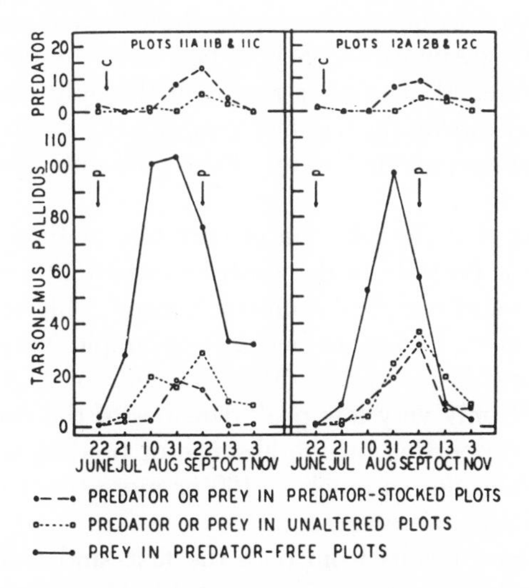

Fig. 11–5. 1953. Second-year plantings: 11A, 11B, and 11C; 12A, 12B, and 12C; Lassen at Cupertino.

Changes in densities of *Tarsonemus pallidus* (= prey) in predator-stocked, predator-free and unaltered plots, and *Typhlodromus* frequencies. Two triplicated series; one plot of each series predator-stocked, one unaltered, and one kept predator-free. The *T. pallidus* in the three plots of a series are plotted on a per-leaflet basis. The predator frequencies in the predator plots express the number of leaflets among thirty which had one or more *Typhlodromus*. "p"s by arrows indicate dates of parathion treatment in C plots. "c"s by arrows indicate the date of unintended grower treatment.

economic control, despite the fact that the control in the unaltered plots was also good, if sometimes more delayed.

Plots 11A, 11B, and 12A, 12B (Fig. 11–5) were disturbed by the grower's use of materials not intended for use in the A plots and not commonly employed by strawberry growers at that season for the B plots. The berries were interplanted between rows of young pear trees. The grower was to cover the A plots at the time that he applied chemicals. However, when he treated the pears he ignored the drift onto the berries, apparently thinking it to be unimportant. The counts showed that on the two dates subsequent to the treatment with DDT and TEPP (shown by an arrow labeled "c" in the illustrations), there was a resultant near-absence of predators in 11B and 12B until late September. This was not the case in other fields. A similar detrimental effect also occurred on the predators in plots 11A and 12A, but the period was not quite as prolonged, possibly because of a more even distribution and the existence of greater numbers of predators in the A, or predator-stocked plots at the time of the treatment. It can be seen that there were very definite spurts in prey population just after the chemical treatment, although the peaks achieved were not as high as those of the predator-free plots 11C and 12C. This condition was obviously due to the effective predation before treatment and during August and September, after the detrimental treatment effect had lessened. The plots represent the only case where predator stocking, either during the late first-year or early second-year, has not resulted in complete economic control of the prey mite, the reason being apparent.

It is surprising that the addition of only two or three predators per plant would make it possible to demonstrate consistently sharp contrasts in prey populations and complete economic control. This occurred regardless of whether a given variety of berry, for example, variety J-7, shown in plots 9A, 9B, and 9C, and 10A, 10B, and 10C of Figure 3, had the capacity to support only very low peak densities of the prey, or whether it was capable of supporting, in the absence of predation (plots 3C and 5C of Fig. 11–4) densities as much as 100 or more times that which occurred when predators were present.

The experimental results from both the first- and second-year fields have been tabulated in combined form in Table 11–2, excluding only the plots of Figure 11–5, those disturbed by unintended chemical application. The prey populations on four successive sampling dates during the mid-season of highest infestation for eleven paired-plot series are presented. Inclusion of data either earlier or later seasonally would serve no purpose and merely complicate the analysis, since this mite is characteristically low in numbers early in the summer and declines again in the fall, irrespective of predator condition. The total for the predator-free condition was 69,630, as compared with 3,075 for the predator-present condition.

PROBLEMS IN COMMERCIAL APPLICATION

Should Plants Be Artificially Infested with Cyclamen Mites?

The results from these studies have been so encouraging as to indicate the possible use of deliberate infestation of new plantings with the pest mite before the periods of highest yield and greatest loss from serious cyclamen-mite infestations, as a means of achieving earlier than normal equilibrium. Smith and DeBach (1953) stated,

> Theoretically it would appear possible, therefore, that when hosts are abnormally scarce artificial infestation of plants in the field with the host insect would serve to synchronize parasite-host or predator-prey populations in such a way as to bring about effective biological control.

This approach is akin to certain types of immunization of animals against losses from disease by controlled vaccination with virulent pathogenic

TABLE 11–2. Total cyclamen mites in predator-stocked and predator-free plots on four successive sample dates during peaks of infestation. Populations on 30 leaflets of second-year strawberry plants in 1953 and 1954*

Sampling period† and predator condition	Data on second-year plants, 1953								Data on second-year plants, 1954			Totals, all plots
	Shasta var. at Morgan Hill		Shasta var. at Edenvale		Lassen var. at Madrone		J-7 var. at Madrone		Experimental var. at San Martin			
	Plot 3	Plot 4	Plot 5	Plot 6	Plot 7	Plot 8	Plot 9	Plot 10	Plot 1	Plot 3	Plot 5	
Period 1:												
Stocked (A) plots	132	136	79	81	132	87	16	15	49	16	59	802
Predator-free (C) plots	1,177	1,404	1,601	1,205	1,066	1,165	384	121	931	1,254	1,387	11,695
Period 2:												
Stocked (A) plots	46	150	36	71	192	48	6	12	92	21	57	731
Predator-free (C) plots	2,194	1,721	2,733	2,039	1,294	1,393	367	185	1,132	2,480	1,392	16,930
Period 3:												
Stocked (A) plots	76	159	93	152	201	86	12	12	40	21	128	980
Predator-free (C) plots	3,970	2,746	3,568	3,440	1,422	951	257	438	2,804	3,860	1,723	25,179
Period 4:												
Stocked (A) plots	40	43	21	43	87	24	12	25	50	21	196	562
Predator-free (C) plots	1,819	1,475	1,122	994	1,038	786	874	655	2,800	2,875	1,388	15,826
Total: four sampling dates												
Stocked (A) plots	294	488	229	347	612	245	46	64	231	79	440	3,075
Predator-free (C) plots	9,160	7,346	9,024	7,678	4,820	4,295	1,882	1,399	7,667	10,469	5,890	69,630

* Plots 11 and 12, Lassen variety at San Jose, were omitted because they were disturbed by unintended chemical applications. All differences between paired predator-stocked and predator-free plots given in the table were highly significant – much greater than necessary for the 1 percent level of significance.

† Actual sampling dates varied for the different plantings but were successive dates for a given planting during midseason peaks of infestation, August through October.

agents, according to a pattern whereby the period of disease is controlled and its economic severity greatly reduced. An example is that of fowl pox of poultry (Bunyea, 1942, p. 981).

Can Predator Stocking Be Done Economically?

Tests are being made on the use of clippings from normal winter pruning of berries. Clippings from fields that have predators always bear predators and cyclamen mites. Growers are reluctant to accept the introduction of cyclamen mites into their fields. It might prove feasible to obtain from these clippings predators that have been separated from the cyclamen mites. However, cyclamen mites must be rather generally present in the new plantings if the predators to be added are to survive and if any good is to be derived. Both components of the ultimate equilibrium should be introduced in order to obtain quick and satisfactory results.

Aramite combines effective control of two-spotted mite with very low toxicity to cyclamen-mite predators. Chemical treatment necessary for aphid control, as well, may prove to be compatible with predator control of cyclamen mite. BHC has shown some possibilities for this purpose. Also, Wilcox and Howland (1955) obtained control of this pest in southern California by use of a 4 percent nicotine dust. Nicotine sprays are known to interfere little, if at all, with *Typhlodromus* and, probably, dusts would not be much different.

Whether this procedure can be economically done on a large scale, whether it will prove compatible with control of other pests, and whether it will be accepted by the industry remain to be seen.

SUMMARY

This study experimentally demonstrates the vital role of predation by a predatory mite, *Typhlodromus cucumeris* or *T. reticulatus* or both, in the control of the cyclamen mite, *Tarsonemus pallidus*, on strawberries. Analyses of population trends of the cyclamen mite and its chief predator were made over a period of five years. By stocking one plot of a pair with predators, and leaving the other unstocked, and by preventing natural invasion of predators in the check plot, it has been possible to demonstrate very marked economic control of the prey species by this predator. During the last two years, a total of eleven such paired series of data was obtained. The total number of cyclamen mites sampled on thirty leaflets from each plot on four different dates in 1953 and 1954 was 69,630 in the predator-free plots, as compared with 3,075 in the predator-stocked plots. The population trends indicated direct *reciprocally dependent oscillations* in numbers of the predator and its prey.

Natural predator invasion of infested fields normally gives sporadic control in second-year berries and good control in third- and fourth-year berries—a result of delay in entrance of the predators. Stocking of fields with predators has given complete economic control, but this has as yet been done only on a trial basis. Deliberate introduction into the fields of the pest mite along with the predator early in each planting is the obvious next step, in order to achieve the earliest possible establishment of equilibrium at low prey densities, thereby effecting economic control.

LITERATURE CITED

Allee, W. C., A. E. Emerson, O. Park, T. Park, and K. P. Schmidt. 1949. Principles of animal ecology. 837 pp. W. B. Saunders Co., Philadelphia, Penn.

Bunyea, H. 1942. Fowl pox (diptheria). Yearbook Agr. 1942:977–86.

Huffaker, C. B., and C. E. Kennett. 1953. Developments toward biological control of cyclamen mite on strawberries in California. Jour. Econ. Ent. 46:802–12.

Huffaker, C. B., and C. H. Spitzer, Jr. 1951. Data on the natural control of the cyclamen mite on strawberries. Jour. Econ. Ent. 44:519–22.

Nicholson, A. J. 1954. An outline of the dynamics of animal populations. Austral. Zool. 2:9–65.

Smith, H. S., and P. DeBach. 1953. Artificial infestation of plants with pest insects as an aid in biological control. Seventh Pac. Sci. Congr. Proc. 4:225–59.

Smith, L. M., and E. V. Goldsmith. 1936. The cyclamen mite, *Tarsonemus pallidus*, and its control on field strawberries. Hilgardia 10:53–94.

Thomas, H. E., and C. P. Marcus, Jr. 1953. Virus diseases of the strawberry. Yearbook Agr. 1953:765–69.

Wilcox, J., and A. F. Howland. 1955. Control of the strawberry aphid in southern California. Jour. Econ. Ent. 48:581–83.

12

The integrated control concept

Vernon M. Stern
Ray F. Smith
Robert van den Bosch
Kenneth S. Hagen

All organisms are subjected to the physical and biotic pressures of the environments in which they live, and these factors, together with the genetic make-up of the species, determine their abundance and existence in any given area. Without natural control, a species which reproduces more than the parent stock could increase to infinite numbers. Man is subjected to environmental pressures just as other forms of life are, and he competes with other organisms for food and space.

Utilizing the traits that sharply differentiate him from other species, man has developed a technology permitting him to modify environments to meet his needs. Over the past several centuries, the competition has been almost completely in favor of man, as is attested by decimation of vast vertebrate populations, as well as populations of other forms of life (Thomas, 1956). But while eliminating many species, as he changed the environment of various regions to fit his needs for food and space, a number of species, particularly among the Arthropoda, became his direct competitors. Thus, when he subsisted as a huntsman or foraged for food from uncultivated sources, early man was largely content to share his subsistence and habitat with the lower organisms. Today, by contrast, as his

Reprinted with permission from *Hilgardia*, 29(2):81–101 (1959). Published by the California Agricultural Experiment Station, University of California, Berkeley, California.

population continues to increase (Hertzler, 1956) and his civilization to advance, he numbers his arthropod enemies in the thousands of species (Sabrosky, 1952).

The increase to pest status of a particular species may be the result of a single factor or a combination of factors. In the last century, the most significant factors have been the following.

First, by changing or manipulating the environment, man has created conditions that permit certain species to increase their population densities (Ullyett, 1951). The rise of the Colorado potato beetle, *Leptinotarsa decemlineata* (Say) to pest status occurred in this manner (Fig. 12–1). When the potato, as well as other solanaceous plants, was brought under widespread cultivation in the United States, a change favorable to the beetle occurred in the environment, which enabled it to become very quickly an important pest (Trouvelot, 1936). Similarly, when alfalfa, *Medicago sativa* L., was introduced into California about 1850, the alfalfa butterfly, *Colias philodice eurytheme* Boisduval, which had previously occurred in low numbers on native legumes, found a widespread and favorable new host

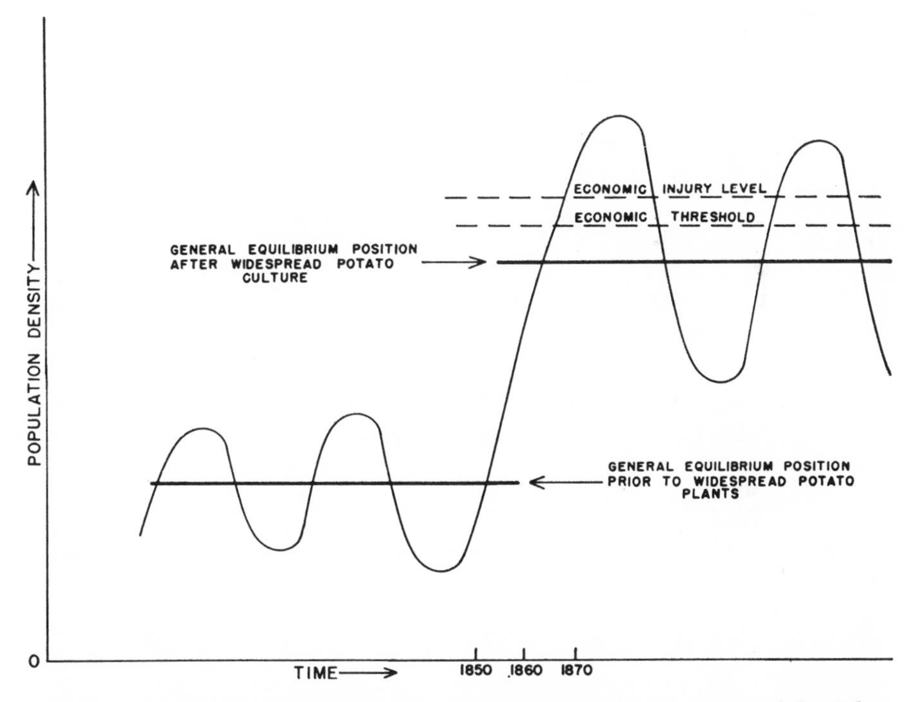

Fig. 12–1. Schematic graph of the change in general equilibrium position of the Colorado potato beetle, *Leptinotarsa decemlineata* following the development of widespread potato culture in the United States. For a discussion of the significance of economic-injury levels and economic thresholds in relation to the general equilibrium position, see p. 89.

plant in its environment, and it subsequently became an economic pest (Smith and Allen, 1954).

A second way in which arthropods have risen to pest status has been through their transportation across geographical barriers while leaving their specific predators, parasites, and diseases behind (Smith, 1959). The increase in importance through such transportation is illustrated by the cottony cushion scale, *Icerya purchasi* Maskell (Fig. 12–2). This scale insect was introduced into California from Australia on acacia in 1868. Within the following two decades, it increased in abundance to the point where it threatened economic disaster to the entire citrus industry in California. Fortunately the timely importation and establishment of two of its natural enemies, *Rodolia cardinalis* (Mulsant) and *Cryptochaetum iceryae* (Williston), resulted in the complete suppression of *I. purchasi* as a citrus pest (Doutt, 1958).

The cottony cushion scale again achieved the status of a major pest when the widespread use of DDT on citrus in the San Joaquin Valley eliminated the vedalia (Ewart and DeBach, 1947).

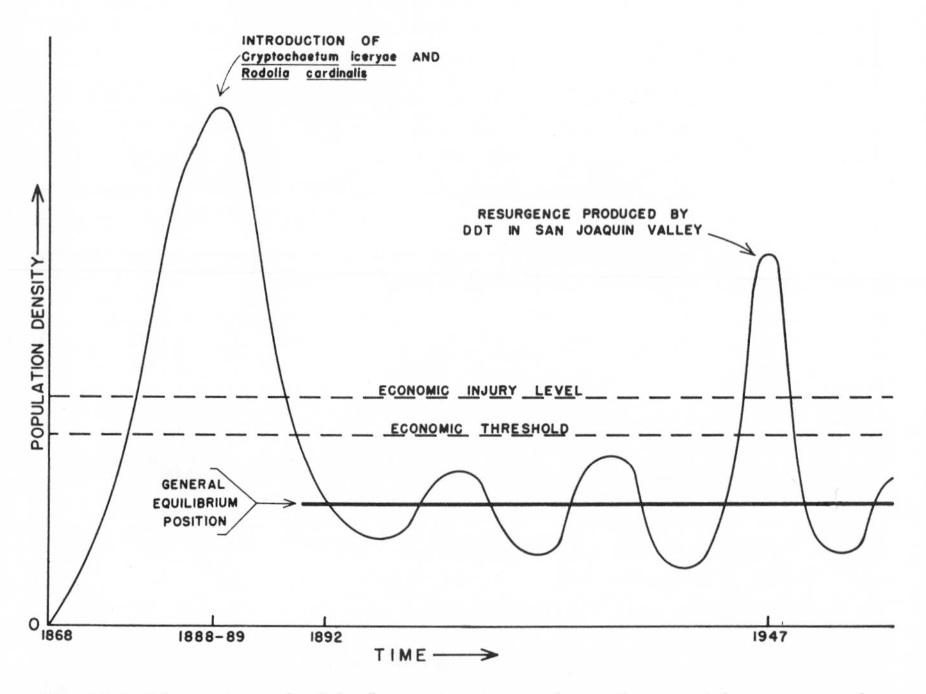

Fig. 12–2. Schematic graph of the fluctuations in population density of the cottony cushion scale, *Icerya purchasi,* on citrus from the time of its introduction into California in 1868. Following the successful introduction of two of its natural enemies in 1888 this scale was reduced to noneconomic status except for a local resurgence produced by DDT treatments.

A third cause for the increasing number of pest arthropods has been the establishment of progressively lower economic thresholds (see p. 89 for definition and discussion). This can be illustrated by lygus bugs *(Lygus* spp.) on lima beans. Not too many years ago the blotches caused by lygus bugs feeding on an occasional lima bean were of little concern, and lygus bugs were considered a minor pest on this crop. However, with the emphasis on product appearance in the frozen-food industry, a demand was created for a near-perfect bean. For this reason, economic-injury thresholds were established and lygus bugs are now considered serious pests of lima beans.

In the face of this increased number of arthropod pests man has made remarkable advances in their control, and economic entomology has become a complex technical field. Of major importance have been new developments in pesticide chemistry and application.

The discovery of the insecticidal properties of DDT, and its spectacularly successful application to arthropod-borne disease and agricultural pest problems, spurred research in chlorinated hydrocarbon chemistry and stimulated the development of other organic pesticides. On a national scale, the experiment stations, state and governmental agencies, and commercial companies, all searching for new or better answers to old insect-pest problems, eagerly accepted the new chemicals. Within a short period many became an integral part of public health and agricultural pest-control programs. Without question, the rapid and widespread adoption of organic insecticides brought incalculable benefits to mankind, but it has now become apparent that this was not an unmixed blessing. Through the widespread and sometimes indiscriminate use of pesticides, the components and intricate relations of crop environments have been drastically altered, and as a result a number of serious problems have arisen (Wigglesworth, 1945; Michelbacher, 1954; Pickett, 1949; Pickett and Patterson, 1953; Solomon, 1953; and others). Among these new problems and old ones which have been aggravated are:

1. Arthropod resistance to insecticides. This phenomenon relating to the genetic plasticity of the arthropods has been reviewed by Metcalf (1955), Hoskins and Gordon (1956), Crow (1957), Brown (1959), and others. In many cases, resistance is already drastic enough to have eliminated certain insecticides from important pest-control programs. There are today in excess of 70 demonstrated cases of arthropod resistance. Actually, a much larger number of pest species exist which are developing resistance or have already done so, but there has not been time to evaluate these cases.

2. Secondary outbreaks of arthropods other than those against which control was originally directed (Massee, 1954; DeBach and Bartlett, 1951; Ripper, 1956; and others). These outbreaks usually result from the interference of the insecticide with biological control (Lord, 1947; Bartlett and

Ortega, 1952; Michelbacher, 1954; Michelbacher and Hitchcock, 1958; and others). This may also occur through the effect of the insecticide on the plant, which, in turn, affects the development of the secondary pest (Fleschner and Scriven, 1957). An example is the increase in mites on plants growing in soil receiving certain chemical treatments (Klostemeyer and Rasmussen, 1953).

3. The rapid resurgence of treated species necessitating repetitious insecticide applications (Holloway and Young, 1943; Bovey, 1955; Schneider, 1955; Stern and van den Bosch, 1959; and others). These flarebacks occur from individuals surviving treatment or from individuals migrating into the treated areas, where they can reproduce unhindered because their natural enemies have been eliminated.

4. The toxic insecticide residues on food and forage crops (Brown, 1951; Linsley, 1956). This problem may result from two sources. First, untimely applications or accidental increases in dosage may result in residues above the tolerance limits. Second, the first three problems mentioned above are interrelated and by aggravating one another may lead to excessive treatment and a residue problem. For example, where the level of resistance is increasing, it requires either more frequent applications or higher insecticide dosages to control the pest, or both. This increased insecticide program may in turn have a drastic effect on the ecosystem, which frequently results in outbreaks of secondary pests or rapid resurgence of the resistant pest for which control was originally intended. Often, under such conditions, where insects threaten the crop or marketability of a crop close to harvest, the grower is faced with the problem of suffering a severe monetary loss or of making an insecticide application closer to harvest than is ordinarily permissible. In many instances, the end result is a residue far above the accepted tolerance limit at harvest time.

5. Hazards to insecticide handlers and to persons, livestock, and wildlife subjected to contamination by drift (Hayes, 1954; Petty, 1957; Upholt, 1955).

6. Legal complications from suits and other actions pertaining to the above problem.

Unquestionably, some of these problems have arisen from our limited knowledge of biological science; others are the result of a narrow approach to insect control. Few studies have included basic investigations on the effects the chemicals might have on other components of the ecosystems to which the pests belong. The entomologist may recognize the desirability of a thorough investigation of these aspects, but because of the need for immediate answers to pressing problems and because of other pressures, he does not have the necessary time. In other instances because fundamental knowledge is lacking, the investigator may be unaware of the intricate nature of the biotic complex with which he is dealing, and of the destructive potential that many chemicals in use today have on the environment of the pests. Finally, and most unfortunately, there are workers who are highly skeptical that biotic factors are of any consequence in the control of pest population densities and thus choose to ignore any approach to pest control other than the use of chemicals.

Whatever the reasons for our increased pest problems, it is becoming more and more evident that an integrated approach, utilizing both biological and chemical control, must be developed in many of our pest problems if we are to rectify the mistakes of the past and avoid similar ones in the future (DeBach, 1951, 1958*a*; Pickett, Putnam, and Roux, 1958; Ripper, 1944; Huffaker and Kennett, 1956; Wille, 1951; Michelbacher and Middlekauff, 1950; and others).

TERMINOLOGY

To clarify the discussion in other parts of this paper some definitions and explanations of terms are here given:

Biological control *The action of parasites, predators, or pathogens on a host or prey population which produces a lower general equilibrium position than would prevail in the absence of these agents.* Biological control is a part of natural control (*q.v.*) and in many cases it may be the key mechanism governing the population levels within the framework set by the environment. If the host or prey population is a pest species, biological control may or may not result in economic control. Biological control may apply to any species whether it is a pest or not, and regardless of whether or not man deliberately introduces, manipulates, or modifies the biological-control agents.

Biotic insecticide *A biotic mortality agent applied to suppress a local insect pest population temporarily.* The effects of the agent usually do not persist and they are similar to those resulting from the use of a chemical insecticide in that they do not produce a permanent change in the general equilibrium position. A polyhedrosis virus applied as a spray to control the alfalfa caterpillar is a typical example of a biotic insecticide. Preparations of microörganisms used in this manner are sometimes referred to as *microbial insecticides.* Predators, such as lady beetles, or parasites, when they are released in large numbers, can also act, in some instances, as biotic insecticides.

Biotic reduction *Deaths or other losses to the population* (e.g., *dispersal, reduced fecundity) caused or induced by biotic elements of the environment in a given period of time.*

Economic control *The reduction or maintenance of a pest density below the economic-injury level* (q.v.).

Economic-injury level *The lowest population density that will cease eco-*

nomic damage. Economic damage is the amount of injury which will justify the cost of artificial control measures; consequently, the economic-injury level may vary from area to area, season to season, or with man's changing scale of economic values.

Economic threshold *The density at which control measures should be determined to prevent an increasing pest population from reaching the economic-injury level.* The economic threshold is lower than the economic-injury level to permit sufficient time for the initiation of control measures and for these measures to take effect before the population reaches the economic-injury level.

Ecosystem *The interacting system comprised of all the living organisms of an area and their nonliving environment.* The size of area must be extensive enough to permit the paths and rates of exchange of matter and energy which are characteristic of any ecosystem.

General equilibrium position *The average density of a population over a period of time (usually lengthy) in the absence of permanent environmental change.* The size of the area involved and the length of the period of time will vary with the species under consideration. Temporary artificial modifications of the environment may produce a temporary alteration of the general equilibrium position (*i.e.*, a temporary equilibrium).

Governing mechanism *The actions of environmental factors, collectively or singly, which so intensify as the population density increases and relax as this density falls that population increase beyond a characteristic high level is prevented and decrease to extinction is made unlikely.* The governing mechanisms operate within the framework or potential set by the other environmental elements.

Integrated control *Applied pest control which combines and integrates biological and chemical control.* Chemical control is used as necessary and in a manner which is least disruptive to biological control. Integrated control may make use of naturally occurring biological control as well as biological control effected by manipulated or introduced biotic agents.

Microbial control *Biological control that is effected by microörganisms (including viruses).*

Natural control *The maintenance of a more or less fluctuating population density within certain definable upper and lower limits over a period of time by the combined actions of abiotic and biotic elements of the environment.* Natural control involves all aspects of the environment, not just those immediate or direct factors producing premature mortality, retarded development, or reduced fecundity; but remote or indirect factors as well. For most situations, governing mechanisms (*q.v.*) are present and determine the population levels within the framework or potential set by the other environmental elements. In the case of a pest population, natural control may or may not be sufficient to provide economic control.

Natural reduction *Deaths or other losses to the population caused by naturally existing abiotic and biotic elements of the environment in a given period of time.*

Population *A group of individuals of the same species that occupies a given area.* A population must have at least a minimum size and occupy an area containing all its ecological requisites to display fully such characteristics as growth, dispersion, fluctuation, turnover, dispersal, genetic variability, and continuity in time. The minimum population and the requisites in occupied area will vary from species to species.

Population dispersion *The pattern of spacing shown by members of a population within its occupied habitat and the total area over which the given population may be spread.*

Selective insecticide *An insecticide which while killing the pest individuals spares much or most of the other fauna, including beneficial species, either through differential toxic action or through the manner in which the insecticide is utilized (formulation, dosage, timing, etc.)*

Supervised insect control *Control of insects and related organisms supervised by qualified entomologists and based on conclusions reached from periodically measured population densities of pests and beneficial species.* Ideally, supervised control is based on a sound knowledge of the ecology of the organisms involved and projected future population trends of pests and natural enemies.

Temporary equilibrium position *The average density of a population over a large area temporarily modified by a procedure such as continued use of insecticides.* The modified average density of the population will revert to the previous or normal density level when the modifying agent is removed or expended (*cf.* "general equilibrium position").

THE NATURE AND WORKING PRINCIPLES OF BIOLOGICAL AND CHEMICAL CONTROL

Biological Control

Biological controls are part of natural control which governs the population density of pest species. On the other hand, with certain exceptions, chemical controls involve only immediate and temporary decimation of localized populations and do not contribute to permanent density regulation. This distinction is not always clearly made, and biological control is often thought of as being similar in its action to chemical control. Perhaps one reason for the misunderstanding is that in spectacular instances a biotic agent may act in the manner of a chemical in eliminating a local pest population. For example, this may occur when weather conditions are favorable and disease pathogens eliminate a localized pest population. Parasites and predators may sometimes act in a similar manner. However, the important prevailing characteristic of biological control is one of permanent population-density regulation. Usually these governing mechanisms occur over such a large area and are so subtle or intricate in their action that they are not easily observed and recorded; thus they tend to be overlooked.

A principal phase of applied biological control is the importation and establishment of natural enemies of pests that accidentally gain entry into new geographical regions. These new pests frequently escape the natural enemies that help to regulate their densities in the areas to which they are indigenous (Elton, 1958). Under satisfactory conditions in the new environment, the pest may flourish and reach damaging abundance. As a counter measure, the natural enemies are obtained from the native home of the pest and transplanted into the new environment to increase the biotic resistance of the environment to the pest.

Biological control is thus utilized to permanently increase environmental resistance to an introduced pest. The hope is that the introduced enemies will lower the general equilibrium position of the pest sufficiently to maintain it permanently below the economic threshold. Most often the introduction of a biotic agent is not so spectacular, and it is an exception when the general equilibrium position of the introduced pest is lowered sufficiently to prevent its occasionally or even commonly reaching economic abundance at certain times or places (Clausen, 1956; Simmonds, 1956). This, of course, is precisely the status of a native pest which, though attacked by a complex of parasites and predators, still has a general equilibrium position high enough to permit it occasionally to cause damage of greater or lesser severity. Thus, in any geographic area the governing mechanisms in the environment are constantly at work to counteract the inherent natality of plant and animal pest species. In terms of crop protection, these regulating factors actually keep thousands of potentially harmful arthropod species permanently below economic thresholds. Moreover, these environmental pressures tend to localize the outbreaks of those forms which on occasion are capable of rising above economic thresholds. A biological control agent is self-perpetuating and capable of response to fluctuations in the population density of the pest it attacks. Biological controls, whether imported or native, are permanent characteristics of a given environment.

Chemical Control

Chemical control of an arthropod pest is employed to reduce populations of pest species which rise to dangerous levels when the environmental pressures are inadequate. When chemicals are used, the damage from the pest species must be sufficiently great to cover not only the cost of the insecticidal treatment but also the possible deleterious effects, such as the harmful influence of the chemical on the ecosystem. On some occasions, the pest outbreak may cover a wide area; in other instances, damaging numbers occur in very restricted locations. These outbreaks occur during the season favorable to the pest, with the relaxed environmental pressures occurring some time before the outbreak. Chemical control is only needed at those times and places where natural control is inadequate. Chemical control should act as a complement to the biological control.

An insecticide must always be manipulated by man, who adds it to a restricted segment of the pest's environment to decimate a localized pest population. Because chemical insecticides are nonreproductive, have no searching capacity, and are nonpersistent, they constitute short-term, restricted pressures. They cannot permanently change the general equilibrium position of the pest population nor can they restrain an increase in abundance of the pest without repeated applications. Therefore, they must be added to the environment at varying intervals of time.

In certain pest-control programs, the insecticide is applied over extensive geographical areas. In some areas, after application, the pest population density may be far below the economic threshold and below its general equilibrium position; but since the insecticide is not a permanent part of the environment, the pest may return to a high level when the effects of the insecticide are gone.

The effectiveness of a chemical insecticide or a biotic insecticide is measured in per cent of kill or in per cent of clean fruits, uninjured cotton bolls, and so forth, in the area of application. Such applications have little influence on the pest in adjoining areas except as localized population depressants. In general, this contrasts sharply with the role of the permanent biotic mortality agent, whose effectiveness is best measured by its influence on the general equilibrium position of the pest species over an entire geographical region or a long period of time.

ECONOMIC THRESHOLDS AND THE GENERAL EQUILIBRIUM POSITION

Chemical control should be used only when the economic threshold is reached and when the natural mortality factors present in the environment are not capable of preventing the pest population from reaching the economic-injury level. The economic-injury level is a slightly greater density than the economic threshold. This difference in densities provides a margin of safety for the time that elapses between the detection of the threatening infestation and the actual application of an insecticide. The economic threshold and the economic-injury level of a pest species can vary depending upon the crop, season, area, and desire of man; the general equilibrium position, on the other hand, barring "permanent" changes in the environment, is a fixed population level (Griffiths, 1951; Strickland, 1954).

A species population is plastic and is undergoing constant change within the limits imposed upon it by its genetic constitution and the characteristics of its environment. Typical fluctuations in population and dispersion are shown in Figure 12–3. The population dispersions shown at the three points in time A, B, and C are not static but rather are instantaneous phases of a continuously changing dispersion.

Thus at point A, when the population is of greatest numerical abundance, it also has its widest distributional range (as depicted by the maximum diameter of the base of the model), and is of maximum economic status (as depicted by the number and magnitude of the blackened pinnacles representing penetrations of the economic threshold). At point B, on the other hand, when the species population is at its lowest numerical abundance, it is also most restricted in geographical range and is of only minor economic status. Point C represents an intermediate condition between points A and B.

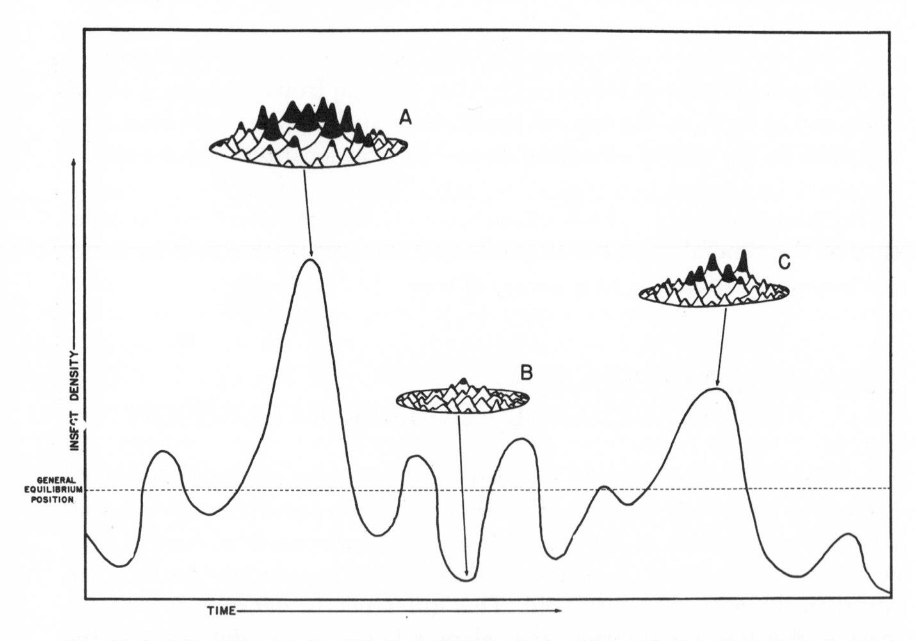

Fig. 12–3. Schematic graph of the population trend and population dispersion of a pest species over a long period of time. The solid line depicts the fluctuations in the population density with time. The broken line depicts the general equilibrium position. The population dispersion is indicated at the specific times *A*, *B*, and *C*. The basal area of these models reflects the distributional range, the height indicates population density. Population densities above the economic threshold are black.

In order to determine the relative economic importance of pest species, both the economic threshold and general equilibrium position of the pests must be considered. It is the general equilibrium position and its relation to the economic threshold, in conjunction with the frequency and amplitude of fluctuations about the general equilibrium position, that determine the severity of a particular pest problem.

In the absence of permanent modifications in the composition of the environment, the density of a species tends to fluctuate about the general equilibrium position as changes occur in the biotic and physical components of the environment. As the population density increases, the density-governing factors respond with greater and greater intensity to check the increase; as the population density decreases, these factors relax in their effects. The general equilibrium position is thus determined by the interaction of the species population, these density-governing factors, and the other natural factors of the environment. A permanent alteration of any factor of the environment—either physical or biotic—or the introduction of new factors may alter the general equilibrium position.

The economic threshold of a pest species can be at any level above or below the general equilibrium position or it can be at the same level. Some phytophagous species may utilize our crops as a food source but even at their highest attainable density are of little or no significance to man (Fig. 12–4, *A*). Such species can be found associated with nearly every crop of commercial concern.

Another group of arthropods rarely exceeds the economic thresholds and these consequently are occasional pests. Only at their highest population density will chemical control be necessary (Fig. 12–4, *B*).

When the general equilibrium position is close to the economic threshold, the population density will reach the threshold frequently (Fig. 12–4, *C*). In some cases, the general equilibrium position and the economic threshold are at essentially the same level. Thus, each time the population fluctuates up to the level of the general equilibrium position insecticidal treatment is necessary. In such species the frequency of chemical treatments is determined by the fluctuation rate about the general equilibrium position, which in some cases necessitates almost continuous treatment.

Finally, there are pest species in which the economic threshold lies below the general equilibrium position; these constitute the most severe pest problems in entomology (Fig. 12–4, *D*). The economic threshold may be lower than the level of the lowest population depression caused by the physical and biotic factors of the environment, *e.g.*, many insect vectors of viruses. In such cases, particularly where human health is concerned, there is a widespread and almost constant need for chemical control. This produces conditions favorable for development of insecticide resistance and other problems associated with heavy treatments.

One solution to pest problems and particularly those in this last cate-

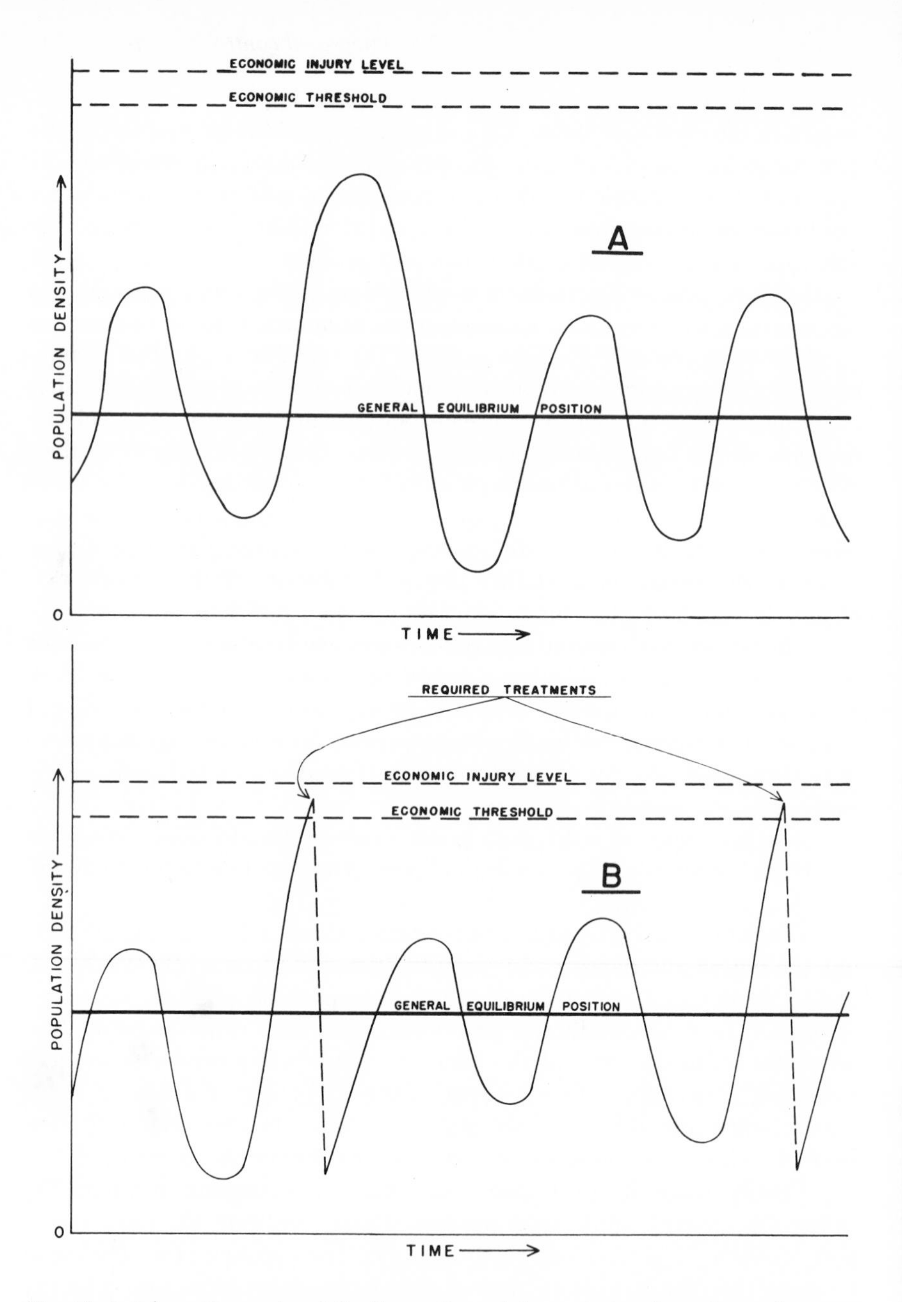

Fig. 12–4. Schematic graphs of the fluctuations of theoretical arthropod populations in relation to their general equilibrium position, economic thresholds, and economic-injury levels. *A,* Noneconomic population whose general equilibrium position and highest fluctuations are below the economic threshold, *e.g., Aphis medicaginis* Koch on alfalfa in California. *B,* Occasional pest whose general equilibrium position is below the economic threshold but whose highest population fluctuations exceed the economic threshold, *e.g., Grapholitha molesta* Busck on peaches in California.

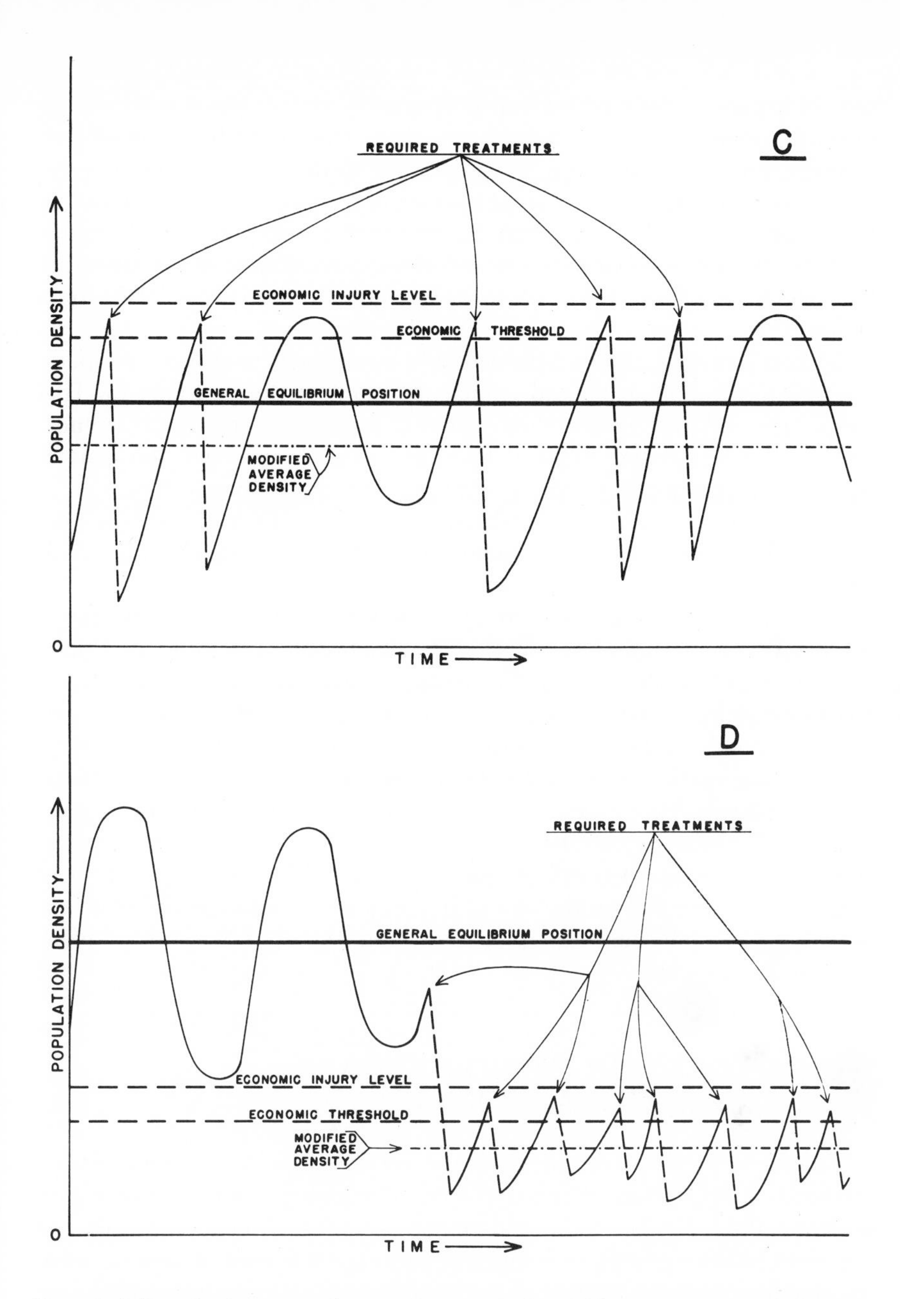

Fig. 12–4 (Cont.). *C*, Perennial pest whose general equilibrium position is below the the economic threshold but whose population fluctuations frequently exceed the economic threshold, *e.g.*, *Lygus* spp. on alfalfa seed in the western United States. *D*, Severe pest whose general equilibrium position is above the economic threshold and for which frequent and often widespread use of insecticides is required to prevent economic damage, *e.g.*, *Musca domestica* in Grade A milking sheds.

gory is to change the environment permanently so that the general equilibrium position will be lowered. For example, this might be accomplished through the introduction of a new biological control agent or through the permanent modification of a large portion of a required habitat. This has been done in certain areas with malaria-vector mosquitoes and similar pests by the draining of swamps and the destruction of other favorable habitats. Such methods may completely eliminate the species from some areas.

Environmental changes unfavorable to the pest may also be made through the use of plants and animals resistant to the pest species. This control method may involve three different aspects—tolerance, preference, and antibiosis (Painter, 1951). If tolerance alone is involved, the general equilibrium position may not be changed but the economic threshold is raised. Where preference or antibiosis is involved, the ability of the pest to reproduce upon the host is reduced, so that the general equilibrium position is lowered.

The lack of a sound measure of economic thresholds, in many cases, has been a major stumbling block to the development of integrated pest-control programs. Our changing economy, variations in natural governing mechanisms from one geographical area to another, differences in consumer demands, and the complexity of measuring the total effect of insects on yield and quality often make the assessment of economic damage extremely difficult. Yet the economic threshold and the economic-injury level must be determined reasonably and realistically before integrated pest control can develop to its fullest. Success in any well-balanced pest-control program is dependent on the aim of holding insect populations below experimentally established economic levels rather than attempting to eliminate all the insects.

THE INTEGRATION OF BIOLOGICAL AND CHEMICAL CONTROL

Biological control and chemical control are not necessarily alternative methods; in many cases they may be complementary, and, with adequate understanding, can be made to augment one another. One reason for the apparent incompatibility of biological and chemical control is our failure to recognize that the control of arthropod populations is a complex ecological problem. This leads to the error of imposing insecticides on the ecosystem, rather than fitting them into it. It is short-sighted to develop a chemical control program for the elimination of one insect pest and ignore the impact of that program on the other arthropods, both beneficial and harmful, in the ecosystem. On the other hand, this approach is no worse than the other extreme which would eliminate chemicals to preserve the
If a chemical treatment destroys the biotic agents without eradicating the

pest, then repeated treatments may become necessary.
biological control even in the face of serious economic damage. For we must recognize that modern agriculture could not exist without the use of insecticides. The evidence that biological and chemical control can be integrated is mounting. It has come from many sources involving many kinds of pests in various situations: see Ullyett (1947), Pickett and Patterson (1953), Ripper (1956), Huffaker and Kennett (1956), DeBach (1958*a*), Stern and van den Bosch (1959), and many others.

In approaching an integrated control program, we must realize that man has developed huge monocultures, he has eliminated forests and grasslands, selected special strains of plants and animals, moved them about, and in other ways altered the natural control that had developed over thousands upon thousands of years. We could not return to those original conditions if it were desirable. We may, however, utilize some of the mechanisms that existed before man's modifications to establish new balances in our favor.

Recognition of the Ecosystem

To establish new, favorable balances, it is first necessary to recognize the "oneness" of any environment, natural or man-made. The populations of plants and animals (including man) and the nonliving environment together make up an integrated unit, the ecosystem. If an attempt is made to reduce the population level of one kind of animal (for example, a pest insect) by chemical treatment, modification of cultural practices, or by other means, other parts of the ecosystem will be affected as well. For this reason, the production of a given food or fiber must be considered in its entirety. This includes simultaneous consideration of insects, diseases, plant nutrition, plant physiology, and plant resistance, as well as the economics of the crops (Forbes, 1880; Ullyett, 1947; Pickett, 1949; DeBach, 1951; Solomon, 1953; Pickett and Patterson, 1953; Glen, 1954; Michelbacher, 1954; Huffaker and Kennett, 1956; Simmonds, 1956; Balch, 1958; Decker, 1958; and others).

In most agricultural ecosystems, some potentially harmful organisms are continually held at subeconomic levels by natural controlling forces. In others, the pests are held below economic levels only part of the time. A pest species may be under satisfactory biological control over a large area or a long period of time, but not in all individual fields or during all periods. In a single field or orchard, or during a portion of a year, the pest population may rise to economic levels, while elsewhere or at other times the pest may be subeconomic. It is in such situations that integrated control programs are especially important. These intermittently destructive populations must be reduced in a manner that permits the biological control which prevailed before or prevails elsewhere to take over again.

Population Sampling and Prediction

The sampling methods utilized by most research investigators for experimental plots are usually too time-consuming and tedious to be of practical value in establishing pest population levels in commercial crops. Special index methods are needed that are rapid and simple to use. Ideally, these should be of such nature that they can be easily utilized by the person examining the crop. But in many cases the grower is not able to evaluate all situations because of the difficulties and complexities involved in determining the status of some pest populations at the times of the year when they must be controlled. Then qualified entomologists will be required to evaluate the populations (Ripper, 1958).

One answer to this problem has been the development of supervised control in California, Arkansas, Arizona, and elsewhere. In a supervised control program the farmer, or a group of farmers, contracts with a professional entomologist who determines the status of the insect populations. On the basis of his population counts, other conditions peculiar to each situation, and his knowledge of the ecology of the pests and their biological controls, the entomologist makes predictions as to the course of the population trends and advises as to when controls should be applied and what kind. For instance, in the case of the alfalfa caterpillar, *Colias philodice eurytheme,* when economic thresholds are reached, the recommended procedure may involve immediate cutting of the hay crop without treatment, application of the polyhedrosis virus (Steinhaus and Thompson, 1949; Thompson and Steinhaus, 1950), or treatment with an insecticide. The course to be taken depends on the characteristics of the particular infestation (Smith and Allen, 1954).

Wherever possible, knowledge must be developed so that we can predict the times when occasional pests will be present in outbreak numbers. This will eliminate unnecessary and environment-disturbing "insurance" treatments. When this is not possible, the treatments can be timed according to the actual pest population levels, as is now done with many field-crop pests.

With those crops that do not yet have fixed chemical control schedules, every effort should be made to plan programs dependent upon pest population levels and to avoid dependence upon insurance and prophylactic treatments. If this is not done there is real danger that on these crops, too, pest-control problems will become increasingly complex.

Augmentation of Natural Enemies

In some situations, the development of integrated control requires the augmentation of the natural-enemy complex (DeBach, 1958*a*). The introduction of additional natural enemies is usually the simplest and best solution. This may not be possible or effective with some pest species, and methods of overcoming the inefficiency of the natural-enemy complex must be sought. This can be done by periodic colonization of parasite or predators (Doutt and Hagen, 1950; DeBach, Landi, and White, 1955), artificial inoculation of the host at times of low density (Smith and DeBach, 1953; Huffaker and Kennett, 1956), modification of the environment, or selective breeding of parasites and predators.

The modification of the environment may involve changes in irrigation, introduction of a covercrop, or development of greater plant heterogeneity. Refuges for beneficial forms can be produced by strip treatments with chemicals (DeBach, 1958*a*) or by the development of uncultivated and untreated areas (Grison and Biliotti, 1953). Modifications of the environment may also involve the control of ants or other organisms which curtail parasite and predator activity (Flanders, 1945; DeBach, Fleschner, and Dietrick, 1951). The selective breeding of parasites and predators may be directed toward increased or modified tolerance ranges to physical conditions (DeBach, 1958*b*) or insecticides (Robertson, 1957).

Where prophylactic treatments are proved to be necessary for a perennial pest, selective materials must be developed and utilized to foster biological control both of other pests and of the pest of direct concern at other times.

Selective Insecticides

Chemical control programs are limited by the nature of the available materials. In the past, nonselective insecticides applied for one insect in a pest complex often have eliminated the biotic factors holding other pests in check. More recently, we have had available a greater variety of materials, some of them selective in their action (Ripper, 1956).

The selective use of insecticides may be accomplished in at least four ways. First, the insecticide itself may be selective in its toxicological action. Narrow-range toxicants may be utilized to reduce a pest of concern and at the same time spare the beneficial forms (Ripper, 1944; Ripper, Greenslade, and Hartley, 1951). A particular material may be selective in one situation and not in another; or it may be selective at low dosages but not at high dosages. Furthermore, the manner of application (Ripper, 1956) and especially the type of carrier and residue deposit may produce differential effects on the insect complex (Flanders, 1941; Holloway and Young, 1943; DeBach and Bartlett, 1951).

Second, we can produce a selective action on a pest-parasite complex by treating only those areas where the pest-parasite ratio is unfavorable. This method is one of the bases of supervised control of the alfalfa caterpillar in California (Smith and Allen, 1954). Population levels of both the host caterpillar and its parasite, *Apanteles medicaginis* Meusebeck, are determined at appropriate intervals in all fields. A prediction of possible damage is made on the basis of these population levels, and only those infestations which are potentially damaging are treated. In this way, on an area-wide basis, the balance is shifted in favor of the parasites, even though parasite adults and parasite larvae within the host caterpillars are often destroyed in the treated fields. The success of such programs will depend on the exact nature of the local problem and the quality of supervision. The rates of dispersal of parasites, predators, and pests are complicating factors.

Third, proper timing of insecticides can produce a selective action on the pest and natural-enemy complex (Ewart and DeBach, 1947; Michelbacher and Middlekauff, 1950; Bartlett and Ewart, 1951; Jeppson, Jesser, and Complin, 1953; Massee, 1954; DeBach, 1955). In such situations, an intimate knowledge of the behavior patterns of the pests and their natural enemies is required.

Fourth, nonselective materials with short residual action may be used if the beneficial forms can survive in a resistant stage or in an untreated reservoir area. Stern and van den Bosch (1959) have demonstrated that parasites of the spotted alfalfa aphid can survive nonselective treatments if they are in the more resistant pupal stage. DeBach (1958*a*) reports successful integration of biological and chemical control of purple scale on citrus where alternate pairs of tree rows were sprayed at 6-month intervals with a nonselective oil treatment.

For some pests a disease pathogen may be used as a selective insecticide (Steinhaus, 1954; 1956). For example, under supervised control in the Dos Palos area of California, the polyhedrosis virus affecting the alfalfa caterpillar has been used successfully either alone or in combination with a selective insecticide to avoid the use of a nonselective treatment. More recently, interest has developed in the commercial use of virulent strains of *Bacillus thuringiensis* Berliner, for the control of certain truck- and field-crop pests in California. The use of disease pathogens as selective insecticides is in its infancy, but can be expected to increase in importance with additional research (Steinhaus, 1951, 1957).

The ideal selective material is not one that eliminates all individuals of the pest species while leaving all of the natural enemies. Use of such a material would force the predators and parasites to leave the treated area or starve (Clausen, 1936; Flanders, 1940). The ideal material is one that shifts the balance back in favor of the natural enemies (Boyce, 1936; Ripper, 1944; Wigglesworth, 1945).

Future of Integrated Control

If our knowledge were adequate today to outline an ideal integrated control program for a crop now utilizing an intensive fixed spray program, it would not be possible to switch to such a program immediately. The effects of the previous treatments may last several years. In some instances, effective biological control no longer exists and would have to be reëstablished. This may be a slow process (DeBach, 1951; DeBach and Bartlett, 1951; Pickett and Patterson, 1953; and others).

It should be emphasized also that the development of integrated control is not a panacea that can be applied blindly to all situations, for it will not work if biotic mortality agents are inadequate or if low economic thresholds preclude utilizing biological control (Barnes, 1959). However, it has worked so well in some appropriate situations that there can be no doubt as to its enormous advantages and its promise for the future.

LITERATURE CITED

Balch, R. E. 1958. Control of forest insects. Ann Rev. Ent. 3:449–68.

Barnes, M. M. 1959. Deciduous fruit insects and their control. Ann. Rev. Ent. 4:343–62.

Bartlett, B., and W. H. Ewart. 1951. Effect of parathion on parasites of *Coccus hesperidum. Jour. Econ. Ent.* 44:344–347.

Bartlett, B. R., and J. C. Ortega. 1952. Relations between natural enemies and DDT-induced increases in frosted scale and other pests of walnuts. Jour. Econ. Ent. 45(5):783–85.

Bovey, Paul. 1955. Les actions secondaires des traitements antiparasitaires sur les populations d'insectes et d'acariens nuisibles. Schweiz. Landw. Monatsh. 33(9/10): 369–79.

Boyce, A. M. 1936. The citrus red mite. *Paratetranychus citri* McG. in California, and its control. Jour. Econ. Ent. 29(1):125–30.

Brown, A. W. A. 1951. Insect control by chemicals. John Wiley & Sons, Inc., New York, N.Y. 817 pp.

1959. Spread of insecticide resistance. Adv. Pest Control Res. 2:351–414. Interscience Publishers, Inc., New York, N.Y.

Clausen, C. P. 1936. Insect parasitism and biological control. Ent. Soc. Amer. Ann. 29:201–23.

1956. Biological control of insect pests in the continental United States. U.S. Dept. Agr. Tech. Bul. 1139. 151 pp.

Crow, J. F. 1957. Genetics of insect resistance to chemicals. Ann. Rev. Ent. 2:227–46.

DeBach, P. 1951. The necessity for an ecological approach to pest control on citrus in California. Jour. Econ. Ent. 44(4):443–47.

1955. Validity of the insecticidal check method as a measure of the effectiveness of natural enemies of diaspine scale insects. Jour. Econ. Ent. 48(5):584–88.

1958a. Application of ecological information to control of citrus pests in California. Tenth Internatl. Congr. Ent. Proc. 3:185–97.

1958b. Selective breeding to improve adaptations of parasitic insects. Tenth Internatl. Congr. Ent. Proc. 4:759–67.

DeBach, P., and B. Bartlett. 1951. Effects of insecticides on biological control of insect pests of citrus. Jour. Econ. Ent. 44(3):372–83.

DeBach, P., C. A. Fleschner, and E. J. Dietrick. 1951. A biological check method for evaluating the effectiveness of entomophagous insects. Jour. Econ. Ent. 44(5): 763–66.

DeBach, P., J. H. Landi, and E. B. White. 1955. Biological control of red scale. California Citrog. 40:254, 271–72, 274–75.

Decker, G. C. 1958. Don't let the insects rule. Agr. Food Chem. 6(2):98–103.

Doutt, R. L. 1958. Vice, virtue and the vedalia. Ent. Soc. Amer. Bul. 4(4):119–23.

Doutt, R. L., and K. S. Hagen. 1950. Biological control measures applied against *Pseudococcus maritimus* on pears. Jour. Econ. Ent. 43(1):94–96.

Elton, Charles S. 1958. The ecology of invasions by animals and plants. 181 pp. Methuen & Co., Ltd., London; John Wiley & Sons, Inc., New York, N.Y.

Ewart, W. H., and P. DeBach. 1947. DDT for control of citrus thrips and citricola scale. California Citrog. 32:242–45.

Flanders, S. E. 1940. Environmental resistance to the establishment of parasitic Hymenoptera. Ent. Soc. Amer. Ann., 33:245–53.

1941. Dust as inhibiting factor in the reproduction of insects. Jour. Econ. Ent. 34(3): 470–1.

1945. Coincident infestations of *Aonidiella atrana* and *Coccus hesperidum*, a result of ant activity. Jour. Econ. Ent. 38:711–12.

Fleschner, C. A., and G. T. Scriven. 1957. Effects of soil-type and DDT on ovipositional response of *Chrysopa californica* (Coq.) on lemon trees. Jour. Econ. Ent. 50(2):221–22.

Forbes, S. A. 1880. On some interactions of organisms. Illinois State Lab. Nat. Hist. Bul. 3:3–17.

Glen, R. 1954. Factors that affect insect abundance. Jour. Econ. Ent. 47:398–405.

Griffiths, J. T. 1951. Possibilities for better citrus insect control through the study of the ecological effects of spray programs. Jour. Econ. Ent. 44:464–68.

Grison, P., and E. Biliotti. 1953. La signification agricole des "stations refuges" pour la faune entomologique. Acad. d'Agr. de France Compt. Rend. 39(2):106–9.

Hayes, W. J. 1954. Agricultural chemicals and public health. U.S. Public Health Reports 69(10):893–98.

Hertzler, J. O. 1956. The crisis in world population. 279 pp. Univ. Nebraska Press, Lincoln.

Holloway, J. K., and T. Roy Young, Jr. 1943. The influence of fungicidal sprays on entomogenous fungi and on the purple scale in Florida. Jour. Econ. Ent. 36(3): 453–57.

Hoskins, W. M., and H. T. Gordon. 1956. Arthropod resistance to chemicals. Ann. Rev. Ent. 1:89–148.

Huffaker, C. B., and C. E. Kennett. 1956. Experimental studies on predation: Predation and cyclamen-mite populations on strawberries in California. Hilgardia 26(4): 191–222.

Jeppson, L. R., M. J. Jesser, and J. O. Complin. 1953. Timing of treatments for control of citrus red mite on orange trees in coastal districts of California. Jour. Econ. Ent. 46:10–14.

Klostermeyer, E. C., and W. B. Rasmussen. 1953. The effect of soil insecticide treatments on mite populations and damage. Jour. Econ. Ent. 46:910–12.

Linsley, E. G., (ed.). 1956. Evaluation of certain acaricides and insecticides for effectiveness, residues, and influence on crop flavor. Hilgardia 26(1):1–106.

Lord, F. T. 1947. The influence of spray programs on the fauna of apple orchards in Nova Scotia: II. Oystershell scale, *Lepidosaphis ulmi* (L.) Canad. Ent. 79:196–209.

Massee, A. M. 1954. Problems arising from the use of insecticides: effect on the balance of animal populations. 6th Commonwealth Entomol. Conf. Rept. pp. 53–57. London, England.

Metcalf, R. L. 1955. Physiological basis for insect resistance. Physiol. Rev. 35:197–232.

Michelbacher, A. E. 1954. Natural control of insect pests. Jour. Econ. Ent. 47(1):192–94.

Michelbacher, A. E., and S. Hitchcock. 1958. Induced increase of soft scales on walnut. Jour. Econ. Ent. 51(4):427–31.

Michelbacher, A. E., and W. W. Middlekauff. 1950. Control of the melon aphid in northern California. Jour. Econ. Ent. 43(4):444–47.

Painter, R. H. 1951. Insect resistance in crop plants. xi + 520 pp. Illus. Macmillan Company, New York, N.Y.

Petty, C. S. 1957. Organic phosphate insecticide poisoning; an agricultural occupational hazard. Louisiana St. Med. Soc. Jour. 109(5):158–64.

Pickett, A. D. 1949. A critique on insect chemical control methods. Canad. Ent. 81(3): 67–76.

Pickett, A. D., and N. A. Patterson. 1953. The influence of spray programs on the fauna of apple orchards in Nova Scotia. IV. A Review. Canad. Ent. 85(12):472–78.

Pickett, A. D., W. L. Putman, and E. J. Le Roux. 1958. Progress in harmonizing biological and chemical control of orchard pests in eastern Canada. Tenth Internatl. Congr. Ent. Proc. 3:169–74.

Ripper, W. E. 1944. Biological control as a supplement to chemical control of insect pests. Nature 153:448–52.

1956. Effect of pesticides on balance of arthropod populations. Ann. Rev. Ent. 1: 403–38.

1958. The place of contracting organizations and professional supervision in the application of pest control methods. Tenth Internatl. Congr. Ent. Proc. 3:93–97.

Ripper, W. E., R. M. Greenslade, and G. S. Hartley. 1951. Selective insecticides and biological control. Jour. Econ. Ent. 44(4):448–59.

Robertson, J. G. 1957. Changes in resistance to DDT in *Macrocentrus ancylivorus* Rohw. Canad. Jour. Zool. 35(5):629–633.

Sabrosky, C. W. 1952. How many insects are there? pp. 1–7 *in:* Insects, 1952 Yearbook of Agriculture. 780 pp. U.S. Dept. Agr., Washington, D.C.

Schneider, F. 1955. Beziehungen zwischen nützlingen und chemischer Schädlingsbekämpfung. Deutsche Gesell. f. Angw. Ent. Verhandl. (13te Mitglied versamm.). 1955:18–29.

Simmonds, F. J. 1956. The present status of biological control. Canada. Ent. 88(9): 553–63.

Smith, H. S., and P. DeBach. 1953. Artificial infestation of plants with pest insects as an aid in biological control. Seventh Pacific Sci. Congr. Zoology (1949) Proc. 4:255–59.

Smith, Ray F. 1959. The spread of the spotted alfalfa aphid, *Therioaphis maculata* (Buckton), in California. Hilgardia 28(21):647–91.

Smith, R. F., and W. W. Allen. 1954. Insect control and the balance of nature. Sci. Amer. 190(6):38–42.

Solomon, M. E. 1953. Insect population balance and chemical control of pests. Chem. and Indus. 1953:1143–47.

Steinhaus, E. A. 1951. Possible use of *Bacillus thuringiensis* Berliner as an aid in the biological control of the alfalfa caterpillar. Hilgardia 20(18):359–81.

1954. The effects of disease on insect populations. Hilgardia 23(9):197–261.

1956. Potentialities for microbial control of insects. Agr. Food Chem. 4(8):676–80.

1957. Concerning the harmlessness of insect pathogens and the standardization of microbial control products. Jour. Econ. Ent. 50(6):715–20.

Steinhaus, E. A., and C. G. Thompson. 1949. Preliminary field tests using a polyhedrosis virus in the control of the alfalfa caterpillar. Jour. Econ. Ent. 42(2): 301–5.

Stern, V. M., and R. Van den Bosch. 1959. Field experiments on the effects of insecticides. Hilgardia 29(2):103–30.

Strickland, A. H. 1954. The assessment of insect pest density in relation to crop losses. 6th Commonwealth Ent. Conf. Rept., pp. 78–83. London, England.

Thomas, W. L. Jr. (ed.). 1956. Man's role in changing the face of the earth. 1193 pp. Univ. Chicago Press, Chicago, Ill.

Thompson, C. G., and E. A. Steinhaus. 1950. Further tests using a polyhedrosis virus to control the alfalfa caterpillar. Hilgardia 19(14):411–45.

Trouvelot, B. 1936. Le doryphore de la pomme de terre (*Leptinotarsa decemlineata* Say) en Amerique de Nord. Ann. Epiphyt. (n.s.) 1:227–336.

Ullyett, G. C. 1947. Mortality factors in population of *Plutella maculipennis* (Tineidae: Lep.) and their relation to the problem of control. South African Dept. Agr. and Forestry Ent. Mem. 2(6):77–202.

1951. Insects, man and the environment. Jour. Econ. Ent. 44(4):459–64.

Upholt, W. M. 1955. Evaluating hazards in pesticides use. Agr. Food Chem. 3(12): 1000–6.

Wille, J. E. 1951. Biological control of certain cotton insects and the application of new organic insecticides in Peru. Jour. Econ. Ent. 44:13–18.

Wigglesworth, V. B. 1945. DDT and the balance of nature. Atlantic Monthly 176(6): 107–13.

13

Screw-worm control through release of sterilized flies

A. H. Baumhover
A. J. Graham
B. A. Bitter
D. E. Hopkins
W. D. New
F. H. Dudley
R. C. Bushland

Screw-worms, *Callitroga hominivorax* (Cqrl.), did not exist in the southeastern United States until about 20 years ago, and it is probable that, if the present infestation could be eradicated, the area might be kept free of infestation through inspection of livestock shipments originating in infested areas.

E. F. Knipling has suggested the possibility that screw-worms might be eradicated from the Southeastern States through the release of sterilized males. In average winters screw-worms survive only in peninsular Florida, and during the summer months infest areas to the north. Flies do not normally migrate from Texas to the Southeastern States.

Laboratory experiments (Bushland & Hopkins 1951, 1953) at Kerrville, Tex., showed that screw-worm flies could be sterilized by irradiation in the pupal stage with x-rays or gamma rays. Under cage conditions it was found that male screw-worm flies mated repeatedly but that females

Reprinted from *J. of Econ. Ent.*, 48:462–466 (1955), by permission of the copyright owner, The Entomological Society of America.

mated only once. If a female mated with a sterilized male she laid only infertile eggs. When mixed populations of sterilized and normal insects were caged together, the ratio of fertile to infertile egg masses deposited by the normal females was about the same as the ratio of normal to sterile males in the cage.

These laboratory experiments were followed by field trials on Sanibel Island off the west coast of Florida. The results supported the laboratory findings. When sterilized males were released at the rate of 100 per square mile per week, the fertility of egg masses laid by the normal flies declined from the 100 percent recorded prior to release to less than 25 percent. After the eighth week of release the original fly population on the island appeared to have been eradicated. However, in the twelfth week a single fertile egg mass was collected.

Since Sanibel Island is only about 2 miles off the coast and is part of a chain of islands, it is within the flight range of insects from untreated areas. Therefore, efforts were made to find a properly isolated screw-worm population for a test to determine whether such a population could be eradicated through release of sterilized flies.

Officials of the Curacao Veterinary Service had sent specimens for identification by the Entomology Research Branch and requested advice concerning the screw-worm problem on the island. Since correspondence indicated that screw-worms were abundant there, the senior author made a trip to Curacao to conduct a survey. As a result of that visit the island administration and the Entomology Research Branch agreed to cooperate in this field experiment.

Test Area

Curacao is an island of approximately 170 square miles about 40 miles off the coast of Venezuela. A constant trade wind blows out of the east. It was believed that the distance over water and the resistance of the wind were sufficient to prevent screw-worms from flying to Curacao from the mainland, and that if proper precautions were taken to prevent accidental importation of flies on cattle boats a valid experiment could be conducted.

Screw-worms were found to be extremely abundant on the island. They probably breed chiefly in wounds on neglected domestic animals. The island is estimated to have about 25,000 goats and 5,000 sheep, which roam through the brush and cactus largely unattended. Other potential hosts are a few hundred roving pigs, some neglected burros, about 300 deer, and numerous rabbits similar to our cottontails. There are also dairy cattle and horses on the island, but those animals, being fairly well cared for, are not considered important in breeding screw-worms.

The island is inhabited by about 115,000 people, most of whom work in town or for the oil refinery. Many of these workers live in the country and own small flocks of goats, which roam about in unfenced pastures. As there are no permanent streams on the island, during dry weather the goats must come to the houses for water except on those plantations having shallow wells in the pastures. Probably half the goats are seen daily when they come near the houses for water, but they do not receive much attention.

As the island is well served by roads, it was convenient to travel by automobile in conducting the field tests. There is an excellent airport. An airplane belonging to a flying club could be rented, and skilled pilots were available for distributing sterilized flies by air.

Estimation of Fly Activity

In this experiment procedures developed in the Florida work were followed. Experience had shown that fairly reliable estimates of fly activity could be made by exposing wounded goats to oviposition by flies in the natural population.

Goat pens were established in 11 locations distributed over the island, as shown in Figure 13–1. Each pen consisted of a fenced enclosure about 20 feet square, in which eight goats were maintained. Each week two goats in each pen were wounded and artificially infested with approxi-

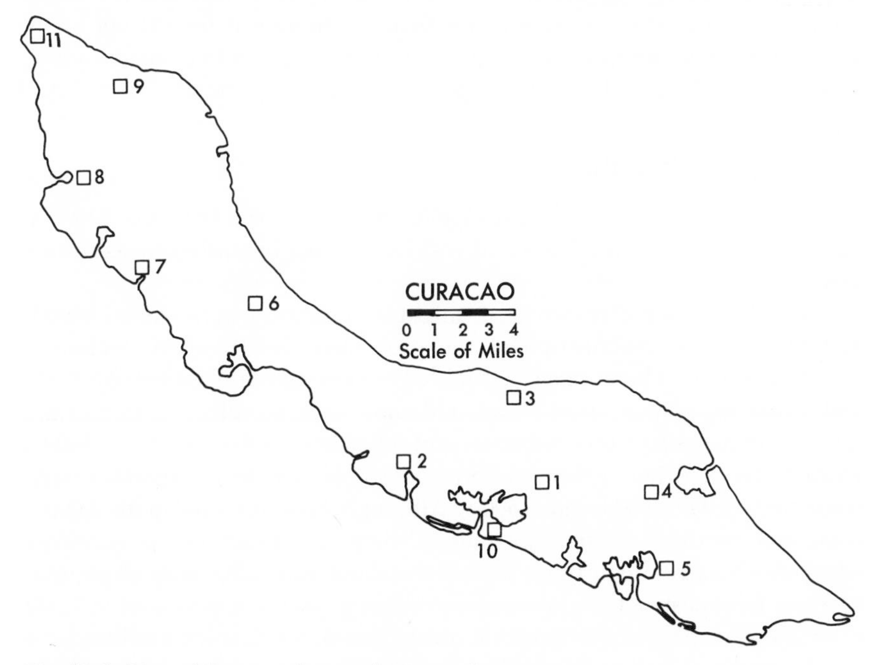

Fig. 13–1. Map of Curacao showing location of goat pens.

mately 100 newly hatched screw-worm larvae. When the larvae were 3 days old, the wounds were fumigated with benzol to destroy the larvae. The wounds were attractive to ovipositing flies from the time of infestation until several days after the benzol treatment.

At the beginning of the experiment in each goat pen there were always at least two fairly fresh wounds and two healing wounds to attract ovipositing flies. Later in the work infested goats were not kept in the pens to avoid the risk of escape. From that time goats were wounded and infested in a locked enclosure and only brought to the field after larvae had been removed from the wounds on the third day.

The goat pens were visited late in the afternoon each day for the purpose of collecting egg masses. Individual egg masses were removed and placed on moist filter paper and held in a petri dish for observations on hatching. Sometimes eggs were laid at a pen after the observer had made his visit. Early in the work, when fertile egg masses were common, such egg masses hatched and the shells usually fell off before the next inspection. The observer could recognize such missed egg masses by the presence of newly hatched larvae in the wound. Infertile egg masses, which did not hatch, remained intact.

Early in the experiment there were undoubtedly errors in estimating egg masses. If eggs hatched prior to a visit, one could only estimate the number of egg masses on the basis of the number of larvae he found in the wound. When egg masses were thickly deposited around a wound, the observer could not always count them accurately if he did not know whether a given clump of eggs represented one large or two small masses. However, it is believed that those sources of error were minor.

Rearing and Sterilization of Flies

Insects for release were reared and sterilized at the Orlando, Florida, laboratory. The flies used were of a strain originally started at Kerrville, Texas.

The larvae were reared on a mixture of ground horse meat, blood, and water by a modification of the procedure described by Melvin & Bushland (1941). The pupae were held at a constant temperature of 80° F., and when the pupae were 5 days old they were sterilized with gamma rays, using the cobalt-60 irradiation unit described by Darden *et al.* (1954). During the experiment three different doses of sterilizing radiation were employed—5,000, 6,000 and 7,500 r. Although flies sterilized with 5,000 r could not produce normal egg masses, they sometimes laid a very few eggs, which caused confusion in record taking when the normal population was near destruction. When the sterilizing dose was increased to 7,500 r, females incapable of ovipositing more than one egg were produced.

Immediately after irradiation the pupae were packed with a little excelsior in No. 1 kraft paper sacks, 130 insects in each sack, and shipped by air freight to Curacao, where they emerged in the sacks and expanded their wings on the excelsior provided.

Within a few hours after emergence was completed, the flies were distributed by airplane. The pilot flew at a height of about 400 feet in flight lanes 1 mile apart. An entomologist riding in the airplane tore open the sacks and dropped them through a pipe which extended below the fuselage. The bottoms of the sacks were smeared with a paste of sand and flour to weight them so that they would drop without drifting unduly. This method of dispersal seemed quite satisfactory, as flies dropped in intact sacks appeared unharmed by the fall.

In the first phases of the experiment the flies were not fed prior to release, but later it was thought desirable to provide the insects with food and water. This was done by packing the insects in sacks with wire-screen bottoms through which they could be fed.

Flies were distributed twice each week. On alternate flights the flight lanes were shifted so that the flies were distributed in lanes ½ mile apart each week.

The packaging rate of 130 pupae per sack was selected on the expectation that out of that number about 100 flies would emerge. Thus, when 50 males and 50 females were expected in each sack, sacks dropped at 1-mile intervals represented a distribution of 50 males per square mile. The actual percentage emergence was determined on sacks retained in the laboratory, and the exact rate of release was calculated subsequently. Thus the rate of fly distribution was regulated by the number of sacks dropped per mile in flight lanes 1 mile apart.

Effects of Releasing 100 Sterilized Males Per Square Mile Per Week

By March 17 ten goat pens were established over the island, and during the following week 133 fertile egg masses were recorded. Just what adult fly density such a figure represents is not known, but that rate of egg-mass deposition was seldom equaled in any experiment in Florida. Beginning on March 26, flies sterilized with 7,500 r of gamma rays were distributed. During the second week 155 egg masses, all of them fertile, were collected. The week of March 31 to April 6 represented a transition period, since some of the ovipositing flies might be young females that had mated since release of sterilized flies began, whereas others would be older insects that had already mated. During that third week 138 egg masses were collected, and of these 4 were sterile. Semiweekly releases of sterilized flies at the rate of 100 per square mile per week were continued until May 7.

TABLE 13–1. Numbers of fertile (F) and sterile (S) screw-worm egg masses collected in goat pens on Curacao from March 17 through May 18, 1954

Date	No. 1		No. 2		No. 3		No. 4		No. 5		No. 6		No. 7		No. 8		No. 9		No. 10		Totals		Per Cent Sterile
	F	S	F	S	F	S	F	S	F	S	F	S	F	S	F	S	F	S	F	S	F	S	
Oviposition before release of sterilized males																							
March 17–23	12	0	37	0	8	0	6	0	13	0	11	0	8	0	14	0	15	0	9	0	133	0	0
24–30	9	0	32	0	8	0	8	0	18	0	20	0	15	0	21	0	16	0	8	0	155	0	0
Transition period																							
March 31–April 6	11	0	48	1	7	1	2	0	13	2	20	0	7	0	19	0	6	0	1	0	134	4	3.0
Oviposition following release of sterilized males																							
April 7–13	5	2	26	2	9	4	12	4	12	1	14	3	4	0	10	0	4	0	8	1	104	17	14.0
14–20	18	6	24	8	30	8	11	2	9	4	25	4	14	4	16	5	22	4	14	1	183	46	20.1
21–27	22	2	8	1	35	5	6	0	6	1	24	1	21	4	30	7	28	4	5	3	185	28	13.1
April 28–May 4	27	1	11	2	23	3	24	3	13	1	15	0	4	1	23	6	14	2	8	4	162	23	12.4
May 5–11	26	6	20	4	39	7	6	3	11	2	24	0	8	4	26	1	33	3	7	9	200	39	16.3
12–18	41	7	29	5	23	8	9	1	11	3	19	1	34	2	40	7	30	3	4	0	240	37	14.4

The records on egg masses are summarized in Table 13–1. The data are remarkably uniform. They definitely show that some of the released males mated with females of the native population, but since egg-mass collections increased rather than declined it appeared that the proportion of sterile matings was not sufficient to cause a decline in the population.

Releasing sterilized males at the rate of approximately 100 (actually 99) per square mile per week along with an equal number of sterile females caused about 15 percent of the native flies to lay egg masses that failed to hatch. It should not be concluded that those sterile males represented only 15 percent of the male population. Undoubtedly many males died soon after distribution before they could find water and food. In observations in the United States normal males lived longer in the laboratory than did sterilized males, and this was probably true in Curacao. Therefore, it should not be assumed that, because only about one-sixth of the Curacao females laid infertile eggs during the period of release at the rate of 100 males per square mile per week, there were about 500 normal males per square mile in the normal population. Probably the normal population was much smaller but merely lived longer and competed more effectively for mates than the released flies.

Effect of Weather on Fly Activity

From the middle of May until July 12 there were no regular releases of sterilized flies on the island. During the first month some special tests were performed which indicated that males sterilized with 7,500 r mated as efficiently as those treated with 5,000 r. During the next 4 weeks no flies were released. Also during this time the eleventh goat pen was established so that the northwest end of the island could be adequately sampled for egg-mass data. From mid-May to mid-July the egg-mass collections indicated a natural decline in the fly population, as collections dropped from 188 to 70 per week.

The decline in the fly population is attributed to the dry weather. After the first week in May there was almost no rainfall until the middle of June. After June 14 rains were frequent, and for the remainder of the year the weather was very favorable for screw-worms, so that any further decline in the fly population could be attributed to sterile matings. In previous years screw-worms had been observed to increase from July to January.

Comparison of 100 and 400 Rates of Release

For 4 weeks beginning July 12 a comparison was made between two rates of releasing males. In the southern half of the island the flies were released twice weekly at the rate of 100 per square mile per week and in the northern half at 400 per square mile per week. When emergence figures were calculated from final laboratory records, the actual releases were found to be 107 and 450 per square mile per week.

From the results of this experiment, shown in Table 13–2, it will be

TABLE 13–2. A comparison of releasing sterilized male screw-worm flies at rates of about 100 and 400 per square mile per week

	100 Males per Week		400 Males per Week	
Date	Average Number of Egg Masses per Pen	Per Cent Sterile	Average Number of Egg Masses per Pen	Per Cent Sterile
July 19–25	10.5	30.2	16.0	37.5
July 26–Aug. 1	9.5	33.3	11.8	55.9
Aug. 2–8	10.0	28.3	6.0	53.3
Average	10.0	30.6	11.3	48.9

noted that the lower rate was more effective than it was in the spring when the natural population was higher. In the first experiment (Table 13–1) releasing at the rate of approximately 100 per square mile caused 22 percent sterility among egg masses collected on the southern half of the island. In the second experiment the sterility from a similar release rate was 31 percent, and from the 400 release rate it was 49 percent.

The record of 49 percent sterility of egg masses collected on the northern half of the island indicated that such a reduction in breeding potential should cause a decline in the fly population.

Eradication Experiment

Beginning on August 9 sterile flies were released over the entire island at the approximate rate of 400 males plus 400 females per square mile per week. Attaining this release rate required the actual release of 68,000 of each sex over the entire island, or 136,000 flies per week. It was necessary to ship twice that many flies to compensate for losses due to conditions beyond our control. Occasionally the insects emerged en route, either because the shipments were delayed or because the packages were exposed to unusually high temperatures. Rearing, irradiating, and packing the insects at the rate of 130 pupae per release sack strained the capacity of the Orlando laboratory. Some weeks it was impossible to meet the release quota because of rearing or shipping mishaps. In order to attain the desired average it was necessary to release practically all the insects that reached Curacao in good condition. For that reason weekly releases varied greatly. During the first 8 weeks, in which screw-worms were practically eliminated from Curacao, the weekly releases ranged from 175 to 701 males per square mile and the average was 435. The figures for each week's release are shown in Table 13–3.

TABLE 13–3. Egg mass records during release of sterilized males over island of Curacao

Date	Males Released per Square Mile	Total Egg Masses in 11 Goat Pens		
		Fertile	Sterile	% Sterile
Aug. 9–15	491	15	34	69
16–22	224	17	38	69
23–29	175	17	36	68
Aug. 30–Sept. 5	381	10	37	79
Sept. 6–12	451	7	42	86
13–19	701	3	23	88
20–26	450	0	10	100
Sept. 27–Oct. 3	607	0	12	100
Oct. 4–10	484	0	0	—
11–17	727	0	0	—
18–24	577	0	0	—
Oct. 25–31	400	0	0	—
Nov. 1–7	400	0	1	100
8–14	485	0	1	100
15–21	393	0	0	—
22–28	400	0	0	—
Nov. 29–Dec. 5	421	0	0	—
Dec. 6–12	498	0	0	—
13–19	313	0	0	—
20–26	265	0	0	—
Dec. 27–Jan. 2	400	0	0	—
Jan. 3–6	188	0	0	—

From October 3 until November 4 no egg masses were collected, but a single infertile egg mass was taken at goat pen No. 6 on November 4 and another at goat pen No. 4 on November 11. After November 11 there was no oviposition on the island detectable by the sampling procedure.

Under weather conditions prevailing on Curacao, about the shortest time for a complete life cycle from egg to egg is 3 weeks, and 4 weeks is about the average. The egg-mass data show that fertile flies were active until mid-September. As there are about 25,000 goats and 5,000 sheep roaming on the island, it would be expected that, if fertile females were ovipositing on observed wounds at the goat pens, many more should be ovipositing on unattended wounds in nature. Flies ovipositing on neglected sheep or goats or on wild mammals should have produced progeny to oviposit a month later in mid-October. The fact that no egg masses were collected after October 3 is a good indication that the native fly population was greatly reduced, but it also suggests that the 11 goat pens under observation did not provide an adequate sample of a declining fly population when the number of ovipositing females was very small.

The females that oviposited on November 4 and 11 probably were the progeny of flies ovipositing early in October. It is possible, but not likely, that the two egg masses were laid by flies 3 or 4 weeks old and hence from parents that oviposited in early September. Since the records were taken at widely separated pens, they probably represented different centers of survival.

Maintenance of the goat pens and systematic releases of sterilized flies were continued through January 6, 1955. Thus, observations were made for 8 weeks after the last record of oviposition. This was ample time for two complete life cycles.

It is quite probable that the two flies that oviposited on November 4 and 11 had brothers and sisters active at the same time. It seems, though, that there were no fertile eggs laid on animals in nature in November, as 6 months have elapsed (May 1955) without further evidence of normal flies.

In addition to taking the egg mass records at goat pens, livestock owners were questioned regarding the occurrence of screw-worm cases in their animals. Very few cases of myiasis were reported after October 1, and every one proved to be due to an infestation of *Callitroga macellaria* (F.) rather than *hominivorax*. This checking with livestock owners is being continued by one of us (B. A. Bitter), but to this writing (May 1955) there has been no evidence of renewed screw-worm activity.

During the course of this experiment there was an obvious increase in the number of surviving kid goats. At the beginning of the study less than half the nannies had kids and twins were seldom seen, but at the conclusion most of the nannies had kids and twins were common. Livestock owners remarked that it had been their practice to seek out newborn kids for treatment with larvicides to prevent almost certain death through navel infestation. As the screw-worm incidence declined the owners abandoned that practice and all remarked at the great increase in the kid crop.

Conclusion

It appears that screw-worms were eradicated from the island of Curacao through the release of sterilized flies. The success of this experiment strongly supports the theory that it may be practical to eradicate screw-worms from the Southeastern States by this method.

Summary

To test whether an isolated population of screw-worms, *Callitroga hominivorax* (Cqrl.) might be eradicated through the release of sterilized flies, experiments were made on the island of Curacao.

Screw-worms were reared in the laboratory at Orlando, Florida. When the pupae were 5 days old they were sterilized with gamma rays and shipped by air freight to Curacao. The adults emerged 2 days after irradiation. Within 24 hours of emergence the insects were distributed by airplane over the island.

When sterilized flies were released at the rate of approximately 100 of each sex per square mile per week, some of the males mated with females of the normal population, as indicated by the deposition of infertile egg masses on wounded goats.

Egg masses collected from wounded goats prior to release of sterilized flies hatched 100 percent.

During April and May, when weather favored fly activity, releasing sterilized males at the rate of about 100 per square mile per week caused 15 percent of the females to lay infertile eggs on wounded goats in pens. This rate of sterility was not sufficient to cause a reduction in the fly population.

At the end of the dry season in July and August releasing at 107 per square mile per week over half the island caused 30.6 percent sterility, and 450 per square mile per week on the other half caused 48.9 percent sterility.

From August 9 to October 3 the entire island was treated at a rate that averaged 435 males per square mile per week. Although frequent rains made conditions suitable for fly increase, the releases were highly effective. In 8 weeks the percentage of sterile egg masses increased from 69 to 100 and the population declined more than 99 percent. The releases were continued. Except for two sterile egg masses collected after 12 and 13 weeks, there was no evidence of normal fly activity after the eighth week of releases. The releases and observations were discontinued 8 weeks after the last egg mass was collected.

The apparent eradication of screw-worms from Curacao supports the theory that screw-worms may be eradicated from the Southeastern States by releasing sterilized flies.

REFERENCES CITED

Bushland, R. C., and D. E. Hopkins. 1951. Experiments with screw-worm flies sterilized by x-rays. Jour. Econ. Ent. 44:725–31.

Bushland, R. C., and D. E. Hopkins. 1953. Sterilization of screw-worm flies with x-rays and gamma-rays. Jour. Econ. Ent. 46:648–56.

Darden, E. B., E. Maeyens, and R. C. Bushland. 1954. A gamma-ray source for sterilizing insects. Nucleonics 12(10):60–2.

Melvin, Roy, and Raymond C. Bushland. 1941. The nutritional requirements of screw-worm larvae. Jour. Econ. Ent. 33:850–2.

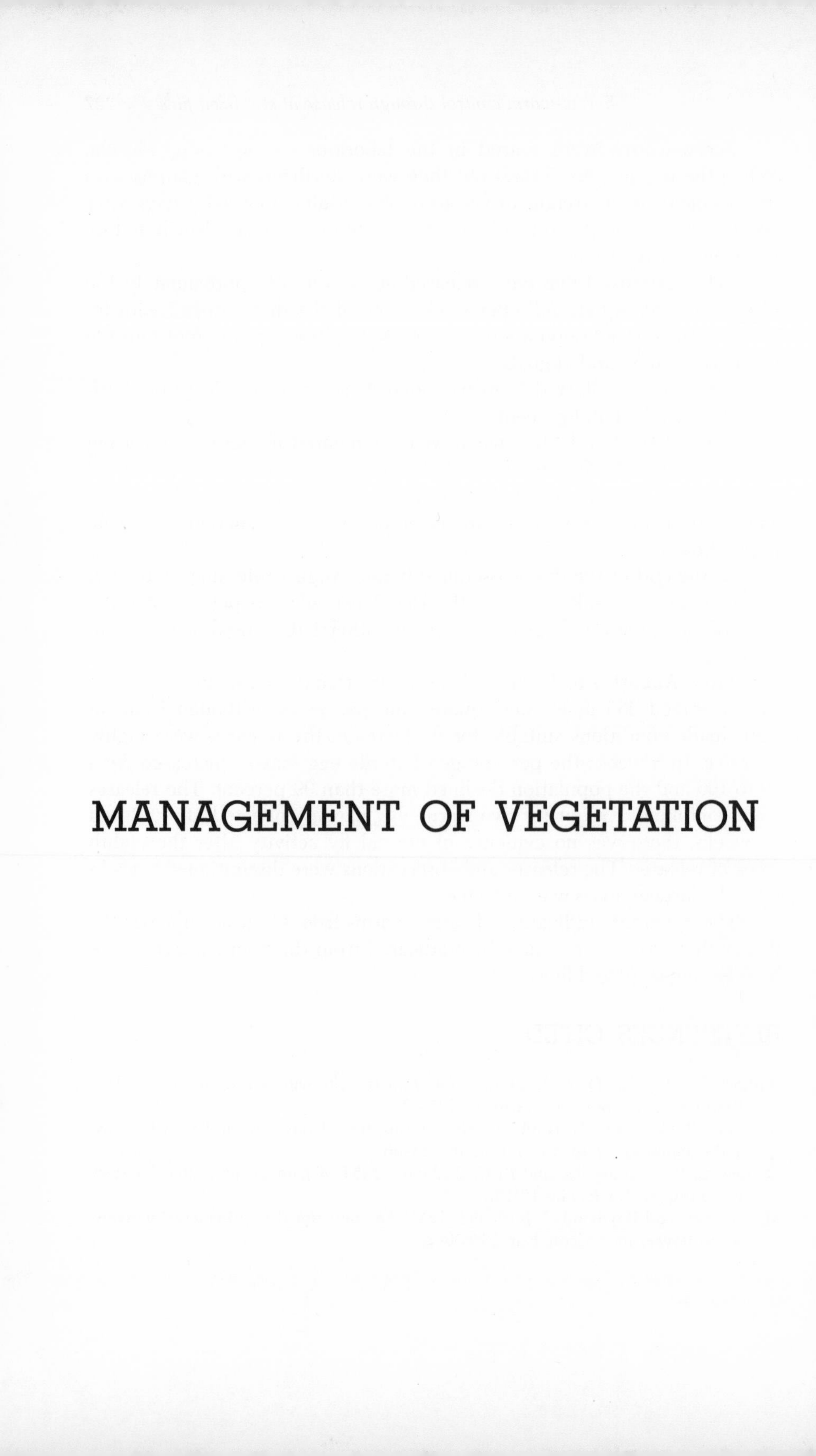

MANAGEMENT OF VEGETATION

14

Preserving vegetation in parks and wilderness

Edward C. Stone

Federal efforts to preserve natural vegetation go back to 1872, when Yellowstone National Park was carved out of the public domain; state efforts go back to 1885, when the New York Adirondack Forest Preserve was established (1). All efforts, however, have been largely unsuccessful because of a failure to appreciate fully that vegetation is a living, dynamic complex and cannot be preserved in the sense in which a building or an archeological site can be preserved. Even the most uniform vegetation is a mosaic created by local variations in the environment and by prior events such as fire, drought, and insect infestation. When a mature plant dies, hundreds of seedlings spring up to take its place, some or all of which may be of different species. Which seedlings survive, and for how long, depends upon their relative growth potential, what effect the dead plant had on its environment before it died, and what kind of environment resulted when it died. Vegetation can only be preserved by controlling the complicated successional forces that have created it and that, if unchecked, will in turn destroy it.

The very efforts made to preserve a natural system of vegetation may bring on unplanned and undesired changes in it. That steps taken to preserve animal wildlife may have this effect is well known to the general public. By 1930 there were overpopulations of elk and bison in Yellowstone National Park, of mule deer in Zion National Park, and of deer and

Reprinted with permission from *Science*, 150(3701):1261–1267 (3 December 1965).

elk in Rocky Mountain National Park, all brought about by control of predators in and around the parks (2). Recognition of the problem led to a reconsideration of these practices, and today, although hampered by a lack of basic data and a restrictive budget, specialists in wildlife preservation are employed in the national parks to plan and apply sounder regulatory methods. While not so dramatic and not so widely publicized as imbalances in wildlife populations, drastic changes in the composition of many of the plant communities in the national parks have occurred during the last 50 years under fire-protection policies and heavy concentrations of use. In a number of cases these changes have progressed so far that even the once dominant plants in a wide variety of plant communities have been replaced, and now trees and shrubs occupy slopes and meadows once clothed in grass and sedge (3).

There are two federal agencies largely responsible for the management of national wildlands, each by charter concerned with conservation of this resource but each with different primary objectives. The Forest Service was organized in 1905 within the Department of Agriculture to manage the forest reserves—later renamed national forests—to secure favorable watershed conditions and to furnish a continuous supply of timber. Shortly thereafter, however, the Forest Service recognized that recreation was an important use of these areas compatible with its other uses, and began developing the recreational facilities that now serve 125 million visitors a year. Some 15 years later it recognized the need for wilderness reserves, and by appropriate administrative action over the next several years set aside almost 12 million acres for this purpose. Subsequently both of these administrative decisions have been sanctioned by congressional action (4).

The Park Service was organized in 1916 within the Department of the Interior to bring together under one administrative head a number of independent national parks formerly administered by several federal departments. It was specifically charged with preserving on these lands plant and animal life, and geological and archeological features of national value, for the enjoyment of the public. Vegetation preservation thus constitutes a minor part of the Forest Service's responsibilities but a major part of the Park Service's responsibilities.

The Forest Service has moved ahead rapidly in meeting its responsibilities as watershed and timber manager and purveyor of recreation facilities. It has been able to do so for a number of reasons. From its inception it was able to staff its key administrative posts with men trained for the job of managing forests for watershed and timber; it could draw upon a wealth of European experience, and it had an excellent research staff engaged in developing workable silvicultural techniques, based upon sound understanding of ecology, for use by its foresters operating in the field. In its minor role as vegetation-preservation manager of 12 million acres of wilderness, the Forest Service has yet to do much of anything.

The Park Service has moved slowly in its major role, that of preservation manager, although it has successfully operated the land under its control for the enjoyment of the public. When established, the Park Service, unlike the Forest Service, had no ready source of professional help to which it could turn. There was no such thing as a vegetation specialist versed in preservation management—that is, a vegetation-preservation manager. Furthermore, administrators could not rely upon European experience for guidance, because there was none. Nor could they turn to a research staff for developing the necessary management techniques, because there was none. To make matters even more difficult, they were forced almost from the beginning to fight a rear-guard action with private companies and government agencies that wanted to open park lands for mining, hunting, water impoundment behind massive dams, logging, and other commercial activities. Thus administrative energies and funds were all but exhausted in maintaining existing park boundaries; and the problem of preserving a variety of undescribed ecosystems, in which changes at the time were not well advanced or readily apparent to the untrained eye, was largely solved as far as the administrator was concerned once an efficient fire-control system had been established, livestock excluded, and insect epidemics brought under control. This does not imply that the Park Service has failed to attract competent biologists to its professional staff and that there has been no effort to stem the successional tide; this is not true. Characteristically, however, individuals in the Park Service who have been trained as biologists have been called upon more for protective and interpretive service than for specialized management of the vegetation complex, because overall park policies have not until recently included the concept of vegetation management except in the narrow aspects of fire, insect, and disease control.

In 1963, public attention was drawn to this state of affairs in the national parks by a report of a committee appointed by the National Academy of Sciences at the request of Secretary of the Interior Udall to aid in "the planning and organizing of an expanded research program of natural history research by the National Park Service." The committee presented "the pressing need for research in the national parks by citing specific examples in which degradation or deterioration has occurred because research on which proper management operations should have been based was not carried out in time; because the results of research known to operational management were not implemented; or because the research staff was not consulted before action was taken" (5). The report stresses the need to develop, by means of extensive research, the ecological basis for managing the preservation of both plant and animal life in the ecosystems involved. "It is inconceivable," the authors remark, "that property so unique and valuable as the national parks, used by such a large number of people, and regarded internationally as one of the finest examples of our national spirit should not be provided adequately with competent research scientists in natural history as elementary insurance for the preservation and best use of the parks." This is an excellent, long overdue report, exhaustive in its treatment of research needs, but it did not in my opinion focus strongly enough on the need for professional vegetation specialists at the operational level, specialists who will be responsible for carrying out the manipulative steps recommended by the research staff, much as the Forest Service professional foresters carry out the cutting practices recommended by its research staff.

The various state agencies that administer park and forest preserves with preservation as the stated objective have come into existence at different times in response to different pressures and differ widely in the composition of their administrative staffs. All engage in protection of some kind, but none, as far as I can ascertain, is involved in manipulative procedures to preserve the integrity of specific types of vegetation. Some ecological research is under way, but again there are no professional vegetation specialists available to carry out the manipulative program that will come out of this research.

THE OBJECTIVES OF PRESERVATION

Since vegetation is never static, preservation must consist, in effect, of managing change. Consequently, it is necessary to determine exactly what the objective is, and thereby to determine how much change, and what kind, can be tolerated.

One of the most common objectives is to keep park lands green or, in the arid West, green and golden. Fire protection is considered the principal means to this end. The fact that vegetation protected from fire may change completely in a relatively short period has rarely been considered, because administrators and the public have not appreciated that this can happen.

Probably the next most common objective is the preservation of certain favored dominant species within the vegetative complex. When this is the objective, the fact that certain successional stages may be fast disappearing and that the overall vegetative structure may be changing within rather broad limits is usually ignored, as long as the dominants remain dominant and the general appearance of the landscape is not altered. Change that does not interfere with the effective display of the dominant species has, in general, been acceptable. Preservation of the redwood tree, for example, has been the sole objective in the world-famous string of redwood parks that extends from south of San Francisco

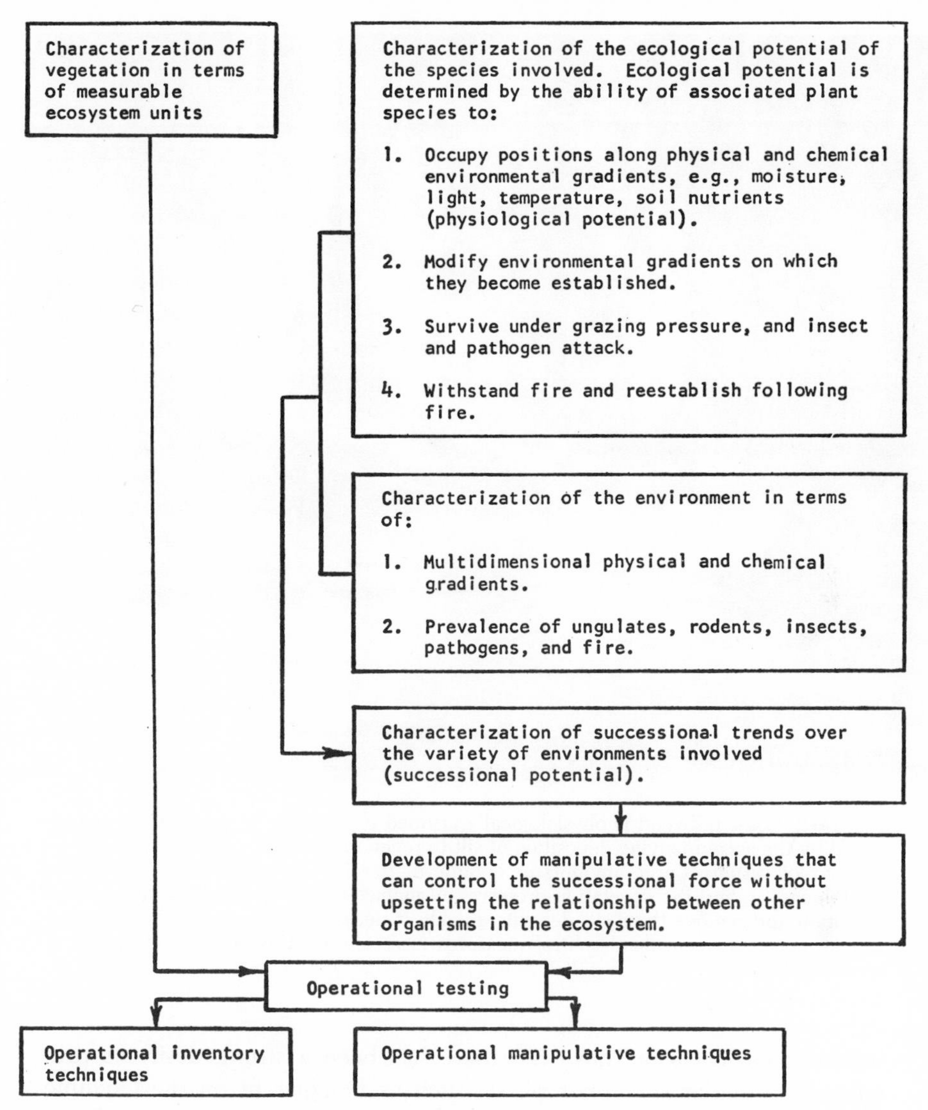

Fig. 14–1. Information required to develop manipulative techniques for vegetation preservation.

to the Oregon border. Consequently, change among associated species has been ignored.

A less common but still popular objective is the preservation of particular successional stages. Most often this objective has reflected a desire to preserve a piece of virgin forest or native grassland and generally has involved a *climax* phase of vegetation, that is, a condition of dynamic equilibrium in which species composition remains more or less

Fig. 14–2. Coast redwood is physiologically attuned to periodic inundation and is provoked by the accompanying deposition of silt to regenerate a new, possibly more active, feeder-root system. Large amounts of organic matter incorporated in the silt can be fatal, however, and the vegetation-preservation manager must be ready to test for this condition and remove these silt deposits quickly if necessary. What steps he will need to take once upstream flood-control measures have altered this peculiar environment are not yet clear.

constant. On occasion, however, there has been a strong desire to preserve a particular *subclimax* phase, such as Douglas fir on the Olympic Peninsula, red pine in the Lake States, and Caribbean pine in the Everglades, where successional change may be proceeding rapidly. In support of this type of preservation, a committee appointed by Secretary Udall to review wildlife-management practices in the national parks has recently recommended that certain successional stages be re-created and maintained in order to present "vignettes" of early America to the park visitor (6). The ease with which various successional sequences can be maintained varies with the area and the type of vegetation involved. Consequently, preservation may not be as cheaply achieved in one area as in another. Provided he is aware of this, the park manager can select appropriate areas and achieve his objectives with a minimum effort.

A closely associated objective is that of slowing succession. Supporters of this objective argue that it is futile to try to stop succession, that if it can be slowed sufficiently a vegetation mosaic containing most of the successional stages could be maintained, and that such a mosaic is what we should strive for in the national parks. Certainly the degree of preservation desired is always an important consideration. Succession can often be slowed for only a few cents per acre, while costs of stopping succession can run to several hundred dollars per acre.

Today, many wilderness supporters argue that we should leave large areas of vegetation alone to change as they may; that man should keep his hands off and let nature run its course, unimpeded by controls against fire, insect, and disease. When pressed on the point of fire control, however, proponents of this policy have usually agreed that some fire control

Fig. 14–3. Fire and browsing by deer (see trees in left foreground) have been important in the creation and maintenance of grass prairies in the redwood-Douglas fir forest along the north coast of California. The vegetation-preservation manager can use fire but is not dependent upon it to maintain this open, park-like intermingling of grassy glades and trees. A variety of selective herbicides – none of which enter the biotic food chain – is available and can be used effectively in conjunction with either spot burning or mowing.

is reasonable, provided it does not interfere with the occurrence of natural fire. Accordingly, lightning fires would be allowed to run unchecked, and if the aboriginal arsonist were alive today he would not be discouraged because he would be a part of the natural environment. Paradoxically, fire started by a careless camper would be dealt with vigorously.

On most of the areas that might be affected by such a program, succession is extremely slow and, because of extensive areas of exposed rock, wildfires soon burn themselves out. With sufficient control of human use these areas will change little in the next 100 or even 200 years, and this is probably what most proponents of a hands-off policy visualize. There are other areas, however, where the understory is now very dense and highly inflammable throughout much of the year and not in the condition that prevailed when the areas were set aside. Uncontrolled wildfire would be catastrophic. Thus, in the absence of fire control, vegetation in one area would be maintained more or less as it is today for many years to come, while in another area it might be violently changed within the next few years.

COMPATIBILITY OF OBJECTIVES

Many preservationists consider management per se to be an unwarranted interference with nature by man. This need not be true. Management consists merely of those actions that are necessary to achieve one or more objectives, whatever they may be, even if the objective is "no management." Management dealing with vegetation may be intensive or extensive, depending upon the objectives, but unless the objectives are thoroughly outlined effective management is impossible. Because vegetation preservation may be only one of several objectives, all must be carefully considered together to determine whether they are compatible. Intensive public use may be compatible with a general policy of keeping certain park lands green, but may be incompatible with a specific policy of preserving dominant species or particular stages in a successional sequence. Probably incompatible objectives are much more widespread in current efforts to preserve vegetation than is generally recognized, because change in vegetation can proceed for many years without detection by the public or even by the park administrator responsible for its preservation.

RESEARCH ON VEGETATION PRESERVATION

The National Academy of Sciences Advisory Committee, in reviewing the research program of the National Park Service, "was shocked to learn that for the year 1962 the research staff (including the chief naturalist and fieldmen in natural history) was limited to ten people and that the Service budget for national history research was $28,000—about the cost of one camp-ground comfort station" (5). If we consider only the magnitude of the research job required to support a realistic vegetation-preservation program, it is easy to understand why the committee was shocked. A million-dollar annual budget and a staff of several hundred scientists, with several times as many supporting personnel, are needed. The Yellowstone National Park staff, for example, has indicated (5) that the research required to support its vegetation-preservation program would entail an analysis of the current climatic trends; a detailed soil survey: and analyses of the vegetative mosaic and the factors creating it; successional patterns in the various biotic communities; the interrelationships of plants and animals, particularly dominant species like ungulates; variations of current ecological conditions from the original; the factors that have caused these deviations; the practicability of re-creating original ecological conditions where ecological damage or deterioration, for instance, soil loss, has occurred; and the direct effect of visitors on important natural features. Thus, dealing effectively with the problem of vegetation preservation in this one park, only one of 31 national parks, will require a dozen or more scientists—climatologists, pedologists, and ecologists of various specializations—with a supporting staff of perhaps a hundred or more.

The Forest Service has not as yet committed its research staff to studying the overall problems of vegetation preservation, but much of what its silviculture and range-management researchers have discovered over the last 35 years is directly applicable. Both the silviculturist and the range manager, like the vegetation-preservation manager, are interested in successional processes and their control. The distinction lies in the end products desired and the tools that can be used to obtain them. The silviculturist is interested in the amount and quality of timber produced. He selects trees to this end and in the process completely alters the structure of the vegetation units involved. His imprint in the form of skid trails, neatly sawn stumps, and extraction roads is everywhere apparent. The range manager is interested in the weight and quality of beef or mutton that vegetation produces. He uses his animals, through rotational grazing schemes of various kinds, to control plant succession. Except for the presence of domestic animals, fences, and occasional scars of a disc harrow on a reseeded, overgrazed range, his mark is not apparent and the structure of the vegetation units involved may not be greatly altered.

What is most needed to get a full-scale vegetation-preservation research program underway by the Forest Service is an administrative decision to do so. Only a few shifts in research emphasis at key points within the present research program, along with a relatively modest augmentation of the basic research staff, are needed. Once embarked upon such a program, the Forest Service soon would be able to develop suitable operational techniques for preserving vegetation on the 12 million acres of wilderness that are its responsibility.

All the state parks involved in vegetation preservation need research support, but few are receiving it. The State Division of Beaches and Parks in California, for example, has an annual support budget of $10,000,000 and is responsible for vegetation preservation on more than 600,000 acres. The size of the individual parks varies from a few hundred to several thousand acres, and the type of vegetation to be preserved varies through cactus and scrub on the Mohave Desert, oak woodland in the Central Valley, mixed conifers in the Sierra, and redwoods along the North Coast. Its annual research budget amounts to only $28,000. A third of this is being spent on a crash program to develop recommendations for preserving redwood groves along the Eel River which are subjected to periodic flooding and to more than 500,000 visitors annually. The rest is being spent by the interpretive-services section. Nothing is being spent on research to determine how the variety of vegetation types that occur in the other parks in the state should be maintained.

The type of information required to develop manipulative techniques for the preservation of vegetation is summarized in Figure 14–1. Some of this information, obtained through the efforts of university-based scientists, their graduate students, and Forest Service researchers, is already available for certain types of vegetation. Rarely, however, will this information be complete from a vegetation-preservation viewpoint, partly because preservation has not generally been the objective of the studies, but mostly because studies of plant succession in this country have closely followed an approach developed by the well-known ecologist F. E. Clements (7). Clements was convinced that successional sequences, which involve changes with time, can be determined by observing changes in vegetative patterns in space, and through his persuasive pen he was able to convince others that this was feasible. The major difficulty encountered in this approach has been that the only method of evaluating the accuracy of a researcher's estimate of successional trends, that is, his first approximation, has been to wait and see.

Where short-lived grasses and herbs have been involved and reestablishment of a dynamic equilibrium following disturbance has been rapid, the Clementsian approach has been reasonably effective. Researchers have been able to modify their first approximation through several subsequent approximations, improving the accuracy of their estimate each time. This is the approach that largely has been used by range-management researchers. Recently, however, they also have experienced its limitations and have begun to turn to detailed environmental analyses and growth-performance studies under controlled environments.

Fig. 14–4. Big cone spruce, shown against the sky line, is not a fire-resistant species but has survived in the chaparral of Southern California because of natural firebreaks created by shallow soils and the regular occurrence of widespread fires in the past. Today these natural firebreaks are overgrown and no longer offer protection. The vegetation-preservation manager must reestablish these firebreaks if stands of big cone spruce are to be preserved.

Where long-lived plants have been involved and reestablishment of a dynamic equilibrium following disturbance has been slow and will not be reached for another hundred years or so, first approximations based on Clements' approach are often little better than educated guesses. In dealing with vegetation of this type a more sophisticated analysis of the ecological potential and the relative magnitude of the environmental factors involved will be required (Fig. 14–1). Studies of comparative growth performance of associated species in controlled environments along the lines suggested by Hellmers (8), and environmental-gradient

Fig. 14–5. Top aerial photograph was taken shortly before the grass-covered slopes in the right foreground were purchased by a local park district and cattle were excluded. Bottom aerial photograph was taken early in 1965 – 35 years later – and shows the extent to which brush has replaced grass within the park boundaries. If the grass cover is to be preserved, the vegetation-preservation manager must remove the brush and take steps to keep it out. Fire as a tool has been ruled out by local smog-control officials. Introducing cattle is difficult because of the number of youth groups using the area. Bulldozers, herbicides, and mowing machines appear to be the alternative tools available.

analyses along the lines suggested by Whittaker (9), Bakuzis (10), and Waring and Major (11), are essential.

SPECIALISTS IN VEGETATION-PRESERVATION MANAGEMENT

The National Academy of Sciences Advisory Committee (5) points out briefly that the Park Service has applied research in a piecemeal fashion and has "failed to insure the implementation of the results of research in operational management." The committee concludes with the comment that "Reports and recommendations on the subject will remain futile unless and until the National Park Service itself becomes research-minded and is prepared to support research and to apply its findings." That the "implementation of the results of research" calls for experts on the management of vegetation has as yet not been recognized. Even preservation-oriented conservationists, who are the backbone of the leading conservation groups in this country, have been slow to perceive this. Many of them still regard vegetation much as they do their own gardens and are quick to suggest how a particular vegetative cover can best be preserved, whether it be in the local nature reserve, in a state park dominated by 1000-year-old redwood trees, or in an untrammeled wilderness.

Obviously the decision as to what should be preserved cannot be left entirely to the specialist. The concerned public, although amateurs in vegetation preservation, must be heard and heeded. But at the same time a realistic assessment must be made of what can be achieved at costs commensurate with public interest, and this depends upon a knowledge of various alternatives and the relative cost and feasibility of achieving each one. Only the vegetation specialist can furnish this kind of information.

The vegetation-preservation specialist must be trained in management, must possess a thorough knowledge of ecology, must be experienced in assessing the relative growth potential of each species in the vegetative mosaic, must be experienced in the use of various manipulative techniques, and must understand research methods. Today there are few men so qualified. There is an impressive number of competent plant ecologists scattered throughout related professions who are oriented toward management, but there are relatively few who have had experience in a detailed assessment of the environmental complex, and even fewer who have had experience in manipulative techniques.

The vegetation-preservation specialist will not replace the research ecologist and to a large extent will be dependent upon him. He must be competent to understand research, to evaluate research findings in terms of his management function, and to translate research into manipulative techniques particularly suited to the specific vegetation he must manage. These manipulative techniques must be based on an understanding of the ecology of the vegetation in question; if such information is not available and ecologists are not employed to develop it, the preservation specialist will be forced to forego his primary responsibility and to spend his time collecting basic ecological data.

Because success in the field of vegetation preservation requires several—usually many—years to evaluate, the vegetation-preservation specialist often will operate in an atmosphere in which unsubstantiated opinions are forcefully urged. Many fire enthusiasts, for example, are convinced that fire protection should be curtailed, and do not recognize that merely because fire control has led to some undesired effects it does not necessarily follow that fire control should be abandoned or prescribed burning introduced. Involved is the whole process of recognizing the management objective, evaluating the ecological forces in play, identifying the conditions which must be achieved to develop the desired vegetation response, and, finally, evaluating all the possible ways of moving toward those conditions economically and with a minimum of unwanted side effects. In all of this the vegetation-preservation specialist will need a fine sense of perspective.

Little can be accomplished in the field of vegetation-preservation management until a source of competently trained specialists has been developed—and perhaps not until considerable numbers of these specialists have infiltrated the various responsible administrative bodies. How can we develop such a source? At the moment I can see only one solution: Ask those universities that have strong programs both in ecology and in land management, for example, those with forestry and range-management curricula, to take on the job. It should be possible to train these specialists by means of a 2-year graduate program, provided it is preceded by an undergraduate degree with a proper emphasis on basic biology and is followed by an appropriate period of apprenticeship. Several universities could readily meet this challenge, provided financial support were assured. The question that remains to be answered is: How soon will the universities that have staffs capable of carrying out this graduate program be asked to join in creating this new profession?

REFERENCES

1. Yellowstone National Park Establishment Act, 17 *Stat.* 32 (1872); S. W. Allen, *An Introduction to American Forestry* (McGraw-Hill, New York, ed. 2, 1938).
2. J. Ise, *Our National Park Policy* (Johns Hopkins Univ. Press, Baltimore, 1961).
3. R. P. Gibbens and H. F. Heady, "The influence of modern man on the vegetation of Yosemite Valley," *Calif. Agr. Exp. Sta. Manual No. 36* (1964); Univ. of Calif. Wildland Research Center, *Outdoor Recreation Resources Rev. Comm. Study Rept. No. 3* (Government Printing Office, Washington, D.C., 1962).
4. Multiple Use Sustained Yield Act, 86 *Stat.* 517 (1960); The National Wilderness Preservation System Act, 78 *Stat.* 890 (1964).
5. National Academy of Sciences–National Research Council, *A Report by the Advisory Committee to the National Park Service on Research* (Washington, D.C., 1963).
6. A. S. Leopold, S. A. Cain, I. N. Gabrielson, C. M. Cottam, and T. L. Kimball, *Living Wilderness* 83, 11 (1963).
7. F. E. Clements, "Plant succession," *Carnegie Inst. Wash. Publ. No. 242* (1916).
8. H. Hellmers and W. P. Sundahl, *Nature* 184, 1247 (1959).

9. R. H. Whittaker, "Vegetation of the Great Smoky Mountains," *Ecol. Monographs* 26, 1 (1956).
10. E. V. Bakuzis, thesis, Univ. of Minnesota (1959).
11. R. Waring and J. Major, "Some vegetation of the California coastal redwood region in relation to gradients of moisture, nutrients, light, and temperature," *Ecol. Monographs* 34, 167 (1964).

15

Vegetation changes on a southern Arizona grassland range

R. R. Humphrey
L. A. Mehrhoff

INTRODUCTION

The problem

As a result of the advent of the white man into southwestern United States and the consequent introduction of cattle and sheep, most of the desert grassland ranges now produce much less forage than they once did. A comparison of past vegetal composition with that of the present shows one startling fact: the almost complete dominance today of noxious shrubs over many millions of acres of range land that were formerly grass.

In this study certain vegetation changes that have taken place on a desert grassland range are discussed and the more important possible causes for these changes are evaluated. The factors evaluated include: climatic changes, grazing by domestic livestock, rodents and fire.

Description of study area

The Santa Rita Experimental Range, where this study was made, contains approximately 53,000 acres typical of the semidesert mixed grass and shrub ranges. It lies along the western base of the Santa Rita Mountains about 35 miles south of Tucson. The area slopes gently toward the

---

Reprinted with permission from *Ecology*, 39(4):720–726 (1958). Study carried out under W-25 Regional Research funds. Arizona Agr. Expt. Sta. Tech. Paper No. 450.

northwest and varies in elevation from 2,900 feet adjacent to the Santa Cruz River valley to 4,500 feet near the mountains.

The plant cover on the experimental range is divisible into 3 major subtypes: semidesert, mesa and foothill. The semidesert, which includes elevations from 2,900 to 3,500 feet, is vegetated largely by shrubs. The mesa, between elevations of 3,500 and 4,000 feet, is largely mesquite with an understory of perennial grasses. The foothill, at elevations of 4,000 to 4,500 feet, is basically a mixed grama grass zone with an intermittent overstory of mesquite and oaks.

Precipitation varies from about 11 inches at the lower elevations to 18 at the upper, with an average of about 14 inches over the entire range. Temperatures tend to be rather high from about March through October. Relative humidity readings are low most of the year.

The principal invading noxious shrubs are burroweed (*Aplopappus tenuisectus* (Greene) Blake), mesquite (*Prosopis juliflora* var. *velutina* (Woot) Sarg.), jumping cholla (*Opuntia fulgida* Engelm.), cane cholla (*Opuntia spinosior* (Engelm. and Bigel.)Toumey), and creosotebush (*Larrea tridentata* (DC.) Coville).

STUDY METHODS

A quadrat frame containing 4 square meters and having an inset frame one-tenth the total area was dropped at 10-pace intervals along each eastwest section line of the Santa Rita Experimental Range. At each drop, the burroweed and cholla plants in the large frame were counted and recorded. All perennials in the small frame were counted. The surrounding vegetation was noted at each drop and the dominant species recorded in estimated order of dominance. At least three species were recorded at each drop. Data were obtained in this manner in 1934,[1] and again in 1954. The vegetation changes indicated by these surveys are herein analyzed. From these analyses the percentage of the perennial vegetation complex comprised by each perennial species for each of the years studied was calculated.

From the dominance listings, maps were prepared showing the range of each important species during the 1934 and 1954 surveys. An abundance scale was established to permit classification of the abundance of each species as heavy, moderate or light. These ratings were based on the number of times the species in question was classed as dominant. If a species was listed as the number one or number two dominant ¾ths or more of the time in any one section of land, it was classed as heavy for that section. If it was listed less than ¾ths but more than ¼th it

[1] 1934 data were collected in part by G. F. Glendening.

would be recorded as moderate; anything less than ¼th was classed as light.

The number of plants per acre of mesquite, cholla, burroweed, and creosotebush in each of these abundance categories was determined by actual counts on acre-size plots. These plots were chosen at random in areas where each of these species had been listed in each abundance class for each of the 4 species involved. The data thus obtained are presented in Table 15–1.

TABLE 15-1. Abundance categories of noxious plants on the Santa Rita Experimental Range in 1954

Species	Heavy stand	Moderate stand	Light stand
	No./acre	*No./acre*	*No./acre*
Burroweed	over 1500	500 to 1500	0 to 500
Cholla	over 350	100 to 350	0 to 100
Creosote-bush	over 300	80 to 300	0 to 80
Mesquite	over 80	40 to 80	0 to 40

Climatic data were obtained from the University of Arizona station of the United States Weather Bureau. Tree-ring data that permitted extension of this analysis into the past beyond recorded Weather Bureau records were furnished by the University of Arizona Laboratory of Tree-ring Research.

VEGETATION CHANGES: 1904-1954

Grasslands

The first detailed description of the vegetation of the Santa Rita Experimental Range was made by Griffiths (1904). From his account a map has been prepared showing the general locations of the species that were dominant at that time (Fig. 15–1). It will be noted that more than half of the range was then classed as grass dominant. By 1954 the area dominated by grasses was largely restricted to the foothill regions and occupied less than ⅕th the total area (Fig. 15–2). As the grasses have receded they have been replaced in large part by shrubs.

Burroweed

Fencing of the Santa Rita Experimental Range in 1903 enclosed some thick stands of burroweed. At that time, the heavy stands were limited to the lower northwest portion of the range (Fig. 15–1). Although

the range was closed to grazing from 1903 until 1915, this weed continued to increase at a rapid rate until by 1910 it occurred throughout most of the lower elevations in the eastern and northern portions of the range (Griffiths 1910). By 1934 the entire range was occupied except for a few areas in the foothill region. From 1934 to 1954 on the other hand, there was no appreciable further increase in total area occupied.
moderate stands 19,500 and light stands 13,400. This represents a 100 percent increase during the last 20 years in heavy stands alone. However,

Analysis of the abundance-class data shows that burroweed increased only slightly in the heavily infested area from 1934 to 1954. In 1934 heavy

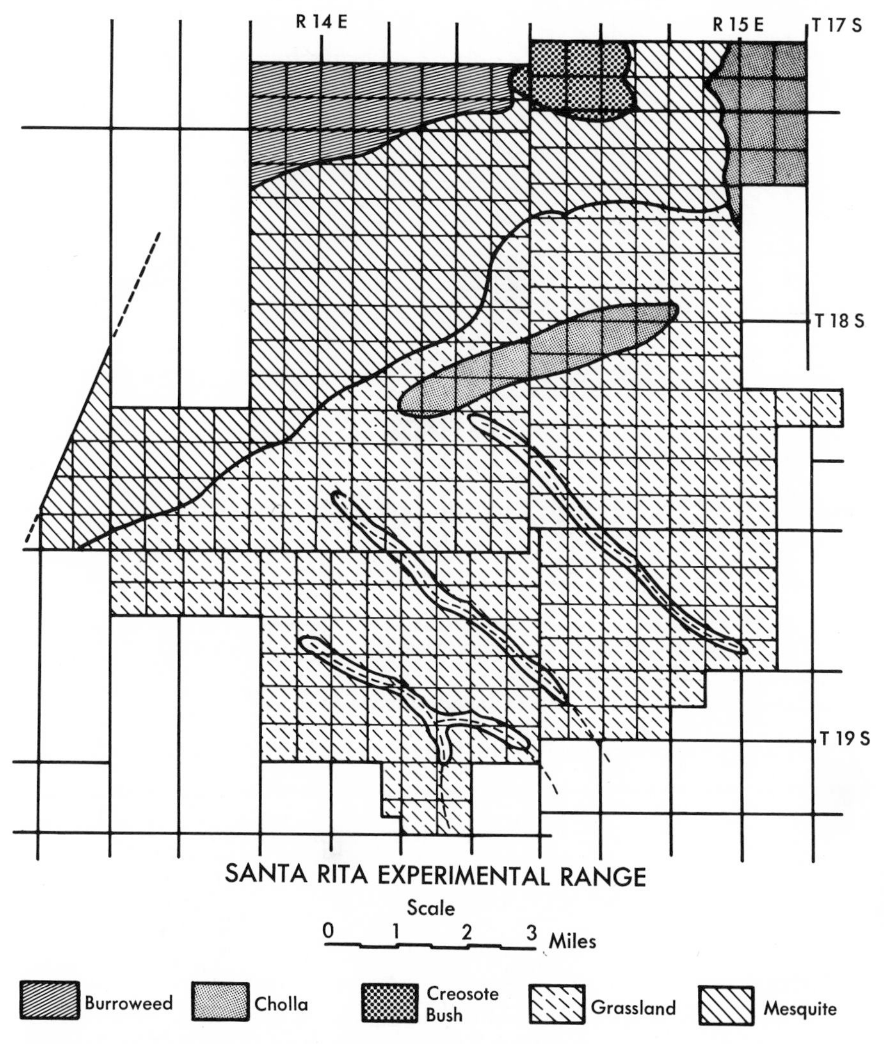

Fig. 15–1. Dominant vegetation on the Santa Rita Experimental Range in 1904.

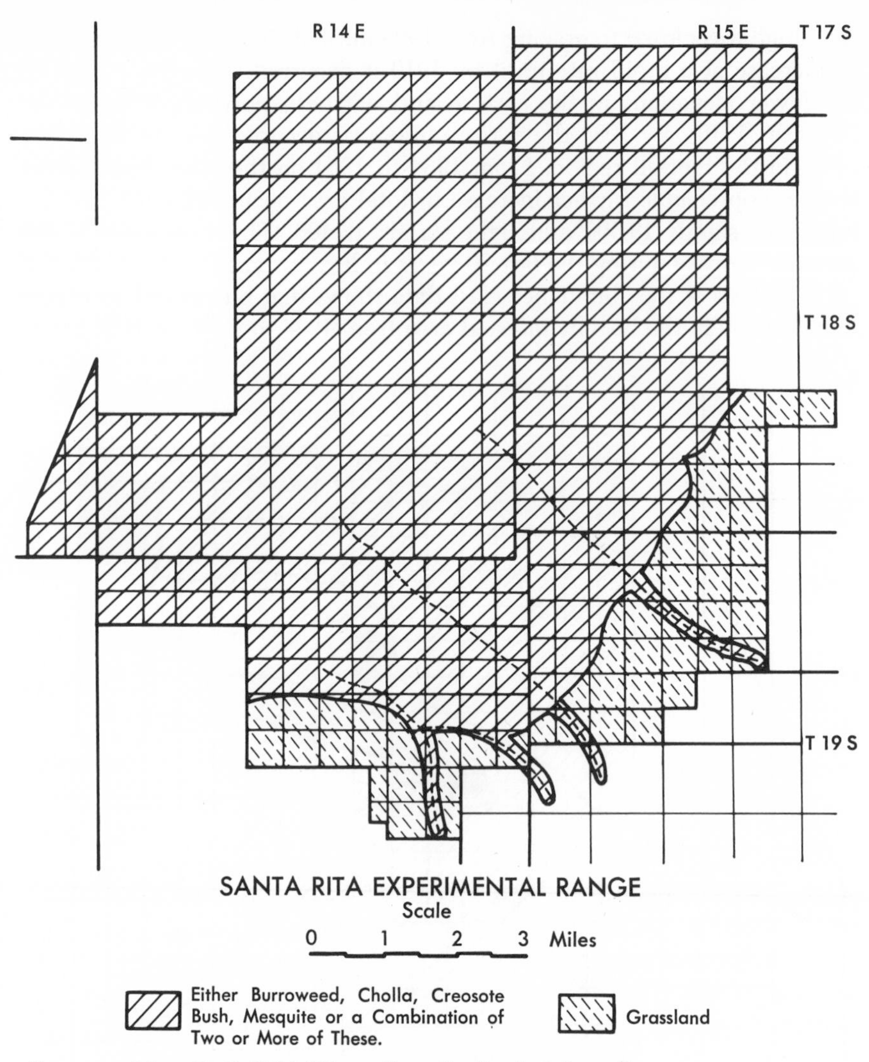

Constructed from D. Griffith's "Range Investigation in Arizona."

Fig. 15–2. Dominant vegetation on the Santa Rita Experimental Range in 1954.

stands covered some 26,500 acres, moderate stands 11,500 and light stands 15,000. By 1954 heavy stands occupied approximately 27,800 acres, moderate stands 10,900 and light stands 14,300.

Cholla

Cholla occurred on the Santa Rita Range in 1904 only in restricted areas in the northwest and central portions of the range (Fig. 15–1). The plant was not considered much of a problem or was even considered desirable because it supplied feed for cattle during drought.

By 1934 the stands of all species of cholla had thickened and increased in area. The spread of these species from about one or two thousand acres in 1904 to about 30,000 in 1934 would seem to indicate that growth conditions during this 30-year period must have been exceptionally favorable. In contrast with burroweed, cholla continued to spread during the 20-year period from 1934 to 1954. In 1934 the cholla infestation was classed as heavy on some 10,000 acres, as moderate on 21,400 and as light on 21,900. By 1954 the heavy stands occupied 20,100 acres, much higher percentages of increase have been observed and recorded for selected sites on the experimental range (Glendening 1952).

The 1954 survey indicated that even the upper foothills had some cholla sprinkled along the ridges. Although these plants did not appear as healthy as those at lower elevations, it is important to note that they had invaded the last stronghold of the grasses.

Mesquite

As early as 1904 the Santa Rita Experimental Range supported a population of mesquite. For the most part the trees were confined to the lower portions of the range (Fig. 15–1). Throughout the area, however, the major drainages contained some trees and these became seed sources for rapid movement onto the intervening mesas.

By 1934, mesquite was firmly established on the experimental range. Infestation on some 14,200 acres was classed as heavy, on 17,000 as moderate, and on 21,700 as light. Since 1934 the plant has continued to spread and increase in density at a rapid rate; by 1954 almost the entire range was covered in some degree. Of the 53,000 acres included in the 1954 study, infestation was heavy on 19,000 acres, moderate on 22,800 and light on 11,200. The areas classed as lightly infested are essentially the foothill grasslands.

Creosotebush

The spread of creosotebush on the experimental range has been small when compared with that of the other dominant shrubs. However, in terms of percent increase this spread looms large and important. In 1904 this shrub was confined to about 950 acres in the northeast corner of the range (Fig. 15–1). By 1934 its range had increased more than 12-fold to include 11,900 acres. At that time infestation on 7,600 acres was classed as light, on 2,600 as moderate and on 1,700 as heavy. The greater portion of the range (41,100 acres) contained no creosotebush. During the 20-year period from 1934 to 1954 there was a continued gradual increase both in total area occupied and in stand density. By 1954 the stand was classed as light on 8,000 acres, moderate on 2,700 and heavy on 2,300. Thus, by 1954, creosotebush occupied an area 73 times as great as it had 50 years before, a truly surprising rate of spread for a plant commonly assumed to be relatively static among semidesert shrubs.

Over-all species analysis

A species analysis of the plants recorded in each of the 2 surveys shows that although the total number of perennials was materially reduced, there was a sharp increase in number of shrubs. In the 1934 survey, perennial plants of all species averaged 56.3 per quadrat drop; in 1954 only 15.6, a decrease of 360 percent. Plants classed as desirable for forage made up 70 percent of the vegetation sampled in 1934 but only 49 percent in 1954, a 70 percent decrease. During this same period, the percentage of undesirable species more than doubled, increasing from 23 to 50 percent.

The change in desirable perennials was most markedly indicated by Rothrock grama (*Bouteloua rothrockii* Vasey), which comprised 30 percent of the perennials in 1934 but only .98 percent in 1954. Marked differences of this sort, particularly when they reflect changes in short-lived grasses or forbs, may be due in part to short-term climatic influences. Rothrock grama reacts rapidly to rainfall conditions and this is doubtless reflected in these widely differing figures. During the 4-year period preceding 1934 above-average rainfall conditions generally prevailed. The period prior to the 1954 survey, on the other hand, was one of severe drought.

Burroweed made up 7.8 percent of the number of perennial plants recorded in 1934; 16.5 percent in 1954. Inasmuch as these figures represent percentages of the total number of perennial plants present, they do not necessarily indicate a proportional increase in numbers of burroweed plants. A reduction of associated perennials results in a relative, but not an actual, increase in the amount of burroweed.

DISCUSSION

Climate as a factor affecting changes in vegetation

The earliest known weather records for the Southwest go back only about 100 years, and the existence of a definite climatic trend cannot be clearly shown in so short a period. As a supplement to recorded climatic data, tree-ring analyses have proved to be a reliable indicator of past growing conditions. For the 60-year period, 1891–1950, unpublished data

(Schulman 1955) shows a correlation coefficient of .7 between representative southern Arizona Douglas fir trees and Tucson rainfall. This high correlation, supplemented by additional study of tree-ring and rainfall records, indicates that tree-rings may be used as reliable indicators of annual rainfall in the Southwest over a period greatly antedating the relatively short period of recorded precipitation.

There is no evidence to indicate that climatic change has been a major factor in effecting the change from grass to brush. Analysis of tree-ring data from 1600 to 1870 (the first period when there was a large influx of white men into the Southwest) does not indicate any appreciable increase or decrease in precipitation. During the 350 years from 1600 to 1950 likewise, there seem to be no growth periods essentially different from those of today. There were periods of above-and below-normal growth, but apparently no consistent climatic change. Although the last 75 years in particular show some extended periods of both drought and above-average precipitation, these deviations from normal are not materially different from others that preceded. Yet, most of the vegetation changes described in this paper have taken place during this 75-year period from 1880 to 1955.

Neither do less extended wet or dry periods appear to have affected the rate of invasion of noxious plants. Tree-ring records and precipitation data both show that during the 19-year period, 1885 to 1904, southwestern ranges were subjected to below-average precipitation and tree growth. During this period, however, the shrubs continued to invade (Griffiths 1904). The period from 1905 to 1921 received above-normal rainfall for almost the entire 18-year span and, although desirable vegetation recovered somewhat, the downward trend of the grasses continued (Griffiths 1910). From 1921 to 1930 drought again plagued the ranges, yet noxious shrubs and weeds continued to increase.

The period 1825 to 1860 is of interest not only from the climatic viewpoint, but also from historical happenings. During most of this 35-year period above-average tree growth was recorded. The favorable moisture conditions indicated by this record must have given a tremendous boost to the grassland vegetation and this, in turn, may have been responsible in part for the reports of unlimited forage credited to the western ranges in the early years (Wislizenus 1848). As a consequence of these reports ranchers and business men visualized an easy fortune from the area and descended upon it with thousands of cattle and sheep.

Grazing as a factor affecting changes in vegetation

When ranges are overstocked, livestock, through their selection of the more palatable plants, influence plant composition on a given area. As grazing removes the more palatable species, the less desirable have more moisture and space available. Clements (1928), generalizing from this, stated that under climax grassland conditions, grasses are able to maintain their dominance in competition with forbs, but that in the event of overgrazing or other disturbance, the forbs gradually predominate. He observed further that when such a range is closed to grazing or the grazing pressure is reduced the grasses again assume dominance, effectively controlling forbs and half-shrubs. Rather different conclusions had been reached earlier by Griffiths (1910), who commented as follows on the reasons underlying shrub invasion, including the half-shrub burroweed, on the Santa Rita Range:

> It will doubtless be impossible to depict all the agencies that are bringing about these changes. It is quite certain that the operations here of the Bureau of Plant Industry have had no influence, for the shrubbery has thickened up on the outside of the inclosure, where the grazing has been very heavy, apparently as much as on the inside. The probability is that neither protection nor heavy grazing has much to do with the increase of shrubs here, but that it is primarily the result of the prevention of fires.

In studies conducted on the Santa Rita Experimental Range by the U.S. Forest Service the rate of spread of mesquite was found to increase regardless of grazing treatment (Glendening 1952). Three different treatments were employed: "(1) Cattle and rabbits excluded–grazing by small rodents, (2) cattle excluded–grazing by rabbits and small rodents, (3) yearlong grazing by cattle, rabbits, and small rodents." Results showed that during the 17-year period, 1932 to 1949, the numbers of mesquite more than doubled on all plots. This increase was greater on the plots closed to grazing than on those that were grazed.

It would seem from this study that the rate of spread of mesquite is not affected by grazing once the seed source is available. In the case of cholla, however, the study showed a direct relationship between grazing activity and plant numbers as shown by average increases of 18, 72, and 149 plants per acre on plots protected from cattle and rabbits, cattle only, and unfenced, respectively. Glendening concluded: "It is improbable that moderation in livestock grazing will prevent the loss of grass cover within mesquite stands where the trees have gained sufficient size and density to completely utilize or materially reduce the moisture supply, and where the population of seed planting rodents is high."

There is little doubt that mesquite invasion has been greatly increased by the ability of the seeds to retain their viability after passing through the digestive tract of cattle and horses (Smith 1899, Griffiths 1904, Fischer 1947, Reynolds and Glendening 1949). This ability is of great importance in the rapid spread of mesquite since it insures a large supply of seed on almost every part of our cattle ranges where mesquite is available as feed.

In addition to mesquite, cattle also spread other southwestern plants. Jumping cholla is spread almost entirely by the action of grazing animals and rodents. The fruit of both cane cholla and jumping cholla is readily taken by cattle and in their eagerness to reach the fruit, the animals often brush against the terminal joints. As a result many of these joints remain hanging to the head, neck, and shoulder regions. The joints eventually fall off, or are rubbed off on bushes and trees and, if they fall in a suitable area, take root and form new plants. In most arborescent Opuntias (often all loosely called chollas) joints commonly fall to the ground as a natural process, and from these new plants arise.

Jumping cholla is not usually propagated by establishment of seedlings (Johnson 1918). Examination of large numbers of the fallen fruits showed no sign of germination. This was established in part from greenhouse studies, in part from range observations. Cane cholla, on the other hand, was found to be largely propagated by seed.

Rodents as a factor affecting changes in vegetation

Many rodents of the Southwest transport seed to unseeded areas. Principal among these is the Merriam Kangaroo rat (*Dipodomys merriami merriami* Mearns). This small rodent buries large quantities of seeds in underground burrows and food caches (Reynolds and Glendening 1949, Reynolds 1950). The ability of most seed-storing animals to relocate their buried seeds is very low (Martin, Zim and Nelson, 1951). As a consequence many seeds remain undisturbed and may eventually germinate. Many of the caches are located at some distance from the seed tree, a fact that facilitates the increase of mesquite onto adjacent areas. Cholla joints also are carried by the woodrat (*Neotoma albigula albigula* Hartley) to their nests and dens, and are often dropped in sites where they take root and grow.

Fire as a factor affecting changes in vegetation

Fires have long been recognized as instrumental in maintaining the former treeless conditions of our grasslands (Stewart 1953). The effects of fires in the southern prairie states were commented on over 100 years ago by Gregg (1844) who noted changes taking place in the grasslands of the great plains.

> It is unquestionably the prairie conflagration that keep down the woody growth upon most of the western uplands. The occasional skirts and fringes which have escaped their rage have been protected by the streams they border. Yet, may not the time come when these vast plains will be covered with timber? . . . Indeed there are parts of the southwest now thickly set with trees of good size that within the remembrance of the oldest inhabitants were as naked as the prairie

plains and the appearance of the timber in many other sections indicates that it has grown up within less than a century. In fact, we are now witnessing the encroachment of the timber upon the prairie wherever the devastating conflagrations have ceased their ravages.

Other travelers commented on the fires which were encountered on their journeys. Cook (1908), a well known agrostologist, commented as follows on the destructive force of the range fires:

Before the prairies were grazed by cattle the luxuriant growths of grass could accumulate for several years until conditions were favorable for accidental fires to spread. With these large supplies of fuel the fires which swept over these prairies were very besoms of destruction not only for man and animals but for all shrubs and trees which might have ventured out among the grass and even for any trees or forests against which the burning wind might blow.

From these reports it is evident that although the fires burned with almost white-hot intensity they moved rapidly over the ground, and did not burn deeply into the soil. It is a matter of common knowledge that grasses are better adapted than shrubs to withstand the effects of fire. This has been pointed out by various writers. Bray (1901) stated:

Apparently under the open prairie regime the equilibrium was maintained by more or less regular recurrence of prairie fires. This, of course, is by no means a new idea, but the strength of it lies in the fact that the grass vegetation was tolerant of fires and the woody vegetation was not. It was only after weakening the grass floor by heavy pasturing and ceasing to ward off the encroaching species by fire that the latter invaded the grass lands. Once the equilibrium was destroyed everything conspired to hasten the encroachment of chaparral—droughts, pasturing, trampling, seed-scattering, etc.

Shantz (1924) in writing about the great plains regions, and giving reasons for their status as grasslands, said:

In the eastern portion of the area fires have in all probability protected the grassland from the encroachment of the forests. Aided by high winds, these fires swept with great rapidity across the grasslands of the prairies and plains. . . . Trees and shrubs are killed by fires, and as a consequence the grasses are able to maintain themselves on land which would support a good forest growth if the trees were adequately protected. Since the settlement of these lands and the consequent checking of the prairie fires, tree growth has been gradually extended, either by planting or natural seeding, and trees now grow throughout the whole prairie region.

Early workers in Arizona reached the same conclusions regarding the treeless nature of our grasslands as those in the great plains regions (Griffiths 1910, Thornber 1910). In commenting on the cause of the encroachment of woody species on our grasslands Griffiths says:

The main factor, though, in the opinion of the writer, has been that of fire. It is firmly believed that were it not for the influence of this factor the grassy mesas would today be covered with brush and trees, the same as the canyons, except that the growth would be smaller, owing to a more limited supply of moisture. In short, the same laws apply here that govern in our great prairie states, . . . where treeless plains were kept so by frequent fires.

Reports from other portions of the Southwest show that the writers and research men from those sections concur with observations made in Arizona. Foster (1917) commenting on the forests of Texas wrote:

The causes which have resulted in the spread of timbered areas are traceable directly to the interference of man. Before the white man established his ranch home in these hills the Indians burned over the country repeatedly and thus prevented an extension of forest areas. With the settlement of the country grazing became the only important industry. . . . Overgrazing has greatly reduced the density of grass vegetation. The practice of burning has during recent years disappeared. The few fires which start are usually caused by carelessness, and . . . burn only small areas. These conditions operated to bring about a rapid extension of woody growth. Almost unquestionably the spread of timbered areas received its impetus with the gradual disappearance of grassland fires and has been hastened by the reduction of the grass cover itself.

A study of the range vegetation in Texas by Buechner (1944) also supports the early observations that grassland fires were the main influence in suppressing the invading woody species.

In most parts of the North American continent, before the advent of the white man, man probably had little or no influence on the vegetation or animal life; indeed, he himself was an integral part of the animal life. . . . But the Edwards Plateau was a notable exception to this general scheme, since the Indian burned the vegetation periodically to facilitate hunting by routing out the game and increasing its visibility. The effect of this practice was to destroy tree and shrub seedlings and produce a grassland in regions that would otherwise have supported arborescent vegetation. . . . As a result of the elimination of fires and the introduction of livestock, profound changes took place in the vegetation. What was once a waving sea of grass as far as the eye could see was changed to a diversified arborescent vegetation.

Recent research on the subject is of interest. Studies indicate that the percentage of survival of older shrubs after grassland fires is much higher than of young seedlings of the same species. Reports from workers on the Santa Rita Experimental Range show on the average a 52 percent kill on mesquite seedlings less than ½ inch in ground diameter but only an 8 to 15 percent kill on large plants (Glendening and Paulsen, 1955).

Humphrey (1949), reporting on fire as a means of controlling burroweed, mesquite, and cholla, noted an approximate 50 percent kill on cholla, mesquite, and bisnaga (*Echinocactus wislizeni* Engelm.) and almost a 100 percent kill on burroweed. In addition to this high kill on the noxious plants, the number of grass plants after burning was considerably higher on the burned area than on the unburned area. An earlier study gave almost a 100 percent kill on burroweed and snakeweed and a 30 to 50 percent kill on cholla (Humphrey and Everson, 1951). Although mesquites were not sufficiently abundant on this area to permit statistical analysis, observations made here and on other burns indicate that individual large mesquites are frequently killed outright by fire. Even though the trees may not be completely killed, they may be reduced to basal sprouts and these are readily injured by browsing animals. In areas swept by frequent fires, the trees may be reduced to this form or may be entirely eliminated.

SUMMARY

Vegetation surveys of the Santa Rita Experimental Range in southern Arizona were made in 1904, 1934 and 1954. The survey data were analyzed to determine changes in area and abundance of the four most common woody plants: creosotebush, burroweed, cholla and mesquite. All of the species increased markedly in area and abundance during the 50-year period 1904 to 1954. Maximum increase had occurred by 1934 but, except for burroweed, continued at a slower rate from 1934 to 1954.

Climate, grazing rodents and fire were evaluated as factors that may have contributed to the vegetational changes noted. Conclusions were: (a) there have been no apparent changes in climate that would result in the vegetational changes noted; (b) grazing by domestic livestock has affected the composition of the vegetation, in part because of seed dissemination, in part because of selective grazing and in part because of the removal of grass by grazing that formerly served as fuel for range fires; (c) rodents bury mesquite seeds and transport cholla cactus joints, thus serving to propagate those plants; (d) fires maintained the desert grassland as such, prior to the introduction of livestock; (e) shrub invasion of southern Arizona semidesert grassland ranges is due primarily to reduction of range fires.

REFERENCES

Bray, W. L. 1901. The ecological relations of the vegetation of western Texas. Bot. Gaz. 32:99–123, 195–217, 262–291.

Buechner, H. K. 1944. The range vegetation of Kerr County, Texas, in relation to livestock and white-tailed deer. Am. Midl. Nat. 31:698–743.

Clements, F. E. 1928. Plant succession and indicators. New York: The H. W. Wilson Co.

Cook, O. F. 1908. Changes of vegetation on the south Texas prairies. USDA Bur. Plant Industry Cir. 14.

Fischer, C. E. 1947. Present information on the mesquite problem. 1056 progress report, Texas Agr. Ex. Sta. A. & M. College of Texas.

Foster, J. H. 1917. The spread of timbered areas in central Texas. Jour. Forestry 15: 442–445.

Gregg, Josiah. 1844. Commerce of the prairies or the Journal of a Sante Fe Trader. New York: G. H. Langley.

Glendening, G. E. 1952. Some quantitative data on the increase of mesquite and cactus on a desert grassland range in southern Arizona. Ecology 33:319–328.

———, and H. A. Paulsen, Jr. 1955. Reproduction and establishment of velvet mesquite as related to invasion of semidesert grasslands. U.S. Dept. Agr. Tech. Bull. 1127. 50 pp.

Griffiths, David. 1904. Range investigations in Arizona. USDA Bur. Plant Industry Bull. 67. 62 pp.

———. 1910. A protected stock range in Arizona. USDA Bur. Plant Industry Bull. 177. 28 pp.

Humphrey, R. R. 1949. Fire as a means of controlling velvet mesquite, burroweed and cholla on southern Arizona ranges. Jour. Range Mgt. 2:175–182.

———, and A. C. Everson. 1951. Effect of fire on a mixed grass-shrub range in southern Arizona. Jour. Range Mgt. 4:264–266.

Johnson, D. S. 1918. The fruit of *Opuntia fulgida*. Carn. Inst. Wash. Pub. 269. 62 pp.

Martin, A. D., H. S. Zim, and A. L. Nelson. 1951. American Wildlife and Plants. 500 pp. New York: McGraw-Hill Book Co.

Reynolds, H. G., and G. E. Glendening. 1949. Merriam kangaroo rat a factor in mesquite propagation on southern Arizona range lands. Jour. Range Mgt. 2:193–197.

———. 1950. Relation of Merriam kangaroo rats to range vegetation in southern Arizona. Ecology 31:456–463.

Schulman, Edmund. 1955. Unpublished data on tree-ring indices in the Southwest.

Shantz, H. L. 1924. Atlas of American Agriculture. Part I sect. E. Natural vegetation.

Smith, J. C. 1899. Grazing problems in the southwest and how to meet them. USDA Div. of Agrostology Bull. 16. 47 pp.

Stewart, O. C. 1953. Why the Great Plains are treeless. Colorado Quart. 2:40–50.

Thornber, J. J. 1910. The grazing ranges of Arizona. Univ. Arizona Agr. Ex. Sta. Bull. 65. 360 pp.

Wislizenus, A. 1848. Memoir of a tour to northern Mexico, connected with Col. Doniphan's expedition, in 1846 and 1847. Sen. Misc. Doc. No. 26, 30th Cong., 1st Sess.

HUMAN POPULATIONS AND FOOD RESOURCES

16

World population growth: an international dilemma

Harold F. Dorn

During all but the most recent years of the centuries of his existence man must have lived, reproduced, and died as other animals do. His increase in number was governed by the three great regulators of the increase of all species of plants and animals—predators, disease, and starvation—or, in terms more applicable to human populations—war, pestilence, and famine. One of the most significant developments for the future of mankind during the first half of the 20th century has been his increasing ability to control pestilence and famine. Although he has not freed himself entirely from the force of these two regulators of population increase, he has gained sufficient control of them so that they no longer effectively govern his increase in number.

Simultaneously he has developed methods of increasing the effectiveness of war as a regulator of population increase, to the extent that he almost certainly could quickly wipe out a large proportion, if not all, of the human race. At the same time he has learned how to separate sexual gratification from reproduction by means of contraception and telegenesis (that is, reproduction by artificial insemination, particularly with spermatozoa preserved for relatively long periods of time), so that he can regulate population increase by voluntary control of fertility. Truly it can be said that man has the knowledge and the power to direct, at least in part, the course of his evolution.

Reprinted with permission from *Science,* 135(3500):283–290 (26 January 1962).

This newly gained knowledge and power has not freed man from the inexorable effect of the biological laws that govern all living organisms. The evolutionary process has endowed most species with a reproductive potential that, unchecked, would overpopulate the entire globe within a few generations. It has been estimated that the tapeworm, *Taenia*, may lay 120,000 eggs per day; an adult cod can lay as many as 4 million eggs per year; a frog may produce 10,000 eggs per spawning. Human ovaries are thought to contain approximately 200,000 ova at puberty, while a single ejaculation of human semen may contain 200 million spermatozoa.

This excessive reproductive potential is kept in check for species other than man by interspecies competition in the struggle for existence, by disease, and by limitation of the available food supply. The fact that man has learned how to control, to a large extent, the operation of these biological checks upon unrestrained increase in number has not freed him from the necessity of substituting for them less harsh but equally effective checks. The demonstration of his ability to do this cannot be long delayed.

Only fragmentary data are available to indicate the past rate of growth of the population of the world. Even today, the number of inhabitants is known only approximately. Regular censuses of populations did not exist prior to 1800, although registers were maintained for small population groups prior to that time. As late as a century ago, around 1860, only about one-fifth of the estimated population of the world was covered by a census enumeration once in a 10-year period (*1*). The commonly accepted estimates of the population of the world prior to 1800 are only informed guesses. Nevertheless, it is possible to piece together a consistent series of estimates of the world's population during the past two centuries, supplemented by a few rough guesses of the number of persons alive at selected earlier periods. The most generally accepted estimates are presented in Figure 16–1.

These reveal a spectacular spurt during recent decades in the increase of the world's population that must be unparalleled during the preceding millennia of human existence. Furthermore, the rate of increase shows no sign of diminishing (Table 16–1). The period of time required for the population of the world to double has sharply decreased during the past three centuries and now is about 35 years.

Only a very rough approximation can be made of the length of time required for the population of the world to reach one-quarter of a billion persons, the estimated number at the beginning of the Christian era. The present subgroups of *Homo sapiens* may have existed for as long as 100,000 years. The exact date is not necessary, since for present purposes the evidence is sufficient to indicate that probably 50,000 to 100,000 years were required for *Homo sapiens* to increase in number until he reached a global total of one-quarter of a billion persons. This number was reached approximately 2000 years ago.

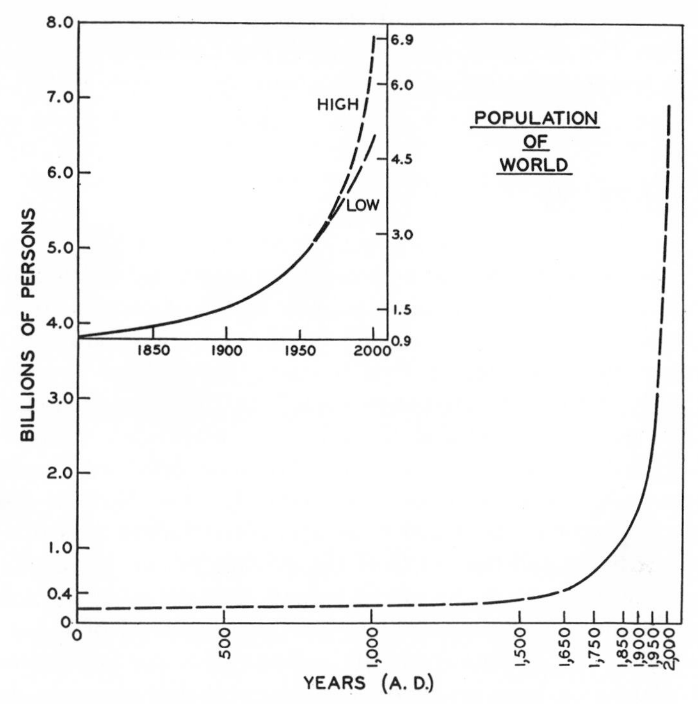

Fig. 16–1. Estimated population of the world, A.D. 1 to A.D. 2000.

TABLE 16–1. The number of years required to double the population of the world. [From United Nations data (9, *14*)]

Year (A.D.)	Population (billions)	Number of years to double
1	0.25 (?)	1650 (?)
1650	0.50	200
1850	1.1	80
1930	2.0	45
1975	4.0	35
2010	8.0*	?

*A projection of United Nations estimates.

* A projection of United Nations estimates.

By 1620, the year the Pilgrims landed on Plymouth Rock, the population of the world had doubled in number. Two hundred years later, shortly before the Civil War, another 500 million persons had been added. Since that time, additional half billions of persons have been added during increasingly shorter intervals of time. The sixth half billion, just added, required slightly less than 11 years, as compared to 200 years for the second half billion. The present rate of growth implies that only 6 to

7 years will be required to add the eighth half billion to the world's population. The change in rate of growth just described has taken place since the first settlers came to New England.

IMPLICATIONS

The accelerating rate of increase in the growth of the population of the world has come about so unobtrusively that most persons are unaware of its implications. There is a small group who are so aroused by this indifference that, like modern Paul Reveres, they attempt to awaken the public with cries of "the population bomb!" or "the population explosion!"

These persons are called alarmists by those who counter with the assertion that similar warnings, such as "standing-room only" and "mankind at the crossroads," have been issued periodically since Malthus wrote his essay on population, about 200 years ago. Nevertheless, says this group, the level of living and the health of the average person has continued to improve, and there is no reason to believe that advances in technology will not be able to make possible a slowly rising level of living for an increasing world population for the indefinite future. Furthermore, the rate of population increase almost certainly will slow down as the standard of education and living rises and as urbanization increases.

A third group of persons has attempted to estimate the maximum population that could be supported by the world's physical resources provided existing technological knowledge is fully utilized. Many of these calculations have been based on estimates of the quantity of food that could be produced and a hypothetical average daily calorie consumption per person.

As might be expected, the range of the various estimates of the maximum world population that could be supported without a lowering of the present level of living is very wide. One of the lowest, 2.8 billion, made by Pearson and Harper in 1945 on the assumption of an Asiatic standard of consumption, already has been surpassed (*2*). Several others, ranging from 5 to 7 billion, almost certainly will be exceeded by the end of this century. Perhaps the most carefully prepared estimate as well as the largest—that of 50 billions, prepared by Harrison Brown—would be reached in about 150 years if the present rate of growth should continue (*3*).

I believe it is worth while to prepare estimates of the maximum population that can be supported and to revise these as new information becomes available, even though most of the estimates made in the past already have been, or soon will be, demonstrated to be incorrect (in most instances too small), since this constitutes a rational effort to comprehend the implications of the increase in population. At the same time it should be recognized that estimates of the world's carrying capacity made in this manner are rather unrealistic and are primarily useful only as very general guidelines.

In the first place, these calculations have assumed that the earth's resources and skills are a single reservoir available to all. In reality this is untrue. The U.S. government attempts to restrict production of certain agricultural crops by paying farmers not to grow them. Simultaneously, in Asia and Africa, large numbers of persons are inadequately fed and poorly clothed. Except in a very general sense there is no *world* population problem; there are population problems varying in nature and degree among the several nations of the world. No single solution is applicable to all.

Since the world is not a single political unity, the increases in production actually achieved during any period of time tend to be considerably less than those theoretically possible. Knowledge, technical skill, and capital are concentrated in areas with the highest level of living, whereas the most rapid increase in population is taking place in areas where such skills and capital are relatively scarce or practically nonexistent.

Just as the world is not a single unit from the point of view of needs and the availability of resources, skills and knowledge to meet these needs, so it also is not a single unit with respect to population increase. Due to political barriers that now exist throughout the entire world, overpopulation, however defined, will become a serious problem in specific countries long before it would be a world problem if there were no barriers to population redistribution. I shall return to this point later, after discussing briefly existing forecasts or projections of the total population of the world.

Most demographers believe that, under present conditions, the future population of areas such as countries or continents, or even of the entire world, cannot be predicted for more than a few decades with even a moderate degree of certainty. This represents a marked change from the view held by many only 30 years ago.

In 1930 a prominent demographer wrote, "The population of the United States ten, twenty, even fifty years hence, can be predicted with a greater degree of assurance than any other economic or social fact, provided the immigration laws are unchanged" (*4*). Nineteen years later, a well-known economist replied that "it is disheartening to have to assert that the best population forecasts deserve little credence even for 5 years ahead, and none at all for 20–50 years ahead." (*5*).

Although both of these statements represent rather extreme views, they do indicate the change that has taken place during the past two decades in the attitude toward the reliability of population forecasts. Some of the reasons for this have been discussed in detail elsewhere and will not be repeated here (*6*).

It will be sufficient to point out that knowledge of methods of voluntarily controlling fertility now is so widespread, especially among persons of European ancestry, that sharp changes in the spacing, as well as in the number, of children born during the reproductive period may occur in a relatively short period of time. Furthermore, the birth rate may increase as well as decrease.

FORECASTING POPULATION GROWTH

The two principal methods that have been used in recent years to make population forecasts are (i) the extrapolation of mathematical curves fitted to the past trend of population increase and (ii) the projection of the population by the "component" or "analytical" method, based on specific hypotheses concerning the future trend in fertility, mortality, and migration.

The most frequently used mathematical function has been the logistic curve which was originally suggested by Verhulst in 1838 but which remained unnoticed until it was rediscovered by Pearl and Reed about 40 years ago (7). At first it was thought by some demographers that the logistic curve represented a rational law of population change. However, it has proved to be as unreliable as other methods of preparing population forecasts and is no longer regarded as having any unique value for estimating future population trends.

A recent illustration of the use of mathematical functions to project the future world population is the forecast prepared by von Foerster, Mora, and Amiot (8). In view of the comments that subsequently were published in this journal, an extensive discussion of this article does not seem to be required. It will be sufficient to point out that this forecast probably will set a record, for the entire class of forecasts prepared by the use of mathematical functions, for the short length of time required to demonstrate its unreliability.

The method of projecting or forecasting population growth most frequently used by demographers, whenever the necessary data are available, is the "component" or "analytical" method. Separate estimates are prepared of the future trend of fertility, mortality, and migration. From the total population as distributed by age and sex on a specified date, the future population that would result from the hypothetical combination of fertility, mortality, and migration is computed. Usually, several estimates of the future population are prepared in order to include what the authors believe to be the most likely range of values.

Such estimates generally are claimed by their authors to be not forecasts of the most probable future population but merely indications of the population that would result from the hypothetical assumptions concerning the future trend in fertility, mortality, and migration. However, the projections of fertility, mortality, and migration usually are chosen to include what the authors believe will be the range of likely possibilities. This objective is achieved by making "high," "medium," and "low" assumptions concerning the future trend in population growth. Following the practice of most of the authors of such estimates, I shall refer to these numbers as population projections.

The most authoritative projections of the population of the world are those made by the United Nations (9, *10*) (Table 16–2). Even though

TABLE 16–2. Estimated population of the world for A.D. 1900, 1950, 1975, and 2000. [From United Nations data (9), rounded to three significant digits]

Area	Estimated population (millions)		Projected future population (millions)			
			Low assumptions		High assumptions	
	1900	1950	1975	2000	1975	2000
World	1550	2500	3590	4880	3860	6900
Africa	120	199	295	420	331	663
North America	81	168	232	274	240	326
Latin America	63	163	282	445	304	651
Asia	857	1380	2040	2890	2210	4250
Europe including U.S.S.R.	423	574	724	824	751	987
Oceania	6	13	20	27	21	30

the most recent of these projections were published in 1958, only 3 years ago, it now seems likely that the population of the world will exceed the high projection before the year 2000. By the end of 1961 the world's population at least equaled the high projection for that date.

Although the United Nations' projections appear to be too conservative in that even the highest will be an underestimate of the population only 40 years from now, some of the numerical increases in population implied by these projections will create problems that may be beyond the ability of the nations involved to solve. For example, the estimated increase in the population of Asia from A.D. 1950 to 2000 will be roughly equal to the population of the entire world in 1958! The population of Latin America 40 years hence may very likely be four times that in 1950. The absolute increase in population in Latin America during the last half of the century may equal the total increase in the population of *Homo sapiens* during all the millennia from his origin until about 1650, when the first colonists were settling New England.

Increases in population of this magnitude stagger the imagination. Present trends indicate that they may be succeeded by even larger increases during comparable periods of time. The increase in the rate of growth of the world's population, shown by the data in Table 16–1, is still continuing. This rate is now estimated to be about 2 percent per year, sufficient to double the world's population every 35 years. It requires only very simple arithmetic to show that a continuation of this rate of growth for even 10 or 15 decades would result in an increase in population that would make the globe resemble an anthill.

But as was pointed out above, the world is not a single unit economically, politically, or demographically. Long before the population of the entire world reaches a size that could not be supported at current levels of living, the increase in population in specific nations and regions will give rise to problems that will affect the health and welfare of the rest of the world. The events of the past few years have graphically demonstrated the rapidity with which the political and economic problems of even a small and weak nation can directly affect the welfare of the largest and most powerful nations. Rather than speculate about the maximum population the world can support and the length of time before this number will be reached, it will be more instructive to examine the demographic changes that are taking place in different regions of the world and to comment briefly on their implications.

DECLINE IN MORTALITY

The major cause of the recent spurt in population increase is a world-wide decline in mortality. Although the birth rate increased in some countries—for example, the United States—during and after World War II, such increases have not been sufficiently widespread to account for more than a small part of the increase in the total population of the world. Moreover, the increase in population prior to World War II occurred in spite of a widespread decline in the birth rate among persons of European origin.

Accurate statistics do not exist, but the best available estimates suggest that the expectation of life at birth in Greece, Rome, Egypt, and the Eastern Mediterranean region probably did not exceed 30 years at the beginning of the Christian era. By 1900 it had increased to about 40 to 50 years in North America and in most countries of northwestern Europe. At present, it has reached 68 to 70 years in many of these countries.

By 1940, only a small minority of the world's population had achieved an expectation of life at birth comparable to that of the population of North America and northwest Europe. Most of the population of the world had an expectation of life no greater than that which prevailed in western Europe during the Middle Ages. Within the past two decades, the possibility of achieving a 20th-century death rate has been opened to these masses of the world's population. An indication of the result can be seen from the data in Figure 16–2.

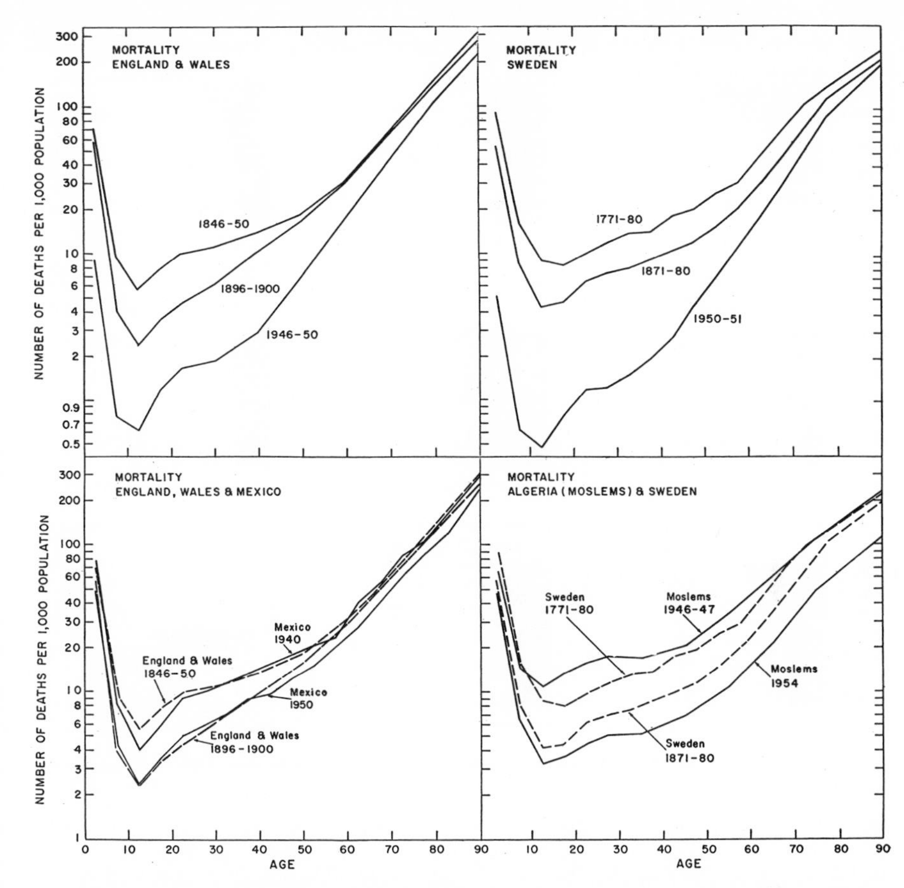

Fig. 16–2. Age-specific death rates per 1000 per year for Sweden, England and Wales, Mexico, and the Moslem population of Algeria for various time periods from 1771 to 1954.

In 1940, the death rate in Mexico was similar to that in England and Wales nearly 100 years earlier. It decreased as much during the following decade as did the death rate in England and Wales during the 50-year period from 1850 to 1900.

In 1946–47 the death rate of the Moslem population of Algeria was higher than that of the population of Sweden in the period 1771–80, the earliest date for which reliable mortality statistics are available for an entire nation. During the following 8 years, the drop in the death rate in Algeria considerably exceeded that in Sweden during the century from 1771 to 1871 (*11*).

The precipitous decline in mortality in Mexico and in the Moslem population of Algeria is illustrative of what has taken place during the past 15 years in Latin America, Africa, and Asia, where nearly three out

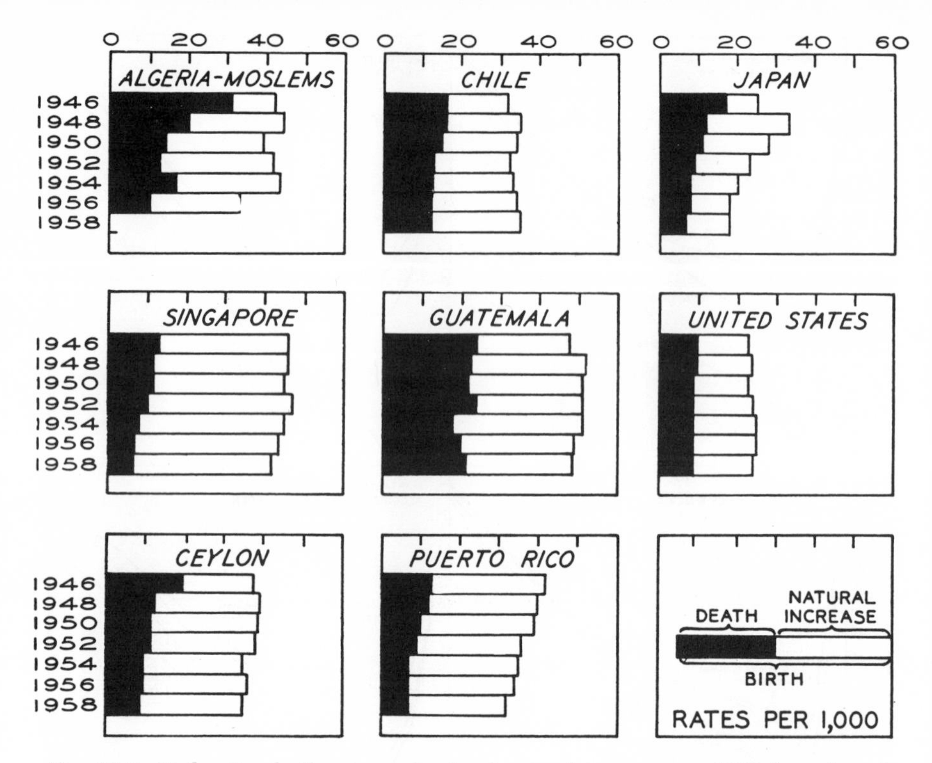

Fig. 16–3. Birth rate, death rate, and rate of natural increase per 1000 for selected countries for the period 1946–58.

of every four persons in the world now live. Throughout most of this area the birth rate has changed very little, remaining near a level of 40 per 1000 per year, as can be seen from Figure 16–3, which shows the birth rate, death rate, and rate of natural increase for selected countries.

Even in countries such as Puerto Rico and Japan where the birth rate has declined substantially, the rate of natural increase has changed very little, owing to the sharp decrease in mortality. A more typical situation is represented by Singapore, Ceylon, Guatemala, and Chile, where the crude rate of natural increase has risen. There has been a general tendency for death rates to decline universally and for high birth rates to remain high, with the result that those countries with the highest rates of increase are experiencing an acceleration in their rates of growth.

REGIONAL LEVELS

The absolute level of fertility and mortality and the effect of changes in them upon the increase of population in different regions of the world can be only approximately indicated. The United Nations estimates that only about 33 percent of the deaths and 42 percent of the births that occur in the world are registered (*12*). The percentage registered ranges from about 8 to 10 percent in tropical and southern Africa and Eastern Asia to 98 to 100 percent in North America and Europe. Nevertheless, the statistical staff of the United Nations, by a judicious combination of the available fragmentary data, has been able to prepare estimates of fertility and mortality for different regions of the world that are generally accepted as a reasonably correct representation of the actual but unknown figures. The estimated birth rate, death rate, and crude rate of natural increase (the birth rate minus the death rate) for eight regions of the world for the period 1954–58 are shown in Figure 16–4.

The birth rates of the countries of Africa, Asia, Middle America, and South America average nearly 40 per 1000 and probably are as high as they were 500 to 1000 years ago. In the rest of the world—Europe, North America, Oceania, and the Soviet Union—the birth rate is slightly more than half as high, or about 20 to 25 per 1000. The death rate for the former regions, although still definitely higher, is rapidly approaching that for people of European origin, with the result that the highest rates of natural increase are found in the regions with the highest birth rates.

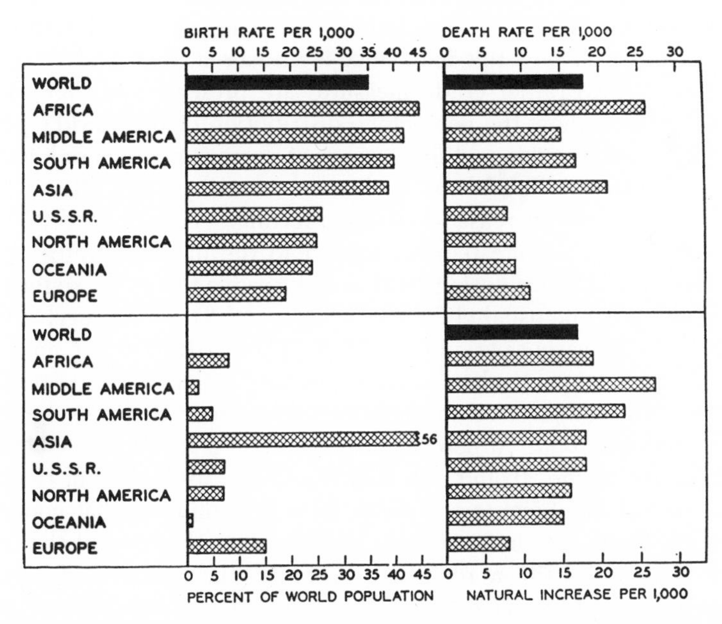

Fig. 16–4. Percentage of the 1958 world population, birth rate, death rate, and rate of natural increase, per 1000, for the period 1954–58 for various regions of the world.

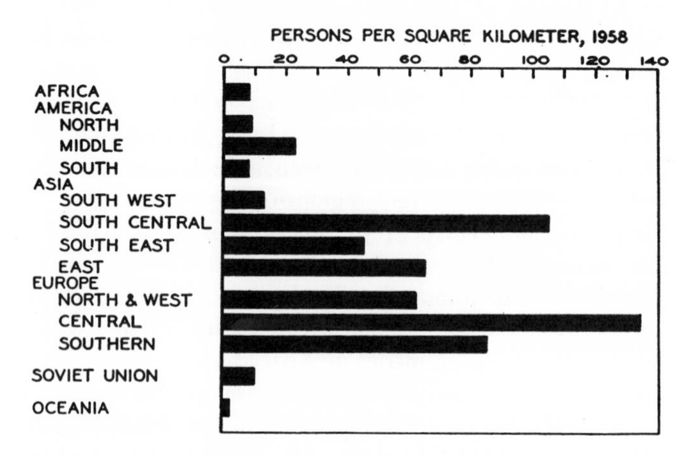

Fig. 16–5. Number of persons per square kilometer in various regions of the world in 1958.

The most rapid rate of population growth at present is taking place in Middle and South America, where the population will double about every 26 years if the present rate continues.

These regional differences in fertility and mortality are intensifying the existing imbalance of population with land area and natural resources. No matter how this imbalance is measured, that it exists is readily apparent. Two rather crude measures are presented in Figures 16–4 and 16–5, which show the percentage distribution of the world's population living in each region and the number of persons per square kilometer.

An important effect of the decline in mortality rates often is overlooked—namely, the increase in effective fertility. An estimated 97 out of every 100 newborn white females subject to the mortality rates prevailing in the United States during 1950 would survive to age 20, slightly past the beginning of the usual childbearing age, and 91 would survive to the end of the childbearing period (Fig. 16–6). These estimates are more than 3 and 11 times, respectively, the corresponding estimated proportions for white females that survived to these ages about four centuries ago.

In contrast, about 70 percent of the newborn females in Guatemala would survive to age 20, and only half would live to the end of the childbearing period if subject to the death rates prevailing in that country in 1950. If the death rate in Guatemala should fall to the level of that in the United States in 1950—a realistic possibility—the number of newborn females who would survive to the beginning of the childbearing period would increase by 36 percent; the number surviving to the end of the childbearing period would increase by 85 percent. A corresponding de-

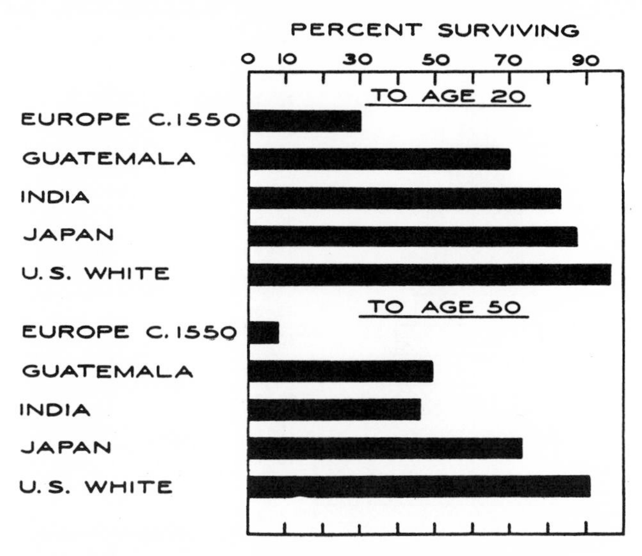

Fig. 16–6. Percentage of newborn females who would survive to the end of the reproductive period according to mortality rates in Europe around A.D. 1500 and in selected countries around 1950.

crease in the birth rate would be required to prevent this increase in survivorship from resulting in a rapid acceleration in the existing rate of population growth, which already is excessive. In other words, this decrease in the death rate would require a decrease in the birth rate of more than 40 percent merely to maintain the status quo.

As can be seen from Figure 16–3, the birth rate in countries with high fertility has shown little or no tendency to decrease in recent years. Japan is the exception. There, the birth rate dropped by 46 percent from 1948 to 1958—an amount more than enough to counterbalance the decrease in the death rate, with the result that there was a decrease in the absolute number of births. As yet there is very little evidence that other countries with a correspondingly high birth rate are likely to duplicate this is the near future.

Another effect of a rapid rate of natural increase is demonstrated by Figure 16–7. About 43 percent of the Moslem population of Algeria is under 15 years of age; the corresponding percentage in Sweden is 24, or slightly more than half this number. Percentages in the neighborhood of 40 percent are characteristic of the populations of the countries of Africa, Latin America, and Asia.

This high proportion of young people constitutes a huge fertility potential for 30 years into the future that can be counterbalanced only by a sharp decline in the birth rate, gives rise to serious educational problems, and causes a heavy drain on the capital formation that is necessary

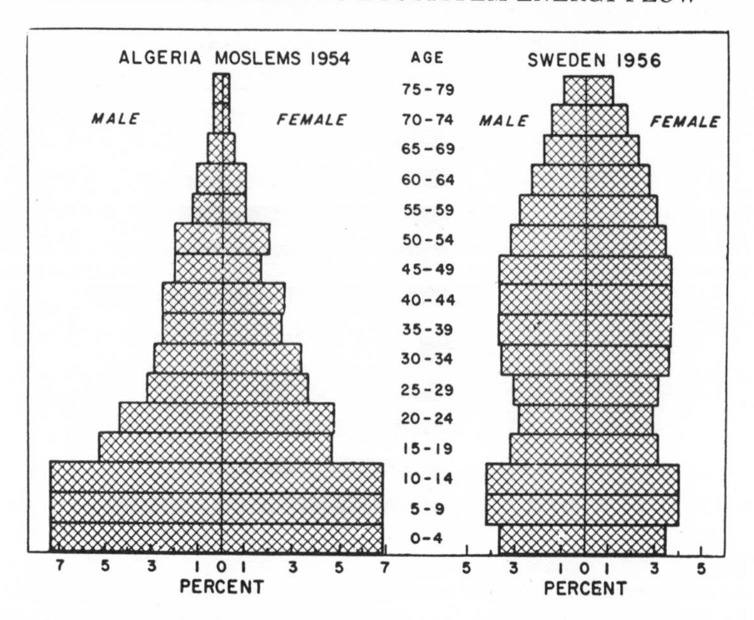

Fig. 16–7. Percentage distribution by age of the population of Sweden in 1956 and the Moslem population of Algeria in 1954.

to improve the level of living of the entire population. A graphic illustration of this may be found in the recently published 5-year plan for India for 1961–66, which estimates that it will be necessary to provide educational facilities and teachers for 20 million additional children during this 5-year period (*13*).

HISTORICAL PATTERN IN WESTERN EUROPE

Some persons, although agreeing that the current rate of increase of the majority of the world's population cannot continue indefinitely without giving rise to grave political, social, and economic problems, point out that a similar situation existed in northwestern and central Europe during the 18th and 19th centuries. Increasing industrialization and urbanization, coupled with a rising standard of living, led to a decline in the birth rate, with a consequent drop in the rate of increase of the population. Why should not the rest of the world follow this pattern?

There is small likelihood that the two-thirds of the world's population which has not yet passed through the demographic revolution from high fertility and mortality rates to low fertility and mortality rates can repeat the history of western European peoples prior to the development of serious political and economic problems. A brief review of the circumstances that led to the virtual domination of the world at the end of the 19th century by persons of European origin will indicate some of the reasons for this opinion.

Around A.D. 1500 the population of Europe probably did not exceed 100 million persons (perhaps 15 to 20 percent of the population of the world) and occupied about 7 percent of the land area of the earth. Four hundred years later, around 1900, the descendants of this population numbered nearly 550 million, constituted about one-third of the world's population, and occupied or controlled five-sixths of the land area of the world. They had seized and peopled two great continents, North and South America, and one smaller continent, Australia, with its adjacent islands; had partially peopled and entirely controlled a third great continent, Africa; and dominated southern Asia and the neighboring islands.

The English-, French-, and Spanish-speaking peoples were the leaders in this expansion, with lesser roles being played by the Dutch and Portuguese. The Belgians and Germans participated only toward the end of this period of expansion. Among these, the English-speaking people held the dominant position at the end of the era, around 1900.

The number of English-speaking persons around 1500, at the start of this period of expansion, is not known, but it probably did not exceed 4 or 5 million. By 1900 these people numbered about 129 million and occupied and controlled one-third of the land area of the earth and, with the non-English-speaking inhabitants of this territory, made up some 30 percent of the population of the world.

This period was characterized by an unprecedented increase in population, a several-fold expansion of the land base for this population, and a hitherto undreamed of multiplication of capital in the form of precious metals, goods, and commodities. Most important of all, the augmentation in capital and usable land took place more rapidly than the growth in population.

A situation equally favorable for a rapid improvement in the level of living associated with a sharp increase in population does not appear likely to arise for the people who now inhabit Latin America, Africa, and Asia. The last great frontier of the world has been closed. Although there are many thinly populated areas in the world, their existence is testimony to the fact that, until now, these have been regarded as undesirable living places. The expansion of population to the remaining open areas would require large expenditures of capital for irrigation, drainage, transportation facilities, control of insects and parasites, and other purposes—capital that the rapidly increasing populations which need these areas do not possess.

In addition, this land is not freely available for settlement. The entire land surface of the world is crisscrossed by national boundaries. International migration now is controlled by political considerations; for the majority of the population of the world, migration, both in and out of a country, is restricted.

The horn of plenty, formerly filled with free natural resources, has been emptied. No rapid accumulation of capital in the form of precious metals, goods, and commodities, such as characterized the great 400-year boom enjoyed by the peoples of western-European origin, is possible for the people of Africa, Asia, and Latin America.

Last, but not least, is the sheer arithmetic of the current increase in population. The number of persons in the world is so large that even a small rate of natural increase will result in an almost astronomical increment over a period of time of infinitesimal duration compared to the duration of the past history of the human race. As was pointed out above, continuation of the present rate of increase would result in a population of 50 billion persons in another 150 years. A population of this magnitude is so foreign to our experience that it is difficult to comprehend its implications.

Just as Thomas Malthus, at the end of the 18th century, could not foresee the effect upon the peoples of western Europe of the exploration of the last great frontier of this earth, so we today cannot clearly foresee the final effect of an unprecedented rapid increase of population within closed frontiers. What seems to be least uncertain in a future full of uncertainty is that the demographic history of the next 400 years will not be like that of the past 400 years.

WORLD PROBLEM

The results of human reproduction are no longer solely the concern of the two individuals involved, or of the larger family, or even of the nation of which they are citizens. A stage has been reached in the demographic development of the world when the rate of human reproduction in any part of the globe may directly or indirectly affect the health and welfare of the rest of the human race. It is in this sense that there is a world population problem.

One or two illustrations may make this point more clear. During the past decade, six out of every ten persons added to the population of the world live in Asia; another two out of every ten live in Latin America and Africa. It seems inevitable that the breaking up of the world domination by northwest Europeans and their descendants, which already is well advanced, will continue, and that the center of power and influence will shift toward the demographic center of the world.

The present distribution of population increase enhances the existing imbalance between the distribution of the world's population and the distribution of wealth, available and utilized resources, and the use of nonhuman energy. Probably for the first time in human history there is a universal aspiration for a rapid improvement in the standard of living and a growing impatience with conditions that appear to stand in the way of its attainment. Millions of persons in Asia, Africa, and Latin America now are aware of the standards of living enjoyed by Europeans and North Americans. They are demanding the opportunity to attain the same standard, and they resist the idea that they must be permanently content with less.

A continuation of the present high rate of human multiplication will act as a brake on the already painfully slow improvement in the level of living, thus increasing political unrest and possibly bringing about eventual changes in government. As recent events have graphically demonstrated, such political changes may greatly affect the welfare of even the wealthiest nations.

The capital and technological skills that many of the nations of Africa, Asia, and Latin America require to produce enough food for a rapidly growing population and simultaneously to perceptibly raise per capita income exceed their existing national resources and ability. An immediate supply of capital in the amounts required is available only from the wealthier nations. The principle of public support for social welfare plans is now widely accepted in national affairs. The desirability of extending this principle to the international level for the primary purpose of supporting the economic development of the less advanced nations has not yet been generally accepted by the wealthier and more advanced countries. Even if this principle should be accepted, it is not as yet clear how long the wealthier nations would be willing to support the uncontrolled breeding of the populations receiving this assistance. The general acceptance for a foreign-aid program of the extent required by the countries with a rapidly growing population will only postpone for a few decades the inevitable reckoning with the results of uncontrolled human multiplication.

The future may witness a dramatic increase in man's ability to control his environment, provided he rapidly develops cultural substitutes for those harsh but effective governors of his high reproductive potential —disease and famine—that he has so recently learned to control. Man has been able to modify or control many natural phenomena, but he has not yet discovered how to evade the consequences of biological laws. No species has ever been able to multiply without limit. There are two biological checks upon a rapid increase in number—a high mortality and a low fertility. Unlike other biological organisms, man can choose which of these checks shall be applied, but one of them must be. Whether man can use his scientific knowledge to guide his future evolution more wisely than the blind forces of nature, only the future can reveal. The answer will not be long postponed.

REFERENCES AND NOTES

1. *Demographic Yearbook* (United Nations, New York, 1955), p. 1.
2. F. A. Pearson and F. A. Harper, *The World's Hunger* (Cornell Univ. Press, Ithaca, N.Y., 1945).
3. H. Brown, *The Challenge of Man's Future* (Viking, New York, 1954).
4. O. E. Baker, "Population trends in relation to land utilization," *Proc. Intern. Conf. Agr. Economists, 2nd Conf.* (1930), p. 284.
5. J. S. Davis, *J. Farm Economics* (Nov. 1949).
6. H. F. Dorn, *J. Am. Statist. Assoc.* 45, 311 (1950).
7. R. Pearl and L. J. Reed, *Proc. Natl. Acad. Sci. U.S.* 6, 275 (1920).
8. H. von Foerster, P. M. Mora, and L. W. Amiot, *Science* 132, 1291 (1960).
9. "The future growth of world population," *U.N. Publ. No. ST/SOA/Ser. A/28* (1958).
10. "The past and future growth of world population—a long-range view," *U.N. Population Bull. No. 1* (1951), pp. 1–12.
11. Although registration of deaths among the Moslem population of Algeria is incomplete, it is believed that the general impression conveyed by Figure 16–2 is essentially correct.
12. *Demographic Yearbook* (United Nations, New York, 1956), p. 14.
13. New York *Times* (5 Aug. 1961).
14. "The determinants and consequences of population trends," *U.N. Publ. No. ST/SOA/Ser. A/17* (1953).

17

The world outlook for conventional agriculture

Lester R. Brown

The problem of obtaining enough food has plagued man since his beginnings. Despite the innumerable scientific advances of the 20th century, the problem becomes increasingly serious. Accelerating rates of population growth, on the one hand, and the continuing reduction in the area of new land that can be put under the plow, on the other, are postponing a satisfactory solution to this problem for at least another decade and perhaps much longer.

Conventional agriculture now provides an adequate and assured supply of food for one-third of the human race. But assuring an adequate supply of food for the remaining two-thirds, in parts of the world where population is increasing at the rate of 1 million weekly, poses one of the most nearly insoluble problems confronting man.

DIMENSIONS OF THE PROBLEM

Two major forces are responsible for expanding food needs: population growth and rising per capita incomes.

Populations in many developing countries are increasing at the rate of 3 percent or more per year. In some instances the rate of increase appears to be approaching the biological maximum. Populations growing

Reprinted with permission from *Science*, 158:604–611 (3 November 1967). Copyright 1967.

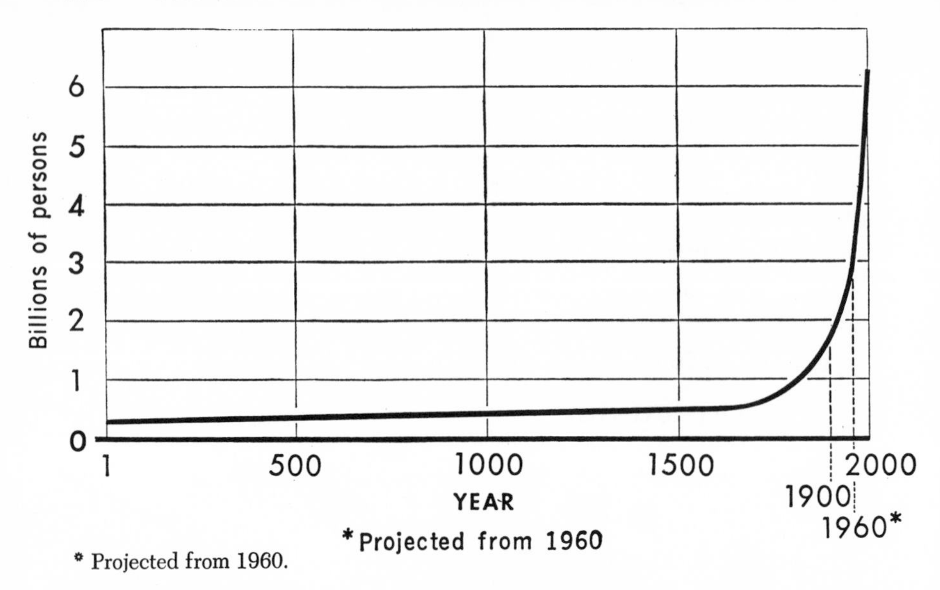

* Projected from 1960.

Fig. 17–1. Twenty centuries of world population growth. [U.S. Department of Agriculture]

by 3 percent per year double within a generation and multiply 18-fold in a century.

According to projections, world population, now just over 3 billion, will increase by another 3 billion over the remaining one-third of this century. (Fig. 17–1). Even with the most optimistic assumptions concerning the effect of newly initiated family-planning programs in developing countries, we must still plan to feed an additional 1 billion people by 1980. The world has never before added 1 billion people in 15 years. More significantly, four-fifths of these will be added to the less-developed countries, where food is already in short supply.

Rising income levels throughout the world are generating additional demand on the world's food-producing resources. Virtually every country in the world today has plans for raising income levels among its people. In some of the more advanced countries the rise in incomes generates far more demand for food than the growth of population does.

Japan illustrates this well. There, population is increasing by only 1 percent per year but per capita incomes are rising by 7 percent per year. Most of the rapid increase in the demand for food now being experienced in Japan is due to rising incomes. The same may be true for several countries in western Europe, such as West Germany and Italy, where population growth is slow and economic growth is rapid.

Comparisons between population growth and increases in food production, seemingly in vogue today, often completely ignore the effect of rapidly rising incomes, in some instances an even more important demand-creating force than population growth.

The relationships between increases in per capita income and the consumption of grain are illustrated in Figure 17–2. The direct consumption of grain, as food, rises with income per person throughout the low-income brackets; at higher incomes it declines, eventually leveling off at about 150 pounds per year.

The more significant relationship, however, is that between total

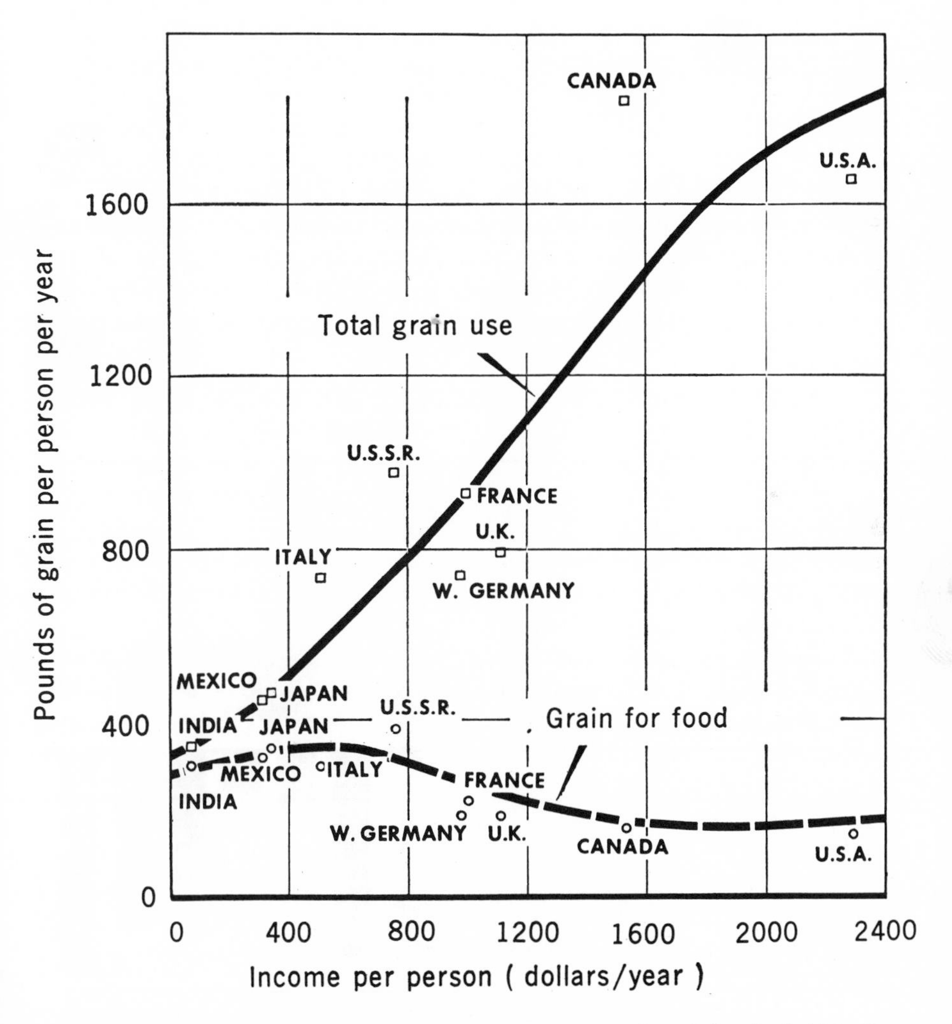

Fig. 17–2. Income and per-capita grain consumption, total and for food (data for 1959–61). [U.S. Department of Agriculture]

grain use and income. Historically, as incomes have risen, the use of grain, both that consumed directly and that consumed indirectly in the form of meat, milk, and eggs, has risen also. The upper curve in Figure 17–2 indicates that every $2 gain in annual per capita income requires one pound of additional grain.

The rapid increases in both population and income are recent phenomena, in historical terms. Both have occurred since the war, and both are gaining momentum on a worldwide scale.

The effect of the resulting explosive increase in the demand for food is greater pressure on the world's food supplies. This rapid expansion of demand, together with the reduction of surplus grain stocks in North America, contributed to a rapid decline in world grain stocks during the 1960's (Fig. 17–3).

Between 1953 and 1961, world grain "carryover" stocks increased each year. The size of the annual buildup varied from a few million tons to nearly 20 million tons. After 1961, however, stocks began to decline, with the reduction or "drawdown" averaging 14 million tons per year.

A stock buildup, by definition, means that production is exceeding consumption; the converse is also true. The trend in grain stocks indicates clearly that 1961 marked a worldwide turning point; as population and income increases gained momentum, food consumption moved ahead of

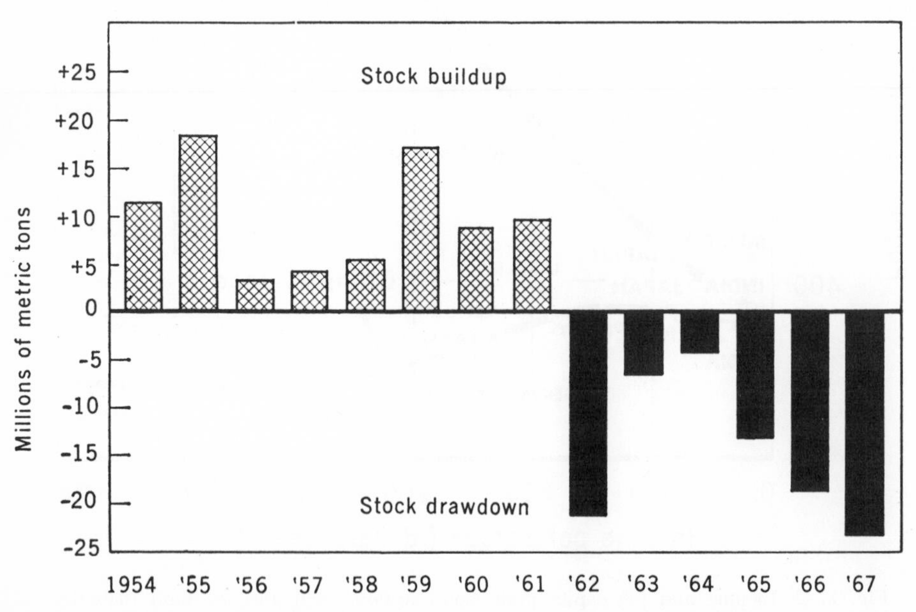

Fig. 17–3. Changes in world grain stocks. [U.S. Department of Agriculture]

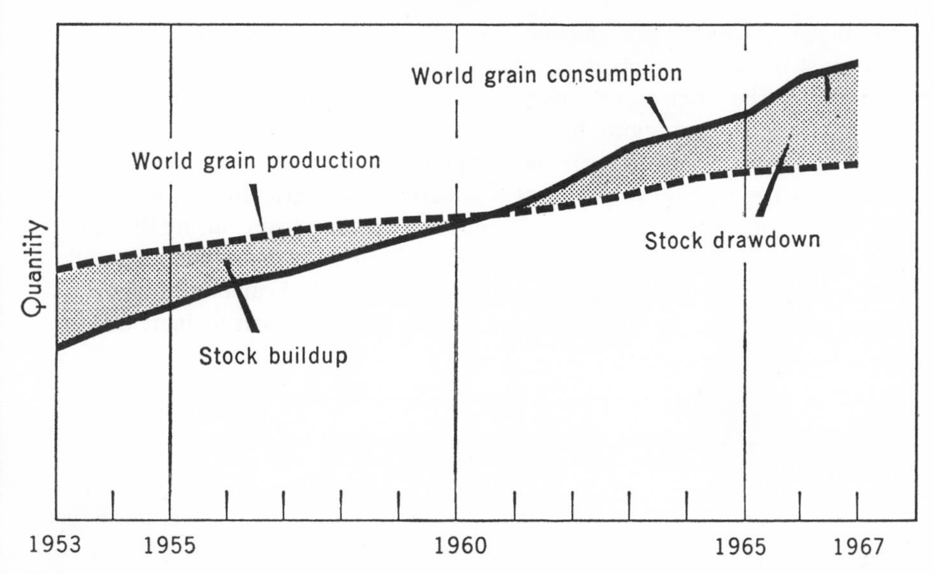

Fig. 17–4. World grain production now lagging behind consumption. (Schematic representation is not drawn to scale.) [U.S. Department of Agriculture]

production. Since 1961, the ever-widening excess of consumption over production has been compensated by "drawing down" stocks. But there is little opportunity for further reductions.

This means that the two lines in Figure 17–4 cannot remain apart much longer. The question is: How will the lines be brought together? Will the production line go up, or will the consumption line come down? What are the implications of recent trends for world food price levels? Rising prices, a possible result, would act both to reduce consumption, particularly among the world's low-income peoples, and to stimulate production. At a time when hunger and, in some cases, severe malnutrition are commonplace in much of the world, reducing consumption is obviously not a desirable alternative. The effect would be to widen the food gap between the world's "haves" and "have-nots."

Meeting future food needs will require immense increases in output. The expected increase of 1 billion in world population over the next 15 years will require expansion of world grain production, now totaling about 1 billion tons, by about one-third, or 335 million tons. Additional demand generated by rising per capita incomes, even if only half as large as the population-generated component, could push the total needed increase toward 500 million tons.

What are the prospects of meeting these future increases in world food needs through conventional agriculture? There are two methods of increasing food production: expanding the cultivated area or raising the productivity (output per unit) of land already under cultivation. Throughout most of history, increases in food production have come largely from expanding the area under cultivation. Only quite recently, in historical terms, have some regions begun to rely on raising output per acre for most of the increases in their food supply (*1*).

Over the past 30 years, all of the increases in agricultural production in North America and western Europe have come from raising the productivity of land. Food output has about doubled in both regions, while the area cultivated has actually declined somewhat. Available technology has made it more profitable to raise output per acre than to increase the area under cultivation.

EXPANDING THE CROPLAND AREA

The world's present cultivated land area totals some 3 billion acres (1.2 billion hectares). Estimates of the possibilities for expanding this area vary from a few hundred million acres to several billion. However, any such estimate of the area of new land likely to be brought under cultivation must, to be meaningful, specify at what cost this is to be accomplished.

Some land which was farmed a few decades ago has now been abandoned because it is no longer profitable. Much of the abandoned farmland in New England and Appalachia in the United States, or in other countries, such as portions of the Anatolian Plateau in Turkey, falls into this category.

In several countries of the world the area of cultivated land is actually declining. Japan, where the area of cultivated land reached a peak in 1920 and has declined substantially since, is a prominent example. Other countries in this category are Ireland, Sweden, and Switzerland.

Most of the world's larger countries are finding it difficult to further expand the area under cultivation. India plans to expand the cultivated-land area by less than 2 percent over its Fourth Plan period, from 1966 to 1971; yet the demand for food is expected to expand by some 20 percent over this 5-year span. Mainland China, which has been suffering from severe population pressure for several decades, has plowed nearly all of its readily cultivable land.

Most of the countries in the Middle East and North Africa, which depend on irrigation or on dry-land farming, cannot significantly expand the area under cultivation without developing new sources of water for irrigation. The Soviet Union is reportedly abandoning some of the land brought under cultivation during the expansion into the "virgin-lands" area in the late 1950's.

The only two major regions where there are prospects for further significant expansion of the cultivated area in the near future are sub-

Saharan Africa and the Amazon Basin of Brazil. Any substantial expansion in these two areas awaits further improvements in our ability to manage tropical soils—to maintain their fertility once the lush natural vegetation is removed.

Aside from this possibility, no further opportunities are likely to arise until the cost of desalinization is reduced to the point where it is profitable to use seawater for large-scale irrigation. This will probably not occur before the late 1970's or early 1980's at best.

The only country in the world which in recent years has had a ready reserve of idled cropland has been the United States. As recently as 1966, some 50 million acres were idled, as compared with a harvested acreage of 300 million acres. The growing need for imported food and feed in western Europe, the Communist countries, Japan, and particularly India is bringing much of this land back into production. Decisions made in 1966 and early 1967 to expand the acreage of wheat, feed grains, and soybeans brought some one-third of the idled U.S. cropland back into production in 1967.

Even while idled cropland is being returned to production in the United States and efforts are being made to expand the area of cultivated land in other parts of the world, farmland is being lost because of expanding urban areas, the construction of highways, and other developments. On balance, it appears that increases in world food production over the next 15 years or so will, because of technical and economic factors, depend heavily on our ability to raise the productivity of land already under cultivation.

INCREASING LAND PRODUCTIVITY

Crop yield per acre in much of the world has changed little over the centuries. Rates of increase in output per acre have, in historical terms, been so low as to be scarcely perceptible within any given generation. Only quite recently—that is, during the 20th century—have certain countries succeeded in achieving rapid, continuing increases in output per acre—a yield "takeoff." Most of the economically advanced countries—particularly those in North America, western Europe, and Japan—have achieved this yield-per-acre takeoff (2).

The first yield-per-acre takeoff, at least the first documented by available data, occurred for rice in Japan during the early years of this century (Fig. 17–5). Yield takeoffs occurred at about the same time, or shortly thereafter, in several countries in northwestern Europe, such as Denmark, the Netherlands, and Sweden. Several other countries, such as the United Kingdom and the United States, achieved yield-per-acre takeoffs in the late 1930's and early 1940's.

Increasing food output per acre of land requires either a change in

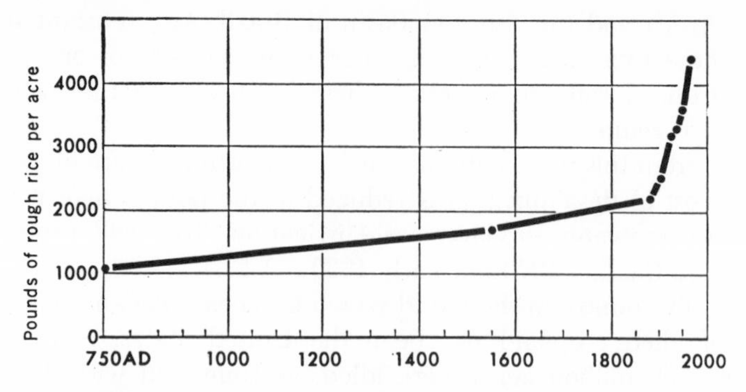

Fig. 17–5. Rice yields in Japan from A.D. 750 to 1960. Historical estimates from Japanese ministry of agriculture. [U.S. Department of Agriculture]

cultural practices or an increase in inputs, or both. Nearly all increases in inputs or improvements in cultural practices involve the use of more capital (*3*). Many (mechanization itself is an exception) require more labor as well (*4*).

A review of the yield trends shown in Figures 17–5 and 17–6, or of any of several others for the agriculturally advanced countries, raises the obvious question of how long upward trends may be expected to continue. Will there come a time when the rate of increase will slow down or cease altogether? Hopefully, technological considerations, resulting from new research breakthroughs, will continue to postpone that date.

Differing sources of productivity

One way of evaluating future prospects for continuing expansion in yields is to divide the known sources of increased productivity into two broad categories: "nonrecurring" and "recurring" sources of increased

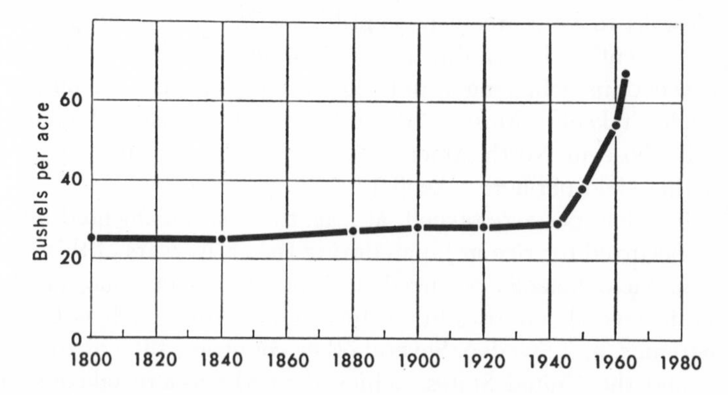

Fig. 17–6. Corn yields in the United States. [U.S. Department of Agriculture]

productivity (5). Nonrecurring inputs are essentially of a one-shot nature; once they are fully adopted, further increases in yields are limited. Recurring inputs, even when fully adopted, offer further annual increases in output through more intensive application.

Corn provides a good illustration. Yields have expanded sharply in the United States (Fig. 17–6). Total production now exceeds 100 million tons of grain annually, or about half the total U.S. grain crop. Much of the increase in corn yields, however, was due to two nonrecurring sources of productivity: the replacement of open-pollinated or traditional varieties with hybrids and, to a lesser extent, the use of herbicides.

Hybrid corn has now replaced open-pollinated varieties on more than 97 percent of the corn acreage in the United States (Fig. 17–7). Further improvements in hybrid varieties are to be expected. (Hybrids in use today are superior to hybrids developed in the mid-1930's.) The big spurt in yields, however, is usually associated with the initial transition from open-pollinated or traditional varieties to hybrids. Consequently, the big thrust in corn yields in the United States resulting from the adoption of hybrids is probably a thing of the past. Likewise, once herbicides are widely used and virtually all weeds are controlled, there is little, if any, prospect of future gains in productivity from this source.

Some sources of increased yields are of a recurring nature. Among these, there is still ample opportunity for further yield increases as a result of the use of additional fertilizer. As plant populations increase, provided moisture is not a limiting factor, corn yields will rise further as more fertilizer is used.

Just how far the yield increase will go in the United States, however,

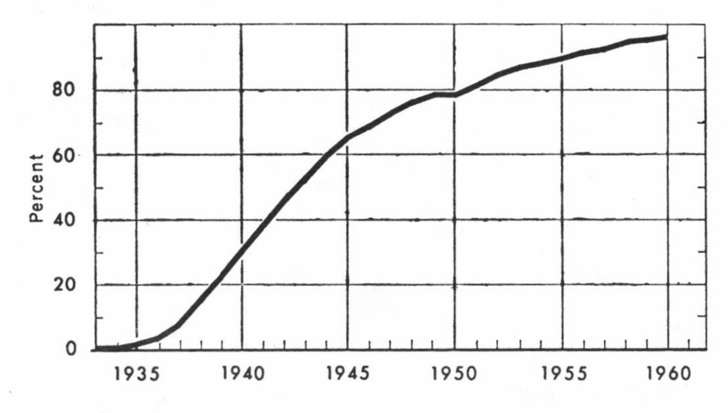

Fig. 17–7. Share of U.S. corn acreage planted with hybrid seed. [U.S. Department of Agriculture]

is not clear. Paul Mangelsdorf of Harvard University, speaking recently at the National Academy of Sciences, asked this vital question (*6*):

With more than 95 percent of the corn acreage already planted to hybrid corn, with the genetic potentials of the hybrids having reached a plateau, with 87 percent of the acreage in the Corn Belt and Lake States already using fertilizer, and with many farmers already employing herbicides, from where will come the future improvements that will allow us to continue our present rate of improvement?

The same question may be asked of other crops in some of the other agriculturally advanced countries.

The S-shaped yield curve

As the nonrecurring sources of productivity are exhausted, the sources of increased productivity are reduced until eventually the rate of increase in yield per acre begins to slow. This might be depicted by that familiar biologic function the S-shaped growth curve (Fig. 17–8). John R. Platt of the University of Chicago recently explained the curve this way (*7*):

Many of our important indices of technical achievement have been shooting up exponentially for many years, very much like the numbers in the biologists' colonies of bacteria, that double in every generation as each cell divides into two again. But such a curve of growth obviously cannot continue indefinitely in any field. The growth of the bacterial colony slows up as it begins to exhaust its nutrient. The exponential curve bends over and flattens out into the more general "S-curve" or "logistic curve" of growth.

We do not know with any certainty when the rate of yield increase for the major food crops on which man depends for sustenance will begin to slow, but we do know that ultimately it will.

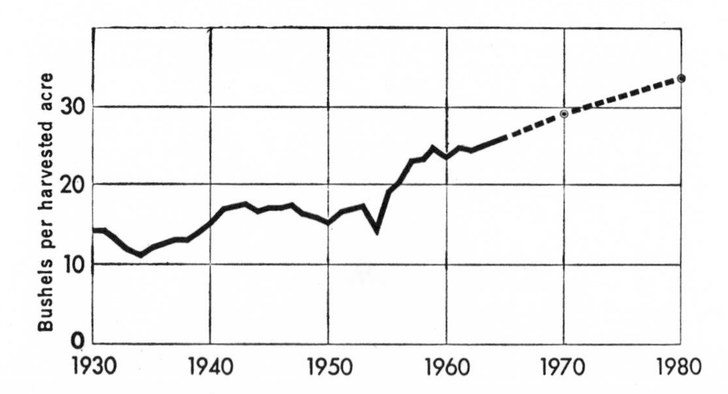

Fig. 17–8. S-shaped yield curve (schematic representation). [U.S. Department of Agriculture]

The key questions are: Is the slowdown near for some of the major food crops in some of the agriculturally advanced countries? Will the slowdown come gradually, or will it occur abruptly and with little warning? Finally, to what extent can the level at which the final turn of the S-shaped yield curve occurs be influenced? Can the level be raised by increasing the prices received by farmers, by adopting technological innovations, and by stepping up investment in crop research?

Most of those countries which have achieved takeoffs in yield per acre are continuing to raise yields at a rapid rate. But there are indications that the rate of gain may be slowing for some crops in some of the more agriculturally advanced countries.

Projected per-acre yield levels for the major grains in the United States show a substantial slowing of the rate of yield increase over the next 15 years as compared with the last 15. The rate of yield increase for wheat, averaging 3.5 percent yearly from 1950 to 1965, is projected to drop to less than 2 percent per year between 1965 and 1980 (Fig. 17–9). Sorghum yields, recently increasing at a rate of nearly 6 percent annually, are projected to increase at just over 2 percent per year between now and 1980 (Fig. 17-10). For corn, the projected slowdown is less dramatic, with yield increases dropping from about 4 percent to 3 percent. Per-acre yields of wheat and grain sorghum have apparently achieved their more rapid gains as the use of nonrecurring technologies becomes almost universal. In Platt's words (7), they may already be "past the middle of the S-curve."

The rate of increase could also be slowing down for certain crops elsewhere in the world. Rice yields in Japan may be a case in point. Yields were relatively static before 1900 but began to rise steadily shortly

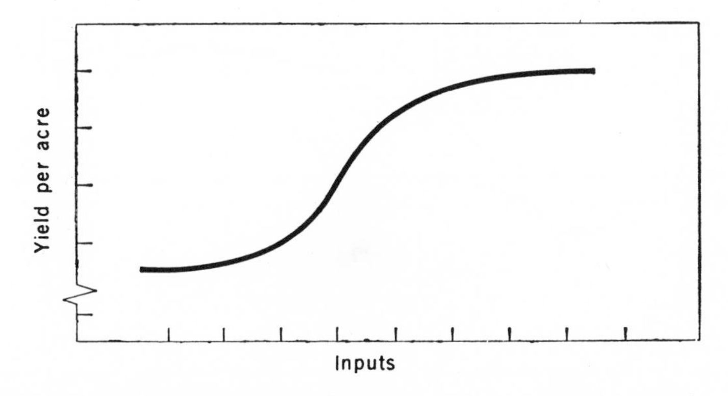

Fig. 17–9. Wheat yields in the United States, with projections. Plotted as a 3-year sliding average. [U.S. Department of Agriculture]

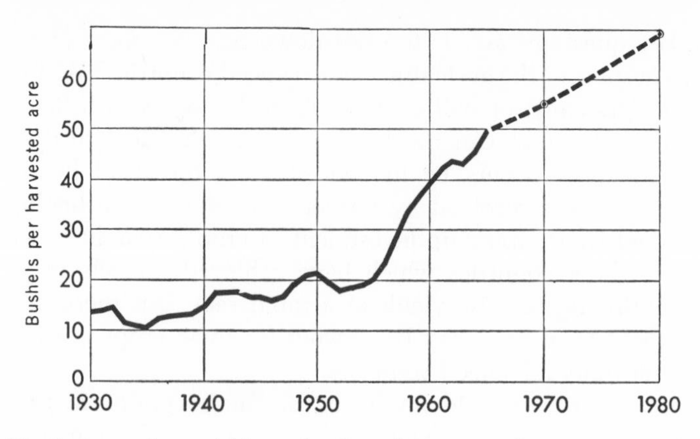

Fig. 17–10. Grain sorghum yields in the United States, with projections. Plotted as a 3-year sliding average. [U.S. Department of Agriculture]

after the turn of the century. This rise continued until about 1959 (except for a brief period around World War II, and a period from 1949 to 1953, when production was disrupted by land reform). Since 1959, U.S. Department of Agriculture estimates (8) indicate, the rate of increase has slowed appreciably and, in fact, has recently nearly leveled off (Fig. 17–11). Whether or not this is a temporary plateau or a more permanent one remains to be seen. Interestingly, projections of per-acre rice yields made by the Japanese Institute of Agricultural Economic Research, using a 1958–1960 base period (9), did not anticipate the recent slowdown in the rate of increase in rice yields.

This recent leveling off of yields, however, may be caused by economic as well as technological factors. One key factor contributing to the very high yields obtained in Japan has been the intensive use of what

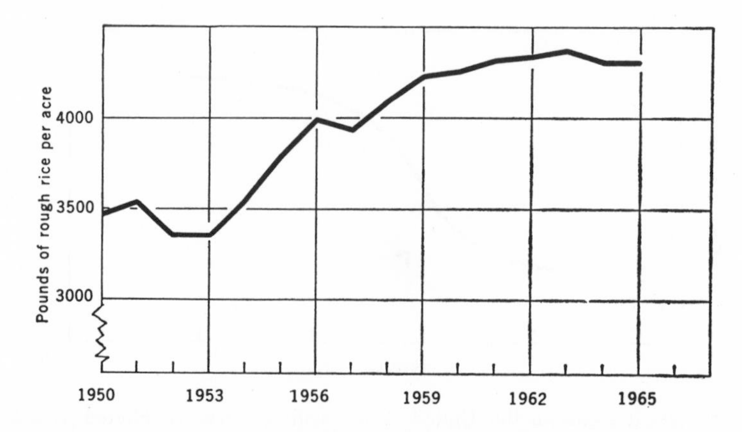

Fig. 17–11. Rice yields in Japan, 1950–1965. Plotted as a 3-year sliding average. [U.S. Agriculture]

was once low-cost labor. In recent years there has been a withdrawal of labor from rice production as rural workers have found more remunerative urban jobs. If economic development continues, it is unlikely that recent trends in labor costs will ever be reversed. Thus, it may well be that per-acre rice yields in Japan are approaching what is, in the immediately foreseeable future at least, a plateau.

A slowdown in the rate of yield increase seems also to be occurring for some of the grain crops in the Netherlands. This is not particularly surprising since yields there are already among the highest in the world. Further yield responses of some grains to the use of additional inputs, such as fertilizer, now seem limited by genetic constraints—the inherent ability of the plant to effectively use additional plant nutrients.

There are, on the other hand, some crops in the agriculturally developed countries which have not yet begun their upward advance on the growth curve. One of the major U.S. crops, the soybean, has thus far stubbornly resisted efforts to generate a yield-per-acre takeoff (*10*). The combination of near-static yields, on the one hand, and the very rapid growth in demand for soybeans, on the other, means that the necessary increases in the soybean supply are obtainable only through a rapid continuing expansion in the area planted to soybeans—an expansion which is steadily reducing the area available for other crops.

During the two decades since World War II, projections of increases in per-acre yields in the United States have invariably underestimated the increases actually achieved. This may be due in part to the yield-raising effect of idling large areas of marginal cropland during this period. There is now a risk that our faith in technology will cause us to overestimate future increases in yields if, in fact, the rate of yield increase ultimately slows as the sources of further gains in productivity diminish.

It is significant that the major sources of increased agricultural productivity—the use of chemical fertilizer; the use of improved varieties, including hybrids; the use of pesticides and irrigation—have all been known for decades, if not longer. The key question now is: Are there any sources of increased productivity in existence or in the process of development comparable to the traditional ones listed above?

The concept of the S-shaped curve is not new, but its implications for future agricultural production have not been fully explored. Although the S-shaped yield curve for crops is, at this point, still an untested hypothesis, it is, in Platt's words (*7*), "at least as plausible as the uncritical assumption that changes like those of the twentieth-century will go on forever."

Photosynthetic efficiency and research

The ultimate factor limiting crop output per acre is the crop's photosynthetic efficiency (*11*). Defined as the percentage of solar energy used relative to that which is available on a given area occupied by a particular crop, photosynthetic efficiency is always quite low, usually less than 3 percent. Density of plant population, actual position of the leaves on the plant, and temperature are key factors accounting for variations within this range.

In 1962, James Bonner of the California Institute of Technology stated (*11*):

> . . . the upper limit of crop yield, as determined by the factors that regulate photosynthetic efficiency, is already being approached today in those regions with the highest level of agricultural practice—in parts of Japan, of Western Europe, and of the United States.

Obviously, research into ways of increasing the upper limit of yield is needed. This increase could be achieved by developing plants which have greater photosynthetic efficiency or by improving present cultural practices so as to increase efficiency per acre, or by both means. The development of smaller and more efficient corn plants, along with reduction in the need for cultivation during the growing season, makes it possible to reduce the width between corn rows—a width that was initially determined by the width of a horse, in the age of the horse-drawn cultivator. The result is a dramatic gain in the number of corn plants per acre, and increased output.

More productive hybrid wheats have been developed, but they are still in the experimental stage and are not yet being grown commercially. Work on breeding new varieties with higher nutritive value—a potentially promising activity—is also under way. The adoption of a new technology takes time, even in an agriculturally advanced country. It took a quarter of a century for U.S. farmers to adopt hybrid corn (Fig. 17–7). Hybrid grain sorghum, introduced in the early 1950's, required about a decade to become widely disseminated.

Both corn and wheat have been the subject of many years of research in the United States and other developed nations. Much less work has been done in rice. To help rectify this situation, the Rockefeller and Ford foundations established the International Rice Institute in the Philippines several years ago. The Institute devotes its efforts not only to the development of new varieties but to the whole range of cultural practices as well.

The need for such research is further emphasized by a recent statement by Harvey Brooks, chairman of the Committee on Science and Public Policy of the National Academy of Sciences (*12*):

> Future food production, even for domestic purposes, will be strongly dependent on the quality and direction of both the basic and applied research undertaken within the next few years. Most of the potential of past basic research has already been realized, and new knowledge will be needed even to maintain present levels of productivity.

Clearly, much more research is essential if we are to (i) get the underdeveloped nations to the yield takeoff point, and (ii) maintain the upward thrust of yields in the developed countries by postponing the final turn on the S-shaped curve (*13*).

RESEARCH AND REALITY

Two groups of factors should be kept in mind in evaluating the real potential of research results for significantly increasing food output on a worldwide basis. The first group centers about the pronounced variations in natural resources and managerial abilities, which can lead to wide differences between record yields and average national yields obtained by individual farmers under localized conditions. The second group concerns the matter of costs and returns, which spells the difference between technical potential and economic reality.

Record yields versus average yields

It is often assumed that record yields attained on experimental plots can be easily and quickly translated into national average yields. Such is not, however, the case. Maximum yields obtained on experimental plots under closely controlled conditions usually far exceed those generally obtained in practice. Average yields of wheat in this country, for example, are far below those attained on experimental plots during the latter part of the last century. The same is true for many other crops.

Equally common and equally unwarranted is the assumption that all countries will eventually attain the average yield prevailing in the nation which now has the highest yield. Potential yield levels attainable by individual countries vary widely with variations in rainfall, temperature, soil types and topography, production costs, managerial abilities of farmers, and other factors.

Wheat yields in the United Kingdom now average about 60 bushels per acre (52 hectoliters per hectare) as contrasted with only 18 bushels per acre in Australia. This does not mean that wheat-production technology is less advanced in Australia than in the United Kingdom. The yield differences do reflect the difference between growing conditions in Australia, where rainfall in the wheat-growing regions averages 12 to 15 inches (30 to 38 centimeters) annually, and those in the United Kingdom, where rainfall may average 40 to 50 inches. Although wheat yields in both the United Kingdom and Australia may continue to rise, there is no reason to assume that the differences in yields between the two countries will narrow appreciably in the foreseeable future.

The average national rice yield in Japan is nearly four times that in

India. A large part of this difference is accounted for by a much greater volume of inputs, including labor as well as modern practices and management. Not to be overlooked, however, is the fact that virtually all of the rice crop produced in Japan is irrigated, whereas only part of India's rice crop is irrigated. A large share of India's rice fields are rainfed, thus the yield levels attained depend greatly on the vagaries of the monsoon.

There are also very wide variations in yield within individual countries. Variations in corn yields within various corn-producing states in the United States are almost as pronounced as variations in corn yields between the various corn-producing countries of the world. Average yields in principal U.S. corn-producing states in 1965, for instance, varied from more than 90 bushels per acre in some states in the Midwest to less than 40 bushels in some states in the southern Mississippi Valley.

It is significant that the leveling off of rice yields in Japan has occurred at a time when average rice yields in the more productive and the less productive prefectures vary widely. Some individual villages in Japan obtain rice yields at least double the national average.

Per-acre yields obtained by individual farmers in the same area may vary even more than do those for various states or prefectures. It is often assumed that the performance of the best farmers can be emulated by all. There are and will continue to be some very basic differences in the innate capacities or motivation of farmers. There is no more reason for assuming that all farmers can or really want to attain a record yield of corn or wheat than to assume that all students can or want to become Harvard Phi Beta Kappas. The distribution of talent and motivation is probably at least as wide within the world's rural communities as in any other area.

Technical potential versus economic reality

The failure to distinguish between the technical potential for expanding food production and the economically profitable possibilities for doing so has resulted in confusing variations in estimates of future food production. The difference between estimates based on these two criteria is often very great. The earlier discussion of the experience in Japan—where rice yields seem to have leveled off in recent years—suggests the importance of economic relationships.

A recent reduction in milk production in the United States closely parallels the Japanese experience with rice yields. Through the early months of 1966, milk production in the United States was 3 to 5 percent below production in comparable months of the preceding year. At prevailing prices it was not profitable for dairy farmers to use some of the existing resources. During 1966, dairy farmers in New York State received scarcely 40 cents an hour for their labor (when allowance is made for interest on their investment), and farmers in Wisconsin received only 50 cents an hour. At a time when slaughter prices were high and there were many job opportunities to choose from—with a 5-day, 40-hour week in industry and a minimum wage of $1.25 per hour (*14*)—it comes as no surprise to learn that many dairy farmers liquidated their holdings and took other jobs. In order to help increase returns to farmers and expand milk production, the Department of Agriculture raised milk support prices twice during 1966, for a total increase of 23 percent.

Both prices received by farmers and costs of production must be taken into consideration in assessing potential increases in production. As farmers move up the per-acre yield curve, the point of diminishing returns is eventually reached. Additional costs begin to exceed additional returns. Thus it is unrealistic to expect farmers to produce up to the full technical potential.

Therefore, while many farmers can produce much more under a given technology, it is sometimes uneconomic, at existing prices and costs, for them to do so. If society is willing to pay higher prices—and it may have to some day—much greater production may be expected.

CONCLUSIONS

1) The worldwide demand for food will continue to be strong in the coming decades. Two forces—rapidly growing population and, in much of the world, rapidly rising incomes—are expected to result in increases in the demand for food even more rapid than those that have occurred during the past.

2) Conventional agriculture has assured an adequate food supply for the economically advanced one-third of the world. The challenge now is to assure an adequate food supply for the remaining two-thirds, where population is now increasing at the rate of 1 million people per week and where malnutrition is already widespread.

3) Economically feasible prospects for significantly expanding the world's area of cultivated land in the 1960's and 1970's are limited and largely confined to sub-Saharan Africa and the Amazon Basin. Even here, agronomic problems will limit the rate of expansion. When the cost of desalting sea-water is substantially reduced—probably not before the late 1970's or early 1980's at best—it may become feasible to irrigate large areas of desert.

4) Given the limited possibilities for expanding the area of land under cultivation, most of the increases in world food needs must be met, for the foreseeable future, by raising the productivity of land already under cultivation. Food output per acre, rather static throughout most of history, has begun to increase rapidly in some of the more advanced countries in recent decades. All of the increases in food production over the

past quarter century in North America, western Europe, and Japan have come from increasing the productivity of land already under cultivation. The area under cultivation has actually declined.

5) Achieving dramatic gains in land productivity requires a massive investment of capital and the widespread adoption of new technology. A similar effort must now be made in the less-developed nations if these nations are to feed their people. The most important single factor influencing this rate of investment is food prices, more particularly the relationship between the price farmers receive for their food products and the cost of modern inputs such as fertilizer.

6) In some of the more-developed countries where per-acre yields have been rising for a long time, there is now evidence that the rate of yield increase may be slowing. Nonrecurring inputs may have made their maximum contribution to output in the case of some crops, pushing yield levels past the middle of the S-shaped logistic curve. Although this cannot be determined with any certainty, the possibility that the middle of the curve has been passed in some instances should be taken into account in viewing the long-term future.

7) If the rate of increase in yield per acre does in fact begin to slow in some of the agriculturally advanced countries, additional pressure will be put on the less-developed countries—which have much of the world's unrealized food-production potential—to meet the continuing future increases in world food needs.

8) Man has not yet been able to by-pass the process of photosynthesis in the production of food. This dependence on photosynthesis plays a significant role in determining the upper levels of the S-shaped yield curve. Additional research is urgently needed to increase the photosynthetic efficiency of crops and to raise the upper levels of economically feasible yields.

REFERENCES AND NOTES

1. I have previously examined these matters in some detail in "Man, Land and Food," *U.S. Dept. Agr. Foreign Agr. Econ. Rep. No. 11* (1963).
2. I have discussed this concept at length in "Increasing World Food Output," *U.S. Dept. Agr. Foreign Agr. Econ. Rep. No. 25* (1965).
3. For further discussion of this point, and the role that may be played by private industry, see L. R. Brown, *Columbia J. World Business* 2, No. 1, 15 (1967).
4. As one leading agricultural economist recently stated, there is considerable evidence that in most low-income countries "technological advance requires a complementary input of labor" [J. Mellor, *The Economics of Agricultural Development* (Cornell Univ. Press, Ithaca, N.Y., 1966), p. 157].
5. The "nonrecurring" concept was introduced by Paul C. Mangelsdorf (see 6).
6. P. C. Mangelsdorf, *Proc. Nat. Acad. Sci. U.S.* 56, 370 (1966).

7. J. R. Platt, *The Road to Man* (Wiley, New York, 1966) [originally published in *Science* 149, 607 (1965)].
8. Estimates published by the Food and Agriculture Organization show a continued increase in rice yields up until 1963–64, followed by successive declines in each of the three following seasons; see annual issues of *Production Yearbook* (Rome) and *Monthly Bull. Agr. Economics Statistics* 15, No. 12, 26 (1966).
9. *Japanese Import Requirement: Projections of Agricultural Supply and Demand for 1965, 1970 and 1975* (Institute of Agricultral Economic Research, University of Tokyo, 1964), p. 84.
10. Soybeans cannot be commercially hybridized and show only limited response to nitrogen: see *The World Food Problem* (Government Printing Office, Washington, D.C., 1967), vol. 2, p. 197.
11. J. Bonner, *Science* 137, 11 (1962).
12. *The Plant Sciences Now and in the Coming Decade* (National Academy of Sciences, Washington, D.C., 1966), p. iv.
13. A detailed discussion of the technical problems and issues faced in intensifying plant production in the developing nations is presented in *The World Food Problem* (Superintendent of Documents, Government Printing Office, Washington, D.C., 1967), pp. 215–233.
14. The minimum wage was recently raised to $1.40 per hour.
15. I am indebted to Dana G. Dalrymple of the U.S. Department of Agriculture for his suggestions and assistance.

18

Prospects for reducing natality in the underdeveloped world

Dudley Kirk

Two-thirds of the world's people live in its "developing" or less industrialized countries.[1] The populations of these countries are growing at an accelerating pace in the so-called "population explosion," thanks to welcome gains in reducing mortality.

The present imbalance between birth and death rates in the developing world is generally recognized. Current rates of population growth cannot continue for any long historical period. Already the growth rate in many developing countries exceeds 3 per cent per year, a rate at which numbers double in twenty-three years and increase twenty-fold in the course of a century. From the mathematics of compound interest, it is clear that in the not too distant future, birth rates must come down or death rates must go up.

Of more immediate concern is the fact that the present and emerging rates of population growth seriously handicap socioeconomic advance in many of the developing countries. In some it has created food shortages and even the threat of famine. A lower rate of growth (that is, a lower birth rate) would facilitate their development.

With continuing decline in death rates throughout the world, the modal European birth rates of some seventeen or eighteen births per

---

Reprinted with permission from *Annals of the American Academy of Political and Social Science*, 369:48–60 (1967).

[1] Africa, Asia excluding Japan and the Soviet Union, and Latin America excluding Argentina and Uruguay.

thousand population would seem to be a reasonable target for the world as a whole by the end of the century. This is what has been achieved in Japan, and this is probably a factor in the spectacular economic success of that country. As in Europe and Japan today, such birth rates would still leave a margin of population growth.

Birth rates in the developing areas still average 40 to 45 per 1,000 population. To bring these rates down to a reasonable balance of natality and mortality (that is, the European level) involves a drop of some 25 points, or more than 50 per cent from the current level. With a present population of 2.3 billion in the underdeveloped world, such a reduction implies the prevention of well over 50 million births per annum at this time. These are the dimensions of the problem of reducing natality in the underdeveloped world.

Fortunately, there are grounds for optimism that birth rate can be brought down in time to avert a rise in death rates, barring some catastrophe such as nuclear war. The prospect for reductions in the birth rate in the less industrialized countries has never been more hopeful than it is today.

(1) Foremost perhaps is the rapid change in climate of opinion regarding family planning. To appreciate the significance of this change it is important to note that the European demographic transition occurred in an atmosphere of overt and covert institutional hostility toward birth control. The European population learned to limit family size, but the process was gradual and stemmed from private decisions made *in spite of* restrictive legislation, religious opposition (Protestant as well as Catholic), and public denunciation of birth control practices in what was generally a "conspiracy of silence" on sexual matters and human reproduction. The prevailing middle-class morality prevented free public discussion, and public authorities harassed militant fringe groups that advocated birth control. With the growing understanding of the impact of high rates of population growth on economic development, the hitherto restrictive atmosphere is rapidly yielding to a climate of public approval and sponsorship of family planning programs which can be expected to accelerate the adoption of contraceptive practice.

(2) In the developing world, except for Latin America, religious doctrine does not oppose family planning. Religions other than the Catholic and the Orthodox do not have clear doctrinal positions that ban the use of contraceptives, and their views on abortion are often more permissive; for example, the Moslem doctrine forbids abortion after the quickening, but it is far less clear on earlier abortions. This is not to suggest that religion does not play a part in natality differentials in the developing countries; it does, but not in the sense of formal opposition to family planning.

(3) New methods of contraception, derived by intensive research,

now offer a wider choice of methods and already include some that seem applicable in poor and peasant cultures. The two major new developments are the oral contraceptives and the intra-uterine devices (IUD's). Even better methods are on the way.

In the Western experience, the decline in the birth rate resulted from later age at marriage; from male methods of contraception, especially *coitus interruptus* and the condom; and from the female method of abortion. These are still much the most common methods of family limitation in the world. The "conventional" chemical and mechanical methods have probably not been of major importance in controlling Western fertility, and seem unsuited to the peoples of the developing world. Continued research will almost certainly produce even more effective, simple, inexpensive, and safe methods, unrelated to the sex act. Not to be overlooked are better techniques of abortion and the general availability and acceptance of abortion in many countries.

(4) The rapid decline of the death rate in today's developing countries is itself an important factor in changing motivations in favor of family limitation. With more children surviving, the pressure of larger families, especially in the context of growing aspirations for the children, is more quickly evident. The change from low to moderate and even high survival rates within a single generation has an impact on parents and grandparents that favors considerably more rapid acceptance of family planning than occurred in the West, where slowly declining death rates were followed by a gradual decline in births.

(5) The traditional view that birth rates and population growth are a "given" not amenable to deliberate social and governmental influence is no longer tenable. Today it is increasingly appreciated among all cultures and strata of people that, within nature's limits, man can control his destiny over births as well as deaths. Numerous surveys in many developing countries on knowledge, attitudes, and practice (the so-called KAP surveys) have shown that the general public, rural as well as urban, is interested in controlling family size. Very few now think it necessary to have as many children "as God wills." On the contrary, almost everywhere there is a substantial "market" for contraceptive knowledge, materials, and services, provided these are made available in a form appropriate to local conditions.

(6) A final point concerns the relationship between population growth and economic progress. Despite impressive aggregate economic gain in many developing countries in recent years (of the order of 4 per cent to 6 per cent per year on growth of real gross domestic product),[2] per capita gains have often been trivial and even negative. In the matter of food and agricultural production, world output since 1958 has grown at about the

[2] United Nations, *Statistical Yearbook, 1965*. Tables 183, 184.

same rate as world population, but only because the developed world has compensated for deficiencies in per capita output in the developing countries.[3] Although primarily agricultural, the developing countries, with over two-thirds of the world's people, account for less than 45 per cent of the world's agricultural output.[4]

POPULATION POLICIES

Confronted with these realities and the failure of their economic development plans to materialize in per capita terms, responsible leaders in developing countries are increasingly concerned with the handicaps of rapid population growth to social and economic progress. Their growing realization that people want and presumably will use available contraceptive information and services has precipitated and reinforced a novel view: that governments can and should do something about high rates of population growth.

The pressures for the adoption of policy are more than a quinquennial affair related to drawing up five-year economic development plans. They are experienced daily in many forms: shrinking plots of land as numerous children inherit their parents' holdings, as in India; in the formidable task of providing schooling for a child population that exceeds 40 per cent of the total population; in crowded urban slums, as rural people migrate to cities to seek escape from the poverty of the countryside; in large numbers of unemployed and many more underemployed; and in growing resort to induced abortion, as in Latin America, where one-fourth of the beds of the major maternity hospitals are occupied by women with complications arising out of illegal abortions.

In response to these and similar forces, population policy has become an accepted part of development programs. By now more than half the people in the developing world live under governments that favor family planning.[5]

Asia

A review of government family planning programs appropriately begins with Asia. In numbers it is the principal home of mankind; it has the most visible population problems; the four largest countries in the developing world are on that continent; and it is the region where national family planning programs under governmetal auspices began.

---

[3] *Ibid.*, Table 6.

[4] *Ibid.*, Table 4.

[5] For official statements of governmental policy on population, see The Population Council, *Studies in Family Planning*, No. 16 (New York: Population Council, in press).

India, Nepal, and Pakistan

The first country to adopt a national policy to control population growth was India. The beginnings, in the mid-1950's, were modest. Major bottlenecks were lack of an administrative structure to reach India's enormous rural population, and acute shortages of medical personnel, especially of women doctors. Except for financial subsidies, authority rests with the individual states. Organization in the Ministry of Health has moved slowly, and the government has been slow to adopt the newer methods of contraception. Nevertheless, the program has gained momentum as indicated by the data on expenditures in Table 18–1.

TABLE 18–1. India's Family Planning Expenditures, 1951–1971

Five Year Plan	Expenditures (Millions of Rupees)
First (1951–1956)	1.5
Second (1956–1961)	21.6
Third (1961–1966)	261.0
Fourth (1966–1971)—Allocation	950.0

Annual expenditures have risen from 13.8 million rupees in 1961–1962 to 60.5 in 1964–1965 and to an estimated 120 in 1965–66.[6] The last two represent per capita expenditures of about two and four cents. The Fourth Plan allocation implies an annual expenditure of $25.4 million or about five cents per person. Food shortages have made India's efforts more urgent. Symbolic of this is the change in name from Ministry of Health to Ministry of Health and Family Planning.

The Indian program has relied chiefly on clinics, which numbered 15,808 in July 1965. Although rhythm and conventional mechanical and chemical methods were first offered, IUD's were introduced in 1965, with one million inserted by mid-1966. Since 1963 over 100,000 male and female sterilizations have been performed per year in "sterilization camps" temporarily set up for this purpose. The cumulative total of sterilizations is soon expected to exceed one million. Oral contraceptives have not yet been approved for general use in India.

Impressive as these recent achievements are, they have not had any measurable influence on the Indian birth rate, except perhaps in the largest Indian cities, where family planning seems to be spreading in the population (but probably as much through private as through government services). The object of the program is to reduce the annual birth rate

[6] B. L. Raina, "India," in Bernard Berelson *et al.* (eds.), *Family Planning and Population Programs: A Review of World Developments* (Chicago: University of Chicago Press, 1966).

from over 40 to 25 per thousand population "as soon as possible." Targets rise annually to 1971 when it is hoped to have 19.7 million IUD users, 4.5 million sterilizations, and 4.7 million condom users.[7] Success in this magnitude is indeed necessary for a major impact on the Indian birth rate. India is a country of almost 500 million people, with some 100 million women in the reproductive ages and some 20 million births annually.

Interested in these developments in India, Nepal incorporated a family planning program in its Third Plan, 1965–1970, but it has yet to be implemented.

Pakistan formulated a population control program in the late 1950's which was allocated 30.5 million rupees in the Second Five Year plan (1960–1965). The objective is to supply family planning services through the existing health services in hospitals, dispensaries and rural clinics for voluntary participation of couples in limiting family size and spacing of children.

These efforts encountered many of the problems noted in India. In the first four years only 9.4 million rupees[8] of the budgeted expenditures of 24.7 million were actually used.[9] The problems related more to administration and organization than lack of funds. The plan for reaching 1.2 million women as contraceptive users and the targets for distribution of contraceptives, specifically condoms and foam tablets, were met only at the level of 17 percent and 15 percent, respectively.[10]

In July 1965 the Family Planning Directorate was upgraded to one of the most ambitious in the world today. Its five-year budget of 300 million rupees represents an average annual budget of about 12 cents per person. A major innovation is to be the insertion of IUD's by midwives, under medical supervision. These midwives will receive incentive payments for referrals and insertions. By 1970 no less than 50,000 village midwives are to be recruited and given a five-week training course.[11]

Korea and Taiwan

More immediate succcess has been achieved in two smaller Asian countries, South Korea and Taiwan. South Korea's Supreme Council for National Reconstruction adopted a national family planning policy in 1961 as an integral part of the development plan. The targets were more specific than in most countries as regards training, use of contraceptives and effects on the rate of growth. By April 1965 some 2,200 full-time field

---

[7] *Ibid.*, p. 119.

[8] One Pakistan rupee equals 21 cents.

[9] E. Adil, "Pakistan," in Berelson *et al.*, *op. cit.*, p. 127.

[10] *Ibid.*, p. 128.

[11] Government of Pakistan, *Family Planning Scheme for Pakistan during the Third Five Year Plan Period, 1965–1970*, pp. 3, 10.

workers had been recruited and trained, one for each 2,500 women in the childbearing ages.[12] Main reliance of the program is now on the IUD. Insertions were 112,000 in 1964, and 233,000 in 1965,[13] against targets of 100,000 and 200,000, respectively. The latter is estimated to be about 15 percent of the target women, that is, those exposed to the risk of unwanted pregnancy. IUD's and conventional contraceptives are being manufactured in Korea.

The official objective is to reduce the rate of population growth from a current estimate of 2.9 percent per year to 1.8 percent in 1971. The Economic Planning Board estimates that by 1980 full implementation of the family planning program will bring down the rate of growth to 1.16 percent compared to 3.15 without a reduction in the birth rates. The difference in growth rates will mean a per capita income 36 percent higher in 1980 that would be the case in the absence of a fertility decline.

These ambitious objectives would sound unrealistic were it not for the fact that the government program is clearly "swimming with the tide" of social change. Attitude surveys show an overwhelming approval of family planning in the Korean population and a rapid increase in contraception and abortion,[14] the latter perhaps because of Japanese influence. Though not yet approved by the government, a bill is currently before the Korean Assembly to legalize induced abortion.

In the absence of accurate vital statistics, it is difficult to measure year-to-year changes in the birth rate. An indirect measure, ratios of children under five to women in the childbearing ages, does strongly suggest a recent rapid decline in the birth rate, especially in 1965. Although this decline was already in process, the greater rate of decline in 1965 might reflect the effects of the government program in 1964, when it first reached mass proportions.

Taiwan does not have an official population policy, but family planning services are now provided throughout the island by the Provincial Department of Health. Impetus for the island-wide program stemmed from the mass action research program in the city of Taichung[15] (in which it was first established that the IUD would be widely accepted in a mass campaign) and from experimental projects begun in 1959 under the euphemistically-called "Pregnancy Health Program." This term paid tribute to the sensitivities of the United States Agency for International

---

[12] Government of Korea, *Korea, Summary of First Five-Year Economic Plan, 1962–1966*.

[13] Ministry of Health and Social Welfare, Government of Korea, *Monthly Report on IUD Insertions* (mimeographed).

[14] According to a survey made in 1964, one out of three pregnancies among married women in Seoul is terminated by induced abortion. See S. B. Hong, *Induced Abortion in Seoul, Korea* (Seoul: Dong-A Publishing Company, 1966), p. 78.

[15] This is fully described in Bernard Berelson and Ronald Freedman, "A Study in Fertility Control," *Scientific American*, 210 [5] (May 1964), pp. 3–11.

Development (AID), which was providing indirect assistance to the health services. The expanded action program for the island as a whole was initiated in 1964, and in 1965 the program effectively achieved the target of inserting 100,000 IUD's.[16]

The principal feature of the Taiwan plan is insertion of 600,000 IUD's within five years.[17] Were there no removals or expulsions this would mean a loop for one-third of the married women of childbearing age, including those marrying in the interim. Since a substantial percentage of the IUD's are not retained, the net effect will probably be less. Oral contraceptives are being introduced on an experimental basis to provide an alternative method for women who cannot or do not wish to use the IUD.

In contrast with most developing countries Taiwan has excellent vital statistics and other methods of evaluating program success. It can be seen on the basis of the information in Table 18–2 that the birth rate in Taiwan has been falling precipitously.

TABLE 18–2. Taiwan: Crude Birth Rate

Year	Rate
1959	41.2
1960	39.5
1961	38.3
1962	37.4
1963	36.3
1964	34.5
1965	32.7

Source: United Nations, *Demographic Yearbook, 1965,* Table 12.

The government program could not have initiated this decline, but the especially large drop in 1964 and 1965 is to be noted. Rising age at marriage, an increase of about two months per year, is known to be a factor. In addition to its direct effects, the government program may operate indirectly to stimulate greater interest in family planning and in induced abortion. Since abortions are illegal, their number is not known. The real tests of the effectiveness of the government program still lie in the future.

Southeast Asia

In Hong Kong, Singapore, and Malaysia, family planning programs were initiated under private auspices and with government subsidies.

[16] Government of Taiwan, *Family Planning in Taiwan, Republic of China, 1965–1966* (1966), p. iii.

[17] Government of Taiwan, *Taiwan, Ten Year Health Program, 1966–1975.*

The private Family Planning Associations in these areas have been among the most successful in the world. In Singapore the government took over the private family planning services in January 1966. These services may well have been a factor in the rapid decline of the birth rate in Singapore from 45.4 per 1000 in 1952 to 29.9 in 1965, although rising age at marriage and changes in age structure were important elements.

Malaysia incorporated a family planning policy in its First Malaysia Plan 1966–1970, adopted in 1965. Voluntary organizations, government departments, and mass communications are to be used for education in and promotion of family planning. The birth rate in Malaysia has been declining, especially among the population of Chinese ancestry. While the role of the private Family Planning Association may have been important in popularizing birth control, its direct services were numerically insufficient to have affected the birth rate.

In Ceylon, the government's national family planning program is being introduced by stages into different sections of the country, first in the Colombo area.[18] Ceylon is a kind of Ireland of Asia—its high age at marriage has led to a lower birth rate than in neighboring India. Swedish technical assistance supports the Ceylon program.

Thailand does not have a population policy, but pilot projects in family planning have been strikingly successful, and family planning is now being introduced as an integral part of health services in major hospitals and in the health units of the northeastern region.

Mainland China

Government interest in family planning in Mainland China goes back to 1956 and 1957 when a birth control campaign was initiated by the government and services were provided in government health clinics. In 1958 a change of policy slowed the campaign to very low gear, but by 1962 renewed governmental interest became evident. In January of that year import regulations were revised to admit contraceptives duty-free. The government advocated later age at marriage. Japanese doctors visiting China in 1964 and 1965 reported that family planning was being advocated as part of maternal and child health programs and that all methods of contraception, including sterilization and abortion, were available. Oral contraceptives and IUD's manufactured in China were apparently becoming increasingly popular.

According to Premier Chou:

> Our present target is to reduce population growth to below 2 percent; for the future we aim at an even lower rate. . . . However, I do not believe it will be

[18] Government of Ceylon, *Provisional Scheme for a Nationwide Family Planning Programme in Ceylon, 1966–1976.*

possible to equal the Japanese rate (of about one percent) as early as 1970. . . . We do believe in planned parenthood, but it is not easy to introduce all at once in China. . . . The first thing is to encourage late marriages.[19]

Many of the 17 million Communist party members and 25 million Young Communists have received birth control instruction, and they, in turn, are expected to become models and teachers. One son and one daughter are now considered an ideal family size.[20]

Far too little is known about the Chinese program, but it may well be the most important national program in the world, if for no other reason than the tremendous numbers involved. The Chinese population may now be as high as 800 million, equal to about one-fourth of the human race.

Middle East and Africa

Four countries in the Middle East and North Africa have adopted national family planning programs. In the United Arab Republic (UAR) the government's interest goes back to 1953, when a National Population Commission was established and a few clinics were opened. Government policy dates from the May 1962 draft of the National Charter, in which President Nasser declared:

Population increase constitutes the most dangerous obstacle that faces the Egyptian people in their drive towards raising the standard of production. . . . Attempts at family planning deserve the most sincere efforts supported by modern scientific methods.[21]

However, a substantial program was not initiated until February 1966, when the government launched a widespread campaign using oral contraceptives.

In Tunisia an experimental program to develop a practical family planning service with IUD's was started in 1964. The success of this experiment led to a national campaign with a goal of 120,000 IUD's.[22] An unusually interesting feature of the Tunisian program has been the use of Destour party members as a major source of information and publicity. The program may have been set back by President Bourguiba's speech on Woman's Day, August 12, 1966, against celibacy and in favor of a young vigorous population.

In April 1965 Turkey repealed an old law against contraception and provided the legal framework and financial basis for a nationwide family

[19] As reported by Edgar Snow in *The New York Times*, February 3, 1964.
[20] *The Sunday Times* (London), January 23, 1966.
[21] UAR, Information Department, *The Charter* (draft presented by President Nasser on May 21, 1962), p. 53.
[22] A. Daly, "Tunisia," in Berelson *et al.*, *op. cit.*, p. 160.

planning program.[23] Full-time family planning personnel are to be trained and added to the Health Ministry, and supplies are to be offered free or at cost. Interesting features of the Turkish program are plans for an informational campaign on birth control in the armed forces (to provide a "ripple" effect when the conscripts return to civilian life) and incorporation of demographic and biological aspects of population into the school curriculum.

Morocco decided in 1966 to adopt a national family program to be introduced by stages through the public health clinics in the various parts of the country. A national sample survey of attitudes on family planning has been started by the government.

Although Algeria has no population policy or program, Dr. Ahmed Taleb, Minister of National Education, stated on the opening of the school year (fall 1966):

> . . . we have to fight an extremely high birth rate. If nothing is done to stop this growth rate through birth control, the problem of educating all the Algerian children will remain unsolved.[24]

Tropical Africa

Although no formal population policies have yet been adopted by countries in Africa south of the Sahara, considerable interest was expressed in population matters at the First African Population Conference held at the University of Ibadan, Nigeria, in January 1966.[25]

Kenya's 1966–1970 Development Plan includes "measures to promote family planning education" through the establishment of a Family Planning Council and by providing services in government hospitals and health centers. In Mauritius the 1966 budget provides funds for family planning services.

Latin America

Latin America has the most rapid rate of population growth of any major region of the world. This has not generally been a matter of much public concern, partly because of the traditional position of the Catholic Church and partly because Latin-American countries have a historical image of themselves as underpopulated.

Two forces are rapidly changing this disinterest. One is the growing recognition that high growth rates are obstacles to achieving planning goals. In many countries population is growing faster than food supply;

---

[23] T. Metiner, "Turkey," in Berelson *et al.*, *op. cit.*, p. 136.

[24] Translation from *La Presse*, Tunis, Tunisia, 30 September 1966.

[25] Proceedings to be published in 1967. Also see J. C. Caldwell, "Africa," in Berelson *et al.*, *op. cit.*

the difficulties in providing public education, health services, and other facilities are formidable; and the intrusive problem of unemployment and underemployment has serious political as well as economic implications as people flock into the cities. A second element is the growing concern, especially in the medical profession, over the problem of induced abortion. The possibility of family planning appears to be gaining favorable attention both among responsible leaders, and among large segments of the population, according to sample attitude surveys in eight Latin-American capitals (coordinated by the United Nations Latin-American Demographic Center in Santiago, Chile).[26]

Latin-American countries may be less likely to adopt formal population policies than other parts of the underdeveloped world owing to the influence of the Church. However, natality regulation has been approved for the public health service in Chile, and several major birth control projects using public health facilities exist in Santiago. In Honduras the Minister of Health recently announced that family planning is to be an integral part of preventive medical services. In Jamaica a Family Planning Unit has been established within the Ministry of Health with administrative costs provided by AID. In Colombia the private Association of Medical Faculties has established a Population Division, which has organized a nationwide program for training health officers in family planning. In October 1966 AID authorized the use of counterpart funds to finance this program. In Peru a government-sponsored population studies center was established to "formulate programs of action with which to face the problems of population and socioeconomic development." Barbados has had an official policy favoring family planning since 1954, expressed chiefly by subsidy of the private Family Planning Association.

EVALUATION OF FAMILY PLANNING PROGRAMS

Most government family planning programs are very new, and it would be unfair to expect major results so soon. Their very existence in so short a time is in itself remarkable. Operating through the existing health network, generally under the Health Ministry, most programs are still largely clinic-oriented despite the common experience that other means may be more effective. Problems of organization; administration; production, distribution and supply of contraceptives; and shortage of skilled personnel are more serious than the question of finance.

---

[26] C. A. Miro and F. Rath, "Preliminary Findings of Comparative Fertility Surveys in Three Latin-American Cities," *Milbank Memorial Fund Quarterly*, Vol. XLIII, No. 4 (October 1965), Part 2, pp. 36–38.

Partly because of their newness, the programs have tended to place great emphasis on the magic of the new contraceptive methods. They all have shied away from abortion, which has been a major factor in reducing birth rates in large populations as in Japan, the Soviet Union, and eastern Europe. The present female-oriented programs minimize the role of male participation, which, if not for presently recommended methods, is nevertheless important for information and motivation. Few of the programs have thus far made use of mass communication, and Ministries of Information and Education have yet to be effectively involved.

National programs may be evaluated at different levels: in-service statistics, as, for example, success in reaching targets in number of clinics, patients, contraceptive users, and the like; more broadly in their effects on knowledge, attitudes, and practices of the general population (measured by the so-called KAP students); and, for the present purpose, the effects on birth rate, the ultimate test of a population control program.

Since very few of the countries of the underdeveloped world have sufficiently accurate vital statistics to measure year-to-year changes in the birth rate, other means must be sought. Censuses can be used to measure natality changes, but these occur too infrequently for the purpose at hand. In the absence of official data, sample registration and periodic sample population surveys are conducted to provide data for measuring year-to-year changes. These have come to be known as Population Growth Estimate (PGE) Studies. Experimental projects of this type are going forward in Pakistan, Turkey, and Thailand.

As programs accelerate, the deficiency in accurate vital data will become increasingly important. In-service data can give good measures of the scope of the programs and their success in achieving targets. However, the number of contraceptive "users," as measured by accepters, can be very deceptive, since failure of the method and, even more important, failure of couples to use methods as required, can significantly reduce "use-effectiveness." This is true of IUD's as well as other effective methods such as oral contraceptives and condoms. Many women do not retain the IUD's or have them removed. In the national program in Taiwan 62 percent of the women still had the IUD in place at the end of twelve months, 52 percent at the end of eighteen months. Experience elsewhere has been better, up to 80 percent retention at the end of a year. In all methods there is substantial shrinkage between ideal use and actual effectiveness in preventing pregnancy.

It should also be noted that family planning programs are most likely to succeed rapidly in countries of greatest socioeconomic advance, where realization of the smaller family ideal has already made some progress. The success of a program in countries like Korea and Taiwan chiefly reflects rapid progress in other ways. In such countries mass acceptance of government services is not an equivalent gain in family planning practice, since many of the couples concerned were already practicing family planning (perhaps by less effective methods or abortion) or would have done so regardless of a government program. In these countries the government program may accelerate a trend already in existence. Indeed, in some countries, the influence of the program on couples to use private sources of supply, methods not requiring supplies, and abortion may well surpass the effect of the direct services offered. Yet this indirect effect is least susceptible to measurement.

SUMMARY AND CONCLUSIONS

By now at least twenty-three nations in the underdeveloped world have explicit official population policies. One is struck by the recency and rapidity with which these programs have come into being. In some instances there is policy, but little if any program; in others, program, but no policy. In most cases neither policy nor program has yet had much opportunity to produce a measureable effect on the national birth rate. Even in those countries with most marked successes, such as Korea and Taiwan, the reduction of the birth rate so far is surely much more the result of general social change than of public policy.

The newness and frailties of family planning programs reflect the tentative approach of governments to their population problems. Thus far, they have involved very small material investments in relation both to economic development plans and to potential economic gains. Some have argued that the "normal" tendency of the birth rate to decline in the course of socioeconomic development will bring about a resolution of present population problems. Government family planning programs now seem to be part of this "normal" development.

With knowledge rapidly becoming available for individual couples to exercise voluntary control over births, fortified by governmental approval and assistance in supplies and medical services, it is quite possible that family planning may progress more rapidly than some other forms of socioeconomic advance. Several non-European areas of different cultures and relatively low per capita incomes (notably Taiwan, Korea, Singapore, and Hong Kong; Soviet Asia and the western provinces of Turkey; and Argentina and Uruguay) already give clear evidence of reductions in birth rates. Given the favorable attitudes found in the KAP surveys, family planning may be easier to implement than major advances in education or the economy, which require large structural and institutional changes in the society as a whole.

The most rapid progress will come in East Asia where the normal demographic transition is already well advanced. While the evidence is scant, family planning may also move ahead rapidly in Mainland China despite its low level of socioeconomic development. The Communist regime has done much to disrupt the ancient pattern of Chinese family life, which, in any case, based on the experience of Chinese outside Mainland China, is less a barrier to the small family norm than had been supposed.

In India, many cultural practices restrain freedom of sexual expression. Widespread acceptance of sterilization (vasectomy), while not yet sufficient to affect the birth rate, is, nevertheless, surprising evidence of a greater concern for family size among men than might have been supposed. Although ten years old, the Indian family planning program has reached significant proportions only in the last year or so. If present trends continue, both in the national program and in independent individual initiative in contraceptive practice, one may expect a sufficient escalation of family planning to produce a decline in the birth rate. That it will achieve the goal of a birth rate of only 25 per 1,000 population by 1971 seems doubtful, but I venture to predict a perceptible drop from the present level of over 40. At the same time, however, the death rate will probably have dropped below 15 so that to reach the target growth rate of one percent will require still further reductions in the birth rate.

In Moslem society, family planning is making headway among the upper classes and among populations closest to European influences, as in Albania, Turkey, and the Central Asia Republics of the Soviet Union. Several Moslem countries have population policies (Turkey, Tunisia, the United Arab Republic, Morocco), but other things being equal, family planning is likely to gain slower popular acceptance among Moslems than among most other cultural groups.[27]

As measured by most indices of socioeconomic advance, tropical Africa would not seem ready for widespread family planning practice. Nevertheless, Africans gladly seize on cultural innovation, and a surprising interest has already been shown. One can scarcely expect any measureable effect on the birth rate in a tropical African country in ten years; but in twenty years I would expect this to have occurred.

In Latin America, urban-rural fertility differentials are not so large as in other parts of the world, partly because of the heavy immigration into the urban shanty towns of rural people who have yet to become integrated into urban life. Nevertheless, studies reveal widespread interest in family planning (at least in the large capital cities), and abortion has become very common. A shift in the position of the Catholic Church toward permissive policies, if not outright acceptance of family planning, will accelerate what is already a major social trend.

In several, perhaps most, Latin American countries I would anticipate a measurable reduction in their very high birth rates within the next ten years. Within twenty years, growth rates should be markedly reduced

---

[27] D. Kirk, "Factors affecting Moslem Natality," in Berelson, *et al.*, *op. cit.*

despite a continuing decline in mortality. Present fertility plus the inertia of age structure, however, will probably double the population of Latin America by the end of the century.

The picture that emerges is a range in birth rates of 20 to 25 per 1,000 population within a decade in the most progressive parts of East and Southeast Asia, and possibly within two decades in India and Mainland China if those countries avoid war, disaster, and social chaos. By that time, one can expect important reductions in the birth rate in all of the larger developing countries in Asia, in the Middle East and North Africa, and in Latin America. Taking the underdeveloped world as a whole, within two decades I expect to see the solution well in sight, though not yet fully achieved.

These conslusions imply great efforts and accomplishments in the face of cultural resistance and inertia. The achievements will not be easily won, nor will they forestall the massive population growth that, in the absence of catastrophe, will be with us at least through the 1970's. The critical problem of world population growth will remain, though in the longer run there is now real hope for its solution.

III

Problems related to ecosystem nutrient cycling processes

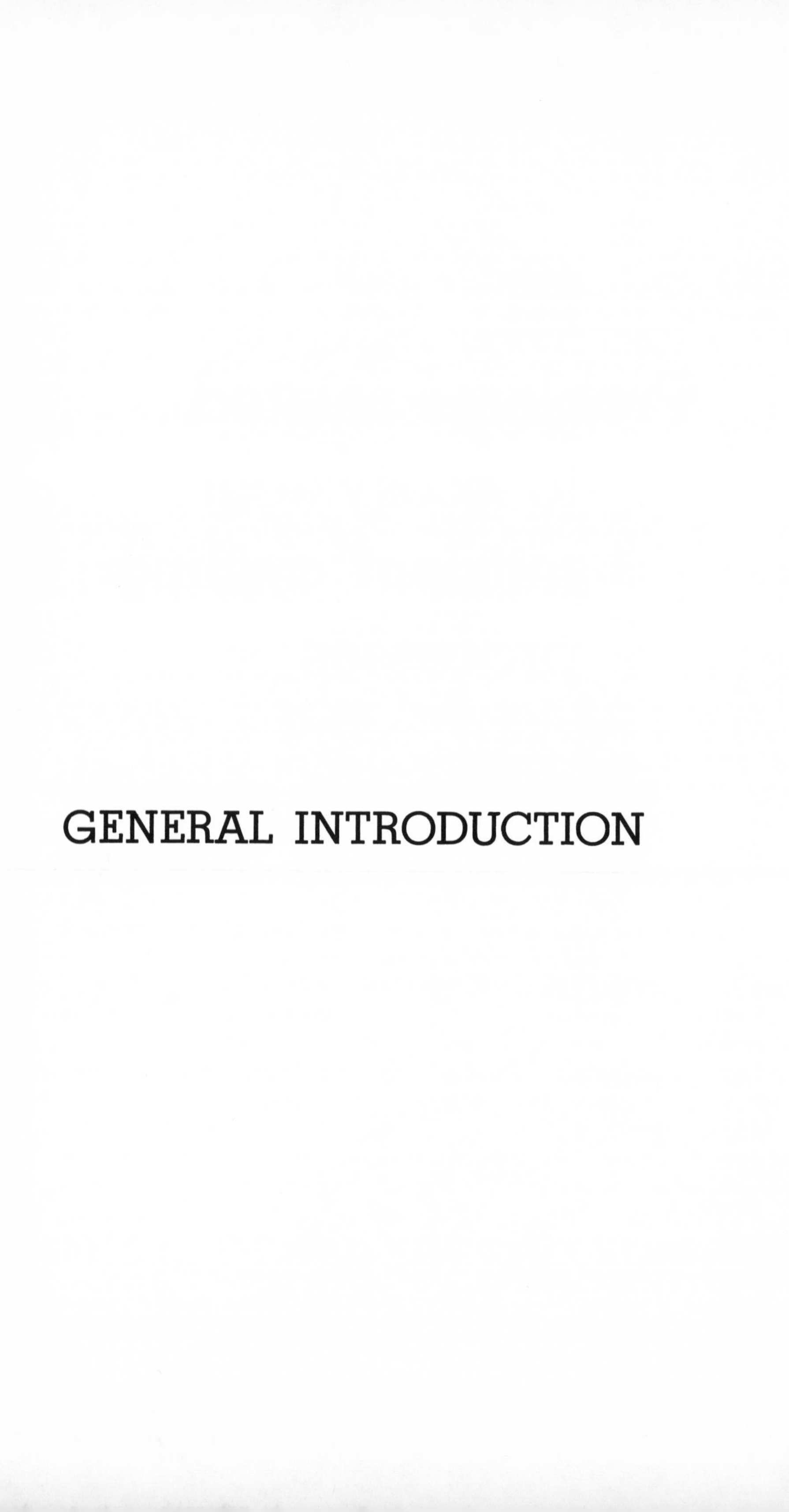

GENERAL INTRODUCTION

19

On living in the biosphere

G. E. Hutchinson

In discussing the subject of "The World's Natural Resources," I want first to make a number of general observations that will provide an intellectual framework into which our developing knowledge, both academic and practical, may be fitted. We live in a rather restricted zone of our planet, at the base of its gaseous envelope and on the surface of its solid phase, with temporary excursions upwards, downwards, or sideways onto or into the oceans. These regions in which we can live and which we can explore are characterized by their temperature, which does not depart far from that at which water is a liquid, and by their closeness to regions on which solar radiation is being delivered. This zone of life is spoken of as the biosphere. Within it, certain natural products can be utilized in both biological and cultural life. It is customary to consider these resources as either material or energetic, but the two categories are not easily separable; contemporary solar radiation is an energetic resource, coal and oil are to be regarded as material resources valuable for their high energy content, which we may call, epigrammatically, fossil solar radiation. There is a third very important though inseparable aspect, namely, the pattern of distribution. Most fossil sunlight, or chemical energy of carbonaceous matter, is diffused through sedimentary rocks in such a way as to be useless to us. Schrödinger says that we feed on negative entropy, and I am almost tempted to regard pattern as being as fundamental a gift of nature as sunlight or the chemical elements.

Reprinted with permission from *Scientific Monthly*, 67:393–397 (December 1948). Also reprinted in *Itinerant Ivory Tower*, Yale University Press, 1953.

The first major function of the sunlight falling on the earth's surface is as the energy of circulation of the oceans and atmosphere. The second is to increase the mobility of water molecules, to become latent heat of evaporation, and so to keep the water cycle operating. The third major function is photosynthesis. Apart from atomic energy, and a little volcanic heat which presumably is actually of radioactive origin, all industrial energy is solar and due to one or the other of these three processes.

The material requirements of life are extremely varied. Between thirty and forty chemical elements appear to be normally involved. Industrially, some use appears to be found for nearly all the natural elements, and some of the new synthetic ones also. Looking at man from a strictly geochemical standpoint, his most striking character is that he demands so much—not merely thirty or forty elements for physiological activity, but nearly all the others for cultural activity. What we may call the anthropogeochemistry of cultural life is worth examining. We find man scurrying about the planet looking for places where certain substances are abundant; then removing them elsewhere, often producing local artificial concentrations far greater than are known in nature. Such concentrations, whether a cube of sodium in a bottle in the laboratory, or the George Washington Bridge, have usually been brought into being by chemical changes, most frequently reductions, of such a kind that the product is unstable under the conditions in which accumulation takes place. Most artifacts are made to be used, and during use the strains to which they are submitted distort them, and they become worn-out or broken. This results in a very great quantity of the materials that are laboriously collected being lost again in city dumps and automobile cemeteries. The final fate of an object may depend on many factors, but it is probable that in most cases a very large quantity of any noncombustible, useful material is fated to be carried, either in solution or as sediment, into the sea. Modern man, then, is a very effective agent of zoogenous erosion, but the erosion is highly specific, affecting most powerfully arable soils, forests, accessible mineral deposits, and other parts of the biosphere which provide the things that *Homo sapiens* as a mammal and as an educatable social organism needs or thinks he needs. The process is continuously increasing in intensity, as populations expand and as the most easily eroded loci have added their quotas to the air, the garbage can, the city dump, and the sea.

The most important general consideration to bear in mind in discussing the dynamics of the biosphere and its inhabitants is that some of the processes of significance are acyclical, and others, to a greater or less degree, cyclical. By an acyclical process will be meant one in which a permanent change in geochemical distribution is introduced into the system; usually a concentrated element tends to become dispersed. By a cyclical process will be meant one in which the changes involved introduce no permanent alteration in the large-scale geochemical pattern, concentration alternating with dispersion. The cyclical processes are not necessarily reversible in a thermodynamic sense; in fact, they are in general no more and no less reversible than the acyclical. Most of the cyclical processes operate because a continuous supply of solar energy is led into them, and sunlight will provide no problems for the conservationist for a very long time. It is important to realize that most of the acyclical processes are so slow that man appears as an active intruder into a passive pattern of distribution. They are safer to disturb because we know what the result of the disturbance will be. If we mine the copper in a given region sufficiently assiduously, we know that ultimately there will not be any more copper available there. Cyclical processes involve complex circular paths, regenerative circuits, feedback mechanisms, and the like. Small disturbances of such processes may merely result in small temporary changes, with a rapid return to the previous steady state. This has been beautifully demonstrated in the experiments of Einsele, who added single massive doses of phosphate to a lake, changing for a time, but only for a time, its entire chemistry and biology. This stability does not imply that if large disturbances strain the mechanism beyond certain critical limits very profound disruption will not follow; in fact, the very self-regulatory mechanisms that give the system stability against small disturbances are likely to accentuate the disruption when the critical limits are transcended. In disturbing cyclical processes we usually do not know what we are doing.

The most nearly perfect cyclical processes are those involving water and nitrogen. Some losses to the sediments of the deep oceanic basins must occur, but they are very small and are doubtless fully balanced or more than balanced by juvenile water and perhaps by molecular nitrogen and ammonia of volcanic origin.

The least cyclical processes are those in which material is removed from the continents and deposited in the permanent basins of the ocean. With one or two exceptions, the delivery to the deep-water sediments of the ocean is of little significance. Most of the mechanical and chemical sedimentation in the oceans takes place in relatively shallow water. The uplifting of shallow water sediments constitutes an important method of completing cycles. One, and perhaps two, exceedingly important elements are, however, sedimented less economically. Calcium, during the Paleozoic, was mainly precipitated in shallow water, but since the rise of the pelagic foraminifera in the Mesozoic, a great deal of calcium, along with an equivalent amount of carbon and oxygen, has been continually diverted to regions from which it is unlikely ever to be removed. At present the sedimentary rocks of the world are an adequate biological and commercial source of calcium, but, with progressive orogenic cycles, less and less of the element will be uplifted (Kuenen), and whatever organisms inherit the earth in that remote future will have to face the problem of the biosphere "going sour on them." For phosphorus, the case is less well established, but there is probably a slow loss in the form of sharks' teeth and the ear bones of whales (Conway), which are very resistant and which are known to be littered about on the floor of the abysses of the ocean.

It is desirable to consider two of the main geochemical cycles in order to gain an idea of the effect of man upon them. It must be admitted that we are ignorant of many matters of importance here. In the cycle of carbon we have a remarkable, possibly a unique, case in which man, the miner, increases the cyclicity of the geochemical process. It is generally admitted that our available store of carbon is ultimately of volcanic origin. A steady stream of carbon dioxide and lesser amounts of methane and carbon monoxide are entering the atmosphere from volcanic vents. Part, probably a major part, of this carbon dioxide is ultimately lost to the marine sediments as limestones; a very small part of it is then returned to the air, wherever lime kilns are in operation. Another part of the CO_2 entering the atmosphere is reduced in photosynthesis. A part of the organic matter so formed is fossilized, and a small part of this fossilized organic carbon is available as fuel, in the form of coal and oil. At the present time it appears that the combustion of coal and oil actually returns carbon to the atmosphere as CO_2 at a rate at least a hundred times greater than the rate of loss of all forms of carbon, oxidized and reduced, to the sediments (Goldschmidt). This particular process obviously cannot go on indefinitely. It concerns only the reduced carbon; to complete the cycle in the case of oxidized carbon, a great deal of energy would have to be supplied. It concerns only the reduced carbon which is aggregated. The poorest sources would be the poorest exploitable oil shales. Most of the reduced carbon is much more dispersed than this; in making an estimate of the total reduced carbon of the sediments, the commercially usable fuels constitute a negligible fraction that need not be considered.

Although the rate at which carbon dioxide is returned to the air by the human utilization of fossil fuels is so very much greater than is the primary production of carbon dioxide from volcanic sources, the rate is evidently a very small fraction—of the order of 1 percent—of the rate of photosynthetic fixation and subsequent respiratory liberation of CO_2 by the organisms of the earth. Since about 1890 a slight increase in CO_2 content in the air, at least at low altitudes over the land surfaces of the Northern Hemisphere, has been noted. This has been attributed to the accumulation of industrially produced CO_2, as the quantity of CO_2 that has appeared in the atmosphere is of the same order of magnitude as the total combustion of fuel (Callendar). In view of the small fraction of the total CO_2 production that industrial output represents, it seems very unlikely that merely adding an extra percent to the natural biological production should overload the cyclic process, so that it rejects quantitatively the additional load. It is known that the air at high altitudes still shows nineteenth-century values. The most reasonable explanation of the observed increase is that the photosynthetic machinery of the biosphere has been slightly impaired, probably by deforestation. It is clear that, in any intelligent long-term planning of the utilization of the biosphere, an extended study of atmospheric gases is desirable, even though for the moment it seems unlikely that the observed change is a particularly serious symptom.

The only other cycle that can be considered in any great detail is that of phosphorus. The chief event in the geochemical cycle of phosphorus is the leaching of the element from the rocks of the continents, and its transport by rivers to the sea. At the present time the rate of this transportation is of the order of 20,000,000 tons of phosphorus per year for the entire earth. Part of this phosphorus, when it enters the sea, will ultimately be deposited in the sediments of the depths of the ocean. Such phosphorus will probably be largely lost to the geochemical cycle, as has just been indicated. The sedimentary rocks of the continents, therefore, will gradually lose phosphorus; there is some evidence that this has actually occurred (Conway). The main return path is by the uplifting of sediments formed in continental seas, which then undergo renewed chemical erosion. Of particular interest are methods by which concentrated phosphorus can be returned to the land surfaces. As far as is known, there are two such methods: The first is the formation of phosphatic nodules and other forms of phosphate rock in regions of upwelling in which water at a low *p*H, rich in phosphate, is brought up to the surface of the sea. The *p*H falls and an apatite-like phosphate is deposited. When the sea floor is later elevated, a commercial deposit may result. The second method is by the activity of sea birds, such as the guano birds of the Peruvian coast. There is little or no unequivocal evidence that guano deposits of great extent were formed prior to the late Pliocene or Pleistocene. Some of the well-known occurrences of rock phosphate, such as that of Quercy, have been explained in this way, but they are certainly not typical guano deposits. During the late Tertiary and Pleistocene, an extraordinary amount of phosphate was deposited on raised coral islands throughout the world, and bird colonies seem to provide the only reasonable agencies of deposition. The great deposits of Nauru, Ocean Island, Makatea, Angaur, the Daito Islands, Christmas Island south of Java, Curaçao, and some of the other West Indies all seem to have been formed in this way. This process is as characteristic of the time as is glaciation, though less grandiose. Its meaning is not clear, but it is probably connected with changes in vertical circulation of the ocean as glaciopluvial periods gave place to interpluvials. Today in certain regions massive amounts of guano are deposited, and it is probable that the oceanic birds of the world as a whole bring out from several tens to several hundreds of thousands of tons of phosphorus and deposit it on land. Only about 10,000 tons of the element are delivered in places where it is not washed away and where it can be carried by man to fertilize his fields.

The main processes that tend to reverse the phosphorus depletion of the continents are, therefore, the deposition of marine phosphorites on the continental shelves and subsequent elevation, and the formation of guano deposits. Both processes are evidently intermittent, and are quantitatively inadequate to arrest deflection of the element into the permanent ocean basins. Man contributes both to the loss and to the gain of phosphorus by land surfaces. He quarries phosphorite, makes super-phosphate of it, and spreads it on his fields. Most of the phosphorus so laboriously acquired ultimately reaches the sea. At present the world's production of phosphate rock is about 10,000,000 tons per annum. This contains from 1,000,000 to 2,000,000 tons of elementary phosphorus. Human activity probably, therefore, accounts for from 5 to 10 percent of the loss of phosphorus from the land to the sea. Man also contributes to the processes bringing phosphorus from the sea to the land. This is done by fisheries. The total catch for the marine fisheries of the earth is of the order of $25\text{–}30.10^6$ tons of fish, which corresponds to about 60,000 tons of elementary phosphorus. The human, no less than the nonhuman, processes tending to complete the cycle seem miserably inadequate. It is quite certain that ultimately man, if he is to avoid famine, will have to go about completing the phosphorus cycle on a large scale. It will be a harder task than that of solving the nitrogen problem, which would have loomed large in any symposium on "The World's Natural Resources" fifty years ago, but possibly an easier problem than some of the others that must be solved if we are to survive and really become the glory of the earth.

The population of the world is increasing, its available resources are dwindling. Apart from the ordinary biological processes involved in producing population saturation already known to Malthus, the current disharmony is accentuated by the effect of medical science, which has decreased death rates without altering birth rates, and by modern wars, which one may suspect put greater drains on resources than on populations. Terrible as these conclusions may appear, they have to be faced. The results of the interaction between population pressure and decline of resource potential are further partly expressed in such wars, which are pathological expressions of attempts to cope with these and other problems and which now invariably aggravate the situation. It is evident that the fundamental causes of war lie in those psychological properties of populations which make them attempt solutions in a warlike manner, and *not* in the existence of the problems themselves. The two problems of war and of resources are, however, at present very closely interrelated; it is probably impossible to find a solution for one without progress in the solution of the other. Any kind of reasonable use of the world's resources involves better international relations than now exist. It is otherwise impossible to operate on a planetary basis, or to avoid the fearful material and spiritual expense of living in a world divided into two armed camps. The lack of international trust is the first difficulty in achieving rational utilization of the resources of the world; the second difficulty for the United States is what may be called the problem of the transition from the pioneering to the old, settled community. For a pioneer, life may be hard, but in good country there is "plenty more where these came from," whether lumber or buffalo tongues or copper. This attitude is incorporated into popular thought very deeply; in a crude form, it is now completely destructive, but it seems possible that attitudes might be developed that could utilize such a point of view. There is at least one thing of which we have plenty more than we use, whether we call it Yankee ingenuity, American know-how, or the human intellect. In some industrial fields there has been a notable series of triumphs of the kind that really will give plenty more of a number of things, and give them in a cyclic manner. The rise of the magnesium industry, and the utilization of the magnesium of sea water as a source of the metal is an interesting example. The production of plastics, though it is probably at the moment not as geochemically economical as it should be, is another. These, along with the development of silicate building materials not involving any particularly uncommon elements, all point the way to the kind of material culture that permits a reasonably high living standard, at least in certain directions, without devastation of the earth.

The future outlook for the world, particularly in food resources, has been put before the public in several recent books, notably those of Fairfield Osborn and William Vogt. Anyone with any technical knowledge understands that the dangers described in these books are real enough, more real and more dangerous perhaps than the threat of an atomic world war. The problem of getting action to forestall such dangers in a culture that has developed under conditions of potentially unlimited abundance just around the corner, is obviously extremely difficult. I do not think it is impossible. The first requirement is a faith that the job can be done. The very difficulty of the task, its apparent impossibility, may here and now prove a challenge that brings the desired response. The difficult we do at once, the impossible takes a little longer. I doubt that a direct appeal to fear will produce any results except a disbelief in the prophets of doom. Cassandra seems even more unpopular in modern America than in ancient Ilium. There would seem to be forces operating in society which tend to reverse the destructive processes, or which could be made to do so. One of the most immediate needs is to find out what they are and do everything possible to strengthen them. I will give one example. A number of industrial concerns—particularly in the chemical and pharmaceutical field, but also some engineering and publishing firms—have used in their advertising a legitimate pride in the learning, skillful research, and development that have gone into the manufacture of their product. It is quite certain that there are many cases in which one particular product, the result of considerable research, and of taking risks in development, actually reduces the drain on the natural resources of the country and the world. I should like to see a small systematic experiment, on the part of some such concern, in advertising in which it is pointed out that by buying this product one is letting the industrial skill behind the product operate for the benefit of one's children. I am fully aware that if this point of view were worked up skillfully enough by a few responsible, public-spirited corporations and put into a form that would pay the corporation, as well as the country and the world, there would be other less responsible concerns who would use the method when their product is actually not one sparing natural resources. This, however, seems to be a lesser evil than the total neglect of the commercial advertising field, which is one of the most potent in determining the values of the public and which at present is largely disruptive. I do not doubt that a professional cultural anthropologist could pick out a great many other fields that could be used to promote the idea of an expanding economy based on an abundance of human ingenuity rather than on an excess of raw materials.

There is one further point that I should like to develop. Though the pursuit of happiness is embedded deeply in the constitutional foundations of the United States, we do not know much about it. It is fairly certain that no metric exists that can be applied directly to happiness, but intuitively we may proceed a little way by arguing as if such a metric could exist. It is obvious that only a very few people, with a genius for sanctity, can be happy if half-frozen and starving. If the temperature be raised, and the food supply and other amenities increased, the possibility of happiness is obviously at first also increased, but beyond a certain increase in the environmental resources available no further increase in happiness would be expected and we might begin to look for a decline. The image of an overheated kitchen, filled with too much electrical equipment and catering to overfed people, will, if adequately evoked, have a nightmarish quality. In more formal language, if we could find a function of the environmental resources that expressed the relationship of happiness to those resources, it is reasonably certain that the function would not be monotonic. For every resource there seems likely to be an optimum level of consumption, but we do not know if this optimum level is, in any particular case, widely exceeded, so that gross overutilization is actually producing avoidable distress.

The problem is not by any means as simple as that of determining discrete optima. In any given society all the cultural values are probably interrelated to form a coherent system, so that the existence of one set of values may greatly modify the others. If, as seems possible, our attitude toward food leads a considerable section of our population to be definitely overweight, it is legitimate to inquire to what extent, by changes in the upbringing of our children, the psychological needs filled by food can be satisfied in other ways so that the psychologically optimum intake falls to a level nearer the psychological optimum for the individual and the moral optimum for the world. We might ask, for instance, how we can substitute the delights of ballet and Mozartian opera, which are geochemically very cheap, for part of those provided by hot dogs or apple pie and ice cream, which may in the long run prove too expensive to use except as a source of energy and essential nutrients. This example is chosen with a view to indicating that the kind of substitutions that might be considered need not be in the least puritanical. It may appear overintellectualized, but that at least is a guarantee that it is not inhuman. Indeed it raises the very interesting problem that those of us in the educational world have to face, namely, why we are raising a generation in the belief that the majority of constructive, complicated, difficult activities are boring duties, when the same generation shows us that in certain specific fields, mainly concerned with electronic amplifiers and with the explosive combustion of hydrocarbons, complicated activity can be very entertaining. What we have to do is to show by example that a very large number of diversified, complicated, and often extremely difficult constructive activities are capable of giving enormous pleasure. This is, in fact, the reason why it is essential that the teachers in our colleges and universities should be enthusiastic investigators in the fields of scholarship or practitioners and critics in their arts. It ought to be possible to show that it is as much fun to repair the biosphere and the human societies within it as it is to mend the radio or the family car.

PROBLEMS OF ENVIRONMENTAL POLLUTION

PESTICIDES

20

Effects of DDT spraying for spruce budworm on fish in the Yellowstone River system

Oliver B. Cope

INTRODUCTION

Widespread mortality among mountain whitefish (*Prosopium williamsoni*), brown trout (*Salmo trutta*), and longnose suckers (*Catostomus catostomus*) in the Yellowstone River below Yellowstone National Park attracted the attention of sportsmen and biologists in the fall of 1955. Thousands of dead fish were seen in the Yellowstone River from October to December, and it was concluded that this mortality must be related to the spruce-budworm spray program that had been conducted in the Yellowstone River drainage during the previous July. Fish studies had not been made in connection with this spray, so it was not possible to ascertain the role of the DDT spray in the destruction of fish. Some collections of aquatic invertebrates had been made before and after spraying in streams in and near the spray area, and these indicated that drastic reductions in invertebrate populations took place immediately after the spraying of DDT.[1]

Plans for spraying an additional 70,000 acres in the Yellowstone

Reprinted with permission from *Trans. Amer. Fish. Soc.*, 90:239–251 (1961). The author is currently associated with the Fish-Pesticide Research Laboratory, Columbia, Missouri 65201.

[1] Cope, Oliver B., and Donald E. Parker. (1956) Survey of conditions attending fish mortalities in Yellowstone River following spruce budworm spray project. Report issued by the U.S. Forest Service, Feb. 3, 1956, 9 pp. + appendices.

The terrain is mountainous, with tributaries draining from altitudes over 9,000 feet into the Lamar River at 6,000–6,500 feet and into the Yellowstone River at 6,000–7,000 feet above sea level. The higher elevations are above timber line. Pines and Douglas fir cover much of the area above drainage were drawn up by the National Park Service and the Forest Service in the spring of 1957. The spray for control of spruce budworm with DDT was scheduled for June and July of 1957 in areas within Yellowstone National Park. This program provided an opportunity to make studies on the effects of the spray on fish and fish-food organisms in the area. Funds for the study were provided by the U.S. National Park Service, the U.S. Forest Service, and the U.S. Fish and Wildlife Service. The Fish and Wildlife Service performed the field work, and the Chemistry Department at Montana State College at Bozeman, Montana, and the Wisconsin Alumni Research Foundation did chemical analyses on materials exposed to DDT.

DESCRIPTION OF THE AREA

The area affected by the 1957 spray program in Yellowstone National Park is in the Yellowstone River drainage in the northern portion of the park and downstream for 65 miles outside the park (Fig. 20–1). the valley floors. The broad valleys of the Yellowstone and Lamar are

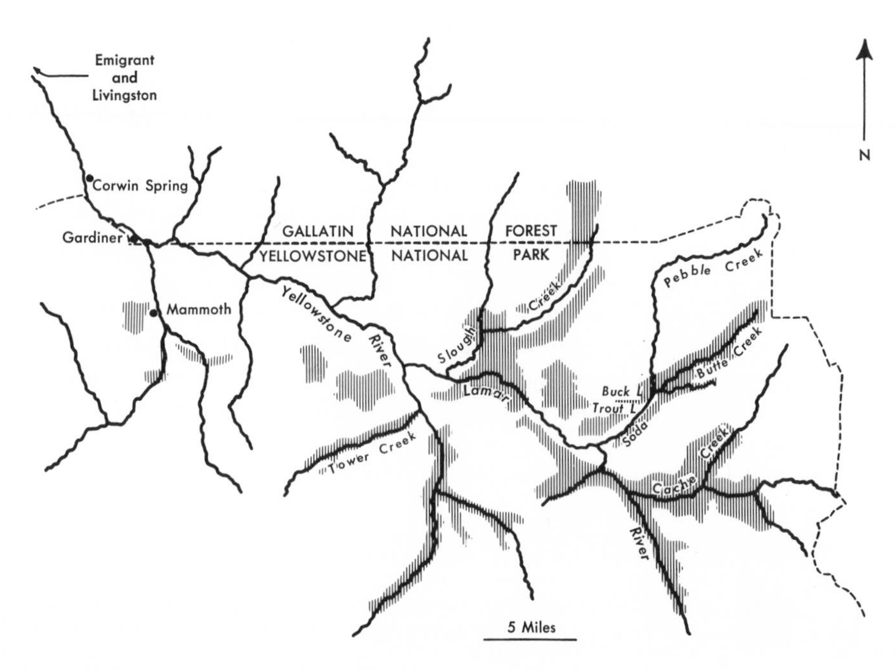

Figure 20–1. Spray areas in Yellowstone National Park – 1957.

open along most of their floors, but forest trees grow close to the water in a few isolated patches. The wide valley floor of the Yellowstone River outside the park is covered with sagebrush, except in the Yankee Jim Canyon area which has sagebrush and sparse Douglas fir close to the water for a few miles.

The tributary streams all have steep gradients along most of their courses. The Lamar and Yellowstone Rivers have very fast currents during flood stage, but even at low water there is sufficient turbulence to cause a constant mixing of waters. Alkalinity, turbidity, and temperature of the streams vary considerably (Table 20–1). The Lamar contains a

TABLE 20–1. Physical and chemical characteristics of sampled waters, 1957

Stream and date	Total alkalinity[1] (p.p.m.)	pH	Water temperature (°F.)	Stream flow (c.f.s.)	Turbidity
Tower Creek (above falls)—July 7	30	7.2	66 (10:25 a.m.)	134	Low
Slough Creek (near mouth)—July 7	34	7.25	44 (12:10 p.m.)	840	Low
Soda Butte Creek (below Pebble Creek)—July 7	54	7.5	52 (1:55 p.m.)	312	Moderate
Trout Lake outlet—July 7	119	8.0	60 (1:40 p.m.)	10	Low
Lamar River (below Slough Creek)—July 7	37	7.8	44 (11:30 a.m.)	1,584	High
Yellowstone River (Corwin Springs)—July 3	41	7.8	58 (3:00 p.m.)	5,500	High

[1] No phenolphthalein alkalinity was found in any samples.

small amount of aquatic vegetation; the Yellowstone has more. The fish fauna of the Lamar consists of cutthroat trout (*Salmo clarki*), longnose sucker, and longnose dace (*Rhinichthys cataractae*), with a few rainbow trout (*Salmo gairdneri*) and mountain whitefish. The Yellowstone River contains the same fauna plus brown trout. Some tributaries of the Lamar and Yellowstone Rivers are rich in aquatic invertebrates used as food by fish. The Lamar River itself is only moderately rich in aquatic invertebrates, but the Yellowstone contains an abundance of animal fish food. A few small lakes are located within the area. They contain cutthroat and rainbow trout and have abundant rooted aquatic vegetation.

THE SPRAY OPERATION

The 1957 spray operation began on July 10 and terminated on July 21 (Table 20–2). The area sprayed consisted of 71,678 acres on 26 plots (Fig. 20–1). Spraying was done from aircraft under a set of standards designed to minimize the deposition of spray on water surfaces. The toxicant was DDT, applied in an oil solution at the rate of 1 pound per acre. The average amount of DDT reaching the ground on six plots was 0.18 pound per acre, as measured by spray cards distributed by U.S. Forest Service entomologists.

TABLE 20–2. Spruce-budworm spraying schedule during July 1957 in the Yellowstone River basin

Spray dates	Area (acres)	Gallons of spray	Pounds per acre		Unit sprayed
			Sprayed	Reaching ground	
12, 13	1,607	1,876	1.17	–	Bunsen Portion and Lava Creek Portion
12	1,395	1,350	0.96	–	Mammoth Hot Springs, Cloggett Butte
18, 21	1,690	1,582	0.93	–	Oxbow–Geode, south of old road
11, 13	2,685	2,690	1.00	–	Crescent Hill–Tower Fall
11	598	560	0.93	–	Deep Creek
13, 14	2,055	2,061	1.00	–	Tower Creek
10, 11	7,045	4,601	0.65	–	Yellowstone Canyon
14, 15, 16, 18	2,688	2,669	0.99	0.26	Specimen Ridge–Jasper Creek
18, 20, 21	3,260	2,875	0.88	–	Slough Creek–Lamar River Canyon
16	3,265	2,100	0.64	–	Buffalo Creek
18	638	525	0.82	–	McBride Lake
11	3,880	3,755	0.96	–	West Slough Creek
14, 16	1,973	1,500	0.76	–	East Slough Creek—lower
16	3,143	2,700	0.85	–	East Slough Creek—upper
15, 16	2,593	2,575	0.99	0.40	Rose Creek
18	2,195	2,200	1.00	–	Chalcedony Creek
18, 20	2,752	2,701	0.98	0.24	Soda Butte
18	1,075	1,030	0.96	0.13	Trout Lake
14, 18	3,113	3,200	1.03	0.03	Druid Peak
10	2,360	2,500	1.05	–	Soda Butte–Amphitheater Creek
16, 17, 18	6,155	6,501	1.06	0.11	Northeast Entrance Road
18, 20, 21	4,048	4,043	1.00	–	Lower Cache Creek
18, 20, 21	3,600	3,677	1.02	–	Lamar River
21	3,742	3,500	0.93	–	South Cache Creek
19, 21	4,123	3,900	0.94	–	Upper Cache Creek
Total or average	71,678	66,671	0.93	–	—

METHODS

The general plan for the field study called for samples and measurements taken 1 week or less prior to spray time, and post-spray samples and measurements taken at intervals following the treatment. Numbers of stream-bottom organisms were determined before and after the spray operation. Samples of water from streams and lakes, samples of fish, and samples of aquatic vegetation for analysis of DDT content were taken throughout 1957 and during 1958 and 1959.

The principal collection stations were located at:

Tower Creek–50 yards upstream from a point opposite the Tower Falls campground;

Lamar River–100 yards upstream from the Cook City Highway Bridge;

Slough Creek–20 yards upstream from the mouth of Buffalo Creek and 100 yards upstream from the mouth of Slough Creek;

Soda Butte Creek–opposite a point 100 yards downstream from the "The Thunderer" sign on the Cook City Highway;

Trout Lake outlet–50 yards downstream from the dam at the outlet;

Trout Lake – east shore of Trout Lake, near the outlet;

Yellowstone River, Corwin Springs–200 yards upstream from the steel bridge at Corwin Springs, 6 miles downstream from Gardiner, Montana;

Yellowstone River, Emigrant–100 yards upstream from Yellowstone River bridge at Emigrant, Montana; and

Yellowstone River, Livingston–50 yards downstream from Yellowstone River bridge, 5 miles upstream from Livingston.

Fish

Collections of fish in affected waters were made near collecting stations and in most tributary streams in the Lamar River, in the Yellowstone River near Emigrant and near Livingston, and in Buck Lake. Fish were collected by shocking and by angling. During 1957 collections were made, whenever possible, before spraying and at 1- to 3-week intervals after spraying. Samples were obtained whenever possible in 1958 and 1959. Fish were frozen for preservation and stored in polyethylene bags. Samples analyzed included rainbow, brown, and cutthroat trout, mountain whitefish, and longnose dace. Most fish samples were about 1 pound in weight. The whole fish were ground and blended and then analyzed for DDT and DDE by the Schechter–Haller method (Schechter and Haller, 1944).

Water

Water samples were collected before and after spraying. Collections were made in most large tributaries, in the Lamar River, in the Yellowstone River as far downstream as Emigrant, and from Trout Lake in the Soda Butte Creek drainage. Each sample consisted of one gallon from the surface of the water and one from 1 to 2 feet below the surface. Each gallon was made up of water from both sides and from the center of the stream. Some of the surface and subsurface samples were combined in the laboratory to form a mixed 2-gallon sample. Samples were acidified with sulfuric acid in the field to avoid the development of high *p*H values and breakdown of the toxicant.

Bottom invertebrates

The several bottom types at nine stations in the tributaries and the Lamar and Yellowstone Rivers made it impossible to use one method for securing indices of abundance of aquatic invertebrates for all stations. The Surber square-foot sampler was used (10 square feet to a collection) wherever the bottom material was small enough. Where this gear was not appropriate, the collections consisted of organisms taken from 20 rocks about 8 inches in diameter. Collections were made at all stations before spraying, just after spraying, and periodically throughout the season into mid-September. Follow-up counts were made in 1958.

Drift samples

A plankton net was used in Tower Creek on July 13 and 14, both before and after spraying, to determine the numbers of aquatic invertebrates drifting downstream in the current. The net had a 9½-inch hoop and 38-mesh bolting cloth. Fishing time was 5 minutes per set.

Aquatic vegetation

Two species of *Potamogeton* were collected from a few streams and lakes before and after spraying. Each sample consisted of two handfuls of plant material which was placed in a plastic bag and frozen before analysis for DDT content.

RESULTS

Fish

No mortality attributable to DDT was found among fish noted in our limited observations. All of the 80 fish samples analyzed contained DDT, and all but four of the 57 tested for DDE, contained that metabolite. Table 20–3 summarizes the results of the analyses on trout, and Table 20–4 lists the results on mountain whitefish. The data suggest certain general patterns with respect to amounts of DDT present, but it is obvious that much variation exists from fish to fish and that caution must be used in interpreting the results.

Lamar River cutthroat trout contained relatively high concentrations of DDT, with the fish collected on August 14 having 11 p.p.m. Certain tributaries of the Lamar River received relatively high dosages of spray, which may account for the high residues measured. Collections made in 1958 in the Lamar showed high residues, but the one collection in 1959 was relatively low. Slough Creek and Cache Creek also had fish with high DDT contents in 1957, but 1958 samples were relatively low. In Buck Lake, cutthroat trout had high levels in 1957 and lower ones in 1958 and 1959.

Downstream in the Yellowstone River the DDT contents of trout were somewhat lower than in the upstream tributaries. At Emigrant, 1957 collections of trout and whitefish had higher residues than did the 1958 and 1959 samples. Whitefish in the Yellowstone had approximately the same DDT levels as cutthroat, rainbow, and brown trout, one whitefish sample containing 6.9 p.p.m. of DDT. There is a suggestion in the data from Corwin Springs and Emigrant that whitefish collected in September 1957 contained more DDT than those collected earlier. It appears that trout collected near Livingston had as much DDT as those captured upstream near Corwin Springs and near Emigrant. The two samples of longnose dace analyzed contained much less DDT than did cutthroat trout collected at the same time and the same place in the Lamar River. Two samples, taken on August 9, 1957, had 0.76 and 0.28 p.p.m., respectively, of DDT. Cutthroat trout from Buck Lake and rainbow trout from Trout Lake contained relatively large amounts of DDT.

TABLE 20–3. DDT and DDE contents of trout collected from the Yellowstone River drainage in 1957, 1958, and 1959

Locality	Species	Date collected	DDT content (p.p.m.)	DDE content (p.p.m.)
Trout Lake	Rainbow	7-30-57	4.9	0.7
Buck Lake	Cutthroat	9-13-57	6.7	1.2
	Cutthroat	9-20-57	5.76	1.26
	Cutthroat	9-22-57	5.1	0.0
	Cutthroat	10-19-57	5.00	–
	Cutthroat	6-27-58	2.10	–
	Cutthroat	8-16-58	0.51	1.90
	Cutthroat	6-16-59	1.18	0.85
Soda Butte Creek	Cutthroat	6-29-58	1.12	–
	Cutthroat	7-27-58	1.09	0.15
Cache Creek	Cutthroat	8- 9-57	14.0	1.6
Lamar River	Cutthroat	8- 9-57	4.2	1.0
	Cutthroat	8-14-57	11.0	3.2
	Cutthroat	9- 4-57	1.8	0.60
	Cutthroat	1-17-58	6.94	–
	Cutthroat	6-26-58	5.04	–
	Cutthroat	8- 5-58	0.85	0.99
	Cutthroat	6-19-59	1.03	0.90
Slough Creek	Rainbow	7-19-57	2.01	0.77
	Rainbow	8- 9-57	10.2	2.5
	Cutthroat	12-11-57	0.44	–
	Rainbow	6-28-58	1.04	–
	Cutthroat	8- 2-58	0.20	1.29
Yellowstone River between Geode and Hellroaring Creek	Cutthroat	8-11-57	0.74	–
Above Yancey's Hole	Rainbow	8-10-57	1.20	–
Above Gardiner	Cutthroat	6-25-58	0.45	–
	Brown	6-28-58	1.09	–
Gardiner	Brown	1-18-58	1.38	6.53
	Brown	7-30-58	0.91	3.74
	Cutthroat	7-30-58	0.37	1.21
	Rainbow	7-30-59	0.38	0.44
Corwin Springs	Brown	8- 3-57	1.1	0.5
	Brown	8-11-57	1.2	0.9
	Cutthroat	8-11-57	2.1	0.2
	Cutthroat	12- 9-59	0.9	–
Emigrant	Cutthroat, rainbow	8- 3-57	1.4	0.69
	Cutthroat	9-25-57	0.9	0.2
	Brown	9-13-57	1.4	0.7
	Brown	9-13-57	0.4	0.0
	Rainbow	9-13-57	0.6	0.05
	Rainbow	9-13-57	1.6	0.05
	Rainbow	9-25-57	0.9	0.1
	Brown	6-21-58	0.56	–
	Rainbow	6-21-58	0.21	–
	Cutthroat	6-21-58	0.18	–
	Rainbow	7-17-58	0.22	0.21
	Brown	7-17-58	0.60	2.79
	Rainbow	8-23-58	0.20	0.40
	Brown	8-18-58	0.62	1.49
	Rainbow	6-19-59	0.68	0.36
	Rainbow	6-20-59	0.24	0.21
Livingston	Rainbow	8-19-57	1.2	0.01
	Brown	8-19-57	1.9	0.4
	Rainbow	10- 9-57	0.13	–
	Brown	10- 9-57	0.76	–
	Brown	11-12-57	1.23	–

Cutthroat trout from Pelican Creek, a tributary to Yellowstone Lake, were selected as control fish for analysis for DDT because no spray program was known to have been carried on within 30 miles. Fish had no opportunity to swim from a sprayed area into Pelican Creek, since impassable falls blocked their way. Analysis of Pelican Creek cutthroat collected on June 15, 1958, revealed the presence of DDT in all six fish

TABLE 20–4. DDT and DDE contents of mountain whitefish collected from the Yellowstone River in 1957, 1958, and 1959

Locality	Date collected	DDT content (p.p.m.)	DDE content (p.p.m.)
Gardiner			
	July 30, 1958	0.62	2.14
	June 17, 1959	1.42	0.15
Corwin Springs			
	Aug. 6, 1957	3.3	0.2
	Aug. 11, 1957	0.06	0.02
	Aug. 11, 1957	2.7	0.0
	Sept. 9, 1957	3.8	0.42
	Dec. 8, 1957	0.77	–
	Jan. 12, 1958	0.91	–
	Dec. 9, 1959	0.7	1.2
Emigrant			
	Aug. 21, 1957	1.1	0.15
	Aug. 22, 1957	3.5	0.0
	Aug. 26, 1957	3.2	0.1
	Aug. 29, 1957	2.0	1.2
	Sept. 13, 1957	6.9	5.6
	Sept. 20, 1957	1.27	2.43
	Jan. 21, 1958	1.52	–
	Feb. 6, 1958	2.02	–
	July 25, 1958	0.74	1.55
	Aug. 4, 1959	1.31	0.59
Livingston			
	Sept. 13, 1957	2.9	1.1
	Oct. 9, 1957	0.85	–

TABLE 20–5. DDT content of cutthroat trout from an apparently untreated area, Pelican Creek, 1958 and 1959

Date collected	DDT content (p.p.m.)	
	Chemical assay	Fly bioassay
June 15, 1958	0.48	0.42
June 15, 1958	0.28	–
June 15, 1958	0.17	0.17
June 15, 1958	0.45	–
June 15, 1958	0.49	0.33
June 15, 1958	0.21	–
June 10, 1959	0.46	–

tested (Table 20–5). In order to test the validity of the chemical method that achieved positive readings in the control fish, fly bioassay tests were run on three of the fish. DDT was found to be present by this test, and in about the same amounts as shown chemically (Table 20–5). Fish collected in Pelican Creek on June 10, 1959, were also found to contain DDT, and in amounts similar to those collected in 1958. We must conclude, therefore, that the cutthroat from Pelican Creek contained DDT, even though we can recognize no source of contamination.

Water

Our interest in exposed waters following the spraying of DDT principally concerned the amount of DDT and the duration of its presence in streams and lakes. The samples analyzed do not show clear relationships between DDT content and the progress of the season in each body of water, because sampling proved to be less frequent than would have been desirable (Table 20–6).

TABLE 20–6. DDT content of water samples collected in the Yellowstone River drainage in July 1957
[Tr. denotes less than 0.01 p.p.m.; N.D. denotes no detectable DDT]

Locality	Date	Hours since last spray	Hour	Depth	DDT content (p.p.m.)
Pebble Creek	12	3	8:30 a.m.	Surface	Tr.
Pebble Creek	12	3	8:30 a.m.	Subsurface	Tr.
Soda Butte Creek	12	3	7:30 a.m.	Surface	Tr.
Soda Butte Creek	12	3	7:30 a.m.	Subsurface	Tr.
Lamar River	15	4	9:00 a.m.	Surface	Tr.
Lamar River	15	4	9:00 a.m.	Subsurface	Tr.
Lamar River	18	4	9:15 a.m.	Combined	0.01
Lamar River	20	1	5:45 a.m.	Combined	0.03
Slough Creek	15	26	8:15 a.m.	Combined	N.D.
Tower Creek	13	1	6:00 a.m.	Surface	0.010
Tower Creek	13	1	6:00 a.m.	Subsurface	0.011
Yellowstone River near Tower Junction	11	3[1]	8:30 a.m.	Combined	0.01
Yellowstone River					
at Corwin Springs	12	5[1]	10:00 a.m.	Combined	N.D.
at Corwin Springs	13	5[1]	12:00 noon	Combined	N.D.
at Corwin Springs	14	9[1]	1:45 p.m.	Combined	N.D.
at Corwin Springs	15	6[1]	11:00 a.m.	Combined	0.01
at Corwin Springs	17	6[1]	11:00 a.m.	Combined	N.D.
at Corwin Springs	18	6[1]	11:00 a.m.	Combined	0.01
at Corwin Springs	19	8[1]	1:30 p.m.	Combined	N.D.
at Corwin Springs	22	8[1]	1:00 p.m.	Combined	N.D.
at Corwin Springs	23	5[1]	10:00 a.m.	Combined	N.D.
Yellowstone River at Emigrant	22	6[1]	11:30 a.m.	Combined	Tr.
Trout Lake	18	3	8:15 a.m.	Combined	0.50

[1] Spraying was done several miles upstream.

The Lamar River contained 0.01 p.p.m. of DDT on the morning of July 18. Many tributaries upstream from the collecting point had been sprayed in the preceding 8 days, but it seems likely from examination of Tables 20–2 and 20–6 that the spray applied on the morning of July 18 in areas on the Lamar River, Lower Cache Creek, Soda Butte Creek, and other places in the Lamar drainage was responsible for the 0.01 p.p.m. of DDT. The reading of 0.03 p.p.m. on the morning of July 20 was probably caused by the spray of that same morning on the Lamar River, since the sample was taken at 5:45 a.m., shortly after the spray had been applied. Slough Creek contained no DDT on July 15 despite the fact that a tributary, West Slough Creek, had been sprayed the day before. The Yellowstone River near Tower Junction had 0.01 p.p.m. of DDT at 8:30 a.m. on July 11. This result can probably be traced to that morning's spray on Tower Falls and the Yellowstone River Canyon, rather than to the previous day's spraying on the Yellowstone River Canyon.

Yellowstone River water at Corwin Springs was sampled on 9 days, from July 12 to 23. Spray was deposited in drainages upstream from this point every day from July 10 to 23, and it is likely that DDT from each spraying reached some stream and was washed into the Yellowstone River. It might be expected that DDT would be found in the Yellowstone River at Corwin Springs during the entire 14-day period, unless it was dissipated, quickly washed away, or adsorbed along the way. The data show that DDT was found in the water at this point on only 2 of the 9 days on which samples were taken, suggesting that either the DDT was usually washed downstream quickly on each occasion or that extremely large amounts of DDT entered the drainage on these two occasions. It would appear that concentrations went downstream in blocks, rather than spreading out and lingering for periods of time.

The Yellowstone River at Emigrant contained a trace of DDT on July 22. It is difficult to assign this to any particular concentration of DDT spray, since the river distances are great. There was widespread spraying on July 21, the last day of the spray period, and this may have been responsible for the presence of DDT in the sample from Emigrant.

The sample of July 18 from Trout Lake, the only lake sample analyzed, shows that there was 0.5 p.p.m. of DDT. This concentration was by far the highest measured and suggests that the lake acted as a reservoir for the DDT and released it slowly through the outlet. Trout Lake was in the spray area on July 18 and could also have received some drifting spray on July 14, 16, and 17.

Bottom invertebrates

Measurements of numbers of bottom invertebrates were made to determine if there were reductions that could be related to the application of DDT spray (Tables 20–7, 20–8, 20–9). Such reductions would be of much importance to fish, since they would significantly reduce the food supply, and since dead or dying organisms containing DDT may be consumed by fish. Studies to be mentioned later have shown these factors to be important to the welfare of trout populations.

The numbers of bottom invertebrates in Tower Creek 2 weeks after spraying were about half those present 1½ weeks before spraying, with the stoneflies and caddisflies suffering the greatest reductions (Table 20–7). Mayflies increased in numbers. Good repopulation had taken place by the middle of 1958. In Soda Butte Creek, where moderate numbers of insects had been counted before the spray, none could be found 1 week after the spray. There was a repopulation later in the season, however, especially of dipterous insects and caddisflies. After 1 year the population numbers were about the same as before the spray, but species composition was different. The number of all forms present in the outlet of Trout

TABLE 20–7. Numbers of bottom invertebrates in Tower Creek, Soda Butte Creek, and the outlet of Trout Lake in 1957–58
[Numbers are the totals found in 10 combined Surber square-foot samples]

Date	Plecoptera	Ephemeroptera	Trichoptera	Diptera	Other	Total
			Tower Creek[1]			
Pre-spray						
7- 2-57	48	10	9	0	0	67
Post-spray						
7-30-57	5	27	2	0	0	34
8- 6-57	2	5	2	0	0	9
8-19-57	7	26	6	19	3	61
8-30-57	26	16	0	63	2	107
9-12-57	23	4	4	2	6	39
7-18-58	3	72	41	0	0	116
8-29-58	52	82	32	4	2	172
			Soda Butte Creek[2]			
Pre-spray						
7- 2-57	8	18	7	0	1	34
Post-spray						
7-26-57	0	0	0	0	0	0
8- 5-57	3	0	1	0	0	4
8-19-57	1	2	1	90	0	94
8-30-57	0	0	14	15	0	29
9-12-57	0	0	20	6	0	26
7-18-58	1	5	2	11	0	19
8-29-58	2	4	4	15	0	25
			Trout Lake Outlet[3]			
Pre-spray						
7- 2-57	750	30	185	35	10	910
Post-spray						
7-25-57	74	24	10	0	0	108
8- 5-57	4	17	2	1	2	26
8-19-57	458	3	25	2	0	488
8-30-57	963	0	17	0	5	985
9-12-57	83	15	19	20	22	159
7-18-58	203	27	17	30	250[4]	527
8-29-58	586	94	6	35	2	723

[1] Last spray July 14.
[2] Last spray July 20.
[3] Last spray July 18.
[4] Hydracarina.

Lake was reduced, with stoneflies, caddisflies, and Diptera showing the greatest changes. There was a further reduction in numbers in this stream by August 5, followed by rapid increases to pre-spray numbers. In August 1958 insect numbers were generally high, due to an abundance of stoneflies. Caddisflies, however, were low in numbers.

In Slough Creek at the campground and at the mouth, large decreases in numbers were measured shortly after spray time (Table 20–8). Stoneflies, caddisflies, mayflies, and dipterans showed extreme changes in numbers. Repopulation during the rest of the season was indicated but was not as extensively measured as in Tower, Soda Butte, or Trout Lake Creeks. In the summer of 1958 numbers of insects at the campground and near the mouth were still below pre-spray numbers. In the Lamar River, a relatively unproductive stream, there appeared to be a significant reduction in bottom organisms after the spray, with small numbers of caddisflies being the only types found until August 5. Repopulation did not reach pre-spray levels by September 12, 1957, although a little comeback was measured. A year later further recovery had been made.

A pre- and post-spray comparison of Corwin Springs, Yellowstone

TABLE 20–8. Numbers of bottom invertebrates in Slough Creek and the Lamar River in 1957 and 1958[1]
[Numbers are the totals found on 20 combined rock samples]

Date	Plecoptera	Ephemeroptera	Trichoptera	Diptera	Other	Total
			Slough Creek at campground			
Pre-spray 7- 2-57	102	45	21	5	0	173
Post-spray 7-18-57	6	0	24	0	6	36
8- 5-57	3	1	1	0	0	5
8-19-57	13	0	9	0	1	21
8-30-57	2	0	6	0	14	22
7-18-58	0	0	11	0	0	11
			Slough Creek at mouth			
Pre-spray 7- 5-57	95	132	63	0	0	290
Post-spray 7-21-57	2	0	9	0	1	12
8- 5-57	8	0	12	0	5	25
8-19-57	1	2	3	0	1	7
8-30-57	1	1	4	0	9	15
9- 2-57	6	1	20	4	1	32
8-29-58	4	86	22	17	52	181
			Lamar River			
Pre-spray 7- 2-57	20	2	9	0	0	31
Post-spray 7-21-57	0	0	5	0	0	5
7-26-57	0	0	0	0	0	0
8- 5-57	3	0	2	0	0	5
8-19-57	2	0	1	0	0	3
8-30-57	0	0	2	0	1	3
9-12-57	3	0	6	0	0	9
8-29-58	3	4	6	2	1	16

[1] Last spray July 21.

River, was made (Table 20–9). The change from July 3 to July 21 was great, with large reductions of stoneflies and mayflies. Further reductions were indicated a week later, and a slow repopulation after that was suggested. Numbers in August 1958 were at pre-spray levels, but the species composition was altered. Post-spray samples at Emigrant and Livingston contained bottom organisms in the same order of magnitude as those at Corwin Springs and showed repopulation in about the same pattern. Repopulation at Emigrant and Livingston was indicated by the appearance in August of many early-instar stonefly and mayfly nymphs, which increased the total numbers over those in earlier samples. The appearance at Emigrant and Livingston of many immature gastropods in August and September accounts for increases in total numbers.

Drift samples

Drift-sample measurements were taken in Tower Creek to obtain an index of dead and dying aquatic invertebrates in the stream following DDT spraying (Table 20–10). A pre-spray sample taken at 4:00 a.m. on July 13 contained 85 organisms. The first post-spray sample, taken at 6:00 a.m., showed an increase in numbers of drifting invertebrates, and one taken at 8:00 a.m. showed a further increase. On July 14 a larger spray application was made in the same drainage, and by 7:30 a.m. the

TABLE 20–9. Numbers of bottom invertebrates in the Yellowstone River in 1957 and 1958[1]
[Numbers are the totals found on 20 combined rock samples]

Date	Plecoptera	Ephemeroptera	Trichoptera	Diptera	Other	Total
			At Corwin Springs			
Pre-spray						
7- 3-57	160	102	17	0	0	279
Post-spray						
7-21-57	14	3	12	0	0	29
7-26-57	9	0	6	0	0	15
8- 5-57	6	0	10	0	0	16
8-23-57	31 (14 large) (17 small)	9	0	1	0	41
9-12-57	10 small	21	35	6	0	72
7-16-58	21	2	8	5	0	36
8-29-58	35	91	103	39	15	283
			At Emigrant Bridge			
Post-spray						
7-29-57	19	0	9	0	13	41
8- 5-57	10	2	3	0	2	17
8-23-57	21 (9 large) (12 small)	22 (9 large) (13 small)	5	2	19	91
9-12-57	10 small	26	17	6	21	70
7-16-58	7	33	6	22	3	71
8-29-58	83	62	245	56	16	462
			Above Livingston			
Post-spray						
7-29-57	17	2	11	0	0	30
8- 5-57	24 (18 small) (6 large)	0	8	0	0	32
8-23-57	27 (19 small) (8 large)	10	9	3	0	49
9-12-57	25 small	60	14	4	85	188
7-16-58	17	7	35	52	0	111
8-27-58	42	45	118	95	8	308

[1] Last spray upstream July 21.

TABLE 20–10. Numbers of aquatic invertebrates collected in 5-minute plankton-net samples in Tower Creek

Date and hour	Hours after spray	Plecoptera	Ephemeroptera	Trichoptera	Diptera	Other	Total
			First spray				
July 13							
4:00 a.m.	Pre-spray	58	14	10	3	0	85
6:00 a.m.	1	90	10	5	4	0	109
8:00 a.m.	3	85	19	54	49	0	207
			Second spray				
July 14							
5:30 a.m.	½	81	19	1	2	1	104
7:30 a.m.	2½	432	91	84	6	0	613

sample contained more than seven times the numbers of invertebrates found in the pre-spray sample.

Aquatic vegetation

The three samples of *Potamogeton* which were analyzed for DDT (Table 20–11) all contained greater concentrations than the water samples from the same localities.

TABLE 20–11. DDT content of vegetation samples (Potamogeton *sp.*) collected from affected waters in 1957

Locality	Days after last exposure	Date collected	DDT content (p.p.m.)	Highest DDT content found in water (p.p.m.)
Lamar River	15	August 5	1.7	0.03
Slough Creek, at campground	15	August 5	2.3	N.D.[1]
Yellowstone River, at Corwin Springs	16	August 6	0.8	0.01

[1] No detectable DDT.

DISCUSSION

The questions as to whether DDT spray moves downstream in a block, and whether DDT may linger for hours in one place in a moving stream are partially answered by this study. It is probable that if the toxicant stayed in the waters of streams for many hours, we would have collected positive samples at Corwin Springs on the Yellowstone River on more than two of the nine occasions. For 11 consecutive days DDT was sprayed in this drainage. The spray that reached streams was funneled into the main stem and flowed past Corwin Springs. It appears that the DDT was either so diluted when it reached this point that it could not usually be measured, or it passed downstream in well-defined slugs that were not sampled in our program. The report by Cope and Park (1957) shows that Canyon Creek, a stream with a discharge of 50 c.f.s. on the Beaverhead National Forest in Montana, had 0.10 p.p.m. of DDT in the water while the spray was being applied, 0.33 p.p.m. a half hour after the end of spray time, and 0.00 p.p.m. 27 hours later. Canyon Creek is similar in size and gradient to many of the tributaries in the Yellowstone River system which were sprayed in 1957, and the data from both studies suggest that the DDT moved downstream in a matter of hours.

Graham and Scott (1959) found on the Ruby River, a meandering stream in Montana, that traces of DDT were in the surface water 32 hours after spraying in a situation in which 0.01–0.02 pound per acre reached the ground at the river's edge.

The dissipation of DDT as it moves downstream is important in the interpretation of data on DDT content of fish in downstream areas. DDT was present in fish below Livingston, which lies about 85 stream miles below the lowermost spray area, and at Emigrant, about 55 stream miles below the lowermost spray area. It has been generally assumed, and reported by Savage (1949), that the effect of DDT rapidly diminishes as it is carried downstream from the spray area. Kerswill and Elson (1955) reported many dead suckers of various sizes having been collected at a counting fence 10 miles below an area sprayed with DDT in the Miramichi River drainage in New Brunswick. Graham and Scott (1959) found 0.01 p.p.m. of DDT in the Ruby River 4 miles below the spray area, 24 hours after the spray was applied.

No unusual numbers of dead fish were found during the present study, and it is concluded that the DDT was not present in these waters in quantities great enough to cause extensive acute mortality of fish.

Relating the DDT spray to the significant reductions in numbers of fish-food organisms in Tower, Soda Butte, and Slough Creeks and the Lamar and Yellowstone Rivers may be difficult in view of the possibility of emergence as adults of some of the aquatic insect populations. A 2- to 3-weeks' time lapse in most cases between pre-spray and post-spray samples was sufficient for some insect populations to emerge in large enough numbers to cause significant reductions of immature forms. The drift samples from Tower Creek, however, appeared to show that reductions in bottom fauna in that stream were caused by the DDT spray and not by emergence. The numbers of dying and dead aquatic insects collected in plankton nets before and after the spray showed large increases within a few hours after spray time and coincided with the reductions measured on the stream bottom. This follows the pattern reported by Cope and Park for Trapper and Canyon Creeks (1957) and by Graham and Scott (1959) in the Ruby River. Here, large volumes of aquatic invertebrates were collected in plankton nets a few hours after spraying. One and 2 days later the drifting organisms were collected in pre-spray numbers and a few weeks later the counts of bottom organisms showed significant losses. Hoffman and Surber (1948) collected drifting insects in Back Creek, West Virginia, after a treatment with wettable DDT. The largest numbers of insects drifting downstream occurred within the first 3 hours after spraying. At the end of 9 hours the number of drifting insects was reduced almost to pre-spray levels.

The interpretation of these bottom-sample numbers in relation to effects of DDT in Yellowstone streams must be carefully studied before cause and effect can be established. The efficiency of sampling was such that only gross changes can be considered. The grouping of organisms into orders, rather than into lower categories, also requires that small differences in total numbers must be disregarded, since changes in species composition are masked. Moreover, insects with short life cycles can pupate and emerge between samplings, giving the effect of changes which might be ascribed to DDT. In other instances, egg-laying between samplings can account for significant increases in population size. For example, in the Trout Lake outlet sample of August 5 there was a large decrease in numbers of bottom organisms, probably caused by DDT toxicity (Table 20–7). This low was followed by a substantial buildup. The increase noted on August 19 consisted of large numbers of early-instar stonefly nymphs which probably came from eggs laid after the spray was deposited. A similar condition was also seen in Tower Creek on August 30 when large numbers of small stonefly nymphs and dipterous larvae appeared, apparently from post-spray egg-laying.

The pattern of bottom-fauna incidence measured on Tower and Soda Butte Creeks, Trout Lake outlet, and the Yellowstone River at Emigrant was characterized by a buildup of organisms followed by a reduction in September. In the Yellowstone River the reduction may have been caused by the emergence of stoneflies since the last August collection showed large and small nymphs, while the September 12 collection had only small ones. In Tower Creek the buildup was caused by stoneflies and dipterans. The last collection had few dipterans, so a fly emergence was probably responsible for the September reduction. The flies in Soda Butte Creek also may have emerged in late August to cause a loss of bottom fauna. The changes in bottom fauna in the various streams all show significant reduction in insect numbers soon after application of the DDT.

It is difficult to relate amounts of DDT in aquatic vegetation in late season to amounts in the water or in the fish, because the vegetation grows after the DDT is adsorbed and aquatic invertebrates graze on the vegetation. It is probable, however, that the DDT contained in aquatic plants enters the bodies of fish, directly or indirectly, and that enough may be ingested to affect the welfare of the fish or be stored in their tissues.

Relatively high levels of DDT were found in the bodies of whitefish and trout during this study compared with the amounts found in the water. Assuming that the fish did not move great distances from the beginning of the season to the end, the manner in which fish acquired DDT in amounts so much greater than any measured in the water or in the limited amount of aquatic vegetation tested is of considerable interest. For example, in the Yellowstone River at Corwin Springs, the largest amount of DDT found was 0.01 p.p.m. in water and 0.08 p.p.m. in aquatic vegetation, while up to 3.8 p.p.m. of DDT was found in whitefish. At Emigrant only a trace of DDT was found in the water, yet 6.9 p.p.m. was found in a whitefish. In Trout Lake 0.50 p.p.m. of DDT was in the water sample analyzed, but rainbow trout contained 4.9 p.p.m. of DDT and 0.7 p.p.m. of DDE 8 days later.

Graham and Scott found from 1.30 to 5.80 p.p.m. of DDT in rainbow and brown trout taken alive from the Ruby River in Montana. These fish were collected about 9 weeks after airplane spraying in the study area. Amounts of DDT in samples of water from the river ranged from 1.35 p.p.m. 5 minutes after the spraying to 0.00 p.p.m. 48 hours after the spraying, with most samples containing less than 0.05 p.p.m. These residues compare favorably with amounts found in the Yellowstone River drainage fish from upstream areas at about the same time after the spraying.

Hunt and Bischoff (1960) reported on a DDD gnat-control project in Clear Lake, California. Treatment was begun in 1949, when 1 part of DDD in 70 million parts of water was dispersed in the lake. Treatments were also made in 1951 and 1954, at the rate of 1:50 million. In 1958 specimens of several species of live fish were collected and their visceral fat analyzed for DDD content. DDD in the fat amounted to 40 p.p.m. in carp and 2,500 p.p.m. in brown bullhead, with other species having intermediate amounts. Later collections in 1958 again showed extremely high DDD levels in the visceral fat and in the fat of edible flesh, especially in black crappie, largemouth bass, and brown bullhead. Visceral fat from white catfish had up to 2,375 p.p.m. of DDD, from largemouth bass, up to 1,700 p.p.m., from black crappie, up to 1,600 p.p.m., and from brown bullhead, up to 2,500 p.p.m. Edible flesh contained DDD up to 221 p.p.m. in white catfish, up to 138 p.p.m. in largemouth bass, up to 115 in black crappie, and up to 79.9 p.p.m. in brown bullhead. Evidently the long exposure at low levels resulted in high levels of DDD in the fish without producing mortality.

DDT enters the fish body from water and in food. Cottam and Higgins (1946) reported on experiments with warm- and cold-water fishes which demonstrated that the most toxic effects were produced when DDT was fed on or incorporated in food, and that DDT was definitely more toxic when dissolved in oil than when given in fat-free carbohydrates or proteins. In view of the experience cited above it is probable that the DDT found in fish in the Yellowstone drainage entered through the ingestion of dead and dying insects. The water contained large numbers of drifting insects affected by the DDT-oil spray. It is not known how much DDT was present in or on the insects, or whether the fish ate them in abundance, but experience indicates that the fish would eat drifting insects. On Canyon and Trapper Creeks in 1956, it was observed that some trout in live-cars ate large amounts of the affected insects and that trout free in the streams gorged themselves on this diet. This engorgement has also been observed elsewhere.

The DDT contents of these fish indicate that an accumulation of DDT took place in the tissues as has been shown by numerous authors reporting on warm-blooded vertebrates. The rate of consumption of DDT by fish in the Yellowstone River is not known, but apparently there was considerable buildup of DDT.

The greater amount of DDT in fish in 1957 than in 1958 and 1959 might be expected. Weighed against the accumulation of residues through ingestion of DDT-bearing animal and plant material are the processes of metabolism and excretion of the toxicant, growth of the fish, and possible mortality of fish with excessive amounts in storage. The data suggest the possibility of an increase in the amount of DDT stored for a few months, followed by a slow process of loss of DDT over a period of many months.

The presence of DDT in fish from an apparently untreated area of Pelican Creek raises questions regarding the interpretation of the positive determinations in fish from treated areas. The DDT levels in the fish from Pelican Creek were somewhat less than 0.5 p.p.m., and 13 of the 80 positive samples from the treated areas had less than this amount. This may suggest that many fish commonly have low levels of DDT in their bodies and that the spruce-budworm spray program may not have been the cause of all of the residues found in fish from the Yellowstone River drainage. In the absence of analyses of water from Pelican Creek, and without zero controls from fish tissue, it appears impossible to resolve the questions concerning these residues. It was during a search for uncontaminated fish for control purposes that the Pelican Creek fish were analyzed and found positive, and all other fish analyzed for this purpose were also found positive.

The presence of DDT and other chlorinated hydrocarbons may not be as unusual as is commonly thought. The Fish-Pesticide Research Laboratory at Denver has frequently found toxicants in control trout and other fish, and it has not always been possible to trace the origins of these materials.

The reliability of the Schechter–Haller test is not questioned, despite the difficulties in consistent extraction and in making valid determinations at low levels. Interfering substances were removed before analysis, so contamination cannot be suspected. The verification of chemical measurements on Pelican Creek fish by fly bioassay lends credence to the reliability of the chemical analyses.

Consideration must be given to the possibility that part of the DDT residues found in and after 1957 may in fact have originated during the 1955 spraying. The 2-year retention of DDT residues from the 1957 treatment forces us to recognize that a long-term storage of DDT in fish tissues could have begun in 1955 and extended beyond the time of the sprayings in 1957.

The comparison of precise amounts of DDT residue with those in other fish from other places at other times is difficult because of the conversion of DDT to DDE and DDA in the animal body. Since this process proceeds at different rates in different individual fish, it appears dangerous to compare residues in the Pelican Creek trout with those of the Yellowstone River trout without knowing when the exposure took place in the Pelican Creek fish.

The Montana Fish and Game Department planted numbers of catchable-sized rainbow and cutthroat trout in the affected area of the Yellowstone River in 1957. The question as to whether these fish contained amounts of DDT different from those of wild fish in the same waters has not been determined. Fish of hatchery origin, recognized through scale growth patterns, contained DDT in amounts similar to those in wild fish. Examination of planting records indicates that the stocking took place before the end of the spray period.

ACKNOWLEDGMENTS

Acknowledgment is made of contributions to the program from the following persons: Elmer E. Frahm, Montana State College, and Francis B. Coon, Wisconsin Alumni Research Foundation, for chemical analyses and consultation; Clarence Hoffman and Milton Schechter, Agricultural Research Service, for advice; M. Barrows, O. Brown, H. Edwards, W. Chapman, J. Packard, O. Bailey, T. Milligan, and R. Burns, National Park Service, for collection of specimens; G. Miller and R. Graham, Montana Fish and Game Department, for collection of specimens and cooperation; David Scott, U.S. Forest Service, for cooperation; and Norman G. Benson, Bureau of Sport Fisheries and Wildlife, for collection of specimens.

LITERATURE CITED

Cope, Oliver B., and Barry C. Park. 1957. Effects of forest insect spraying on trout and aquatic insects in some Montana streams. U.S. Forest Service, Missoula, Montana, 56 pp.

Cottam, Clarence, and Elmer Higgins. 1946. DDT: Its effect on fish and wildlife. U.S. Fish and Wildl. Serv., Circ. 11, 14 pp.

Graham, Richard J., and David O. Scott. 1959. Effects of an aerial application of DDT on fish and aquatic insects in Montana, U.S. Forest Service, Missoula, Montana. 35 pp.

Hoffman, Clarence H., and Eugene W. Surber. 1948. Effects of an aerial application of wettable DDT on fish and fish-food organisms in Back Creek, West Virginia. Trans. Am. Fish. Soc., 75:48–58.

Hunt, Eldridge G., and Arthur I. Bischoff. 1960. Inimical effects on wildlife of periodic DDD applications to Clear Lake. Calif. Fish and Game, 46:91–106.

Kerswill, C. J., and P. F. Elson. 1955. Preliminary observations on effects of 1954 DDT spraying in Miramichi salmon stocks. Prog. Repts., Atlantic Coast Stations, Fish. Res. Bd. Canada, No. 62:17–24.

Savage, James. 1949. Aquatic invertebrates: Mortality due to DDT and subsequent re-establishment (1944–45). *In:* Forest spraying and some effects of DDT. Ontario Dept. Lands and Forests, Biol. Bull. 2:39–47.

Schechter, M. S., and H. L. Haller. 1944. Colorimetric test for DDT and related compounds. Jour. Amer. Chem. Soc., 66:2129–2130.

21

Inimical effects on wildlife of periodic DDD applications to Clear Lake

Eldridge G. Hunt
Arthur I. Bischoff

INTRODUCTION

The indirect effects of pesticides on wildlife are of growing concern to conservationists. These effects are insidious and are often entirely unnoticed or are not discernible for a long period after initial contact with a toxic material. The materials involved in this type of poisoning are usually of low acute toxicity; many are accumulative in action and are stored in animal flesh. These properties are found especially in certain members of the chlorinated hydrocarbon family of insecticides. The amount of toxic material accumulated in animal tissue may be increased over a period of years as a result of multiple contacts. Clinical symptoms or deleterious effects may appear at any time during this period. The level of accumulation of toxic material that can be tolerated before clinical symptoms occur may be different for the various animal species or individuals of the same species. Some animals are able to store large amounts of certain toxic materials in tissue with no apparent ill effects. However, continued accumulation of these materials usually affects vital functions and may eventually result in death (Wallace, 1959; Rudd, 1958).

The literature provides examples of the effects of this type of poison-

Reprinted from *California Fish and Game*, 46(1):91–106 (1960), by permission from the California Department of Fish and Game. This work was performed as part of Federal Aid in Wildlife Restoration, California Project W-52-R, "Wildlife Investigations Laboratory." The authors are affiliated with the California Department of Fish and Game.

ing on animal populations. A few are decreased fecundity, interference with normal food chain activities, and upset of interdependent relationships of one animal species with another (Rudd, 1958; DeWitt, 1957; Genelly and Rudd, 1956; Springer, 1956).

The use of a chlorinated hydrocarbon insecticide DDD (dichloro diphenyl dichloroethane) in gnat control programs at Clear Lake, Lake County, California, resulted in a wildlife-pesticide problem (Fig. 21–1). Involved were the effects of this insecticide on mammal, amphibian, and fish populations at the lake. A study on the effects of the gnat control programs on wildlife at Clear Lake was begun in March, 1958, by the California Department of Fish and Game and is scheduled to continue while present hazards to wildlife exist. This paper presents the results of chemical analyses of specimens collected at Clear Lake and information obtained from investigations in the area.

Fig. 21–1. Western Grebes—A species involved in a wildlife-pesticide problem at Clear Lake. *Photograph by William Anderson.*

ACKNOWLEDGMENTS

The effects of the control programs on wildlife could not be determined conclusively without analyses of animal tissues for the presence and amount of DDD. These analyses were made by chemists of the Food and Drug Laboratory, California Department of Public Health; and the Bureau of Chemistry, California Department of Agriculture. We wish to thank the personnel from these laboratories not only for the actual chemical analyses, but also for their assistance in assessing the significance of the analyses. Guidance was provided by staff members of the Department of Fish and Game, Department of Public Health, and Department of Agriculture in planning various phases of the study. The collection of specimens was made by Game Management, Inland Fisheries, and Wildlife Protection personnel of the Department of Fish and Game. Assistance with these collections was also provided by employees of the Lake County Mosquito Abatement District. The study was enhanced by the co-operative attitude of the agencies concerned with this problem. Critical review of this manuscript was made by Dr. Robert L. Rudd.

PHYSICAL DESCRIPTION OF CLEAR LAKE

Clear Lake is an irregularly shaped body of fresh water approximately 19 miles long and seven miles wide (Fig. 21–2). The lake is about 1,325 feet above sea level and has a surface area of approximately 41,600 acres. It has an upper and lower portion connected by a narrows that is less than one mile wide. Clear Lake is relatively shallow with maximum depths of 30 feet in the upper portion and 50 feet in the lower portions. The marginal slope of the lake bottom is gradual in most places, especially in the larger upper portion where it is saucer shaped. The deposits on the bottom are predominately soft, deep, black ooze except for a few beaches and a portion of Konocti Bay, which are largely volcanic gravel.

Mountains of the Coast Range surround the lake and slope to the water's edge along most of the shoreline. The peaks of some of the highest mountains in the area are over 4,000 feet in elevation. Several alluvial plains fan out into the lake on the west side.

An estimate of the average amount of water during late summer is 300 billion gallons or 850,000 acre feet. The lake level fluctuates with seasonal precipitation and the demands for water from agricultural and domestic interests. Occasionally the level ranges from 8 or 10 feet above to one to two feet below the zero reading on a scale indicator used to measure variation in surface level (Lindquist and Deonier, 1943).

The annual bottom temperatures range from 45 to 78 degrees F.

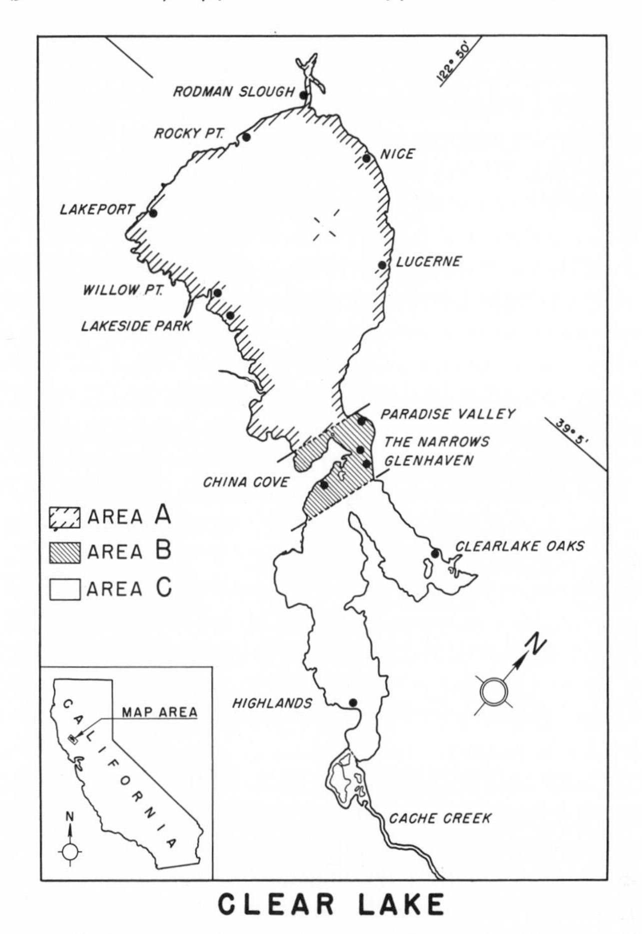

Fig. 21–2. Map of Clear Lake showing locations of animal collections. *Drawing by Cliffa Corson.*

Compared with other mountain lakes, Clear Lake is rather turbid most of the time. The distance at which Secchi Disks can be seen may vary from 5 to 75 inches (Walker, 1949).

HISTORY OF GNAT CONTROL AT CLEAR LAKE

The periodic appearance of large numbers of gnats, *Chaoborus astictopus*, during the summer has presented a problem to Clear Lake residents for many years and has had an adverse effect on the large resort business. The annoyance caused by these insects is due wholly to their extreme abundance and attraction to light. These gnats are closely related to mosquitoes but are not bloodsuckers and probably do not feed as adults

(Herms, 1937). Concern over the outbreaks of gnat populations led to occasional studies over a period of years to determine methods of control. The study of the biology and methods of controlling this insect has been extensive. Adverse effects on fish life have been considered in the planning of chemical control programs. The following is a brief chronological account of activities relating to gnat control at Clear Lake from 1916 until the initiation of this study in 1958.

Between 1916 and 1941 several studies were made on the biology of the gnat. Experiments were conducted to determine the effectiveness of several larvicides and ovicides. No large scale control program developed from these experiments. The possibility of using native fish to control the gnat also was explored and examinations made to determine the food habits of 10 fish species. This food habit study indicated that nine of the species examined consumed gnat larvae or pupae and/or adults at some time during the year. Although fish were found to consume great quantities of gnats, it was apparent that some other method of control was required.

During the war years all work was suspended, but in 1946 research to develop a satisfactory control program was resumed. Laboratory experiments with newly-developed chlorinated hydrocarbon insecticides indicated that these materials could be used to control the gnat. Two chemicals, DDT and DDD, were found to have desirable properties and were selected for further study.

Studies made on the toxicity and effect of these chemicals on fish and gnat larvae in aquaria and in smaller bodies of water in the Clear Lake area indicated that DDD would provide the satisfactory gnat control with less hazard to the aquatic environment and its inhabitants than DDT. The final selection of DDD as the larvicide to be used in Clear Lake was made after test applications of this material to nearby Blue Lake in 1947 and Detert Reservoir in 1948. DDD was reported as causing relatively low fish mortality when used at a dilution of one part of insecticide to 70 million parts of water (Lindquist and Roth, 1950). The effects of the DDD treatment of Blue Lake on bottom organisms were reported by Lindquist and Roth (1950), and on insects, fish, and plankton by Murphy and Chandler (1948). A hydrographic survey of Clear Lake was made in the winter of 1948 and 1949 by the U.S. Geological Survey to provide necessary information on the volume of water and the physical properties of the lake. Results of this survey were used in computing desired insecticidal dosages and in planning the actual DDD application.

The first large scale DDD treatment of Clear Lake was made in September, 1949; 14,000 gallons of chemical concentrate were used. The insecticide was prepared by a chemical company according to the following formula: DDD (TDE) 30 grams, emulsifier 30 ml. and xylene 72 ml. (Lindquist *et al.* 1951). The material was applied from six barges towed by tug boats. Final dilution in the lake water was estimated at one part of active insecticide to 70 million parts of water. Additional treatment of 20 small lakes and reservoirs within 15 miles of Clear Lake was made. It was determined that a 99 percent kill of gnat larvae resulted from this treatment. As a result very few gnats were observed for two years.

In July, 1951, gnat larvae were found by a night plankton tow, for the first time since the 1949 treatment. The numbers of gnat larvae and adults increased; and a second DDD treatment of the lake was made in September, 1954. The concentration of active insecticide in the lake water shortly after this treatment was greater than the first treatment and was estimated to have been one part of active insecticide to 50 million parts of water. Rate of larval kill was again estimated at 99 percent. All mud samples taken later that year were negative for *Chaoborus* (Brydon, 1955).

In December, 1954, 100 western grebes, *Aechmophorus occidentalis*, were reported dead, and specimens were sent to the Department of Fish and Game Disease Laboratory. (Any mention of grebes in the text unless otherwise designated refers to western grebes.) Infectious disease was not detected in these specimens. In March, 1955, more dead grebes were reported, and results of autopsies on specimens submitted to the disease laboratory were also negative for infectious disease.

Gnat populations increased during 1955-56; and a third treatment of Clear Lake was made in September, 1957. DDD was again applied at the rate of one part of active insecticide to 50 million parts of water. Gnat control was considered not to be as successful as after previous treatments. The lower rate of success was attributed by some observers to inimical weather conditions during application and by others to a buildup of resistance to DDD by the gnats. During December, 1957, approximately 75 grebes were reported dead on the shores of Clear Lake. A few weeks after this report two sick grebes were submitted to the disease laboratory. No infectious disease was found. Two sections of fat from these birds were submitted to the Bureau of Chemistry, California Department of Agriculture, for toxicological examination. Results of chemical analysis of the fatty tissue indicated DDD was present at the unusually high concentration of 1,600 p.p.m. (parts per million). Contaminated food was suspected to be the cause of death.

Accordingly in March, 1958, a fish collection was made by personnel of the Lake County Mosquito Abatement District. Samples were submitted for toxicological examination. They included single specimens of carp, *Cyprinus carpio*, white catfish, *Ictalurus catus*, black crappie, *Pomoxis nigromaculatus*, brown bullhead, *Ictalurus nebulous*, largemouth bass, *Micropterus salmoides*, and Sacramento blackfish, *Orthodon microlepidotus*. Visceral fat was analyzed by Department of Agriculture chemists. The amount of DDD found in the fat ranged from 40 p.p.m. in the carp to 2,500 p.p.m. in the brown bullhead. After the results of these analyses were received by the Department of Fish and Game the present study was initiated and is being continued.

METHODS

Animals collected for chemical analyses were taken by conventional methods. Fish were either netted in gill nets, seines, or fyke nets, or were taken by hook and line. Birds collected for testing were shot; and bull-frogs, *Rana catesbeiana,* were provided by the local warden. Samples of fish for chemical analysis were usually skinned and filleted. Some fish samples were submitted whole except for being eviscerated, but they proved difficult to grind in preparation for chemical analysis. Visceral fat was dissected from the internal organs of certain fish. Fish from two of the three collections made were classified according to brood year and age groups by accepted methods (Marzolf, 1955; Lagler, 1952). No fish in the second collection were aged. However, an attempt was made to collect older fish of similar size.

The quantitative analysis for DDD in grebe flesh was made from composite samples of visceral and subcutaneous fat. Samples of frog tissues were analyzed from visceral fat only. Flesh samples that could not be analyzed in a fresh condition were frozen and stored until analysis could be made.

The analytical procedure used for detection of DDD was a modification of the Schechter-Haller Method for color development (Pontoriero and Ginsburg, 1953). Removal of interfering substances such as fat was accomplished by use of the Davidow exchange column following extraction of the sample with solvent (Davidow, 1950), or with the use of dimethylformamide (Burchfield and Sterrs, 1953). The presence of DDD was determined by specific color development. The only commonly encountered compounds that give the same color reactions are nontoxic isomers and metabolites of DDT and DDD. To be sure that isomers or metabolites were not involved in the color determination, bio-assays using house flies, *Musca domestica*, were done on the first fish samples. Results of these bio-assays established the presence of a toxic substance, and no additional bio-assays were made. Since no appreciable amount of DDT had been used in the area, the toxic substance in all analyses was assumed to be DDD.

PROCEDURE AND RESULTS

Samples of flesh tissue for chemical analyses were limited to specimens collected in the Clear Lake area. Some life history data on bird and fish species were also obtained. The original selection of species to be sampled was based on both availability and suspected DDD contamination. Animals believed to have come in contact with DDD and which were readily available were collected first. Information derived from analyses of samples of the first two collections indicated that certain species could be used as indicators of maximum DDD concentrations. Subsequent collections were planned to include primarily these species.

The number of animals taken during each collection was determined by one or more of the following: (1) the total number of samples that could be readily analyzed by co-operative agencies, (2) the availability of the animal to be sampled, and (3) the number needed to provide an adequate sample. The number of analyses varied with each collection but was always less than 25. Analyses were made of composite as well as single samples. Samples were considered adequate when the data obtained satisfactorily clarified specific problems. All fish collected were apparently healthy and exhibited no symptoms of being affected by the insecticide. Symptoms of fish poisoned by chlorinated hydrocarbons were described by Henderson *et al.* (1958).

First Collection

Samples obtained during the first collection, May, 1958, were composed entirely of fish taken from two locations, one at each end of the lake. The basic information desired from these samples included: (1) the distribution and comparative levels of contamination of DDD in fish, (2) the effect of cooking on breakdown of this chemical, (3) the amount of DDD in different fish species, and (4) the percentage of the fish population that was contaminated with DDD. The results of chemical analysis of this collection are given in Table 21–1. For purposes of simplification only common names will be used in the tables.

Results of these analyses regarding the distribution and comparative levels of contamination of DDD in fish flesh agreed with findings of other investigations that most of the insecticide is accumulated in fatty tissue (U.S. Dept. Pub. Health, Educ. and Welfare, 1955; Bann *et al.* 1956; Metcalf, 1957). The amount of chemical in visceral and flesh fat from the same samples was similar. Two samples of edible flesh were wrapped in aluminum foil and baked at 400 degrees F. for one hour prior to analysis. There was very little breakdown of DDD, and the major change due to cooking was loss of water and transfer of some fat containing DDD from the flesh to the cooking container. Those species containing the highest concentration of DDD in edible flesh were brown bullhead, largemouth bass, and black crappie. It was apparent that larger samples taken from additional locations would be necessary to obtain more precise information regarding the extent of DDD contamination of the fish at Clear Lake.

Consideration was also given to possible contamination of amphibians with DDD. Visceral fat from nine bullfrogs was analyzed and found to contain 5 p.p.m. of DDD. This level of accumulation was considered low, and no further examinations of frog tissue were made.

TABLE 21–1. Analysis of Fish Flesh From Collection Made in May, 1958

Species	Year class	Age group	Parts per million of DDD				
			Visceral fat	Edible flesh		Fat of edible flesh†	
				Raw	Cooked	Raw	Cooked
Bluegill	1955	III	–	9.52	–	255	–
	1956	II	–	5.27	–	293	–
	1957	I*	175	6.75	–	278	–
	1957	I*	254	7.14	–	350	–
Black crappie	1956	II*	2,690	115.00	–	2,840	–
Brown bullhead	1954	IV	548	79.90	60.3	1,110	1,010
	1955	III	342	11.80	–	218	–
	1955	III	650	24.60	–	912	–
Largemouth bass	1954	IV	1,550	115.00	133.0	1,360	1,310
Sacramento blackfish	1956	II	983	10.90	–	773	–
	1956	II	–	17.60	–	475	–
Hitch	1956	II	–	10.90	–	106	–
	1957	I	–	28.10	–	763	–

* Composite of two fish; all other samples of individual fish.
† Fat between the muscle tissues.

TABLE 21–2. Analysis of Fish Flesh From Collection Made in July, 1958

Species	Area	Number fish in sample	PPM of DDD edible flesh
Black crappie	A	6	5.4
	B	5	61.0
	C	1	5.8
Brown bullhead	A	25	15.5
	B	1	32.2
	C	32	24.8
Bluegill	A	24	7.3
	B	49	10.0
	C	20	6.6
Carp	A	10	51.3
	B	35	62.3
	C	4	63.8
Sacramento blackfish	A	1*	–
	B	16	20.4
	C	–	–
Hitch	A	–	–
	B	52	18.5
	C	–	–
Largemouth bass	A	1*	5.0
	B	1	97.3
	C	2	19.7
White catfish	A	11	30.4
	B	50	81.6
	C	6	129.0
Tagged catfish 10-year-old	B	1	133.0
Total	A	79	–
	B	210	–
	C	65	–

* These fish were of insufficient size to constitute a satisfactory sample for analysis.

Second Collection

The second collection, July, 1958, was also composed entirely of fish. An attempt was made to obtain a representative sample of the more numerous fish species from several key areas along the lake shore. Because of the large number collected, fish were segregated into groups of similar size and appearance instead of aging them by conventional methods. This proved unfortunate because data collected from other samples regarding age and DDD accumulation were not directly comparable. One tagged white catfish 10 years old was collected and was the only sample of known age in this collection. Results of analysis of the second collection are given in Table 21–2. The area designation used in this table of A, B, and C was established arbitrarily. Area A corresponds to the upper section of the lake, Area B to the central section of the lake, and Area C to the lower sections of the lake. All three areas of the lake contained DDD contaminated fish.

Of the nine fish species in the second collection the highest concentration of DDD was found in white catfish and largemouth bass. These species were chosen for sampling on all subsequent collections. Sampling of any other fish species was to be intermittent.

Third Collection

The third collection was made in October, 1958, and included samples of largemouth bass, white catfish, Sacramento blackfish, grebes, and plankton. Results of the analysis of these samples are presented in Table 21–3. The year-old group of Sacramento blackfish and largemouth bass included in this sample was hatched seven to nine months after the last DDD treatment. Flesh samples of both species, one a plankton eater (Sacramento blackfish), the other a carnivore (largemouth bass), contained significant amounts of DDD. Grebes collected were believed to be from a large population of winter visitants. At the time of the grebe collection several thousand grebes were on the lake. In the same general area four weeks prior to the collection fewer than 50 grebes were seen. The plankton samples were too small for accurate chemical analysis. Due to the inadequacy of the sample the presence or absence of DDD in plankton at Clear Lake could not be established.

Table 21–4 presents data on the amount or range of DDD found in all samples of animal flesh analyzed. The amount of DDD found in all flesh samples exceeded by many times the specified rate of dilution of active insecticide in the lake on a p.p.m. basis. All fish, bird, and frog samples contained DDD. Because of a difference in food habits it was believed that carnivorous fish accumulated a greater amount of DDD than did plankton eating fish of similar size and age.

TABLE 21–3. Analysis of Animal Flesh From Collection Made in October, 1958

Species	Year class	Age group	Percent fat in edible flesh	Parts per million DDD: Edible flesh	Parts per million DDD: Visceral fat
White catfish	1956	II	–	22	–
	1955	III	–	26	–
	1953	V	–	64	–
	1953	V	3.0	109	–
	1953	V	–	113	–
	1953	V	4.0	142	–
	1953	V	–	178	2,110
	1953	V	9.5	196	2,375
	1950	VIII	2.0	106	–
	1950	VIII	–	111	–
	1950	VIII	–	221	–
	1949	IX	–	162	–
	1949	IX	2.1	174	–
Largemouth bass	1958	0	6.0	22*	–
	1958	0	–	24*	–
	1958	0	–	25*	–
	1957	I	–	30	–
	1957	I	–	42	–
	1951	VII	2.0	138	–
Sacramento blackfish	1958	0	1.5	7*	–
	1958	0	–	9*	–
Western grebe	Mature	–	†	–	723

* Composite samples not filleted; all other samples were fillets.
† Composite sample of subcutaneous and visceral fat.

TABLE 21–4. Range of DDD Contamination of All Specimens Collected at Clear Lake

Species	Specimens analyzed	Number of analyses	Visceral fat: Analyses	Visceral fat: Specimens	Visceral fat P.P.M. of DDD: Single samples	Visceral fat P.P.M. of DDD: Composite samples	Edible flesh: Analyses	Edible flesh: Specimens	Edible flesh Parts per million DDD: Single samples	Edible flesh Parts per million DDD: Composite samples	Cooked samples	Fat of cooked samples
White catfish	82	20	3 –	1* –	1,700-2,375 – –	– –	14 3	14 67	22.0–221.0 – –	– 30.4–129.0	– –	– –
Largemouth bass	19	12	2 –	1* –	1,550-1,700 – –	– –	6 4	6 12	5.0–138.0 – –	– – 19.7–25.0	133.0† –	1,310† –
Brown bullhead	62	10	4 –	1* –	342–2,500 – –	– –	4 2	4 57	11.8–79.9 – –	– – 15.5–24.8	60.3† –	1,010† –
Black crappie	15	6	1 1	1 2	1,600 – – –	– 2,690 –	1 3	1 11*	5.8 – – –	– – 5.4–115.0	– –	– –
Bluegill	100	9	2 –	2 –	175–254 – –	– –	3 4	3 95	5.3–8.5 – –	– – 6.6–10.0	– –	– –
Hitch	54	3	– –	– –	– – – –	– –	2 1	2 52	10.9–28.1 – –	– – 18.5–	– –	– –
Sunfish	1	1	– –	– –	– – – –	– –	1 –	1 –	5.4 – – –	– – – –	– –	– –
Sacramento blackfish	32	7	2 –	1* –	700–983 – –	– –	2 3	2 29*	10.9–17.6 – –	– – 7.0–20.4	– –	– –
Carp	50	4	1 –	1 –	40 – – –	– –	3 –	49 –	– – – –	51.3–62.3 – –	– –	– –
Total fish	415	72	–	–	– –	–	–	–	– –	– –	–	–
Frogs	9	1	1	9	– –	5 –	–	–	– –	– –	–	–
Grebe	7	2	2	7	– –	723 –	–	–	– –	– –	–	–
Total	431	75	–	–	– –	–	–	–	– –	– –	–	–

* Includes specimens used in other analyses.
† Single analysis of one fish.

Two additional collections have been made at Clear Lake, and the results of analysis have not been received at the time of this writing. Present plans call for the contamination of sampling at four-month intervals until the DDD continuation in animal flesh is considered by the investigators to be insignificant.

Several observations indicate a decrease in the nesting grebe population at Clear Lake since the first DDD treatment. Prior to this treatment the nesting population of grebes was in excess of 1,000 pairs (A. H. Miller, pers. com., 1959). During the nesting seasons of 1958 and 1959 less than 25 pairs were seen during surveys made by Spruill (pers. com., 1959) and the authors. Separate surveys in 1959 by Miller and Spruill each reported that nests of grebes were not found in areas where they had been in previous years. As many as 15 pairs of grebes were reported on the lake during this past nesting season but no nests or young have been found. It is not known whether these birds constructed nests that were not found or they made no attempt to nest. A collection of five adult grebes was made in August, 1959. There were an estimated 60 grebes on the lake at that time, and although the normal nesting season had been completed no immature birds of this species were seen. These data, although fragmentary, indicate little or no nesting success of this species during 1959. However, nests of several pied billed grebes, *Podilymbus podiceps*, and coots, *Fulica americana*, were found; and mallards, *Anas platyrhynchos platyrhynchos*, and cinnamon teal, *Anas cyanoptera* nested successfully in the area. Reasons for this reported difference in nesting success are not known.

After the 1954 and 1957 applications of DDD dead grebes were reported. Based on examination of specimens submitted to the laboratory, disease was not believed to have caused the death of these birds. Toxicological examination was not made on dead grebes in 1954. Two grebes were submitted for toxicological examination in January, 1958. Fat samples from these birds contained 1,600 p.p.m. of DDD. According to chemists of the Department of Agriculture the probable cause of death of these birds was chronic DDD poisoning. As further indication that they were poisoned, both birds exhibited nervous tremors comparable to those characterizing chlorinated hydrocarbon poisoning as described by Radeleff and Woodard (1955). Several other grebes observed on Clear Lake showed similar symptoms (Speth and Taylor, pers com., 1958).

No attempt was made to determine acute or chronic toxicity of DDD to grebes, and no reference to toxicity of any chlorinated hydrocarbon to this or related species was found in literature. However, strong circumstantial evidence indicates that grebe losses occurring after DDD treatments were caused by chronic poisoning from DDD. Differences in DDD concentrations in the fat from fish and grebes may also be interpreted to suggest that grebes show a higher susceptibility to DDD than do many fishes.

DISCUSSION

The use of DDD has affected certain wildlife species in the Clear Lake area. Dead grebes were observed following each DDD application. Various amounts of the toxic material were found in all animals examined. From what source the animals accumulated DDD and to what extent animal populations were affected by this chemical is not fully known.

At present there is no significant public health hazard involved in consumption of Clear Lake fish according to toxicologists of the State Department of Public Health. However, to assure that safe levels for human consumption of Clear Lake fish are not exceeded and to prevent the possibility of increasing the present hazard of DDD poisoning of wildlife, no further treatment of Clear Lake with DDD will be made. It is probable, therefore, that a complete understanding of this particular pesticide-wildlife problem will never be obtained. However, some of the data presented in this paper may be applicable to other problems resulting from similar pesticide applications.

Conclusive results based on chemical analyses showed that all organisms contained DDD. These results established which types of tissue had the greatest concentration of DDD, the amount of chemical found in tissues of various species, and the extent of DDD contamination of fish in the lake.

Results based on circumstantial evidence are believed to be correct. The information collected pertaining to the chronic poisoning of grebes is not complete. It is known that these birds are subject to periodic die-offs, and such an occurrence could have reduced the nesting population at Clear Lake. Additional information will be required to relate accurately the results of chemical analyses of tissues and the cause of death of the grebes. The use of fat analysis to establish a diagnosis of poisoning can be misleading. Apparently, high levels of insecticides can be built up without harm by certain animals (Radeleff and Woodward, 1955).

Although the data are in part circumstantial, the following items indicate poisoning rather than other causes of mortality: (1) the decline in the grebe population corresponded with the period in which pesticide applications were made, (2) the absence of any known infectious disease in autopsied grebes picked up after two of the chemical treatments of the lake, (3) clinical symptoms common to poison victims were exhibited by some grebes from the lake, and (4) an abnormally high concentration of DDD was found in fatty tissue of dead grebes. Observations of dead grebes were made following each DDD application. These die-offs began one or two months after pesticide application and usually lasted several weeks. The fact that die-offs were noted during these periods only, indicates the possibility of chronic poisoning of grebes.

Assuming that grebes were poisoned by DDD, it is important to know how they obtained the toxicant. It is believed that most of the toxic materials was assimilated from ingested contaminated fish and insects. The food of grebes at Clear Lake has been reported to consist primarily of fish, with some insects included (Lawrence, 1950). The chemical could also have been absorbed through the skin or picked up from contaminated feathers consumed by the birds or from other less conspicuous sources. A more logical assumption is that the control programs resulted in the creation of "poisonous fish." The toxicant accumulated in the grebes as a result of ingesting these "poisonous fish" and other contaminated food and resulted in the death of grebes.

The accumulation of DDD in fish flesh could be correlated with food habits exhibited in normal food chain relationships. Generally speaking smaller fish accumulated less DDD on a p.p.m. basis than did larger fish. Plankton eaters accumulated less of the toxic material than carnivorous fish species of the same size. For example, flesh samples of one-year-old largemouth bass contained more than twice the amount of DDD than flesh samples of a one-year-old Sacramento blackfish. Flesh samples from three- to seven-year-old bass contained up to five times as much DDD as a one-year-old bass and 20 times as much as a one-year-old Sacramento blackfish. The theory of DDD transmission through a food chain would be more acceptable if the presence of DDD in plankton organisms could have been established.

It is assumed that certain species of Clear Lake fish such as white catfish and largemouth bass possess a greater tolerance for DDD than do grebes. This is based on results of analysis of fatty tissue. Analysis of visceral fat from apparently healthy largemouth bass and white catfish indicated accumulation of DDD at levels as high as 2,275 p.p.m. and 1,700 p.p.m., respectively. The highest concentration of DDD found in grebe tissue—1,600 p.p.m.—was from birds believed killed by chronic DDD poisoning.

Several theories may be offered regarding further aspects of the Clear Lake problem. Two of these theories pertain to possible consequences of further DDD treatments. The first relates to the establishment of resistance to DDD by various animals. After animals build up a resistance to DDD, or other chlorinated hydrocarbon insecticides, a relatively greater degree of exposure may be endured without perceptible adverse effects. Examples of resistance to chlorinated hydrocarbons in insects is well documented by Hammerstrom (1958) and Shepard (1958), but this phenomenon has not been reported in warmblooded animals. Fish might accumulate DDD in small increments and might not be affected by small additional dosages beyond the normal toxic threshold. Presumably these fish would not be affected by additional DDD treatments.

A second theory expresses an opposing idea. This is based on the fact that certain species have already accumulated and stored large amounts of the toxic chemical, and one or more additional treatments would result in the loss of many fish. DDD already accumulated might not break down and be eliminated, and that accumulation of DDD would continue until lethal dosages were reached. In experiments with DDT conducted by Tarzwell (1950), continued applications of sub-lethal dosages of this material resulted in fish losses.

There may be other effects of these insecticide programs that will become evident in the future, and there are probably some effects that will never be measured. An example of this pertains to the effect of insecticides on reproduction. The accumulation of sub-lethal dosages of DDD by various animal species might reduce fecundity. It has been shown by other investigators that feeding DDT to quail and pheasants resulted in lower than normal rate of hatch and a greater number of malformed progeny than was found in control birds (Genelly and Rudd, 1956; DeWitt, 1955).

If it were true that certain fish have at present accumulated DDD to a level slightly below the critical toxic level, then such fish could be endangered by further consumption of other contaminated fish. This would mean that a die-off of fish caused by DDD poisoning may yet occur without further treatment of the lake.

SUMMARY

1. A study is being made of the effects on wildlife of DDD used in gnat control programs at Clear Lake, Lake County, California. The study began in May, 1958, and is being continued at the date of this writing.
2. This paper is based on chemical analysis of animal tissue of specimens collected at Clear Lake and on information obtained during field investigations in that area.
3. Conclusive results obtained during the study were:
   (a) All fish, bird, and frog samples analyzed contained DDD.
   (b) The amount of DDD found in all flesh samples exceeded the specified rate of dilution of active insecticide in the lake water on a p.p.m. basis.
   (c) Flesh samples of largemouth bass and Sacramento blackfish hatched between seven and nine months after the last DDD application contained 22 to 25 p.p.m. and 7 to 9 p.p.m. of DDD, respectively.
   (d) All areas of the lake contained DDD contaminated fish.
4. Conclusions based in part on circumstantial evidence were:
   (a) Grebe losses occurring after DDD applications were caused by chronic DDD poisoning.

(b) The nesting population of grebes at Clear Lake has declined as a result of DDD treatment of the lake.

(c) Certain species of fish possess a greater tolerance for DDD than do grebes.

(d) Because of a difference in food habits carnivorous fish accumulated a greater amount of DDD than did plankton eating fish of the same size.

LITERATURE CITED

Bann, J. M., T. J. DeCino, N. W. Earl, and Y. P. Sun. 1956. The fate of aldrin and dieldrin in the animal body. Agr. Food Chem., Vol. 4, No. 11, pp. 937–941.

Brydon, Harold W. 1955. The 1954 control treatment of the Clear Lake gnat, *Chaoborus astictopus* D. & S., in Clear Lake, California. Proc. and Papers, 23rd Ann. Conf. Calif. Mosq. Control Assoc., pp. 108–110.

Burchfield, H. P., and E. E. Sterrs. 1953. Partition of insecticides between N, N-Dimethylformamide and hexane. Contrib. Boyce Thompson Inst., Vol. 17, pp. 333–334.

Davidow, B. 1950. Isolation of DDT from fats. Jour. A.D.A.C., Vol. 33, No. 1, pp. 130–132.

DeWitt, J. B. 1955. Effects of chlorinated hydrocarbon insecticides upon quail and pheasants. Jour. Agr. Food Chem., Vol. 3, No. 8, pp. 672–676.

1957. H-bomb in the pea patch. Wildlife in North Carolina, Vol. 21, No. 9, pp. 4–6.

Genelly, R. E., and R. L. Rudd. 1956. Effects of DDT, toxaphene and dieldrin on pheasant reproduction. Auk., Vol. 73, No. 4, pp. 529–539.

Hammerstrom, R. J. 1958. Insect resistance to insecticides. Publ. Health Repts., Vol. 73, No. 12, pp. 1126–1131.

Henderson, C., Q. H. Pickering, C. M. Tarzwell. 1958. The relative toxicity of ten chlorinated hydrocarbon insecticides to four species of fish. Trans. Am. Fisheries Soc., Vol. 88, No. 1, pp. 23–32.

Herms, W. B. 1937. The Clear Lake gnat. U.S. Agr. Exp. Sta. Bull., No. 607, 22 pp.

Lagler, F. K. 1952. Freshwater fishery biology. Wm. C. Brown, Publ., Dubuque, Iowa, 360 pp.

Lawrence, E. G. 1950. The diving and feeding activity of the Western grebe on the breeding grounds. Condor, Vol. 52, No. 1, pp. 3–16.

Lindquist, A. W., and C. C. Deonier. 1943. Flight and oviposition habits of the Clear Lake gnat. Jour. Econ. Ento., Vol. 35, No. 3, pp. 441–415.

Lindquist, A. W., and A. R. Roth. 1950. Effect of dichlorodiphenyl dichloroethane on larva of the Clear Lake gnat in California. Jour. Econ. Ento., Vol. 43, No. 3, pp. 328–332.

Lindquist, A. W., A. R. Roth, and John R. Walker. 1951. Control of the Clear Lake gnat in California. Jour. Econ. Ento., Vol. 44, No. 4, pp. 572–577.

Marzolf, C. R. 1955. Use of pectoral spines and vertebrae for determining age and rate growth of the channel catfish. Jour. Wildl. Mgt., Vol. 19, No. 2, pp. 243–249.

Metcalf, R. L. 1957. Advances in pest control research. Vol. 1, Intersci. Publ. Inc., New York, 514 pp.

Murphy, G. I., and H. P. Chandler. 1948. The effects of TDE on fish and on the plankton and litoral fauna in lower Blue Lake, Lake County, California. Calif. Fish and Game, Inland Fisheries Admin. Rept., No. 48–14, June 1948.

Pontoriero, L. P., and J. N. Ginsburg. 1953. An abridged procedure in the Schechter Method for analyzing DDT residues. Jour. Econ. Ento., Vol. 46, No. 5, pp. 903–904.

Radeleff, R. O., and G. T. Woodard. 1955. The diagnosis and treatment of chemical poisoning of animals with particular reference to insecticides. Proc. Am. Vet. Med. Assn. 92nd meeting. pp. 109–113.

Rudd, R. L., and R. E. Genelly. 1956. Pesticides: their use and toxicity in relation to wildlife. Calif. Fish and Game, Game Bull. No. 7, 209 pp.

Rudd, R. L. 1958. The indirect effect of chemicals in nature. Talk presented at 54th Conv. Nat. Audubon Soc., N.Y., Nov. 10, 1958.

Shepard, H. H. 1958. Methods of testing chemicals on insects. Burgess Publ. Co., Minn. 356 pp.

Springer, Paul F. 1956. Insecticides boon or bane. Audubon, Vol. 58, No. 3, pp. 128–130; No. 4, pp. 176–178.

Tarzwell, C. M. 1950. Effects of DDT mosquito larviciding on wildlife. V. Effects on fishes of the routine manual and airplane application of DDT and other mosquito larvicides. Publ. Health Repts., Vol. 65, No. 8, pp. 231–255.

U.S. Dept. Health, Education and Welfare. 1955. Clinical memoranda on economic poisons. (Revised April 1, 1955), 56 pp.

Walker, John R. 1949. The Clear Lake gnat, *Chaoborus astictopus* D. & S. Staff Communication, Calif. Dept. Publ. Health. mimeo 15 pp.

Wallace, George J. 1959. Insecticides and birds. Audubon, Vol. 61, No. 1, pp. 10–12–13–35.

22

DDT residues in an East Coast estuary: a case of biological concentration of a persistent insecticide

George M. Woodwell
Charles F. Wurster, Jr.
Peter A. Isaacson

DDT residues (1) have become an intrinsic part of the biological, geological, and chemical cycles of the earth (2) and are measurable in air (3), water (4), soil (5), man (6), and even in animals from the Antarctic, many hundreds of miles from places where DDT has been applied (7, 8). While the presence of residues does not prove an effect on living systems, the worldwide distribution of a substance as persistent and broadly toxic as DDT is itself reason to question whether residues are accumulating to toxic levels in certain populations. Accumulation occurs either through direct absorption from the environment or by concentration along food chains, and this latter phenomenon has been documented in several aquatic situations (9). Such accumulations have been correlated with recent declines in populations of carnivorous birds (8, 10) and other organisms. Residues of persistent pesticides are now so widespread that they must be considered as potentially aggravating the problems of eutrophication by degrading populations of consumers.

We measured residues of DDT in the soils of a brackish marsh on

Reprinted with permission from *Science*, 156(3776):821–824 (12 May 1967). Work carried out at Brookhaven National Laboratory under the auspices of the USAEC.

the south shore of Long Island, New York, and in various organisms from the area. Results showed a high concentration of residues in the marsh and a systematic increase in DDT residues with increase in trophic level, thus providing an especially clear example of what has been called "biological magnification." In many cases the concentrations approached those in organisms known to have died of DDT poisoning, which suggests that DDT residues are currently reducing certain animal populations within the estuary.

The marsh from which soil samples were taken was the mouth of Carmans River at the eastern end of Great South Bay, Long Island. The area was selected as representative of relatively undisturbed marsh. The marsh along the western side of the river was sampled by a modification of the technique described by Woodwell and Martin (5). A "sample" consisted of six subsamples, each subsample being a core 4.8 cm in diameter and either 20 or 40 cm long, taken with a sharpened aluminum tube pressed into the soil. In each area of interest, subsamples were collected systematically about 10 m apart. Seven such samples were taken, four in the *Spartina patens* marsh, two along the margins of drainage ditches dug for mosquito control, and one from the bottom of the estuarine bay a few meters from the edge of the marsh.

For DDT analyses, 2-cm increments from equal depths among the six subsamples were pooled. Analyses were performed on the increments from 0 to 2, 4 to 6, 8 to 10, and 18 to 20 cm, and where deeper samples were taken, 38 to 40 cm. Total residues were calculated on a weight-per-acre basis by integrating the area under the curve expressing residues per square meter at the various depths.

Plankton were collected in a No. 6 (0.239-mm mesh) plankton net. All organisms were living when taken except as indicated; fish were netted; birds were shot. Samples were stored frozen until analyzed, and mud samples were oven-dried before analysis. In most cases, whole organisms were analyzed, but feathers, beaks, feet, and wing tips of birds were discarded. Analyses were on 1-g samples of the homogenized organism. Extraction was from Florisil with petroleum ether–diethyl ether as described by Cummings *et al.* (11). Analyses of samples to which measured quantities of DDT, DDE, and DDD, individually, had been added prior to extraction indicated recoveries averaging 96 percent. Analyses for DDT, DDE, and DDD were by electron-capture gas chromatography; certain identifications were confirmed by thin-layer chromatography (11).

Residues of DDT in the *Spartina* marsh varied widely from less than 3 to more than 32 lb./acre (Table 22–1). The mean concentration of the four samples from this marsh (each a composite of 6 cores) was 13.1 lb./acre. Slightly lower quantities occurred along the ditches, but total residues were still 1 to 5 lb./acre; submerged bay bottom contained 0.28 lb./acre. In all of these samples most of the residues (approximately 90 per-

TABLE 22–1. DDT residues (DDT + DDE + DDD) (*1*) in Carmans River marsh and in the bottom mud of Great South Bay, N.Y., August 1966. Each sample was a composite of six subsamples, taken to the depths indicated

Zone	Sample No.	Depth (cm)		Total residues	
				Lb/ acre	Kg/ ha
Spartina mat	1	0–20		2.69	3.01
	2	0–40		9.23	10.3
	3	0–20		7.86	8.81
	4	0–40		32.6	36.5
			Mean	13.1	14.7
Drainage ditch	1	0–20		4.63	5.19
	2	0–40		1.10	1.23
			Mean	2.87	3.21
Bay bottom (submerged)		0–40		0.28	0.31

cent) occurred in the upper 4 cm of the profile. Residues were highly variable in relative proportions of DDT, DDE, and DDD. In general there was an increase in the proportion of DDE, and a decrease in DDT, with increase in depth. In the 0- to 2-cm samples, the mean DDE content was about 25 percent; in the 18- to 20-cm samples it was about 60 percent. Residues in the bottom of the bay contained only traces of DDT and DDD, the principle residue being DDE.

Thirty-nine samples of plants and animals from the vicinity were analyzed. Arrangement of the samples in sequence according to increasing concentration of DDT residues (Table 22–2) shows a progression according to both size and trophic level, larger organisms and higher carnivores having greater concentrations than smaller organisms and organisms at lower trophic levels. Total residues ranged through three orders of magnitude from 0.04 part per million (p.p.m.) in plankton to 75 p.p.m. in a ring-billed gull. Shrimp contained 0.16 p.p.m.; eels, 0.28; insects from the marsh, 0.30; and mummichogs (*Fundulus*), 1.24 p.p.m. Among fish, the needlefish, a carnivore, had the highest content, 2.07 p.p.m., about twice that of *Fundulus*, which forms part of its food. In general, the concentrations of DDT residues in carnivorous birds were 10 to 100 times those in the fish on which they feed. Concentrations of DDT in the waters of Great South Bay must be assumed to be less than the 0.0012-p.p.m. saturation limit, a reasonable estimate probably being closer to 0.00005 p.p.m. (4, 12). Based on this estimate, birds near the top of these food chains have concentrations of DDT residues about a million times greater than the concentration in the water.

The shift in relative proportions of DDT, DDE, and DDD with progression in trophic level is also conspicuous. Organisms containing high

TABLE 22–2. DDT residues (DDT + DDE + DDD) (*1*) in samples from Carmans River estuary and vicinity, Long Island, N.Y., in parts per million wet weight of the whole organism, with the proportions of DDT, DDE, and DDD expressed as a percentage of the total. Letters in parentheses designate replicate samples

Sample	DDT residues (ppm)	Percent of residue as		
		DDT	DDE	DDD
Water*	0.00005			
Plankton, mostly zooplankton	.040	25	75	Trace
Cladophora gracilis	.083	56	28	16
Shrimp†	.16	16	58	26
Opsanus tau, oyster toadfish (immature)†	.17	None	100	Trace
Menidia menidia, Atlantic silverside†	.23	17	48	35
Crickets†	.23	62	19	19
Nassarius obsoletus, mud snail†	.26	18	39	43
Gasterosteus aculeatus, threespine stickleback†	.26	24	51	25
Anguilla rostrata, American eel (immature)†	.28	29	43	28
Flying insects, mostly Diptera†	.30	16	44	40
Spartina patens, shoots	.33	58	26	16
Mercenaria mercenaria, hard clam†	.42	71	17	12
Cyprinodon variegatus, sheepshead minnow†	.94	12	20	68
Anas rubripes, black duck	1.07	43	46	11
Fundulus heteroclitus, mummichog†	1.24	58	18	24
Paralichthys dentatus, summer flounder‡	1.28	28	44	28
Esox niger, chain pickerel	1.33	34	26	40
Larus argentatus, herring gull, brain (d)	1.48	24	61	15
Strongylura marina, Atlantic needlefish	2.07	21	28	51
Spartina patens, roots	2.80	31	57	12
Sterna hirundo, common tern (a)	3.15	17	67	16
Sterna hirundo, common tern (b)	3.42	21	58	21
Butorides virescens, green heron (a) (immature, found dead)	3.51	20	57	23
Larus argentatus, herring gull (immature) (a)	3.52	18	73	9
Butorides virescens, green heron (b)	3.57	8	70	22
Larus argentatus, herring gull, brain§ (e)	4.56	22	67	11
Sterna albifrons, least tern (a)	4.75	14	71	15
Sterna hirundo, common tern (c)	5.17	17	55	28
Larus argentatus, herring gull (immature) (b)	5.43	18	71	11
Larus argentatus, herring gull (immature) (c)	5.53	25	62	13
Sterna albifrons, least tern (b)	6.40	17	68	15
Sterna hirundo, common tern (five abandoned eggs)	7.13	23	50	27
Larus argentatus, herring gull (d)	7.53	19	70	11
Larus argentatus, herring gull§ (e)	9.60	22	71	7
Pandion haliaetus, osprey (one abandoned egg)‖	13.8	15	64	21
Larus argentatus, herring gull (f)	18.5	30	56	14
Mergus serrator, red-breasted merganser (1964)‡	22.8	28	65	7
Phalacrocorax auritus, double-crested cormorant (immature)	26.4	12	75	13
Larus delawarensis, ring-billed gull (immature)	75.5	15	71	14

* Estimated from Weaver *et al.* (*4*). † Composite sample of more than one individual. ‡ From Captree Island, 20 miles (32 km) WSW of study area. § Found moribund and emaciated, north shore of Long Island. ‖ From Gardiners Island, Long Island

proportions of DDT, as opposed to its metabolites, are common at lower trophic levels; at upper levels, most of the residue is DDE. In most organisms, DDD and DDE are somewhat less toxic than DDT.

The secondary effects of applications of DDT to marshes, streams, and forests are the subject of an extensive literature. Single applications in the range of 0.1 to 0.3 lb./acre have repeatedly caused drastic reductions in populations of crayfish, shrimp, amphipods, isopods, annelids, fish, fiddler crabs, blue crabs, and others, sometimes with no recovery for years (13). Aerial spraying with 0.5 lb. of DDT per acre in New Brunswick, Canada, caused extremely high mortality of young salmon, reduced salmon food organisms, and was correlated with reduced reproductive success in woodcock (14, 15). Applications of 1 to 5 lb./acre are known to have serious long-term effects on amphibians, fish, and birds (16–18).

While it is not true that residues distributed through 4 cm of highly organic soil are continuously available to the biota in the same degree as immediately after spraying, there is little question that residues in soil are leached by water, moved by erosion, and absorbed by mud-dwelling and mud-scavenging organisms. As a result of such processes, DDT residues in a marsh inevitably enter environmental cycles. Deleterious effects on wildlife from 13 lb./acre on the Carmans River marsh might therefore be expected. Detailed long-term observations have shown substantial reductions during the past decade in local populations of shrimp, amphipods, summer flounder, blue crab (*Callinectes sapidus*), spring peeper (*Hyla crucifer*), Fowler's toad (*Bufo woodhousei fowleri*), woodcock (*Philohela minor*), and various other species, known to be sensitive to DDT, that are indigenous to this area (13–19). Other aspects of human disturbances unquestionably contribute to degradation of estuaries, but do not offer adequate explanations for all of these declines. This is especially true for declines in populations indigenous to marshes that have been remote from other disturbances (19).

The concentration of DDT residues that affect animals in nature is difficult to appraise. Analyses of whole organisms, rather than of specific organs, are most representative of the degree of exposure to DDT, although correlation of such measurements with death is not precise. Nevertheless, broad correlations exist between whole-body concentrations and mortality (20) and are useful in appraising the hazards of residues in the estuary we sampled. For instance, fish of several species, known to have been killed by DDT, contained whole-body concentrations of 1 to 26 p.p.m., commonly averaging 4 to 7 p.p.m. (18, 21); the concentrations reported for living fish in Table 2 lie between 0.17 and 2.07 p.p.m., within and somewhat below this apparently lethal region. Birds known to have died of DDT poisoning contained 30 to 295 p.p.m. of DDT residues when analyzed as whole birds, the average for several species being about 112 p.p.m. (17); the birds in our sample contained residues ranging from 1 to 75 p.p.m.

Living birds and fish that were analyzed in this study contained DDT residues that exceed one tenth of the mean concentrations in organisms known to have died of acute exposures to DDT. This in itself implies that concentrations of DDT in this food web are approaching the maximum levels observable in living organisms and now occasionally reach acutely lethal levels in both birds and fish. Two observations lead to this conclusion: (i) the trophic-level effect has been shown to produce concentrations in carnivorous birds and fish that are 10 to 1000 times the concentrations lower in the food web. We must assume that, although we have sampled fish-eating carnivorous birds such as the merganser, carnivorous or scavenging birds feeding on birds would have even higher levels, probably by as much as another 10-fold. Such birds probably would not survive. (ii) Perhaps even more important because it occurs at all trophic levels, great variability characterizes the total amounts of residues in animals from the wild. Differences in the range of 5- to 10-fold can be expected between the minimum and maximum concentrations in birds of a single species killed under similar conditions by DDT (17). While the causes of such variability are not clear, its existence implies that mortality from DDT residues is now occurring in these populations through elimination of individuals at the upper end of the range of concentrations. This constant attrition or "cropping" process would be extremely difficult to detect, since it does not produce spectacular "kills," and dead individuals, widely scattered, tend to disappear rapidly (17, 22). The probability appears high that not only are the populations of many of the organisms of this sampling now being affected by accumulation of DDT residues, but that other species in the area have already been depleted to the point where study is difficult or impossible.

Acute mortality, however, is but one effect of DDT. Sublethal concentrations, although studied less, may actually have more important effects on populations in nature. In the laboratory, sublethal amounts of DDT reduce reproductive success in bobwhite, ring-necked pheasants, and mice (23). Evidence has linked chlorinated hydrocarbon residues with reduced reproduction in field populations of trout (24), osprey, woodcock, bald (25) and golden eagles, peregrine falcon, and others (8, 10, 15). Serious population declines have occurred for some of these species. Minute quantities of chlorinated hydrocarbons affect the patterns of behavior of goldfish and upset temperature-regulating mechanisms in salmon (26). Such sublethal effects might be expected within the populations represented in Table 22–2.

One important conclusion is that analyses of water have limited meaning when evaluating the effects of DDT residues on animal populations. Water can be expected to contain a lower concentration of DDT than other components of an ecosystem—quantities that are usually vanishingly small or "nondetectable" (4). Even these very low concentrations may be important because natural mechanisms can concentrate residues many thousands of times. A better criterion of hazard from DDT pollution would be analyses of carnivores or other organisms that concentrate the residues.

Concentrations of DDT residues reported here are not unique to this marsh or even to Long Island. Observations from widely scattered fish and bird populations in North America show concentrations approximating those reported here (8, 9), which suggests that DDT residues are moving through the biological, geological, and chemical cycles of the earth at concentrations that are having far-reaching and little-known effects on ecological systems.

REFERENCES AND NOTES

1. DDT residues include DDT and its decay products (metabolites), DDE and DDD; DDT, 1,1,1-trichloro-2,2-bis(*p*-chlorophenyl)-ethane; DDE, 1,1-dichloro-2,2-bis(*p*-chlorophenyl)ethylene; DDD, also known as TDE, 1,1-dichloro-2,2-bis(*p*-chlorophenyl)ethane.
2. G. M. Woodwell, *Sci. Amer.* 216, 24 (March 1967).
3. P. Antommaria, M. Corn, and L. DeMaio, *Science* 150, 1476 (1965); G. M. Woodwell, *Forest Sci.* 7, 194 (1961).
4. L. Weaver, C. G. Gunnerson, A. W. Breidenbach, and J. J. Lichtenberg, *Public Health Rep.* 80, 481 (1965).
5. C. A. Edwards, *Residue Rev.* 13, 83 (1966); G. M. Woodwell and F. T. Martin, *Science* 145, 481 (1964).
6. G. E. Quinby, W. J. Hayes, Jr., J. F. Armstrong, and W. F. Durham, *J. Amer. Med. Assn.* 191, 175 (1965).
7. W. J. L. Sladen, C. M. Menzie, and W. L. Reichel, *Nature* 210, 670 (1966); J. L. George and D. E. H. Frear, *J. Appl. Ecol.* 3 (suppl.), 155 (1966).
8. "The effects of pesticides on fish and wildlife," *U.S. Fish and Wildlife Service Circ. 226* (1965); "Pesticides in the environment and their effects on wildlife," *J. Appl. Ecol.* 3 (suppl.) (1966).
9. E. G. Hunt and A. I. Bischoff, *Calif. Fish and Game* 46, 91 (1960); J. J. Hickey, J. A. Keith, and F. B. Coon, *J. Appl. Ecol.* 3 (suppl.) 141 (1966); J, O. Keith, *ibid.*, p. 71; E. G. Hunt, *Scientific Aspects of Pest Control* (Nat. Acad. Sci.-Nat. Res. Council, Publ. No. 1402, 1966), p. 251.
10. P. L. Ames, *J. Appl. Ecol.* 3 (suppl.), 87 (1966); I. Prestt, *ibid.*, p. 107; S. Cramp, *Brit. Birds* 56, 124 (1963); J. D. Lockie and D. A. Ratcliffe, *ibid.* 57, 89 (1964); D. A. Ratcliffe, *ibid.* 58, 65 (1965); D. A. Ratcliffe, *Bird Study* 12, 66 (1965); A. Sprunt, *Audubon Mag.* 65, 32 (1963).
11. J. G. Cummings, K. T. Zee, V. Turner, F. Quinn, and R. E. Cook, *J. Assn. Offic. Anal. Chemists* 49, 354 (1966); M. F. Kovacs, Jr., *ibid.*, p. 365.
12. M. C. Bowman, F. Acree, Jr., and M. K. Corbett, *J. Agr. Food Chem.* 8, 406 (1960).
13. H. H. Ross and W. Tietz, Jr., *Illinois Nat. Hist. Surv.* (1949) (mimeographed); P. F. Springer, *Proc. 50th Annu. Mtg. N.J. Mosquito Extermination Assn.* (1963), pp. 194–203; P. A. Butler and P. F. Springer, *Trans. 28th North American Wildlife and Natural Resource Conf.* (1963), pp. 378–390; P. F. Springer, *Diss. Abstr.*

22, 1777 (1961); P. F. Springer and J. R. Webster, *Mosquito News* 11, 67 (1951); R. A. Croker and A. J. Wilson, *Trans. Amer. Fish. Soc.* 94, 152 (1965); A. D. Hess and G. G. Kenner, Jr., *J. Wildlife Management* 11, 1 (1947); J. L. George, R. F. Darsie, Jr., and P. F. Springer, *ibid.* 21, 42 (1957).

14. M. H. A. Keenleyside, *Can. Fish. Cult. No. 24*, 17 (1959); F. P. Ide, *Trans. Amer. Fish. Soc.* 86, 208 (1956).
15. B. S. Wright, *J. Wildlife Management* 29, 172 (1965).
16. B. A. Fashingbauer, *Flicker* 29, 160 (1957); R. J. Barker, *J. Wildlife Management* 22, 269 (1958); P. Goodrum, W. P. Baldwin, and J. W. Aldrich, *ibid.* 13, 1 (1949); C. S. Robbins, P. F. Springer, and C. G. Webster, *ibid.* 15, 213 (1951); L. B. Hunt, *ibid.* 24, 139 (1960); G. J. Wallace and E. A. Boykins, *Jack-Pine Warbler* 43, 13 (1965); R. B. Anderson and W. H. Everhart, *Trans. Amer. Fish. Soc.* 95, 160 (1966); R. L. Rudd, *Pesticides and the Living Landscape* (Univ. of Wisconsin Press, Madison, 1964).
17. C. F. Wurster, Jr., D. H. Wurster, and W. N. Strickland, *Science* 148, 90 (1965); D. H. Wurster, C. P. Wurster, Jr., and W. N. Strickland, *Ecology* 46, 488 (1965).
18. K. Warner and O. C. Fenderson, *J. Wildlife Management* 26, 86 (1962).
19. Long-term field studies by D. Puleston and A. P. Cooley, many of them specifically covering the Carmans River marsh and documented in personal notes kept over more than 20 years, show the decline or disappearance of these and many other populations, including American bittern, *Botaurus lentiginosus*, least bittern, *Ixobrychus exilis*, green heron, and marsh hawk, *Circus cyaneus*. During this period the physical characteristics of the marsh have remained largely unchanged.
20. DDT and its residues are nerve toxins; when DDT is suspected as a cause of death in vertebrates, the best appraisal is by analyses of residue concentrations in the brain [W. E. Dale, T. B. Gaines, W. J. Hayes, Jr., and G. W. Pearce, *Science* 142, 1474 (1963); L. F. Stickel, W. H. Stickel, and R. Christensen, *ibid.* 151, 1549 (1966)]. However, DDT residues can be stored in adipose tissues for long periods without conspicuous effects; symptoms occur when fat reserves are utilized, redistributing the toxin [R. F. Bernard, *Mich. State Univ. Mus. Publ. Biol. Ser.* 2(3), 155 (1963)]. In many species of birds, for example, fat reserves are utilized during reproduction and migration. Because of the accumulation of residues in other tissues, analyses of whole bodies appears to be a better criterion of exposure to DDT and of its hazard to the organism than analyses of brain tissues alone.
21. A. V. Holden, *Ann. Appl. Biol.* 50, 467 (1962); F. H. Premdas and J. M. Anderson, *J. Fish. Res. Board Can.* 20, 827 (1963); D. Allison, B. J. Kallman, O. B. Cope, and C. C. Van Valin, *Science* 142, 958 (1963).
22. W. Rosene, Jr., and D. W. Lay, *J. Wildlife Management* 27, 139 (1963).
23. J. B. DeWitt, *J. Agr. Food Chem.* 3, 672 (1955); *ibid.* 4, 863 (1956); R. E. Genelly and R. L. Rudd, *Auk* 73, 529 (1956); R. F. Bernard and R. A. Gaertner, *J. Mammal.* 45, 272 (1964).
24. G. E. Burdick *et al.*, *Trans. Amer. Fish. Soc.* 93, 127 (1964).
25. On the basis of 61 specimens taken from many parts of the United States, the bald eagle averages 11 p.p.m. of DDT residues in breast muscle, a quantity probably approaching lethal concentrations (8).
26. R. E. Warner, K. K. Peterson, and L. Borgman, *J. Appl. Ecol.* 3 (suppl.), 223 (1966); D. M. Ogilvie and J. M. Anderson, *J. Fish. Res. Board Can.* 22, 503 (1965).
27. Supported in part by a grant from the Research Foundation of the State University of New York and in part by Brookhaven National Laboratory under the auspices of AEC.

23

DDT residues in the eggs of the osprey in the north-eastern United States and their relation to nesting success

Peter L. Ames

INTRODUCTION

The Osprey (*Pandion haliaëtus*) is a large fish-eating bird of prey found in nearly every continent in the world, always associated with large bodies of water. In many parts of the United States it may be seen in considerable numbers, nesting aggregations of several hundred pairs being spread over a few square miles. In common with nearly all other birds of prey, the Osprey has suffered a gradual decline since the early years of the twentieth century (Bent, 1937). This decline, which cannot have averaged more than 2–3 percent per year during the first five decades of the century, has recently attained the rate of about 30 percent annually.

In 1957 I began an ecological study of a large breeding concentration of Ospreys near the mouth of the Connecticut River, where it empties into Long Island Sound. The colony extended over an area of about 10 square miles (25.9 km^2) and numbered at least 100 pairs. About one-third of these were nesting on Great Island, a low salt marsh area of about 300 ac (120 ha), where many nests were on low structures and some on the ground. Almost from the start of the study it was evident that the production of young birds was extremely low. In 1960, the first season for which good data are available, the number of young Ospreys fledged was seven from

Reprinted with permission from *Journal of Applied Ecology*, 3(Supplement):87–97 (1966). The author is currently associated with the Museum of Vertebrate Zoology, University of California, Berkeley.

seventy-one active nests, a rate of less than 0.1 young per nest. Nestling production in normal Osprey populations is 2.2–2.5 young per nest. The low production in 1960 was due in part to a high rate of egg predation, which was not repeated the following year, when thirty-one pairs produced twelve young (average: 0.4 young per nest). Even this rate is extremely low.

The five years from 1957 to 1961 were spent largely in gathering data on population trends and the factors affecting them. The situation was discussed by Ames & Mersereau (1964) and may be summarized as follows:

1. The number of nesting pairs of Ospreys in the Connecticut River colony has declined at an average rate of 31 percent annually since 1960. The maximum known population was 200 pairs in 1938 (J. Chadwick, personal communication) and this had dropped to twelve pairs by 1965.

2. The production of nestlings in the colony has been 0.1–0.4 fledglings per pair, as compared with 2.2–2.5 for healthy colonies in New York and Virginia.

3. Poor fledgling production, in this colony at least, is certainly due to poor hatching, for the observed fledgling rates have been very close to hatching rates.

4. Courtship, nest-building, and other aspects of breeding behaviour appear normal. The usual clutch of three eggs is laid and incubation is continuous. There is no evidence of egg-eating nor other abnormal behaviour.

5. There is no correlation between the degree of isolation of the nest site and nesting success. In order to analyze the role of human interference, tidal flooding and egg predation, nesting platforms were provided for the marsh-nesting birds. By 1963 nearly half of the colony (then down to twenty-four pairs) nested on these elevated platforms. The rate of nesting success of the platform-nesting birds was about equal to that of those in trees.

6. By 1964 it was evident that nesting success is not randomly distributed, nor is it greater in one region or one nest site type. Certain pairs have produced young with nearly normal regularity; others have not produced within the 9 years of the study.

7. There do not appear to be any significant differences between successful and unsuccessful pairs in the number and species of fish eaten. Moreover, successful pairs under observation seem to have no difficulty in providing fish for three large young. The males at productive nests spend much time perched near their nests, suggesting that food pressures are not acute.

The situation found in Connecticut is mirrored in other parts of the Atlantic Coast, but the human variables are quite different. In eastern Maine Ospreys suffer very little harassment by man, but their population decline has paralleled that in Connecticut. In south-eastern Massachusetts the major drop apparently was between 1890 and 1930, and there are now only about thirty-five pairs in an area which once held about four times that number. 1965 production was about 0.15 young per nest (Memorandum, Massachusetts Audubon Society, 2 August 1965). Members of the Rhode Island Audubon Society have documented a similar situation in that state (Emerson & Davenport, 1963) and have noted the low rate of fledgling production. Eight nests which produced only three young among them in Rhode Island in 1963 were located in a nearly impenetrable swamp, in which human interference seems unlikely. Elsewhere in Rhode Island the shooting of adults may have been a factor. A large Osprey colony on Gardiners Island, New York, directly across Long Island Sound from the Connecticut River colony, declined from about 300 pairs in 1900 to about twenty-one pairs in 1963. Wilcox, who has followed this colony closely for about 25 years, notes (personal communication) that the drop in the last 15 years has been about 50 percent. The fledgling rate in 1963 was about 0.3 young per nest, the hatching rate about 0.4 eggs per nest.

In southern New Jersey the loss of nest trees through land development and other forms of human activity has greatly reduced the numbers of nesting Ospreys, a decline which continues. I know of no recently published reports on the population changes or rates of fledgling production. The present population appears to be about 5 percent that of 50 years ago.

A portion of the large Osprey population of Chesapeake Bay (Maryland and Delaware) is now being studied by Fred C. Schmidt of the United States Fish and Wildlife Service. The work is in too early a stage for a detailed evaluation of population trends, but the pesticide levels and rates of nestling production are discussed below.

From Virginia to Florida, Ospreys are found nesting in somewhat lower densities than in some of the northern states. The Florida birds do not appear to be in trouble (Alexander Sprunt, personal communication) but I know of no quantitative study of the species in any of the southeastern states.

PESTICIDE STUDIES

In June 1962, six eggs were taken from Osprey nests on Great Island, Connecticut, when about 1 week past the expected hatching dates. All were in advanced state of decomposition. The eggs were analyzed at the Connecticut Agricultural Experiment Station by paper chromatography and the well-known method of Schechter and Haller. As reported by Ames & Mersereau (1964), DDT residues (mostly DDE) averaged 555 μg per egg. Three small samples of fish tissues, taken from Osprey nests, were also analyzed and found to contain 0.7–1.8 p.p.m. of DDT and 1.8–7.4 p.p.m. of its metabolites.

In 1963 we felt that conclusive results might be gained by comparing the egg-pesticide levels and hatching data in the declining Connecticut Ospreys with the values obtained from the population in Chesapeake Bay, Maryland, where the species appeared to be maintaining its numbers. Other factors, such as human activity, nest site availability, and food fish species had not been analyzed in the Maryland population, so Schmidt undertook to study various aspects of the nesting success, as well as to collect a series of eggs for analysis. Each of us, in his respective area, was to take for analysis one egg from each of a number of clutches. The remaining eggs in the nest (usually two) were to be checked for hatching and fledgling rates. Because of the relative accessibility of Osprey nests in both colonies, there seemed little danger that our visits would prejudice the results of the remainder of the clutch. In June some overdue eggs were collected from the same nests as the early samples. In addition, a number of eggs were collected and analyzed from Maine, Rhode Island and New Jersey, usually without complete nest histories.

During the first week of May 1963, thirty-eight eggs were taken from Osprey nests on the lower Potomac River, Maryland, at the western edge of the Chesapeake Bay area. Not all of the known nests in the area were sampled. Single eggs were taken from twenty-seven nests, two from one nest, and three (complete clutches) from three nests. The complete clutches were taken in order to determine the variability of pesticide distribution within a clutch. In the Connecticut colony fifteen eggs were taken in late April and early May, relatively earlier in the incubation period, as nesting is about 1 week later in Connecticut than in Maryland. As soon as possible after collecting, and in all cases within 24 h, each egg was weighed, its volume was measured by immersion in a graduated beaker (accurate to about ± 3 ml), and it was opened to determine the developmental condition. All eggs (less shells) were stored in glass jars at about $-18°$ C until analyzed. They were analyzed in the laboratory of the Patuxent Research Center of the Fish and Wildlife Service by thin layer (aluminum oxide) chromatography for the DDT group of chlorinated hydrocarbons only.

EGG FERTILITY AND NESTING SUCCESS

When examined fresh from the nests, most of the eggs were found to contain live, normal-looking, embryos. Of the fifteen Connecticut eggs taken early in the season, eleven had embryos (both alive and dead), and only one of the remaining four eggs had been incubated long enough for infertility to be positively established. Of the eleven fertile eggs, three contained decomposing embryos, indicating death prior to collecting. Even when an egg is found to be 'clear' (i.e. with no sign of development) after a week or more of incubation, one cannot be sure of its infertility; the embryo may have died at an extremely early age. The term 'infertile' is best applied to eggs which show no sign of development, not, as is often the case, to eggs which, for one reason or another, fail to hatch. In this sense, the fertility of the twelve Connecticut eggs which provide sure evidence was 92 percent and the minimum fertility of the fifteen eggs, assuming the three uncertain eggs to be infertile, was 73 percent.

Of the thirty-one Maryland eggs, twenty-nine were fertile, three with decomposing embryos. One of the remaining two eggs was crushed in handling, making accurate determination of fertility impossible. If present, the embryo must have been very small. The other egg appeared freshly laid. If these two are omitted as uncertain, fertility was 100 percent; if they are treated as infertile, fertility was 93 percent. Observed embryonic death in the Connecticut series was about twice as high as in the Maryland series, but the former were collected later in the incubation period than the latter, so the differences are probably not significant.

All four of the Maine eggs contained large, healthy-appearing, embryos. The Rhode Island and New Jersey eggs were collected when past due to hatch and their contents were too decomposed for fertility evaluation.

The 1963 breeding results of the Connecticut and Maryland Ospreys are summarized in Table 23–1. The upper half of the table shows the

TABLE 23–1. Observed and theoretical nesting success in Connecticut and Maryland Ospreys, 1963

	Connecticut		Maryland	
	Total population	Sampled	Total population	Sampled
No. of nests	22	15	35	25
Total eggs (observed)	63	42	99	75
Eggs sampled	15	15	38	25
Eggs remaining after sampling	48	27	61	50
Observed hatch, eggs	9	4	27	27
Observed hatch (%)	18·8	14·8	44·3	54·0
Theoretical hatch, without sampling	12	6	41	41
Theoretical hatch, eggs per nest	0·54	0·41	1·25	1·60
Observed fledge, individuals	9	4	24	24
Percentage of eggs producing fledglings	18·8	14·8	39·3	36·0
Theoretical fledge, without sampling	12	6	39	36
Theoretical fledge, young per nest	0·54	0·41	1·1	1·4

observed results in terms of hatching and fledging; the lower half gives an estimate of what the results would have been had not about one-third of the eggs been taken for analysis. For each area data are given for the entire number of nests under observation and for that portion of the nests from which single eggs were taken. In Connecticut only single eggs were taken, so the table indicates all of the direct effects of sampling. In Maryland, however, the figure of thirty-five nests includes five in which no eggs remained after sampling (three clutches of three eggs each and two clutches of one egg) and one nest from which two out of three eggs were taken. A more meaningful evaluation of the possible effects of pesticides may be gained by considering only the twenty-five nests from which sample eggs were taken and in which some eggs were left for observation. It is interesting to note that none of the eight eggs in the four unsampled nests hatched. Three of these unsampled clutches were smaller than normal (one, two and two eggs), suggesting that predation may have taken place.

The estimates of the hatching and fledgling production which would have been observed in the absence of sampling were arrived at by applying to the total number of pre-sampling eggs the observed hatching and fledging rates of the eggs remaining after sampling. This method is probably more accurate in projecting hatching than fledging rates, for the survival of nestlings might be affected by lowering clutch-size.

It appears that the Maryland birds were about 2–2.5 times as successful as the Connecticut birds in hatching their eggs. The corrected hatch of about 0.40–0.54 eggs per nest in Connecticut was about equal to that observed in 1961 and 1962, when no sampling was performed during the incubation period. The hatchability of the Maryland eggs, 1.3–1.6 eggs per nest, appears to be lower than that observed in most vigorous Osprey populations, and one may well question whether the Maryland Ospreys are maintaining their numbers as well as is generally believed.

RESULTS OF CHEMICAL ANALYSES

The amounts of DDT residues found in Osprey eggs from the various regions sampled are shown in Table 23–2. To aid in comparisons with other Osprey populations and with other species, the values are given in average microgrammes (10^{-6} g) per egg and as average microgrammes per cubic centimetre of egg volume (including shell). The usual method of stating the amounts of pesticides in eggs, 'parts per million' (microgrammes per gramme), leads to erroneous conclusions when the various samples have undergone different amounts of drying, through decomposition or in the normal course of incubation. Stickel, Schmidt, Reichel & Ames (1965) discuss various methods of stating egg pesticide levels, using as examples early and late Osprey eggs from the Connecticut series. They conclude that one of two methods should be employed. Either: (1) the fresh weight of the egg contents should be determined as a function of

TABLE 23–2. DDT and its metabolites in the eggs of Ospreys from the north-eastern United States

Locality	Year	No. of eggs	Average volume (ml)	DDE		DDD		DDT		Total residues		DDD (% of total)
				μg	μg/ml	μg	μg/ml	μg	μg/ml	μg	μg/ml	
Maine	1963	3	72	120	1·7	7	0·1	5	0·06	130	1·8	5
Rhode Island	1963	1	68	500	7·4	100	1·5	ND	ND	600	8·8	17
Connecticut	1962	6	68	450	6·7	100	1·5	Trace	Trace	550	8·1	22
Connecticut	1963	15	68	320	4·7	20	0·3	10	0·1	350	5·1	5
New Jersey	1963	2	Not measured	350	5·1	40	0·6	10	0·1	400	5·9	10
Maryland	1963	25	70	160	2·3	40	0·6	5	0·07	205	3·0	18

ND = None detected.

the entire egg volume and the pesticide amounts stated in parts per million (fresh weight), or (2) the amount of pesticide should be stated in microgrammes per millilitre of total egg volume. The latter of these methods is employed here because it allows ready comparison with the eggs of other raptors, and because the conversion factor from egg volume to weight of contents is not known for Ospreys.

The average microgramme amounts of residues listed in Table 23–2 are derived from single-egg estimates based on chromatograph spots. The values were stated by the chemist with an accuracy of ±25–50 percent. The values are statistically valid when a good number of samples is analyzed, as with the Connecticut and Maryland eggs. The values obtained from the Maine, Rhode Island and New Jersey eggs may be taken only as an indication of the widespread nature of the contamination. The amounts of all DDT residues in the individual eggs from Connecticut and Maryland are indicated in the histogram (Fig. 23–1), expressed in microgrammes per millilitre. The mean of the fifteen Connecticut eggs (5.1) is 1.6 times that of the twenty-five Maryland eggs on which the hatchability data are based (3.0), but both values are very small. It is interesting to note that DDD (=TDE) averages 18 percent of the total DDT residues in the Maryland eggs, but only 5 percent in the Connecticut eggs. This difference, which is also found in the fish samples, probably reflects the greater agricultural use of Rhothane (DDD) in Maryland. The amount of DDD in the Maryland eggs varies from 0 to 44 percent of the total residues; that in Connecticut eggs varies from 0 to 16 percent. What part of this DDD is produced by the metabolism of DDT cannot be determined on the basis of our present knowledge.

In 1964 we took several whole clutches of eggs from both Osprey colonies, in an attempt to learn whether there were differences which could be examined by artificial incubation. The hatching success of both groups of eggs was so low that no comparisons could be made. Most of the clutches were taken early in the season and, as expected, most of the

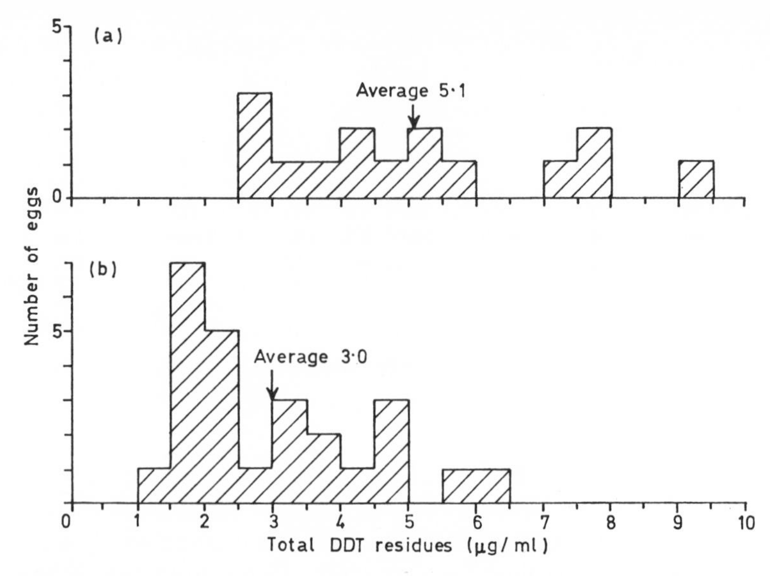

Fig. 23–1. Comparison of residue levels in eggs of Ospreys from (a) Connecticut, and (b) Maryland, in 1963.

pairs laid replacement clutches. The hatching rate of the second clutches was about the same as the first clutches, 11 percent. We are still investigating methods of artificially incubating Osprey eggs and feel that this may prove a fruitful avenue of investigation.

RESULTS OF ANALYSIS OF OSPREY FOOD FISHES

In conjunction with the programme of egg analyses, fifteen samples of Osprey food fishes were collected from nests in Connecticut and thirteen from nests in Maryland. Most of the Connecticut samples were obtained at two nests with one and three young, respectively, and from which egg samples had been obtained. The Maryland samples were also obtained from nests with young. The results, shown in Table 23–3, must be evaluated with some caution, for several reasons:

1. Most of the samples were taken early in the season, when the young were small. For about 2 weeks after hatching the young and the adult female did not consume all of the fish supplied by the male, with the result that we were able to acquire uneaten parts of the prey, sometimes as much as 75 g, without interfering with nesting success. As the nestlings grew larger there was seldom any uneaten prey in the nests. Attempts to frighten the adult off the nest while she was feeding young resulted in her taking the fish with her. The single specimen of Shad from

Connecticut was obtained on 25 August by frightening one of the grown young, who was feeding at the nest.

2. The samples represent the part of the fish normally eaten by the adult female and the young. The head of the fish, usually eaten by the male during the early part of the nesting season, was rarely included in the sample.

3. The basic validity of using late spring fish samples to evaluate early spring pesticide intake might well be questioned, for the species of food fish utilized change with the shifts in abundance of one species or another. Early in the season Connecticut Ospreys were found to feed on Eels far more frequently than later in the season. Black-backed Flounder were eaten from March through June. For this reason the Eel sample known to be locally caught was purchased in a fish store. The rapid assimilation of pesticide by our experimental quail (Table 23–3) suggests that

TABLE 23–3. DDT residues in fish samples from Connecticut and Maryland

Species	No. of individuals	Total wet weight (g)	DDE		DDD		DDT		Total residues		DDD (% of total)
			μg	ppm	μg	ppm	μg	ppm	μg	ppm	
CONNECTICUT											
Black-backed Flounder	6	376	160	0·4	30	0·1	300	0·8	490	1·3	6·1
Windowpane Flounder	2	70	50	0·7	10	0·1	140	2·0	200	2·9	5·0
Alewife	4	60	20	0·3	10	0·2	100	1·7	130	2·2	7·7
Shad	1	70	80	1·1	40	0·6	100	1·4	220	3·1	18·2
Cunner	1	19	Trace		Trace		60	3·1	60	3·1	5
Eel	1	40	80	2·0	40	1·0	100	2·5	220	5·5	18·2
MARYLAND											
Eel	4	572	60	0·1	110	0·2	60	0·1	230	0·3	48
Yellow Perch	3	256	20	0·1	10	0·04	30	0·1	60	0·2	17
White Perch	2	93	Trace		Trace		Trace		5	0·05	–
Striped Killifish	1	22	Trace		Trace		Trace		5	0·1	–
Menhaden	2	125	Trace		Trace		Trace		5	0·05	–
Toadfish	1	140	20	0·1	10	0·1	10	0·1	40	0·3	25

Scientific names of fish species: Black-backed Flounder, *Pseudopleuronectes americana*; Windowpane Flounder, *Lophosetta maculata*; Alewife, *Alosa pseudoharengus*; Shad, *A. sapidissimus*, Cunner, *Tautogolabrus adspersus*; Eel, *Anguilla rostrata*; Yellow Perch, *Perca flavescens*; White Perch, *Roccus americanus*; Striped Killifish, *Fundulus majalis*; Menhaden, *Brevoortia tyrannus*; Toadfish, *Opsanus tau.*

residues in Osprey eggs are the result of feeding during the few weeks or even days before the eggs are laid.

4. Smaller individuals and species are less represented in our fish samples than they are in the whole of the Ospreys' food, because they are more frequently eaten entirely.

5. The values obtained, although stated as 'p.p.m., wet weight', are often higher than would have resulted if the fish had been truly fresh, due to the unmeasured amount of drying before collecting. The Black-backed Flounders, Shad and Eel from Connecticut were nearly fresh

when acquired, but the Windowpane Flounders, a species with a high surface-to-volume ratio, had lost perhaps half of their fluid weight when collected.

Despite the errors inherent in sampling, the values of DDT residues given in Table 23–3 probably show the relative residue intakes of the Ospreys of Connecticut and Maryland. The data are in the immediate range of residue found by Tompkins (1964) in fish from the upper waters of the Connecticut River (about 1.2 p.p.m. in 1963; 3.5 p.p.m in 1964) and those reported by Ames & Mersereau (1964) for three samples of fish from Osprey nests (1.7, 3.9 and 7.4 p.p.m.). Taken collectively, the data suggest that the amount of DDT and its metabolites in fish eaten by Connecticut Ospreys is 5–10 times higher than that in the food of Maryland Ospreys. Although these differences are easily sufficient to explain the differences in the residue content of the eggs from the two populations, neither group of fish samples contained residues sufficient to cause systemic poisoning in adults.

As indicated in Table 23–3, there are major differences between Connecticut and Maryland Ospreys in the fish species utilized for food, but the differences are not as profound as the table suggests. At a nest under observation in Connecticut at least one Yellow Perch was brought in by the male. Maryland Ospreys are sometimes seen carrying flatfishes. Flounders, which make up about half of the observed prey of Connecticut Ospreys throughout the season, are certainly less often utilized by the Maryland birds. Extensive knowledge of the species and age classes of fish eaten by Ospreys would greatly facilitate determining the rate of pesticide intake. The current increase in the study of pesticides in fishes will certainly provide a better basis for evaluating intake levels in fish-eating birds.

DDT RESIDUES IN THE EGGS OF JAPANESE QUAIL

It is difficult to evaluate the effects of DDT residues in the eggs of Ospreys, mainly because of the paucity of data relating the residue levels to hatchability in any species under laboratory conditions. An attempt in this direction was made by Victor J. Hardaswick and myself in the winter of 1964–65 with the experiment described below. The analysis of eggs is not yet complete, but preliminary results are worth noting.

Japanese Quail (*Coturnix coturnix japonica*) were chosen because of their small size, high rate of egg production and polygamous breeding habits. Twenty cages of Quail were set up, in five racks of four cages each, with one male and three females in each cage. All were kept on a 16-h 'day', produced by fluorescent lights. Four racks were test groups; the fifth was controls. One group was dosed with technical Rhothane (DDD) at 10 p.p.m. in the food, another at 50 p.p.m., a third group with *pp'*-DDE at 10 p.p.m., and a fourth with *pp'*-DDE at 50 p.p.m. The appropriate amount of pesticide was introduced dry into 10 kg of Purina Game Bird Layena, a commercial food prepared in 1 mm pellets. The entire 10 kg was slowly tumbled for 40 min in a cylindrical container on a skewed axis. The food was provided to the birds in gravity feed containers loaded from the top. Both food and water were continuously available.

Analysis of food samples has indicated that dry-mixing pesticide with pelletized food does not provide the anticipated dose rate. In our gravity feed containers, and probably also in the initial tumbling, the abrasion of food pellets results in food material being lost from the outsides of the pellets in the form of a coarse powder. This powder collects in the bottom of the feeding containers and is not consumed by the birds unless the feeders are allowed to become empty. Analysis of the powder from the bottom of a feeder of DDD at 10 p.p.m revealed a concentration of 27 p.p.m. of DDD, while pellets from the surface of the feeder, where they were being consumed by the Quail, showed only 3 p.p.m. It is apparent that less pesticide than planned was entering the birds.

Eggs were collected from all cages at noon daily. Each egg was numbered and weighed shortly after collecting. For 2 consecutive days all of the eggs would be incubated, then for the following 2 days all would be saved for analysis, then 2 days' eggs would be incubated. The analyses were performed on the gas chromatograph at the Connecticut Agricultural Experiment Station by Mr. Lloyd Keirstead. Incubation was performed in a 'Humidaire' forced-air incubator at 37.5° C and approximately 85 percent relative humidity. The eggs were automatically rotated 90 degrees around the short axis every 2 h. Hatches were scored daily by noting the empty shells, no attempt being made to separate the chicks or to study chick survival. Eggs which failed to hatch were removed when 2 days overdue and were examined for gross defects.

Dosing was not begun until fertile eggs had been obtained from all of the cages. During the acclimatization period several males were killed by females and there was some cannibalism among females, necessitating some substitutions in order to get workable cage groups. The shuffling of birds resulted in some loss of randomness, but definitely reduced the death rate during the experiment. The administration of pesticides was continued for 60 days, at the end of which time the birds were sacrificed and weighed. During the 60-day period 2881 eggs were laid by the entire flock, an average of about 580 eggs per test rack. About half of these, or about 290 eggs per rack, were incubated. The remainder were pooled by sample periods, providing about fifteen samples of sixteen to twenty-two eggs for each chemical at each dose. At the time of writing, four samples from each chemical/dose have been analyzed, providing a basis for tentative conclusions regarding the deposition of DDE and DDD in eggs and their effects on hatchability.

1. During the test period deaths in the test groups were not significantly different from those in the control group; one or two birds per test rack. Only one male died, but data from his cage had to be discarded after his death, which was, fortunately, late in the experiment.

2. The weights of test birds remained as high as those of the controls, and all groups gained weight during the test period.

3. The amounts of DDE in the eggs of birds on the nominal 10 p.p.m. rose from 0.27 μg/ml on the 4th day to 5.7 μg/ml on the 29th day, and 8.5 μg/ml on the 50th day. Apparently the increase was considerably more rapid at the beginning of the period than at the end. At 50 p.p.m. of DDE the amount in the eggs rose linearly from 7.7 μg/ml on the 5th day through 24.9 μg/ml on the 24th day, to 57.5 μg/ml on the 60th day, when the experiment ended.

4. The amounts of DDD were consistently much lower than those of DDE at the same dose and time. The nominal dose of 10 p.p.m. produced 0.15 μg/ml in the eggs after 7 days, 0.40 μg/ml after 48 days, and 0.90 μg/ml after 64 days. DDD at 50 p.p.m produced a steady level of about 3.5 μg/ml in the eggs from the 6th to the 57th day.

5. At the egg pesticide levels attained in the 60-day period of the quail study there appears to have been no lowering of hatchability in any of the test groups. Other effects may turn up when the post-mortem examinations are evaluated. It is noteworthy that the hatching rates in all groups, including the controls, were less than 50 percent. Most of the losses occurred at a time close to hatching, when eggs would have benefited from higher humidity. We avoided the use of a separate, high-humidity hatching chamber because we did not wish to introduce an additional variable. It may prove difficult to separate eggs which may have died from the effects of the pesticides from those which died from desiccation at the critical point just before hatching. Further work is planned to elucidate this point.

6. There are a number of interacting factors in the cage behaviour of Japanese Quail which may influence or even counteract the effects of pesticides on fertility and hatchability. It is likely that some of the very attributes which make these quail (and many other galliform birds) good subjects for controlled studies also limit the applicability of the results. High egg production, large egg size, early maturation, polygamy and a good 'cage disposition' make the Japanese Quail appear an ideal experimental bird, but I urge other investigators to consider carefully all aspects of the behaviour and physiology of the species before using it for pesticide research.

ACKNOWLEDGMENTS

For the last four years I have been assisted in the Osprey study by Mr. Gerald S. Mersereau, who has done much of the arduous field work. I am grateful to Dr. and Mrs. Roger T. Peterson, Mr. William H. Stickel, Dr. Lucille F. Stickel and Mr. Frederick C. Schmidt for much material and intellectual aid. I am indebted to Mr. Allen H. Morgan, Mr. Robert C. Woodruff, Mr. LeRoy Wilcox and Mr. Alexander Sprunt, IV, for information on the status of the Osprey on various parts of the Atlantic Coast.

I must express my appreciation to Mr. Victor J. Hardaswick for permission to discuss the preliminary results of the quail study on which we worked jointly. The National Audubon Society deserves our gratitude for financial support of the study. DDT and DDD were generously provided by the manufacturers, the Geigy Chemical Corporation and the Rohm and Haas Company. For the analysis of quail eggs we are indebted to the Analytical Laboratory of the Connecticut Agricultural Experiment Station and, in particular, to Mr. Lloyd G. Keirstead, who made a special effort to fit our samples into a pressing schedule. Mr. Roland C. Clement deserves thanks for many helpful suggestions in planning the quail study.

REFERENCES

Ames, P. L. and G. S. Mersereau. (1964). Some factors in the decline of the Osprey in Connecticut. *Auk*, 81, 173–85.

Bent, A. C. (1937). Life histories of North American birds of prey. Pt. 2. *Bull. U.S. Natn. Mus.* 195, 352–79.

Emerson, D. and M. G. Davenport (1963). Profile of the Osprey. *Naragansett Naturalist*, 6, 56–8.

Stickel, L. F., F. C. Schmidt, W. L. Reichel, and P. L. Ames. (1965). Ospreys in Connecticut and Maryland. I: Effects of pesticides on fish and wildlife. *Circ. Fish Wildl. Serv., Wash.* 226, 4–6.

Tompkins, W. A. (1964). *A pesticide study of the Westfield, Farmington and Connecticut watersheds.* Connecticut River Watershed Council, Greenfield, Mass.

24

Organochlorine pesticides in Antarctica

J. O'G. Tatton
J. H. A. Ruzicka

Considerable interest was aroused in 1965 by news from the United States that DDT had been discovered in Antarctic wildlife. Details of this finding were given later in two papers. The first of these by Sladen *et al.* (1) described the detection of DDT and two of its toxic metabolites, DDE and TDE, in six Adèlie penguins (*Pygoscelis adeliae*) and one crab-eater seal (*Lobodon carcinophagus*). All these specimens were taken at Cape Crozier on Ross Island (Fig. 24–1). The second paper by George and Frear (2) described a more ambitious study of samples taken on Ross Island and in the nearby McMurdo area. DDT only was detected in four out of sixteen Weddell seals (*Leptonychotes weddelli*) and four out of sixteen Adèlie penguins. Of sixteen skuas (*Catharacta skua maccormicki*) examined, nearly all contained DDT and DDE. No pesticides were detected in water and snow samples, nor in samples of four phyla of marine invertebrates and an Emperor penguin (*Aptenodytes forsteri*). Of three species of fish taken in the Ross Sea, only one specimen of *Rhigophila dearborni* contained DDT but no DDE.

These discoveries suggested that the contamination of the environment by the new persistent organochlorine pesticides had spread to what is usually regarded as the most remote and isolated part of the Earth. The nearest land mass to Antarctica is Cape Horn in South America (Fig.

Reprinted by permission from *Nature*, 215(5099):346–348 (22 July 1967). Reproduced by permission of the authors and the Government Chemist, Laboratory of the Government Chemist, London.

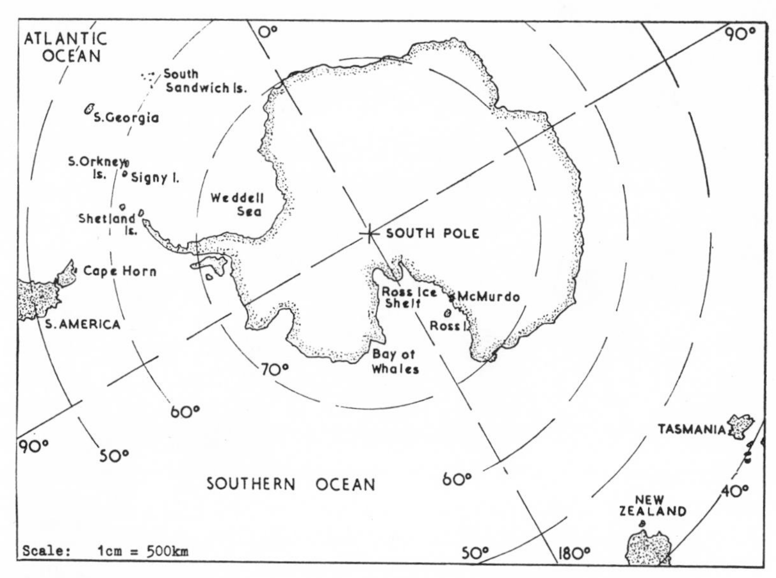

Fig. 24–1. Antarctica.

24–1), about 1,000 km distant, and most other land masses are at least 2,000 km away. The character of Antarctica is such that the only human residents on the continent are explorers and research teams, but the traffic of men and supplies brings visiting airmen and sailors to the region. Although the number of men on the continent has grown in recent years (3) and at times there may be as many as 4,000, the population is still minute relative to the land mass of about 13 million sq. km. There are no insect pests in the Antarctic and therefore no need for insecticides. Moreover, there is a general international agreement, under the Antarctic Treaty of 1959, that Antarctica shall remain a faunal and a floral preserve and that the existing ecosystems shall not be disturbed by, for example, the introduction of alien species or such powerful influences as pesticides. The detection of DDT in Antarctic wildlife therefore raised immediately the problem of its origin.

Both of the above reports referred to samples taken in the area served by the large United States base at McMurdo and it is perhaps significant that only DDT, and its first two metabolites DDE and TDE, were detected. It is now well known that the air and coastal waters of, for example, the United States and the British Isles contain traces of organochlorine pesticides (4–9) so that pesticides could have reached Antarctica by way of these two media. Air, rain and coastal waters, however, usually show traces not only of DDT and its metabolites but also of other organochlorine pesticides in common use, such as BHC and dieldrin. None of these other pesticides was detected in the samples from the McMurdo area.

Therefore, there was always the possibility that no matter how vigilant the authorities concerned may have been, DDT had reached McMurdo simply because man himself had transported it there in his ships, food, clothing or stores.

If this latter supposition were correct, then other areas of Antarctica outside the environment of McMurdo might still be free from contamination. It was considered that this could be investigated by sampling in the area of the British Antarctic Survey Base at Signy Island in the South Orkney Islands (Fig. 24–1) some 4,500 km from the American base at McMurdo. If no DDT were detected there, then it would suggest that the contamination in the McMurdo area was directly due to local human activities. If DDT were found on Signy Island then again it might be attributed to human activities but it would also reinforce the argument that there was general contamination of all Antarctica whether sea-borne or air-borne or both. The British Antarctic Survey generously agreed to co-operate in obtaining suitable samples from the Signy Island area and to transport them to England in a permanently frozen condition.

The samples included the livers, blubbers, abdominal fat and stomach contents of a number of chinstrap penguins (*Pygoscelis antarctica*). Most of these were adult birds, 3–8 years old, at the end of the breeding season when they were moulting and very fat; three of the birds were described as immature but at least 1 year old. The samples also included three penguin eggs, probably chinstrap, the fat and livers of two brown skua and two blue-eyed shags (*Phalacrocorax atriceps*), and the livers of four fish (*Notothenia neglecta*) about 3 years old. All these specimens, except for the fish, were taken on Signy Island, as far removed as possible from the Survey's base. The fish are the most abundant species in local waters and were taken in Borge Bay on the east side of the island. They are bottom feeders and are preyed on by shags and seals. The birds were killed by pithing or with a blunt instrument, except for the shags which were shot with a rifle. Dissecting instruments were washed in acetone which was also sampled as a precaution and later found to be completely free from pesticide residues. The items for analysis were placed in glass containers with caps lined with metal foil. These containers had been specially cleaned and prepared in this Laboratory. The samples were then maintained in a continuously frozen state during the journey to London where they were stored at $-25°$ C pending analysis.

The samples were all in an excellent state of preservation when examined and showed no signs of decomposition. They were extracted with hexane or an acetone–hexane mixture, and these extracts cleaned up by the method of de Faubert Maunder *et al.* (10) which includes a dimethylformamide–hexane partition followed by passage through a column of prepared alumina. The solutions which resulted were then examined by gas–liquid chromatography using silicone, Apiezon and cyanosilicone columns, with electron-capture detection (11, 12). The solutions were further examined by a thin-layer chromatographic method (13, 14), by which the solutions were spotted on to silica gel plates and developed with a 1 percent solution of acetone in hexane. After development, the appropriate areas of the thin-layer, corresponding to concurrently developed standard solutions of pesticides, were removed and extracted with hexane. These extracts were examined by gas–liquid chromatography and the identity and proportions of each of the compounds detected by the initial examination confirmed. The presence of *pp'*-DDT in the samples was further confirmed qualitatively and quantitatively by warming the appropriate thin-layer extract with ethanolic sodium hydroxide. The *pp'*-DDE formed by hydrolysis of the *pp'*-DDT was then determined by gas–liquid chromatography as already described. The presence of *pp'*-TDE was confirmed in a similar manner by hydrolysis to *pp'*-DME (1-chloro-2,2-di-4-chlorophenylethylene).

The results of the analyses are set out in Table 24–1. They give clear indication that all the penguins examined contained small amounts of BHC isomers, dieldrin, *pp'*-DDT and *pp'*-DDE in their liver, blubber and fat. Most of these samples also contained small amounts of heptachlor epoxide and *pp'*-TDE. Not shown in the table, but also detected in some of the samples, were small amounts, generally less than 0.005 p.p.m., of *pp'*-DME, a further breakdown product of DDT and generally regarded as non-toxic. Most of the pesticide residues detected were of the order of 0.001–0.010 p.p.m. but some of the samples contained nearly 0.050 p.p.m. of *pp'*-DDE in the blubber and fat. There was no apparent relation between the age of the birds and the distribution of results. The three penguin eggs, each weighing about 100 g, also contained BHS isomers, dieldrin, *pp'*-DDT and *pp'*-DDE in similar proportions to those found in the birds.

The stomach contents of the birds consisted almost wholly of krill (*Euphausia sp.*). The Antarctic seas are very rich in krill which form the main food of these penguins. Analysis of these krill samples indicated that here at least was one source of the pesticides found in the birds. Precisely the same compounds, but in lower proportions, were detected in the krill as in the birds, except that heptachlor epoxide was not detected in the krill, probably because the concentrations were less than those capable of being detected, allowing for the size of the samples and the methods used.

All the fish livers showed BHC isomers, heptachlor epoxide, dieldrin, DDT and its metabolites, in small amounts similar to those found in the penguin livers.

The skua is known to nest in Antarctica and the specimens taken for this study came from nests on Signy Island. Outside the breeding season, the birds range widely and often appear in the tropics. They are scavengers and predators and such birds have often been shown to contain large

TABLE 24–1. Organochlorine pesticide residues in Antarctic wildlife (p.p.m.)

	No. of samples	Alpha-BHC	Beta-BHC	Gamma-BHC	Heptachlor epoxide	Dieldrin	*pp'*-DDE	*pp'*-TDE	*pp'*-DDT
Penguin liver	11	0·000–0·002	0·002–0·008	0·000–0·003	0·000–0·006	0·001–0·006	0·001–0·018	ND	0·001–0·010
		Mean 0·001	Mean 0·005	Mean 0·002	Mean 0·002	Mean 0·002	Mean 0·006	—	Mean 0·005
Penguin blubber	10	0·001–0·002	0·000–0·006	0·001–0·002	0·001–0·003	0·004–0·010	0·013–0·048	0·000–0·006	0·005–0·012
		Mean 0·002	Mean 0·003	Mean 0·001	Mean 0·002	Mean 0·007	Mean 0·032	Mean 0·003	Mean 0·008
Penguin abdominal fat	5	0·001–0·002	0·000–0·004	0·000–0·002	0·001–0·002	0·006–0·010	0·029–0·048	0·000–0·004	·0·006–0·011
		Mean 0·001	Mean 0·001	Mean 0·001	Mean 0·002	Mean ·0·008	Mean 0·039	Mean 0·001	Mean 0·008
Penguin stomach contents	2	0·001, 0·001	0·005, ND	0·001, 0·001	ND, ND	0·001, 0·001	0·001, 0·001	ND, ND	0·001, 0·001
Penguin eggs	3	0·003–0·005	ND	0·003–0·005	ND	0·003–0·008	0·014–0·032	0·003–0·005	0·005–0·012
		Mean 0·004	—	Mean 0·004	—	Mean 0·005	Mean 0·021	Mean 0·004	Mean 0·008
Notothenia (fish)	4	0·001–0·007	0·002–0·008	0·001–0·004	0·002–0·004	0·001–0·009	0·002–0·013	0·000–0·018	0·006–0·020
		Mean 0·003	Mean 0·005	Mean 0·003	Mean 0·003	Mean 0·003	Mean 0·007	Mean 0·007	Mean 0·011
Brown skua liver	2	ND	ND	ND	0·100, 0·035	ND	4·00, 0·890	ND	0·33, 0·23
Brown skua fat	2	0·004, ND	0·040, ND	0·006, ND	0·120, 0·730	ND	5·80, 26·0	ND	0·890, 2·50
Blue-eyed shag liver	2	0·001, 0·002	0·006, 0·003	0·001, 0·003	0·002, ND	0·002, 0·001	0·011, 0·015	ND	0·003, 0·009
Blue-eyed shag fat	2	ND	0·009, 0·010	ND, 0·002	ND	0·004, 0·006	0·051, 0·140	0·009, 0·018	0·012, 0·023

ND, not detected.

amounts of pesticide residues (17, 18). Nevertheless, 0.73 p.p.m. of heptachlor epoxide, 26 p.p.m. of *pp′*-DDE and 2.5 p.p.m. of *pp′*-DDT are, by any standard, very high concentrations to find in the fat of a wild bird.

Blue-eyed shags nest in rookeries and are not wanderers. The specimens examined here were from a shag rookery on Signy Island and had probably never left the local waters. As might be expected, therefore, both the identity and proportion of the pesticides found in these birds are the same as those found in the penguins except that the *pp′*-DDE and *pp′*-DDT contents appear to be a little higher.

Shortly after receiving and analyzing the samples described above, we also received the livers of three sheathbills (*Chionis alba*). These had been taken on a separate occasion by the British Antarctic Survey, on Signy Island following the sudden deaths of a number of these birds from unknown causes. All of these livers contained *pp′*-DDE (0.030, 0.100 and 0.014 p.p.m.) and two of them contained dieldrin (0.015 and 0.012 p.p.m.) but no other pesticide residues.

The picture revealed by this present investigation shows that all the wildlife sampled on Signy Island and in its surrounding waters was contaminated by the persistent organochlorine pesticides. The great majority of the samples contained small amounts of four of the most commonly used of these compounds or their toxic metabolites—BHC isomers, heptachor expoxide (a very toxic metabolite of heptachlor), dieldrin, *pp′*-DDT, and two of its toxic metabolites, *pp′*-DDE and *pp′*-TDE. The pesticides themselves have now been in large scale agricultural and veterinary use for about 15–20 years. In this time, substantial amounts would have found their way down rivers to the oceans and have been carried away by deep sea currents. Large amounts would also have been lost to the atmosphere by volatilization and some of this would have returned to some other part of the land or sea by precipitation (6). It is clear from the analysis of the krill in the stomachs of the penguins and of the fish livers that the Antarctic waters contain not only traces of DDT but also traces of other organochlorine pesticides. The analysis of the penguins and their eggs and of the blue-eyed shags shows that this contamination is also present in the avian wildlife of the region. It thus seems possible that the DDT found in the early studies may, in fact, have been the first signs that contamination had reached Antarctica through the media of the seas and air.

The penguins of Antarctica have the auk family (*Alcidae*) as their ecological equivalent in the northern hemisphere. Some measure of the degree of contamination of Antarctic waters might be gained from a comparison of the results from the present study and those from studies already made of the auk family. Around the shores of the United Kingdom this family is represented chiefly by razorbills (*Alca torda*), puffins (*Fratercula arctica*), guillemots (*Uria aalge*) and black guillemots (*Cepphus grylle*). The pesticide residues occurring in the eggs of the first three of these species have been studied closely for the past 4 years so that the general concentrations of these compounds in their eggs are well established (5, 15, 16). The concentrations are of the order of fifty times higher than were contained in the penguin eggs studied here which would suggest that the waters of Antarctica are still far less contaminated than those around the British Isles.

We thank the British Antarctic Survey for their co-operation in this enterprise and, in particular, Dr. J. R. Brotherhood, of the Survey, who took all the samples and arranged their dispatch to England.

REFERENCES

1. Sladen, W. J. L., C. M. Menzie, and W. L. Reichel, *Nature,* 210, 670 (1966).
2. George, J. L., and D. E. H. Frear, *J. Appl. Ecol.,* 3 (suppl.), 155 (1966).
3. Stonehouse, B., *New Sci.,* 27, 273 (1965).
4. *Use of Pesticides, Rep. President's Sci. Adv. Comm.* (U.S. Government Printing Office, Washington, 1963).
5. Moore, N. W., and J. O'G. Tatton, *Nature,* 207, 42 (1965).
6. Abbott, D. C., R. B. Harrison, J. O'G. Tatton, and J. Thomson, *Nature,* 208, 1317 (1965).
7. Abbott, D. C., R. B. Harrison, J. O'G. Tatton, and J. Thomson, *Nature,* 211, 259 (1966).
8. West, I., *Arch. Environ. Health,* 9, 626 (1964).
9. Ruzicka, J. H. A., J. H. Simmons, and J. O'G. Tatton, *J. Sci. Fd. Agric.* (in the press).
10. De Faubert Maunder, M. J., H. Egan, E. W. Godly, E. W. Hammond, J. Roburn, and J. Thomson, *Analyst (Lond.),* 89, 168 (1964).
11. De Faubert Maunder, M. J., H. Egan, and J. Roburn, *Analyst (Lond.),* 89, 157 (1964).
12. Simmons, J. H., and J. O'G. Tatton, *J. Chromatog.,* 27, 253 (1967).
13. Abbott, D. C., H. Egan, and J. Thomson, *J. Chromatog.,* 16, 481 (1964).
14. Harrison, R. B., *J. Sci. Fd. Agric.,* 17, 10 (1966).
15. *Report of the Government Chemist, 1965,* 75 (HMSO, London, 1965).
16. *Report of the Government Chemist, 1966* (HMSO, London, in the press).

PROBLEMS OF ENVIRONMENTAL POLLUTION

RADIATION AND RADIOISOTOPES

25

Radiation and the patterns of nature

George M. Woodwell

The partial answers we have to the question of what radiation does *in* and *to* nature are revealing not only of the effects of radiation on living systems, but also of the architecture of the systems themselves. My object is to show the patterns of the effects of radiation on natural communities, and how the patterns parallel and help to explain the normal patterns of structure, function, and development of these communities. It is important in this discussion to remember that most life as we know it has evolved in environments in which total exposures to ionizing radiation have amounted to less than a few tenths of 1 roentgen per annum, and that ionizing radiation is generally thought to have played a very minor role among the selective processes of evolution. It is somewhat surprising therefore that the effects of radiation on natural communities follow predictable patterns apparently related to the evolution of life.

The significance of natural communities to biology and to man is not immediately apparent. For my purposes it is important to recognize that all organisms have evolved as functional units in communities of organisms, and that the structure and function of these communities have determined in some measure the structure and function of the organisms themselves. So we can think of Darwin's struggle for existence as operative in the evolution of not only species but also groups of species and whole communities. This is not a new concept; it was set down by Darwin in his *Origin of Species*, published in 1859.

Reprinted with permission from *Science,* 156(3774):461–470 (28 April 1967). Work carried out at Brookhaven National Laboratory under the auspices of the USAEC.

The evolutionary implications of Darwin's struggle for existence at the community level are shown most clearly by a simple example, which is based rather freely on Darwin's own studies in the Galápagos Islands. Let us assume a small group of islands in the tropics, volcanic, and therefore young in a geologic sense, but supporting the limited flora and fauna that have arrived from the mainland some 1000 kilometers away. The climate is diversified, ranging from desert to moist forest. The islands have trees, grasses, and shrubs, but no mammals and few birds.

Over the years, probably hundreds of years, chance, possibly in the form of westering storms, brought small flocks of birds. From among these flocks at various times some finches survived and found a favorable habitat, rich in a diversity of foods and free of both mammalian and avian predators; reproducing rapidly, each new immigrant population became a plague, much as the Japanese beetle, the sparrow, the starling, the gypsy moth, and a host of other introductions have become plagues in our own experience. Food, although at first abundant, quickly became limiting, and the struggle for existence intensified. Competition for food was fierce and a premium attached to any ability to exploit new food supplies—foods different from those exploited by competitors. Small differences in behavior or in size or shape of beak resulted in small differences in survival and in ability to rear young. These differences, when hereditary and useful, were passed on and amplified in the population, and on each island there developed a population of finches peculiarly adapted to that environment and different from populations on other islands.

There was one additional complication. Exchanges of individuals or small groups of individuals occurred occasionally among the islands, continually testing the degree of genetic isolation achieved by the evolution of different races. Frequently these transported populations failed on the new island or were absorbed into the now-indigenous population; occasionally, however, a small one found itself partially isolated ecologically, by behavior, food supply, or local preference of habitat, from the indigenous population and survived as a distinct population, competition and evolution tending to accentuate the isolating mechanisms. Thus the islands gradually acquired a diverse bird fauna consisting largely of races of finches: ground finches, tree finches, a warbler finch, a woodpecker finch —each race using a set of resources used elsewhere in the world by a totally different species. Ecologists call the resources used by any one species a niche; where niches overlap and resources are shared, they say that competition occurs.

We see from this example, which is a grossly simplified version of *Darwin's Finches* (1), that the evolution of life proceeds toward reduction of competition, toward utilization of space and other resources, toward diversity in form and function, toward the filling of niches. We see, moreover, that the evolution of a race is affected not only by its physical environment, but also by the evolution of other races whose evolutions are in turn affected. Thus the chain of cause and effect here becomes entangled in bewildering ways. The product is a complex and, in some degree, mutable array of plants and animals which, itself, has clear and predictable patterns of structure, function, and development; these are "natural communities." Thus physical environments that are similar tend to support organisms that are similar in form and function, if not in species. So certain climates support forests the world over; others, grasslands; others, desert; and these words—forest, grassland, desert, and tundra—have meaning for us in terms of climate and flora and fauna.

Thus, where environments are similar, we find organisms that may have little or no common genetic past performing parallel functions. In Australia the marsupials, for instance, fill the grazing niches filled by placental mammals elsewhere; and the genus *Eucalyptus* has filled the tree niches occupied elsewhere by a score of other genera. The communities in which these organisms participate are one answer, tested through millions of years of evolution, to the very fundamental question: How can the resources of environment be used to perpetuate life? This is, of course, a fundamental objective of man; the use of environment to best advantage.

The evolutionary answer is a magnificently durable one and, in terrestrial communities, usually a surprisingly stable one, free of plagues or rapid changes in sizes of population. By this I mean that controls of population size have evolved, building stability into these complex biological systems—putting the "balance" into nature.

Now let us consider for a moment certain other characteristics of natural communities. It is clear that the communities have developed over long periods and are very much a product of the evolution of life; and that they vary in a spatial sense with geography, climate, topography, and a host of other environmental factors. They also vary with time.

To show the variation with time, let us assume for a moment that after we harvest our corn crop in the eastern United States we simply abandon the land. The weeds of the garden take over; crabgrass, at first; later, grasses; then, pine forest; and finally, after 100 years or so, an oak forest. The general pattern is familiar; environmental circumstance may modify details. The change from herbaceous weed field to forest involves not only changes in the species forming the communities, but also changes in the total weight of living matter on a unit of land, in the total amount of essential nutrients available, in the total amount of water used, in the total amount of niches available, and probably in the rates of biologic evolution itself. This process—succession—becomes one of the great central principles of biology.

We can examine one succession, from abandoned field to forest, most easily by considering stored energy in plants over time. By plotting such

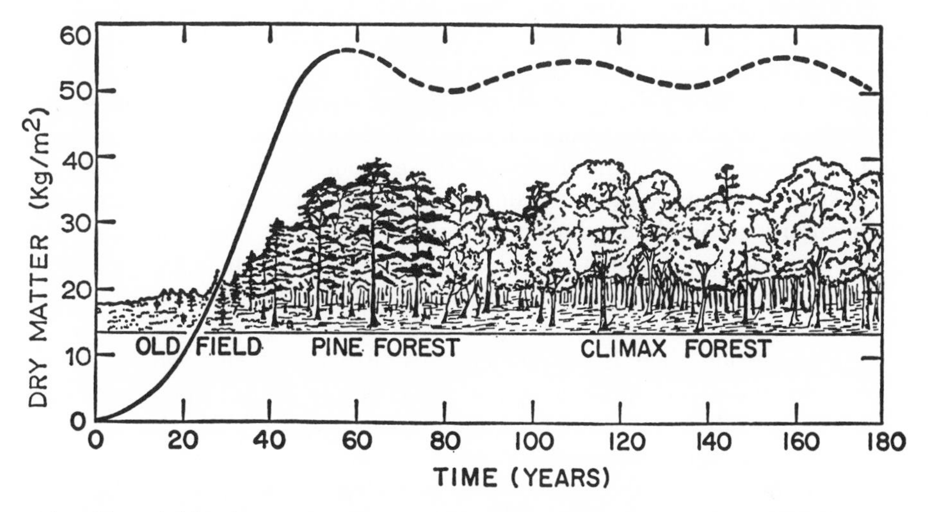

Fig. 25–1. Field-to-forest succession in the eastern United States. The oscillations of climax are assumed.

data (my own, and those produced by workers elsewhere in eastern North America) we obtain an S-shaped curve similar to the growth curve of a single organism (Fig. 25–1). It rises slowly during the early herbaceous stages, rises much more rapidly during the pine-forest stage, and levels during the deciduous-forest stage as the degree of stability increases. Along this curve several very fundamental changes occur in community structure and function. There is, of course, a shift in species from herbaceous plants to trees. But there is also thought to be increase in diversity —total numbers of species present—from the few of the abandoned field to the many of the ultimate forest. There is change in degree of stability from the field, with its patches of ragweed and crabgrass which may be rapidly replaced by any of several species, to the forest with its spatial uniformity and slow replacement. There is increase in the total pool of minerals within the system: small amounts held within the herbaceous communities, large amounts in the forest. Total respiration and total photosynthesis increase, but at different rates, producing a regular change with time in the ratio of photosynthesis to respiration during the course of succession. We assume in addition that the total amount of water used increases along this succession (2).

If, in the course of such a succession, one or more factors essential to the system become exhausted or available only in short supply, the rate of succession is slowed and the climax is diverted, possibly by as much as from forest to grassland. Thus, in areas of low rainfall, succession ends in a stable grassland or woodland; where little mineral nutrient is available, whatever the reason, the succession is slowed and the S-shaped curve levels.

If, on the other hand, the environment is changed drastically by erosion or by sudden change in climate, or catastrophically by fire or windstorm or even by fallout from a bomb, then the changes that occur in these arrays tend to be just the reverse of those occurring during a normal succession: the communities are simplified, niches are opened, the nutrient inventory accumulated during succession is lost at least partially, the community becomes less stable, and a new succession begins, possibly marked by large fluctuations in populations that reproduce rapidly (such as insects) and can exploit the open niches.

Succession, then, is such a fundamental part of biology that it forms the logical core for appraisal of the effects of any change in environment, most especially a change that has such far-reaching and basic implications for life as ionizing radiation.

At first glance the problems in appraisal of the effects of ionizing radiation on the communities along a successional gradient seem so complex as to be impossible. But we can borrow a trick from the mathematicians and examine the effects on the extremes: we can use a gradient of exposures from very high to very low and examine the early stages of succession, which, in eastern North America, are abandoned fields, and the later stages, which are forests. The question we ask is, in each of these stages: What are the effects of irradiation on the community? In the forest, for instance, we need to know what exposure to radiation changes the composition of the plant community. When the composition does change, how does it change? Do species behave individually, or are there groups of species having similar characteristics? After what exposures do we expect insect populations to change? Do we affect metabolism, use of water? How do we affect them? Are there any patterns of radiosensitivity that may be useful for prediction of effects of radiation or for interpretation of the structure and function of unirradiated communities? The overriding question is: What are the patterns of radiation effects on the structure, function, and development of natural communities? This was the question posed in 1961 when the work at Brookhaven, which I shall discuss, was started.

We had then considerable information on radiation effects on many species of plants (3). It was known that the amount of damage caused by any exposure was related to the size and number of chromosomes in the cell nucleus (4). Sparrow had observed that certain species of pine trees are killed by exposures in the same general range as those killing man. Other data had shown a very great range, more than 1000-fold, in the sensitivity of plants to damage by radiation (5). The sensitivity of pines had been confirmed (6), and it had been shown that forests are generally more sensitive than had been known (7). Field observations, however, were most limited, and there was good reason to explore the problems experimentally and in detail.

Our approach entailed the establishment of two experiments, in each of which we used a single large source of γ-radiation (equivalent to about 9500 curies of Cs^{137}), arranged in such a way that it could be lowered into a shield (for safe approach) or suspended several meters above ground to provide radiation over a large area. The sources were large enough to administer several thousand roentgens per day within a few meters, the dose approaching background levels beyond 300 meters. The two experiments were conducted in an irradiated old field in the now-well-known γ-radiation field established in 1949 (8), and in an irradiated forest—a completely new installation (9, 10); thus they gave us a sample from each end of the successional curve that I have discussed.

A section of the γ-radiation field was abandoned in the fall, after harvest, and the herbaceous communities common to abandoned gardens were allowed to develop. On Long Island about 40 herbaceous species participate in colonizing land prepared in this way; one of the most conspicuous is pigweed (*Chenopodium album*) because of its height (up to 1 meter) and abundance. During the 2nd year, horseweed (*Erigeron canadensis*) is the most conspicuous and one of the most abundant. In subsequent years, grasses such as broom sedge (*Andropogon* spp.) and asters (*Aster ericoides*) become dominants, to be followed by pine, and oak-hickory forest (11, 12).

Irradiation produced striking changes in the communities of the early stages of the succession. Although we have studied several of these communities over five summers at Brookhaven, I shall discuss here only the 1st-year communities. The most conspicuous change was drastic simplification at high exposures. We can measure simplification as a reduction in numbers of species per unit area, or in "diversity." Figure 25–2 is a plot of diversity along the radiation gradient. Irradiation at 1000 roentgens per day reduced diversity to about 50 percent of that of the unirradiated community, another field 2 kilometers distant. This decrease was continuous along the radiation gradient and was not marked by any abrupt decline indicating exclusion of several species in a narrow range of rates of exposure. Certain species survived daily exposures that exceeded 2000 roentgens.

The pattern of distribution of standing crop, or total weight of plants, at the end of the growing season, was strikingly different (Fig. 25–3). Total standing crop along the radiation gradient ranged between about 400 grams per square meter in the control community and 800 grams at 1000 roentgens per day, with a consistent increase with increase in exposure between these extremes. While the significance of this increase is not entirely clear, it is plain that, at exposures exceeding 1000 roentgens

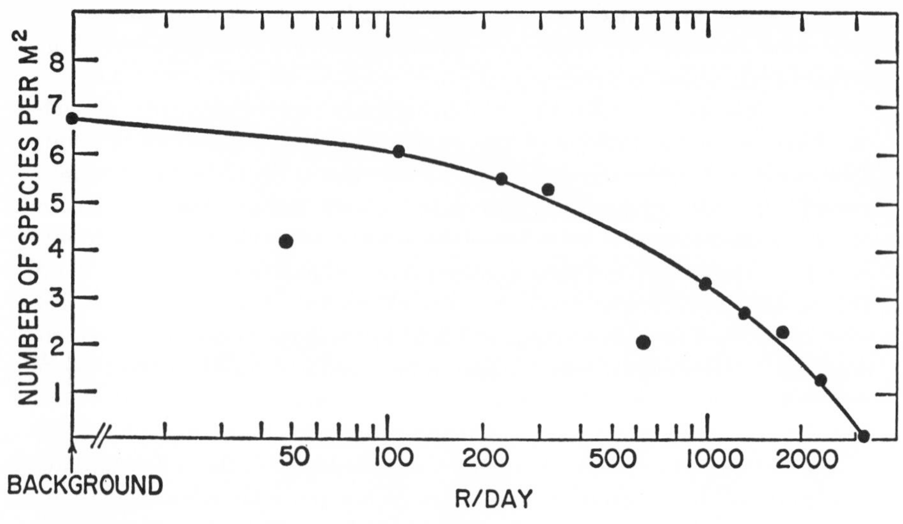

Fig. 25–2. Diversity in the 1st-year-old field.

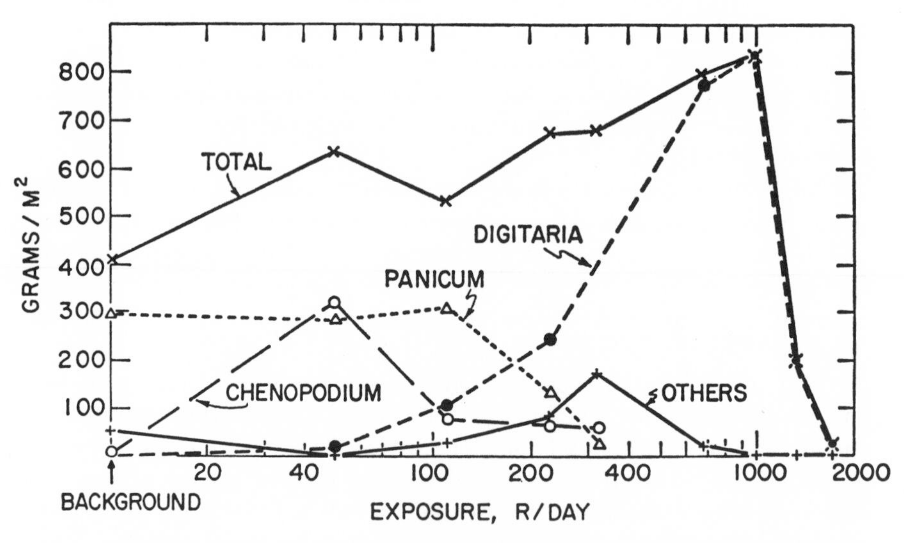

Fig. 25–3. Total dry weight of plants, by species, in the irradiated old field. Dry weights were measured at the end of the season and do not represent total production.

per day, total standing crop dropped abruptly to a few grams per square meter and, although some species survived even higher exposures, production of plant mass was very low indeed. There is clear evidence that at intermediate exposures exclusion of one species freed resources for others, crabgrass being by far the most benefited; at exposures exceeding

200 roentgens per day it was the major contributor to the total standing crop. Thus these old-field communities appear to be plastic, maintaining and possibly even increasing the total amount of energy fixed, despite a reduction in diversity of up to 50 percent. It also appears that diversity of species is more sensitive to radiation effects than is organic production. This relation is borne out by a brief consideration of coefficient of community, and percentage similarity.

The coefficient of community is simply the total number of species common to two communities, expressed as a percentage of the total number of species in both communities; Figure 25–4 shows an approximately linear relation between the coefficients of community along the radiation gradient, calculated for the control community, and the logarithm of radiation-exposure rate (11). There appears to be no threshold for effects on composition by species at exposures as low as 50 roentgens per day. If

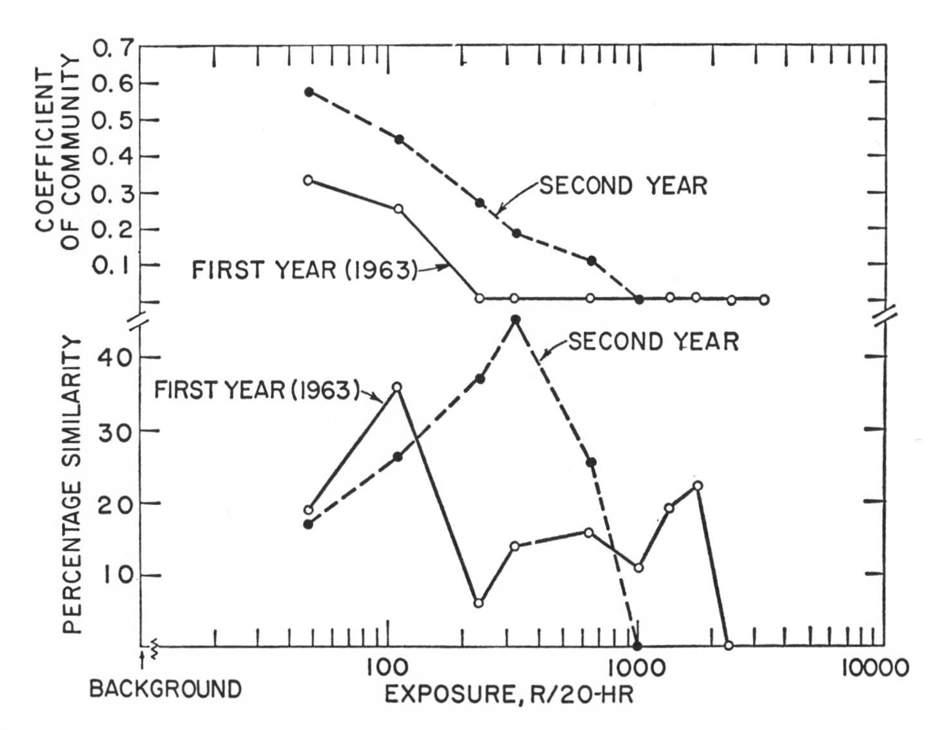

Fig. 25–4. Coefficient of community, and percentage similarity for irradiated communities compared with the control community (2 kilometers distant). The linear relation between coefficient of community and the logarithm of exposure rate shows that species composition, alone, independent of density, is a useful criterion of the severity of disturbance by radiation.

we weight the coefficient of community with a measure of abundance of each species, we can calculate what is called percentage similarity, and Figure 25–4 shows that there is no simple relation between these figures and radiation exposure, an observation that seems to confirm the earlier observation that the relative abundance of any species, however measured, is primarily controlled by competition with other species. Thus diversity and coefficient of community (and probably any other index of species diversity) emerge as relatively sensitive measures of radiation damage—and probably of any type of environmental change; abundance, density, and standing crop are insensitive.

Let us examine somewhat more closely the characteristics of plants that survive high rates of exposure. Two characteristics seem particularly significant: first, at high exposures the incidence of species that normally grow close to the ground [prostrate, decumbent, or depressed (13)] is substantially higher than in unirradiated communities (Fig. 25–5); second, there appears to be sorting on the basis of chromosome size, plants with large chromosomes being excluded from the areas receiving high exposures (Fig. 25–6). While it is difficult to venture a reason for apparent correlation between small size of chromosomes and a prostrate or decumbent growth habit among plants of old-field communities, these observations suggest that such a pattern may exist.

Thus the first year of succession is characterized by a loose array of herbaceous plants, most of them annuals or biennials, of varying life-forms and physiologies. Diversity in form and function allows rapid colonization of a wide variety of disturbed areas, and contributes toward making the community as a whole resilient in the face of disaster—such as a gardener's hoe or a gradient of ionizing radiation. The primary effect of stresses, including irradiation, is reduction of diversity. In the case of radiation, the

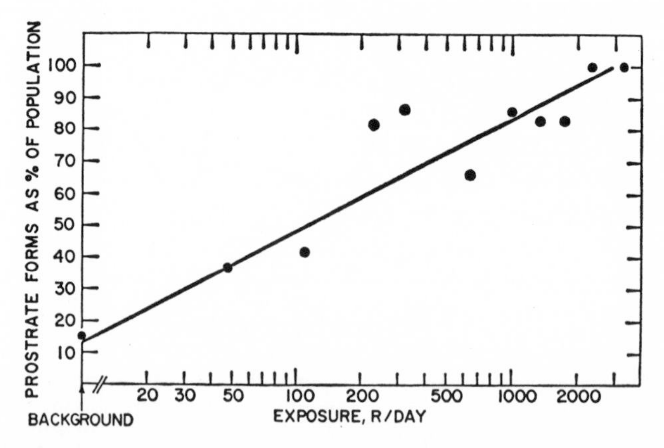

Fig. 25–5. Life-forms in an irradiated field (1963, 1st year); "prostrate" forms include forms labeled normally prostrate, decumbent, or geniculate by Fernald (13).

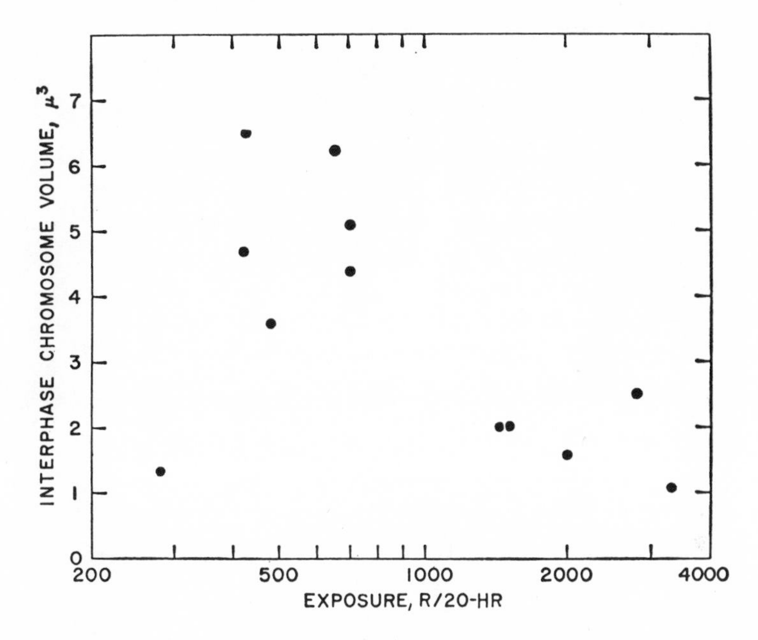

Fig. 25–6. Average interphase chromosome volumes of 12 species of plants, showing maximum exposure at which any individual lived [from Wagner (25)].

reduction is continuous along the radiation gradient and not characterized by simultaneous exclusion of two or more closely associated species, an observation testifying to the looseness of the community organization. Although it is true that the plasticity of the community as a whole makes it resistant to radiation damage, it is certainly not true that all species in the community are equally resistant. Daily irradiation at 50 or more roentgens produced continuous sorting of species according to life-form and according to the average volumes of their chromosomes.

Irradiation of the forest commenced in November 1961 after a detailed series of preirradiation studies. The approach was to make a case-history study of one relatively complex ecological system by examining as many aspects as possible of its structure and function, both normal and pathological (10). Six months after installation of the source the forest appeared as in Figure 25–7; my data, with few exceptions, apply to the forest as it was in the summer of 1962, after approximately the same period of exposure as the old field.

Five zones were apparent along the radiation gradient: a central zone in which no higher plants survived; a sedge zone containing *Carex pensylvanica* and a few sprouts of the heathshrub layer; a shrub zone where the two blueberries and huckleberry survived; an oak-forest zone at daily

Fig. 25–7. Effects of 6-month exposure to gamma radiation ranging in intensity from several thousand roentgens per day near the center of the circle to about 60 roentgens at the perimeter of the defoliated area. The experiment is part of a study at Brookhaven National Laboratory of the effects of chronic exposure to ionizing radiation.

exposures less than about 40 roentgens; and the oak-pine forest in which radiation effects on growth were apparent, without change in species composition (14).

The zoning of vegetation reflected the decline in diversity along the gradient (Fig. 25–8). If the normal "plot" in this forest be accepted as having 5.5 species, then 50-percent diversity occurred at 160 roentgens per day, or less than one-fifth the exposure to reduce diversity by 50 percent in the herbaceous community. Shielding by the stems of large trees in the forest allowed survival by species at average exposures substantially greater than the normally lethal exposures. Therefore the differential is probably even greater, and the forest may have its diversity depressed by 50 percent at exposures as low as one-tenth of those required in the herb field (15).

Unlike the old field, standing crops in the forest declined at approximately the same rate as diversity (Fig. 25–9). This relation between diver-

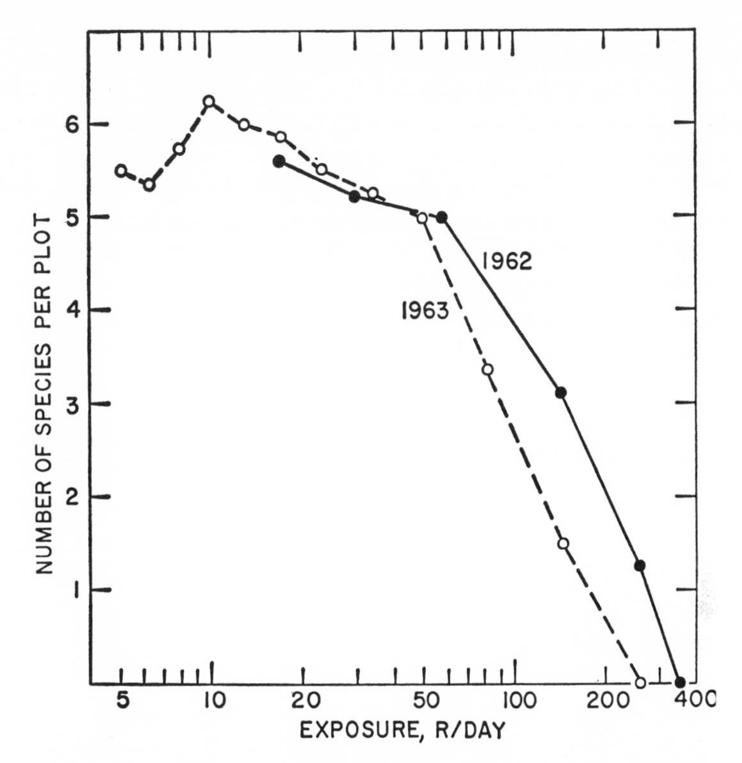

Fig. 25–8. Species diversity along the radiation gradient in the irradiated forest in 1962 and 1963. Measurement of diversity in a forest requires differently sized samples for differently sized plants; thus the unit of diversity here is "species per plot" [from Woodwell and Rebuck (*15*)].

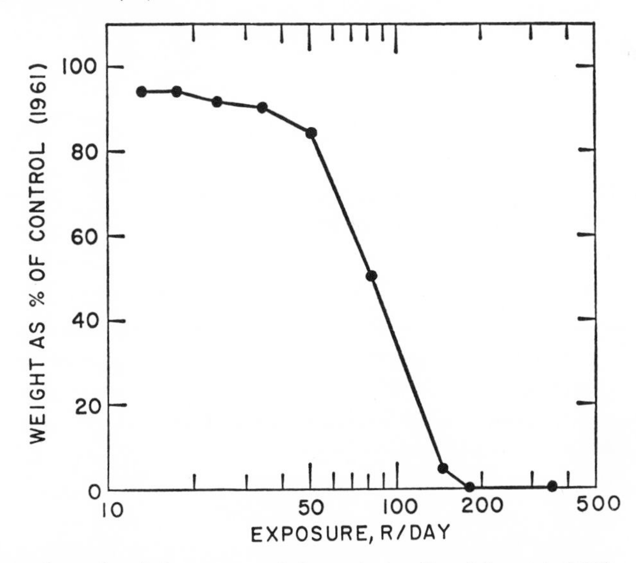

Fig. 25–9. Total weight of above-ground shoots in irradiated forest in 1962.

sity and abundance measured by standing crop is, of course, to be expected, since there is no possibility of a population of oak trees, 9 meters in height, expanding within a year to fill a niche vacated by pine. Nor was there invasion by any of the herbaceous species more resistant to radiation. There was, however, expansion of the population of *Carex*, a plant that normally occurs as an ubiquitous but very sparse herb, to cover as much as 20 percent of the total ground surface. This expansion was in response to the demise of the tree and shrub cover; it points to the potential importance of rare, or at least inconspicuous, species, capable of rapid regeneration, in maintaining certain aspects of function in disturbed communities.

Other examples of rapid response to the changed resources in the damaged community abound, especially among insect populations. In general these populations have followed quite closely change in food supply (16). Populations that utilize dead organic matter and decay organisms increased in the central zone of high mortality to the vegetation; bark lice are a good example (Fig. 25–10). While this type of change seems quite straightforward and predictable, all changes in insect populations were not: during the 2nd year of the experiment, for instance, there was an unexpected and still-unexplained population explosion of aphids on white

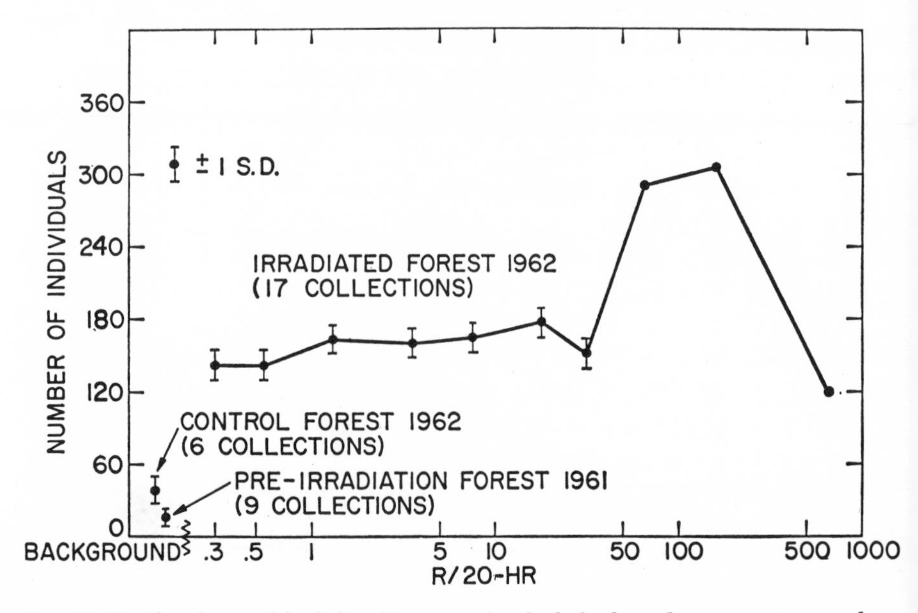

Fig. 25–10. Abundance of bark lice (*Psocoptera*), which feed on decay organisms and dead organic matter, along the radiation gradient [from Brower (*16*)].

Fig. 25–11. Abundance of aphids (*Myzocallis* sp.) on oak leaves in 1963 along the radiation gradient. At 9.5 roentgens per day, populations were more than 200 times normal.

oaks exposed to 5 to 10 roentgens per day (Fig. 25–11). Aphids share with certain fungi, such as wheat rust, ability to reproduce asexually very rapidly to exploit any available resource. Although mobile, they are not strong fliers and do not migrate far; it is unlikely that the high populations resulted from migration from neighboring forests. It seems much more probable that leaves of trees exposed to 5 to 10 roentgens per day differed qualitatively from leaves of unirradiated trees sufficiently to support large populations of aphids; the difference appears to be not in either total sugars or total proteins, but in some more subtle factor detectable by aphids but not yet by man (17).

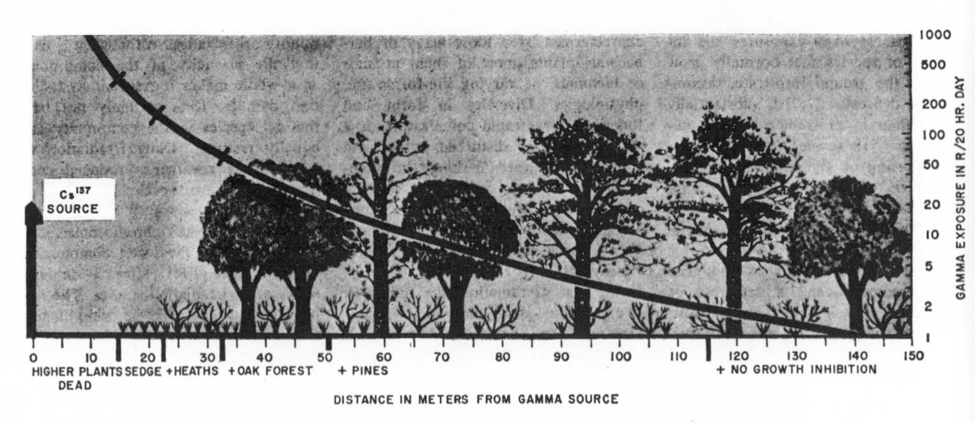

Fig. 25–12. Pattern of radiation damage to oak-pine forest in 1962 after about 6 months' exposure.

The effects of irradiation on the forest are best summarized by the profile (Figure 25–12) showing the five vegetation zones and their approximate distribution along the radiation gradient in 1962. The most striking observation is the relative sensitivity of all the higher plants. No higher plant indigenous to the forest survived the 1st year of exposures exceeding 350 roentgens per day; in the old field, certain species survived more than 3000 roentgens per day. The 50-percent diversity point occurred in the forest at less than 160 roentgens per day; in the field, at 1000 roentgens per day. It seems abundantly clear that the forest as a unit is substantially more sensitive than the herb field. A second important relation is that there is sorting by size along the radiation gradient, smaller forms of life being generally more resistant than trees; this relation also extends to mosses and lichens.

This sorting by size, which now seems to be a well-established characteristic of radiation damage, has interesting parallels elsewere in nature. It occurs along gradients of increasing climatic severity, such as the transition from forest to tundra in the north, and on mountain slopes. At such transitions, forest is replaced by low-growing shrubs, frequently blueberries and other members of the heath-plant family. In more extreme environments the heath shrubs are replaced by a sedge mat formed by a species of *Carex*; in the most extreme, the *Carex* is restricted to protected spots, and mosses and lichen are the only vegetation. The parallel with the irradiated forest is quite remarkable, holding even to genera and species, in certain instances. The conclusion to be drawn from this relation is merely that characteristics that confer resistance to certain types of environmental extremes also, curiously enough, confer resistance to damage by radiation.

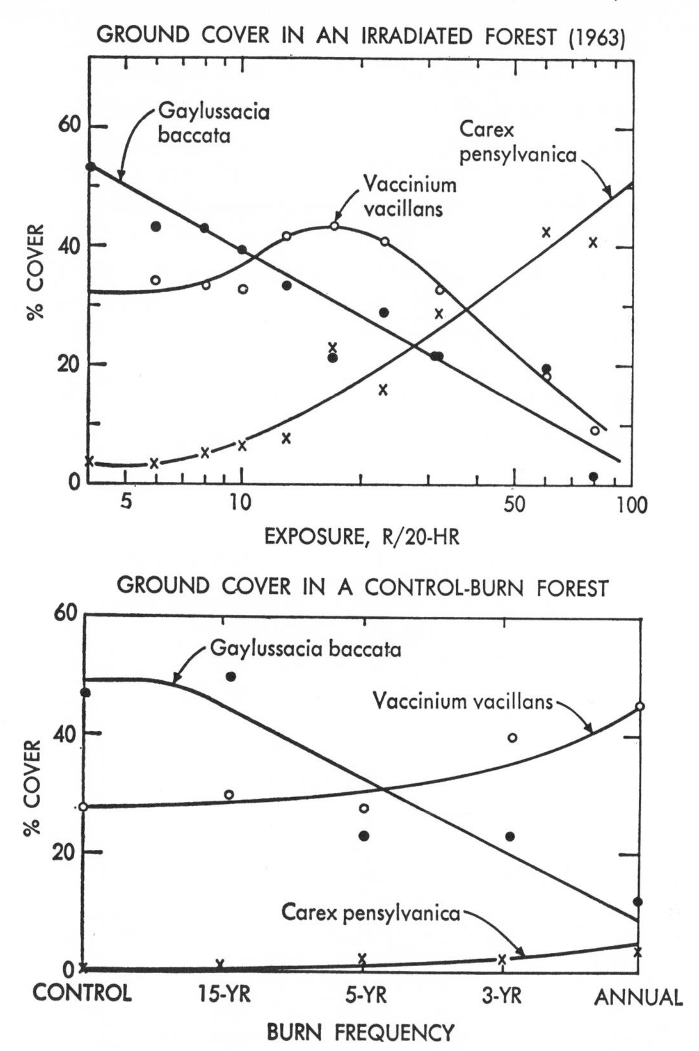

Fig. 25–13. Comparison of effects of ionizing radiation and different frequencies of burning on the shrub-and-sedge community [from Brayton and Woodwell (*19*)].

We can test the hypothesis a little more rigorously by examining in detail the shrub layer of the forest, which is itself a small community containing two species of blueberries, the huckleberry, and the sedge. Changes in this community after burning have been studied intensively (18); their general pattern appears in Figure 25–13: with increased frequency of fire, the huckleberry populations decline, the blueberries increase, and the sedge increases. Under irradiation the pattern is strikingly similar until the point at which radiation kills the blueberry. The parallelism between the effects of fire and of radiation should not be expected to be universal, for many factors are implicated. Nonetheless there seems to be a strong parallel between the effects of radiation and the effects of another extreme; and in both instances, as well as in the herb field, the correlation between durability and small stature applies (19).

While there is no completely satisfactory explanation of the parallels, one important contributory factor may be simply the size of the plant. Perennialism, height and complexity of structure all represent investments of energy in nonphotosynthesizing tissue, tissue that requires energy for maintenance. We might think of this tissue as a mortgage that must be paid off with income from photosynthesis. As the size of a plant increases, both mortgage and total income increase, but at different rates. In Figure 25–14 are plotted the total weights of trees against $h \times d^2$, a measure of size (20). Since total weight of the tree is not a proper measure of total

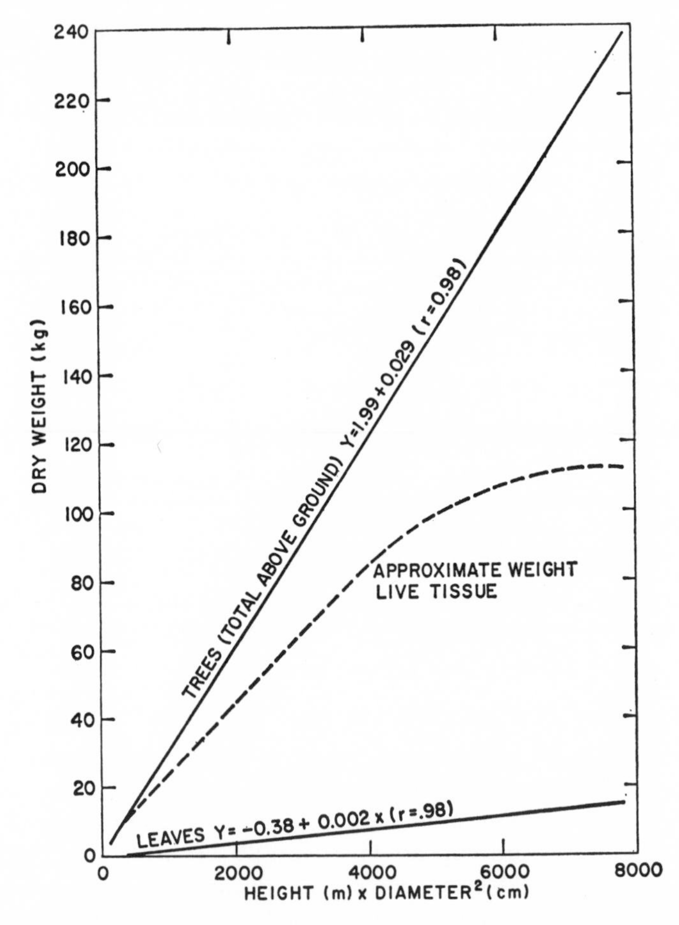

Fig. 25–14. Relation between weights of leaves, weights of trees, and tree sizes for an oak-pine forest on Long Island.

living tissue (there being considerable nonliving tissue in a tree), we have also plotted an estimate of the weight of tissue that may normally be considered living. It seems clear that in small trees leaves represent a substantially larger fraction of the total weight of live tissue than in large trees. An increase in size thus puts greater demands on the photosynthetic mechanism simply for maintenance, leaving less for growth and repair.

A similar relation applies along the successional gradient that we have discussed (Fig. 25–15). In the early stages of succession most of the tissue produced is green, and the mortgage payments to support respiration are small. As succession progresses, the complexity of structure increases, but total living tissue increases more rapidly than the weight of

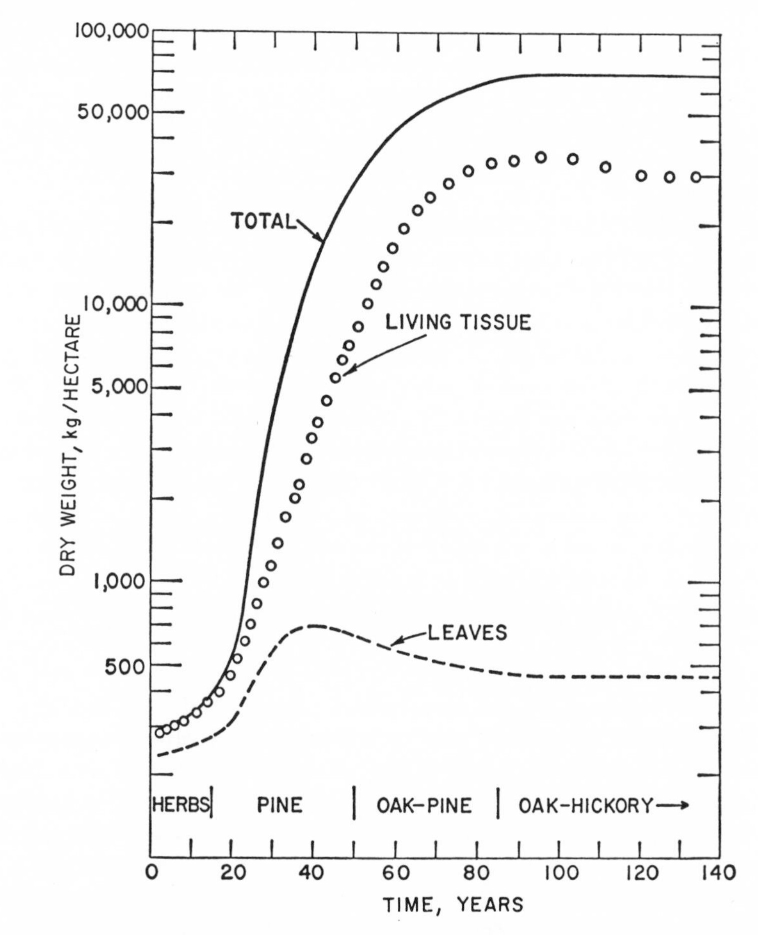

Fig. 25–15. Approximate relations between total above-ground standing crop of plants, weight of living tissue, and weight of leaves in a normal field-forest succession of eastern North America.

leaves, which supply the energy for respiration; the mortgage increases, but income does not increase proportionally. It is true that the existence of the structure allows greater use of space, greater total photosynthesis up to a point, and greater diversity of species. But it is also true that the maintenance of the structure and diversity hinges on the annual interest paid from photosynthesis into the structural mortgage, and, if for any reason the interest is not paid, the structure begins to decay.

And here lies the crux of the matter: the mechanisms related to energy fixation—bud set, bud burst, leaf production, photosynthesis—are at much greater hazard than mechanisms related to energy use. Almost any disturbance of a forest may reduce its capacity for fixing energy; it either increases respiration or reduces it relatively slightly. If the disturbance is chronic, the vegetation comes to a new equilibrium, supporting a less complex structure. For this reason we might expect a forest to be more sensitive to disturbance than is an herb field because the forest is less plastic in species composition and because its capacity for fixing energy must remain substantially intact or it will burn up more than it fixes and deteriorate. And that is exactly what happens, but it is far from the whole explanation.

We have shown for the herbaceous field that there was sorting along the radiation gradient, dependent on chromosome volume: plants with large average chromosome volumes are sensitive; those with small volumes, resistant. The pattern in the forest was similar. If we plot the average chromosome volumes (8, 21) against the daily exposure required to inhibit growth to 10 percent of growth of unirradiated plants (Fig. 25–16), it is abundantly clear that radiosensitivity correlates with size of the chromosomes, and that this correlation applies to populations in nature as well as to cultivated populations. Also, the larger plants tend to have larger chromosome volumes; the smaller plants, smaller. Clearly, chromosome volume has played a role in the persistence of plants along the radiation gradient in the forest as well as in the field.

Now let me recapitulate briefly: the successional gradient we have used to explore effects of radiation on natural communities is characterized in the early stages by a loosely structured community or series of communities shifting in species composition, diversity, dominance, density, total mass of living matter, and probably in every other measurable parameter, within relatively broad limits, in response to disturbance. It is also true that the species of the early successional communities, including mosses and lichens, tend generally to be resistant to radiation. The forest does not share the plasticity of communities having simpler structure; in this respect the forest is more sensitive to any disturbance. In plants, large size alone, because of its effects on the ratio of photosynthesis to respiration, contributes to this type of sensitivity; but, more importantly, plants of the forest are inherently sensitive to radiation damage because they

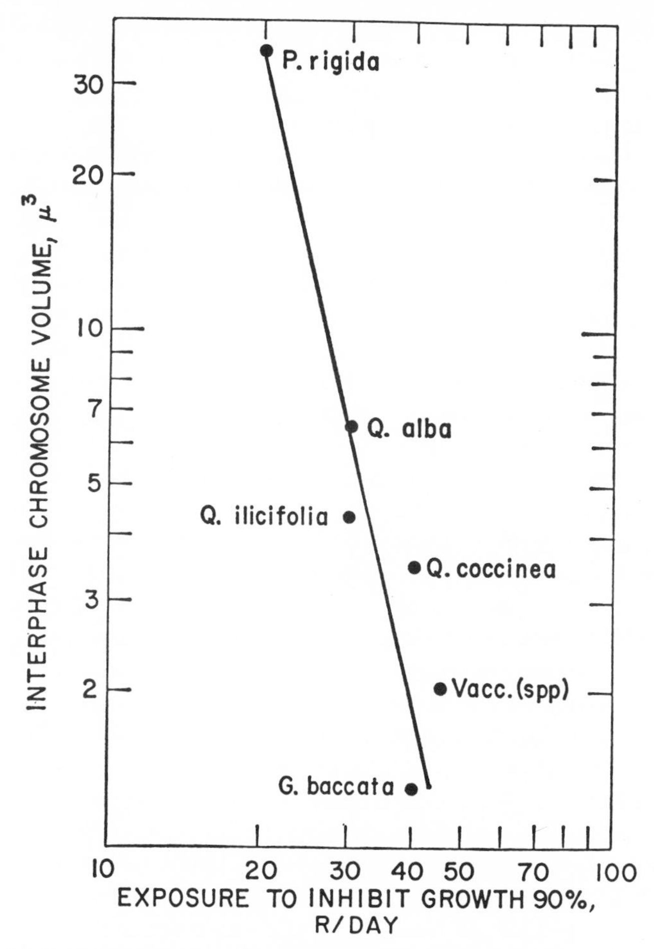

Fig. 25–16. Relation between radiosensitivity, measured as inhibition of growth (90 percent), and interphase chromosome volume in irradiated forest (*21*).

have large chromosomes and because woody species in general are more sensitive than herbaceous plants having the same-sized chromosomes (22).

Thus there seems to be a shift toward greater sensitivity to radiation as succession progresses. The shift is due to at least three factors: (i) what I term the relative plasticity of the communities; (ii) increase in the amount of structure in the communities, with its implications for the photosynthesis:respiration ratio; and (iii) changes in the intrinsic characteristics of the plants participating in these communities, including changes in size of chromosomes. All these factors work in the same direction, contributing toward greater sensitivity to radiation and probably to other types of disturbance later in succession.

What do the patterns of radiosensitivity mean? Could they be sheer coincidence, on the one hand, or a useful new clue to the mechanisms of evolution on the other?

It is difficult to discard them as mere coincidence: true, they are imperfect: there are radiation-resistant plants in the forest and radiosensitive plants in the field; furthermore, the pines, the most sensitive of all, are a minor part of the mature forest, and the pattern of increasing radiosensitivity along succession is imperfect in detail. Yet the difference in sensitivity between field and forest spans a factor of nearly 10; if we include lichen communities, which sometimes precede herbaceous communities in succession, there is a factor of 10 of additional resistance beyond that of the herbs (23). And the correlations between life-form and size and radiosensitivity, and the parallels between radiation effects and known effects of environmental gradients are too strong to be set aside lightly. There is no evidence at all that the enormous range of radiosensitivities among higher plants correlates in any way with the distribution of radioactivity in nature; nor is there reason to believe that radiation levels have changed appreciably during the quarter-billion years or so of existence of the higher plants. Rather, it seems that we must look further for other environmental factors or combinations of factors that have affected the evolution of that constellation of characteristics we measure when we measure radiosensitivity, including especially chromosome size.

It is an intriguing if somewhat over-simplified hypothesis that sensitivity to radiation damage is a measure of sensitivity to environmentally induced mutation (mutation is used in its broadest sense). It seems reasonable to accept the concept that rates of mutation tend toward some optimum, which is under hereditary control (24). If the rate were too high, there would be reduction in reproductive success; if too low, there would be insufficient variability to meet the evolutionary demands of constantly changing environments.

Certainly it is conceivable that environments vary in capacity to produce mutations. If, on the basis of current evidence about mutations, one were to seek a mutagenic natural environment (independent of radiation intensity), he would probably seek one characterized by extremes: extremes of temperature, moisture availability, and solar radiation. One thinks immediately of surfaces exposed to the sun: soil, rock, bark. The evidence that I report suggests strongly that plants that normally inhabit such surfaces—algae, lichens, mosses, and prostrate-growing vascular plants—are more resistant to ionizing radiation (and doubtless to many other stresses) than plants of more ameliorated environments such as forests.

Whether this suggestion will prove to be true when examined in a larger context than has yet been possible remains to be seen. Nonetheless, we now infer that ability to survive such rigorous environment conditions also confers in some degree, at least, resistance to ionizing radiation. The factors that confer resistance involve growth form, length of life cycle, regenerative capacity, and cytological characteristics, especially average interphase chromosome volume. Experimental examination of this question is a current challenge to radiation research. Only by willingness to look at such really tough questions will we gain further insight into both radiation and the patterns of nature.

REFERENCES AND NOTES

1. D. Lack, *Darwin's Finches* (Cambridge Univ. Press, London, 1947).
2. This series of general statements is a synthesis of concepts current in ecology; for further discussion see R. Margalef, *Amer. Naturalist* 97, 357 (1963); F. E. Clements, *Plant Succession and Indicators* (Wilson, New York 1928).
3. Largely because of the efforts of A. H. Sparrow and colleagues at Brookhaven National Laboratory.
4. This work has been recently summarized: A. H. Sparrow, R. C. Sparrow, K. H. Thompson, and L. A. Schairer, in *Proc. Use of Induced Mutations in Plant Breeding* (Pergamon, Oxford, 1965), pp. 101–32.
5. A. H. Sparrow and G. M. Woodwell, *Radiat. Bot.* 2, 9 (1962).
6. By a 3-year study (Emory Univ., Atlanta; directed by R. B. Platt) of the effects of radiation on vegetation surrounding an unshielded reactor in Georgia.
7. R. B. Platt, in *Ecological Effects of Nuclear War, BNL 917 (C-43)*, G. M. Woodwell, Ed. (Brookhaven National Laboratory, Upton, N.Y., 1965), pp. 39–60.
8. By A. H. Sparrow and colleagues, Brookhaven National Laboratory.
9. A. H. Sparrow, in *Large Radiation Sources in Industry* (Intern. Atomic Energy Agency, Vienna, 1960), Vol. 3, pp. 195–219; ——— and W. R. Singleton, *Amer. Naturalist* 87, 29 (1953).
10. G. M. Woodwell, *Radiat. Bot.* 3, 125 (1963).
11. ——— and J. K. Oosting, *ibid.* 5, 205 (1965).
12. G. E. Bard, *Ecol. Monographs* 22, 195 (1952); H. J. Oosting, *Amer. Midland Naturalist* 28, 1 (1942).
13. M. L. Fernald, *Gray's Manual of Botany* (American Book, New York, 1950).
14. G. M. Woodwell, *Science* 138, 572 (1962).
15. ——— and A. L. Rebuck, *Ecol. Monographs,* 37, 53 (1967).
16. J. H. Brower, dissertation, Univ. of Massachusetts, 1964.
17. G. M. Woodwell and J. H. Brower, *Ecology,* in press.
18. M. F. Buell and J. E. Cantlon, *ibid.* 34, 520 (1953).
19. R. D. Brayton and G. M. Woodwell, *Amer. J. Bot.* 53, 816 (1966).
20. G. M. Woodwell and P. Bourdeau, in *Proc. Symp. Methodol. Plant Ecophysiol.* (UNESCO, 1965), pp. 519–27.
21. G. M. Woodwell and A. H. Sparrow, *Radiat. Bot.* 3, 231 (1963).
22. R. C. Sparrow and A. H. Sparrow, *Science* 147, 1449 (1965).
23. G. M. Woodwell and T. P. Gannutz, *Amer. J. Bot.*, in press.
24. J. F. Crow, *Sci. Amer.* 201, 138 (1959); M. Kimura, *J. Genet.* 52, 21 (1960).
25. R. H. Wagner (Brookhaven National Laboratory), unpublished data.

26. Research carried out at Brookhaven National Laboratory under the auspices of the AEC. Many have contributed for years to this work. I thank especially F. H. Bormann of Yale University, R. H. Whittaker of the University of California, Irvine, and A. H. Sparrow of Brookhaven National Laboratory for many long discussions.

26

Bioaccumulation of radioisotopes through aquatic food chains

J. J. Davis
R. F. Foster

INTRODUCTION

With an increasing number of atomic energy installations and their associated problems of disposal of liquid wastes, we recognize that more and more aquatic environments are going to be exposed to at least low concentrations of radioactive materials. For the safety of human populations who may be drinking water which contains such radioactive materials, a set of maximum permissible concentrations has been recommended (International Commission on Radiological Protection, 1955). By themselves, however, such recommendations are inadequate to define completely the radiological hazard which may develop through aquatic food chains. Where biological systems are involved, the organisms may accumulate certain isotopes to many times the initial concentrations in the water. There are many radioisotopes, however, that apparently are not biologically concentrated.

This paper describes some of the mechanisms involved in the accu-

Reprinted with permission from *Ecology*, 39(3):530–535 (1958). R. F. Foster is now Manager, Water and Land Resources Section, Battelle-Northwest, Richland, Washington. This paper is based on work performed under contract No. W-31-109 Eng-52 for the U.S. Atomic Energy Commission. The authors wish to express their gratitude to Dr. H. A. Kornberg for stimulating interest in the mathematical expression of some of the basic concepts, and for his general guidance of the biological studies at Hanford which have furnished most of the examples used here. We are also grateful to the members of the Aquatic Biology Operation, particularly Mr. P. A. Olson and Mr. D. G. Watson, who contributed data which has not been published elsewhere.

mulation of radioisotopes by aquatic organisms, with special reference to food webs and metabolic rates, and presents some examples of how the concentration of radioisotopes in organisms can be used to measure relationships between different species.

THE ACCUMULATION OF RADIOACTIVE MATERIALS

In order to interpret the reasons for, or to predict the concentration of, radioactive substances in aquatic forms, the biologist must appreciate that several basic processes are involved. The most important are: (1) the mode of uptake, which includes adsorption to exposed areas, absorption into tissues, and assimilation of ingested material; (2) retention, which is a function of the biochemistry of the particular elements and components involved, the site of deposition, the turnover rate, and the radioactive half-life; and (3) the mode of elimination, which may involve ion exchange, diffusion, excretion, and defecation.

Mode of Uptake

The metabolism of the different radioelements and the relative importance of the different modes of uptake will fluctuate widely between different species, environments, and seasons. While this paper is principally concerned with assimilation through food chains, the processes of adsorption and absorption of radioactive substances directly from the water cannot be neglected. They are primary mechanisms by which inorganic materials are acquired by aquatic plants which are the food sources of the animals. The absorption of radioisotopes of strontium, barium-lanthanum and sodium by fresh-water fish has been demonstrated by Prosser *et al.* (1945). Absorption of radiocalcium has been demonstrated by Lovelace and Podoliak (1952) and by Rosenthal (1956). Chipman (1956) showed that cesium readily passed through excised pieces of tuna skin but that there was little absorption of strontium or ruthenium from sea water. Fish immersed in effluent from the Hanford reactors concentrated Na^{24} in the tissues about 130-fold. Direct absorption of other isotopes which are dominant in the effluent, including Cr^{51}, Cu^{64}, P^{32}, As^{76}, and rare earths, appeared to be inconsequential, however. In fish that live downriver from the Hanford reactors, sorption of radioactive materials directly from the effluents accounts for only about 1.5 percent of the total radioactivity. Consequently, sorption is of much less importance than ingestion in the uptake of radioactive materials by Columbia River fish.

Adsorption occurs almost instantaneously, as has been demonstrated with yttrium on cells of the marine alga *Carteria* by Rice (1956), while equilibrium by absorption is usually reached by algal cells (Whittaker, 1953) and by vascular aquatic plants (Hayes, *et al.*, 1952) within a few hours. Because of the rapid uptake of radioisotopes by these mechanisms, Columbia River plankton, composed almost entirely of diatoms, appears to reach equilibrium about one hour after floating into the zone containing effluent from the Hanford reactors (Foster and Davis, 1955).

Assimilation of ingested materials is the dominant means by which many radioactive materials become accumulated in animals since the bulk of their essential elements is obtained from their food. The contribution of food webs to the concentration of radioisotopes in aquatic animals was apparent from samples collected from the Columbia River soon after the first Hanford reactors began operation. Fish collected downriver from the reactors were approximately 100 times as radioactive as laboratory fish that were exposed to equivalent mixtures of the effluent, but fed uncontaminated food. Bottom animals, particularly herbivorous insect larvae, were found to be even more radioactive than the fish. The concentrations of radioactive materials in Columbia River organisms have never approached hazardous levels, however.

Differences Between Species

The relative concentrations of beta emitters in various Columbia River organisms are shown in Figure 26–1. There are several reasons for the differences which occur between these species.

(1) Several radioisotopes are involved and their relative proportions

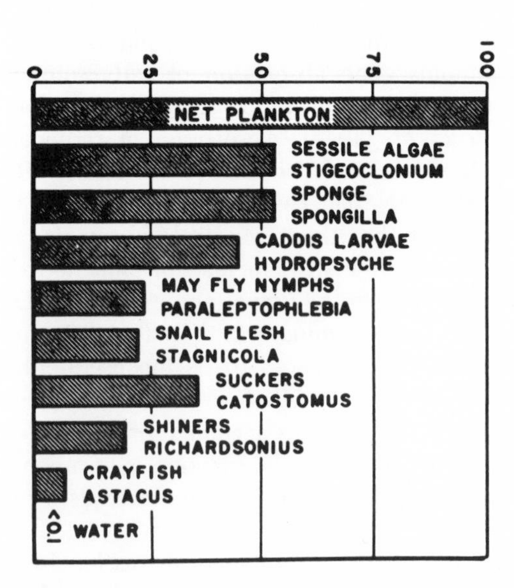

Fig. 26–1. Radioactivity in different Columbia River organisms.

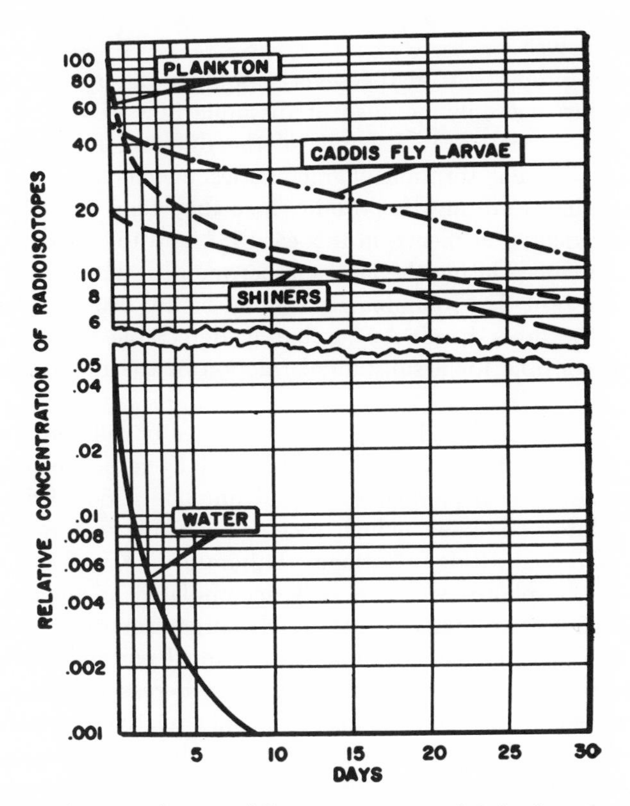

Fig. 26–2. Radioactive decay in different organisms and Columbia River water.

are different in the various organisms. A good indication of the proportions of the several isotopes can be obtained from curves like those in Figure 26–2 which show the characteristics of the radioactivity decay of the isotope mixtures peculiar to each species. The positions of the curves in Figure 26–2 at zero time approximate the relative concentrations of radioisotopes in the water, small fish [*Richardsoninus balteatus* (Richardson], caddis larvae [*Hydropsyche cockerelli*, Banks], and plankton of the Columbia River during the late summer months. The predominance of short-lived isotopes in the water is shown by the steep slope of the bottom curve. Short-lived emitters also contribute most of the radioactivity in the plankton but these have virtually disappeared by the fifth day. The remaining activity in the plankton, which is only about 20 percent of that originally present, emanates from P^{32} and other isotopes with half-lives greater than two weeks. In the caddis larvae and fish, only about 5 percent of the initial radioactivity originates from short-lived emitters. After the first day the rate of decay is quite uniform and characteristic of P^{32} (half-life 14.3 days). The dominance of the P^{32} has been confirmed by radiochemical analysis.

The relative proportions of the several isotopes differ from one organism to another not only because of dissimilarities in the chemical composition and physiological demands of the different forms but also because of the different sorption characteristics which vary with morphology. Food chains are also important since they tend to "select for" isotopes of the essential elements, in this case P^{32}, and to "select against" nonessential elements. During the late summer months, the concentration of P^{32} in small fish of the Columbia River may be 165,000 times that of the water. On the other hand, As^{76} is barely detectable in the fish although it is responsible for a substantial fraction of the radioactivity in the water.

The marked variation in the relative abundance of different isotopes which can occur at different trophic levels and even between similar species has recently been pointed out by Krumholz (1956). From data collected at White Oak Lake, which received a variety of radioactive wastes from the Oak Ridge National Laboratory, Krumholz states: "Although radiophosphorus was generally accumulated in much greater amounts than any other radioelement by the organisms that served as food for the fish, that element made up only a small portion of the total radiomaterials concentrated in the fish tissues; whereas radiostrontium, which was present in the food organisms in only relatively small quantities, was accumulated in high concentrations in the fish skeletons. Furthermore, although the contents of the bluegill stomachs contained more radioactivity, on the average, than those of the black crappies, the crappies accumulated considerably greater amounts of radiomaterials in the hard tissues than the bluegills did. The bluegills, on the other hand, accumulated more radiomaterials in the soft tissues than the black crappies. Both species concentrated radiostrontium in quantities 20,000 to 30,000 times as great as those in the water in which they lived."

(2) Variation in moisture content between different organisms is a second reason for the differences in concentration of radioisotopes shown in Figure 26–1. Chemical composition is also a factor since we are actually concerned with the quantity of a particular element in a unit mass of live tissue. The percentage of the live weight of the Columbia River plankton, caddis larvae, and minnows which is contributed by the inorganic ash is respectively 16, 2.2, and 3.0; and the concentration of phosphorus in the living organisms is about 150 p.p.m. for plankton, 2,000 p.p.m. for caddis larvae, and 6,000 p.p.m. for minnows. Even greater differences may occur between the different tissues of an individual. Figure 26–3 shows how the concentration of radioactive materials varies between different tissues of whitefish in the Columbia River. Since virtually all of

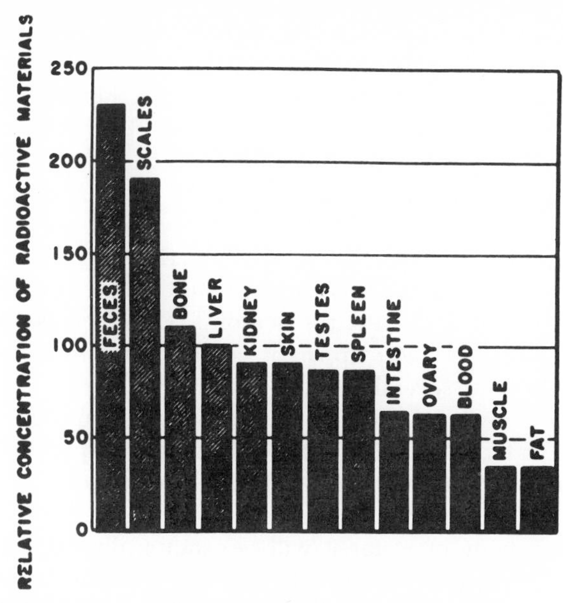

Fig. 26–3. Radioactivity in different tissues of Columbia River fish.

the activity is from P^{32}, this gives a good indication of the relative concentration of phosphate in the different tissues.

(3) A third reason for differences in concentrations of radioisotopes between different organisms is their relative position on the food pyramid. Although elements are exchanged continuously between the water and the organisms of a food web, there is a mean retention time for each element in each organism. Each trophic level thus serves as a kind of pool or reservoir in which essential elements are retained for some mean length of time before they are passed on to the next level. The size of each pool will be governed by the total amount of an element held by the entire biotic mass making up the particular trophic level. A major fraction of most radioactive contaminants accumulated by aquatic life will be held by the plankton and benthic algae because of their relatively large total mass. Rigler (1956) found that over 95 percent of the P^{32} added to a lake was taken up by plankton (including bacteria) within 20 minutes. But retention time is not necessarily a function of the size of the pool. Indeed it is more apt to be inversely related since most elements will remain for a longer time in the larger organisms than in the small plant forms, although the small plants constitute the largest pool. Since, in the Columbia River, we are dealing with a flowing stream where isotopes are added at a more or less constant rate, much of the mineral exchange system can be considered as a once-through process rather than a cycle. Some radioactive decay will occur while the isotopes are retained in each

trophic level. This decay, and thus the effective retention period, should be measurable by a progressive decline in specific activity—the concentration of an isotope per unit mass of the element. For example, under certain conditions midge larvae in the Columbia River may contain on the order of 4 μc P^{32}/g of P and the small fish which eat the midge larvae about 0.5 μc P^{32}/g of P. Since the half-life of P^{32} is two weeks, the phosphorus deposited in the fish must be, on the average, about six weeks "older" than that in the midge larvae. The relative "age" of the isotope will differ between species ar . will change with the age, size, and growth rate of the individual and with the seasons. The decrease in specific activity will, of course, be more apparent for short-lived isotopes than for those with half-lives of several weeks or more.

The specific activity of the river biota should be appreciably lower than that of the water not merely because of the time required to incorporate the isotope into the organisms but also because of the "pools" of elements fixed in the biota and sediments. When a radioisotope is first introduced into a body of water it will be isotopically diluted with the stable form of the element which is dissolved in the water. Soon, it also will become isotopically diluted by exchange with the stable form of the element which has not been in solution. With a single addition of isotope into a "static" environment, the specific activity will eventually become uniform throughout the biota. Reservoirs of phosphates in the solids of lakes have been described by Hutchinson and Bowen (1950), who have studied phosphorus exchange with the use of P^{32}.

Rate of Accumulation by Aquatic Organisms

The nearly instantaneous uptake of isotopes by adsorption and the rapid uptake by absorption have been mentioned. When animals are chronically feeding on radioactive materials, the rate at which their concentration of the isotopes approaches equilibrium will be a function of the radioactive and biological half-lives of the particular isotope involved.

Figure 26–4 shows the rate at which caddis fly larvae (*Hydropsche cockerelli*) accumulated radioactive materials (mostly P^{32}) when fed filamentous algae (mostly *Spirogyra*) that had been cultured in reactor effluent. If there was no biological turnover of phosphorus in the caddis larvae, the time required to reach some fraction of the equilibrium level would be a function of the radioactive decay constant and could be predicted from the equation:

$$\frac{Q_t}{Q_e} = 1 - e^{-\lambda t}$$

where Q_e is the amount of the isotope present at equilibrium, Q_t is the amount present at some time (t) before the equilibrium is reached, and λ is the radioactive decay constant.

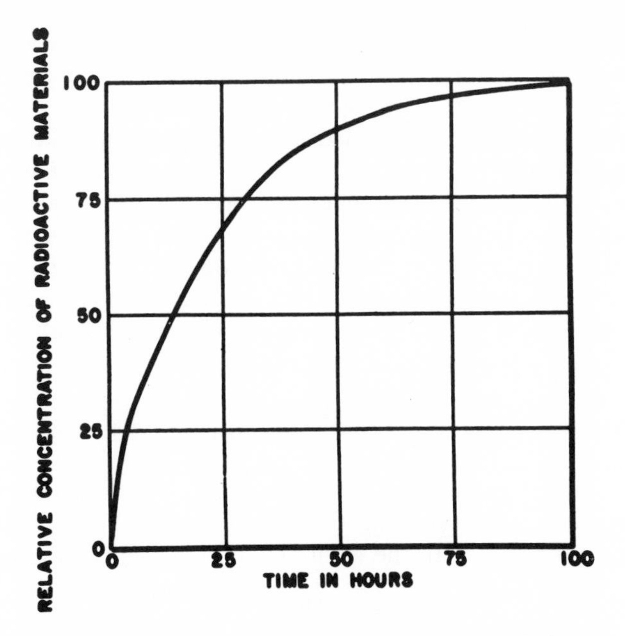

Fig. 26–4. Rate of accumulation of effluent isotopes by caddis fly larvae.

Since true equilibrium will only be reached after infinite time, we can consider practical equilibrium to occur when $Q_t = 0.9\ Q_e$, and solve the equation for t. For any isotope, t will be equal to the half-life multiplied by $\frac{-\ ln\ 0.1}{.693}$. For P^{32} it is approximately 47 days.

The curve presented in Figure 26–4 shows a much shorter time which indicates that significant biological turnover is present. The equation is easily modified to take this into account:

$$\frac{Q_t}{Q_e} = 1 - e^{-\ \delta t}$$

where δ is the sum of λ and β, where β is the constant for biological half-life. At 0.9 of equilibrium,

$$t = \frac{-\ ln\ 0.1}{\beta + \lambda}$$

From Figure 4, t is about 50 hours and

$$50 = \frac{2.302}{\frac{.693}{T_b} + \frac{.693}{343}}$$

(343 is the half-life of P^{32} in hours).

The biological half-life, T_b, is about 16 hours. Under such conditions the specific activity of the P^{32} will not diminish appreciably at this trophic level.

If laboratory tests can be carried out in conjunction with field observations, some interesting ecological relationships can be deduced. For example, we might measure the concentration of P^{32} in small minnows collected from a contaminated environment and find this to be 2×10^{-3} μc/gram. If the average size of the minnows was 5 grams, then each fish would have a body burden of about 10^{-2} μc of P^{32}. From a laboratory test, which duplicated field conditions as closely as possible, we might find that 0.9 of Q_e was reached in 20 days. Then

$$\delta = \frac{-\ln 0.1}{t} = \frac{2.302}{20} = 0.115$$

The same test might show that half of the ingested P^{32} was assimilated and deposited ($a = 0.5$). Assuming the concentration of P^{32} in the river fish to be in equilibrium with the environment and neglecting growth,

$$Q_e = \frac{aq}{\delta}$$

where q is the quantity of P^{32} ingested per unit time—in this case each day. Then,

$$10^{-2}\ \mu c = \frac{0.5\, q}{0.115}$$

and

$$q = 2.3 \times 10^{-3}\ \mu c/\text{day}.$$

In order to have reached the observed concentration of P^{32}, each minnow must have consumed about 2.3×10^{-3} μc of P^{32} each day. If, from stomach analyses, we have found that the fish feed predominantly on midge larvae and from field collections we have found that the midge larvae have a concentration of P^{32} of about 10^{-2} μc/g, then we can surmise that each minnow has been eating about 0.23 grams of the midge larvae each day.

Seasonal Variations

Since most aquatic animals are poikilothermic, their metabolic rates, and thus their feeding rates, change with variations in temperature and so with the seasons. For those aquatic forms that accumulate radioactive substances principally via ingestion, the concentration of radioisotopes fluctuates with metabolic rate. Figure 26–5 shows the seasonal fluctuations which occur in the radioactivity of plankton (diatoms) and minnows (*Richardsonius balteatus*) in the Columbia River. Fluctuations in plankton are quite similar to those in the water since the radioisotopes are

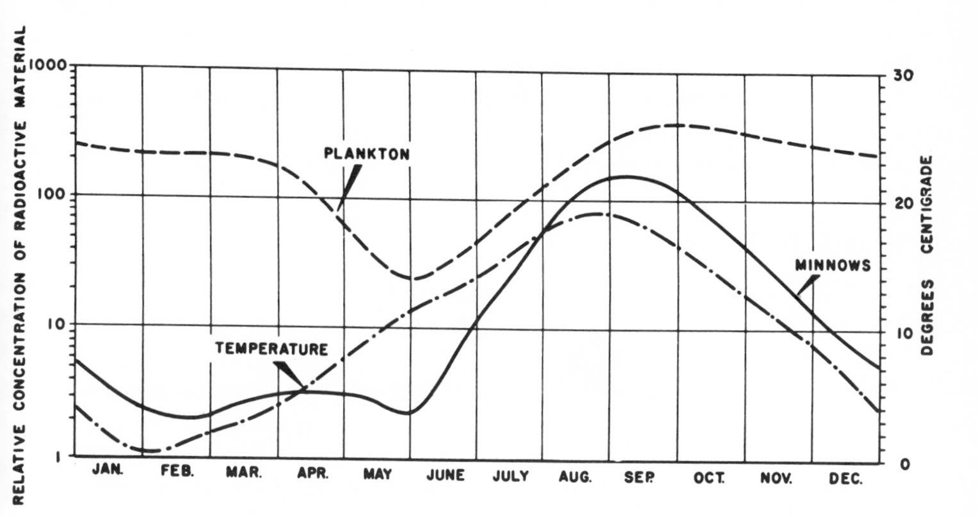

Fig. 26–5. Seasonal fluctuations in radioactivity of Columbia River organisms.

acquired by direct absorption and adsorption (Foster and Davis, 1955). On the other hand, fluctuation in the radioactivity of the minnows is more closely related to the temperature. The 75-fold increase in concentration of radioisotopes in the fish between winter and late summer does not mean simply that the fish are eating 75 times as much food. The seasonal fluctuations result from the interaction of all of the factors mentioned above which influence the accumulation of radioactive materials. As the feeding rate increases for each organism, its intake of radioisotopes may be disproportionately large. The consumer is not only eating more grams of food, but each food organism has become more radioactive, and the effective time intervals between trophic levels have become less. Possibly the food habits of the species in question have also changed. A complete evaluation of the seasonal fluctuations in any one species would require an immense amount of work, not only on the food habits of the species but also on its physiology and on the radioactive contamination of its food organisms.

Not all seasonal variations are associated merely with temperature since deviations may occur where complex life cycles are involved. This occurs in immature insects which are less radioactive during quiescent periods than when the larvae or nymphs are feeding. It is also true of salmon that return to the Columbia River to spawn. The adult salmon virtually stop feeding when they enter fresh water, and consequently pick up very little radioactive material. Krumholz (1956) also observed definite seasonal changes in the accumulation of radiomaterials by fish of White

Oak Lake. These corresponded to some extent with seasonal changes in temperature. He noted, however, that the accumulation of radioisotopes in black crappie and bluegills stopped at the first of August when the temperature reached about 80° F. He attributes the rapid loss of radioactive materials during August and September to a period of summer dormancy for these species.

SUMMARY

Some radioactive materials introduced into aquatic environments may be accumulated by the organisms. The amount of accumulation will vary over many orders of magnitude depending upon the kinds of isotopes involved and many physical, chemical, and biological factors. Such concentration is of considerable importance in the control of radiological hazards and the aquatic biologist has definite responsibilities in this area.

The processes of adsorption and absorption are of major importance in the uptake of radioisotopes by plants but appear to be of less importance than the food chain in the uptake by aquatic animals. The concentration of radioactive substances will vary between species and tissues and will fluctuate according to food habits, life cycles, and seasonal changes.

Within the biotic mass, a major fraction of most radioactive contaminants will be held by the organisms which make up the primary trophic levels. In a flowing stream, the specific activity of a radioisotope will diminish along the food chain. Where the turnover rates of certain isotopes can be measured, inferences can be drawn on feeding habits.

REFERENCES

Chipman, Walter A. 1956. Passage of fission products through the skin of tuna. U.S. Fish and Wildlife Service Special Scientific Report—Fisheries No. 167.

Foster, R. F. and J. J. Davis. 1955. The accumulation of radioactive substances in aquatic forms. Proceedings of the International Conference on the Peaceful Uses of Atomic Energy, 13(P/280):364–367.

Hayes, F. R., J. A. McCarter, M. L. Cameron, and D. A. Livingstone. 1952. On the kinetics of phosphorus exchange in lakes. Jour. Ecol. 40:202–216.

Hutchinson, G. E. and V. T. Bowen. 1950. Limnological studies in Connecticut. IX. A quantitative radiochemical study of the phosphorus cycle in Linsley Pond. Ecology 31:194–203.

International Commission. 1955. Recommendations of the international commission on radiological protection. British Jour. of Radiology, Suppl. No. 6, 92 pp.

Krumholz, L. A. 1956. Observations on the fish population of a lake contaminated by radioactive wastes. Bull. Am. Mus. Nat. Hist. 110:277–368.

Lovelace, F. E. and H. H. Podoliak. 1952. Absorption of radioactive calcium by brook trout. The Progressive Fish-Culturist 14:154–158.

Prosser, C. L., W. Pervinsek, Jane Arnold, G. Svihla, and P. C. Thompkins. 1945. Accumulation and distribution of radioactive strontium, barium-lanthanum, fission mixture and sodium in goldfish. USAEC Document MDDC-496:1–39.

Rice, T. R. 1956. The accumulation and exchange of strontium by marine planktonic algae. Limnology and Oceanography 1:123–138.

Rigler, F. H. 1956. A tracer study of the phosphorus cycle in lake water. Ecology 37: 550–562.

Rosenthal, H. L. 1956. Uptake and turnover of calcium-45 by the guppy. Science 34: 571–574.

Whittaker, R. H. 1953. Removal of radiophosphorus contaminant from the water in an aquarium community. *In:* Biology Research—Annual Report 1952. USAEC Document HW-28636:14–19.

27

Cesium-137 in Alaskan lichens, caribou and eskimos

W. C. Hanson

Cesium-137 is generally recognized as a most important constituent of worldwide radioactive fallout resulting from nuclear weapons tests. Its importance is nowhere greater than in the arctic and subarctic regions of the northern hemisphere, where the unique features of the lichen-caribou (reindeer)-man food chain have created a situation of increasing interest to workers in the fields of radiation ecology and radiological health.

Our radiation ecology studies in Alaska began in 1959 and have progressed from a description of the general spectra of fallout radionuclides in arctic biota to a definition of seasonal cycles, rates and routes of radionuclide movement within the ecosystems, with particular attention to ^{90}Sr and ^{137}Cs. Cesium-137 has been of greatest importance in these studies, not only because of its public health aspects, but also because its convenient gamma emission and biological concentration permits effective study of its transfer within the natural systems. The results of these studies and those of other northern regions have been remarkably consistent. The ^{137}Cs situation in the arctic regions may be briefly described as depending upon (1) the extremely effective retention of ^{137}Cs by lichens, (2) the importance of lichens as a winter food for caribou and reindeer, and (3) the dependence upon caribou and reindeer for food by many northern peoples.

Reproduced from *Health Physics*, 13:383–389 (1967) by permission of the Health Physics Society. This paper is based on work performed under United States Atomic Energy Commission Contract AT(45–1)–1830.

The transfer of ^{137}Cs up the food chain in this series of steps or stages, known as trophic levels, illustrates the simple and thus delicately balanced structure of arctic ecosystems.

CESIUM-137 IN LICHENS

It is apparent that lichens represents a most important reservoir of ^{137}Cs and other fallout radionuclides because of their longevity (decades, even one century), persistence of aerial parts, and their dependence upon nutrients dissolved in precipitation (1). These properties result in a tenacious holding of ^{137}Cs. Recent field experiments in Alaska utilizing ^{134}Cs

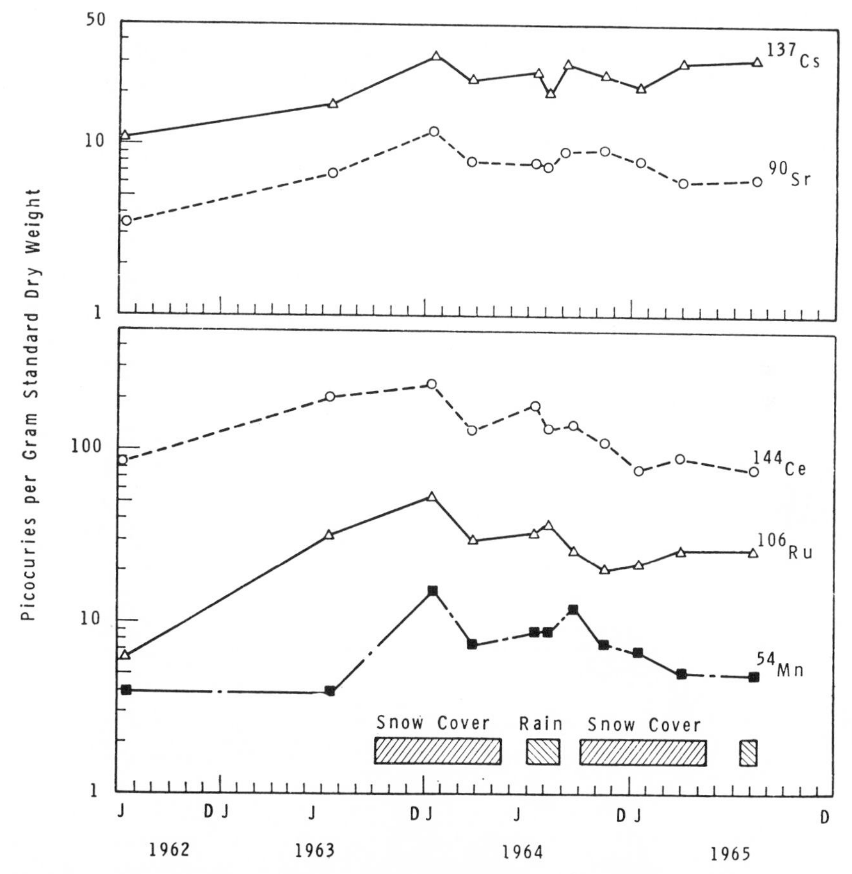

Fig. 27–1. Concentrations of ^{54}Mn, ^{90}Sr, ^{106}Ru, ^{137}Cs and ^{144}Ce in *Cladonia-Cetraria* lichens at Anaktuvuk Pass, Alaska during 1962–1965.

as a tracer suggest a biological half-time of about 13 years for cesium in lichens (2). This value is median to half-time estimates of 17 years (3) and 6–10 years (4) obtained by comparison of changes in ^{137}Cs levels in lichens and ^{137}Cs deposition in northern Scandinavia.

Further evidence recently reported (5) showed that ^{137}Cs applied to the green living top part of lichens remained there rather than being translocated throughout the plant, as was ^{90}Sr. The continuing long-term accumulation of ^{137}Cs in Alaskan lichens is shown in Figure 27–1, which illustrates that concentrations of other fallout radionuclides ^{54}Mn, ^{90}Sr, ^{106}Ru and ^{144}Ce have decreased since early 1964, particularly during periods of snow cover when the lichens were shielded from direct fallout deposition. A comparison of ^{137}Cs concentrations in eight lichen species suggests that ^{137}Cs levels tend to be ranked according to a species' ecological "niche," structural diversity and growth pattern. This is illustrated here by three examples (Fig. 27–2).Highest concentrations occur in *Cornicularia divergens,* which grows on windswept ridgetops that are nearly free of snow during winter and are first snow-free in the spring; intermediate levels were found in *Cetratia cuculata* collected from the sides

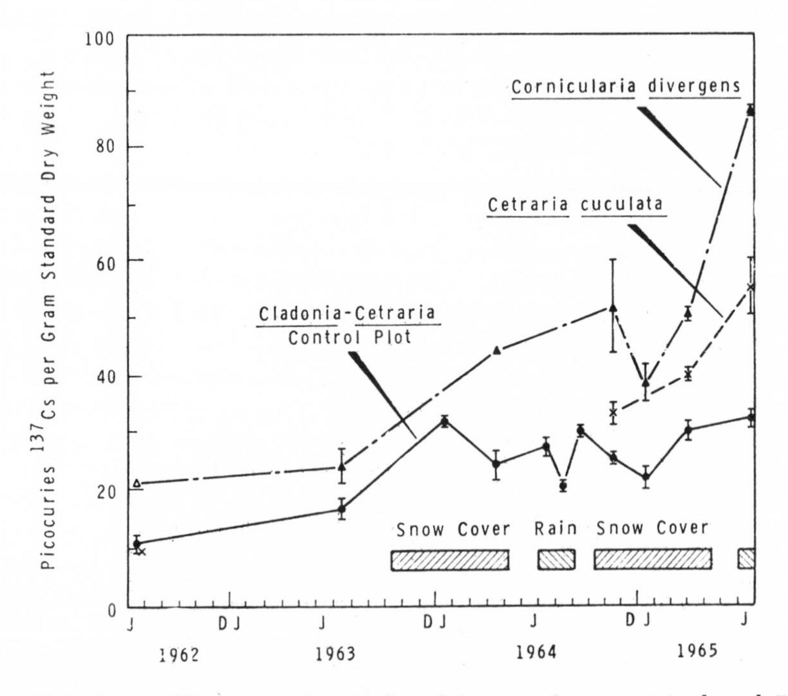

Fig. 27–2. Cesium-137 concentrations in three lichen sample types at Anaktuvuk Pass, Alaska during 1962–1965. Values shown are mean ± one standard error.

of the ridges and which were covered by an average of 5 cm of snow for about 6 months each year, and least amounts were in mixed *Cladonia-Cetraria* samples which were snow-covered to a depth of 2–3 dm during winter.

The quantities of fallout ^{137}Cs per square meter of *Cladonia-Cetraria* lichens at a site in Anaktuvuk Pass, Alaska during the month of July of the years 1962–1965 were 18, 28, 41, and 48 nCi, respectively. These values are nearly identical to those reported from northern Sweden (3) at comparable times and are slightly greater than cumulative ^{137}Cs fallout estimates obtained from HASL soil and precipitation samples from other Alaskan locations (6, 7). These data reinforce other observations that lichen mats are efficient accumulators of fallout and may well serve as natural fallout meters (8).

CESIUM-137 IN CARIBOU AND REINDEER

The heavy utilization of lichens by caribou and reindeer as a basic winter food results in the rapid accumulation of ^{137}Cs levels in flesh to levels that exceed those of other arctic herbivores.

Within any one herd of Alaskan caribou there has usually been a coefficient of variation of about 30 percent in concentrations of ^{137}Cs (9). Finnish investigators have recently reported greater variability among ^{137}Cs concentrations in male reindeer flesh compared to females due to physiological parameters (4). Controlled metabolic studies of orally administered ^{137}Cs in reindeer suggest that muscle physiological activity may influence accumulation and distribution (10).

Caribou and reindeer samples collected from several locations in northern Alaska suggest that animals from the Noatak and Kobuk River drainages may contain higher ^{137}Cs concentration than do animals from other locations, although it is difficult to characterize ^{137}Cs concentrations among the several caribou and reindeer herds in Alaska because of the extensive migrations of the animals and large area involved. There is, however, a definite seasonal cycle of ^{137}Cs concentrations in caribou and reindeer, with maximum values during the winter period when lichens are their main food and minima during the summer pasturing period (3, 5, 8, 11, 12).

The rapid decrease of flesh ^{137}Cs concentrations with the shift to summer diets, as well as other experimental data, indicates a biological half-time of about 3–5 weeks in caribou and reindeer. Recent experiments in the USSR have been cited (5) in which 50 percent of the ^{137}Cs body burden had a biological half-time of 1.5 days and 50 percent had a half-time of 20 days. Thus, ^{137}Cs concentrations in Alaskan caribou flesh often have varied over a threefold range from winter high to summer low, while ^{137}Cs levels in lichens were essentially unchanged. Maximum ^{137}Cs values in caribou flesh have been consistently associated with animal samples obtained during April or early May (the spring kill of animals migrating from the winter range). These values usually have been slightly greater than those in caribou collected throughout the winter period near Anaktuvuk Pass, and many represent animals moving northward from ranges containing greater ^{137}Cs concentrations than those of the Anaktuvuk Pass area. The ratio of summer range:winter range caribou flesh values generally has been about 1:3 to 1:10, and is near that reported in Swedish and Russian reindeer herds individually identified and related to their respective ranges (3, 5).

A sharp increase of ^{137}Cs concentrations in caribou flesh sampled during April and May was apparent in 1964 and 1965, and emphasized the possible importance of caribou migration patterns in regulating the ^{137}Cs body burdens in Anaktuvuk Pass Eskimos. These spring-killed animals are the main food source during the summer months when few other game animals are available. Although we have no conclusive proof the increases were due to sampling of animals from distant ranges, we assume they were associated with the caribou migration patterns.

CESIUM-137 IN ESKIMOS

The dependence upon wildlife and other natural resources for substantial amounts of food results in major differences in seasonal and geographic utilization of animals in various villages (13). These differences appear to control the ^{137}Cs levels in the natives, as shown by our initial surveys of radionuclides in northern Alaskan natives and their foods (14, 15, 16). In general there was a linear relationship between reported caribou or reindeer consumption and the ^{137}Cs body burdens. The higher levels usually have been found at the inland village at Anaktuvuk Pass. These people have a very limited economic base and are thus more dependent upon their own resources than other native Alaskan villages. Caribou meat forms about one-half of the total diet at Anaktuvuk Pass, because of necessity and preference to other foods. The Anaktuvuk Pass people represent the last major remnant of nunamiut Eskimos that were formerly nomadic and now remain at their mountain village site. Their culture has always centered on the caribou and the welfare of the people is still highly dependent upon the local environment. Average adults estimate they eat about 5–6 kg of caribou meat per week. Body burdens are about 50–100 times those of persons on an average temperate zone diet.

The seasonal pattern of ^{137}Cs concentrations in Anaktuvuk Pass Eskimos is inversely related to the concentration in the caribou because of the timing of the kill and the stockpiling practices. Animals killed during the fall season are returning from summer ranges and contain low levels of ^{137}Cs. This meat supply forms the base of the winter diet of the Eskimos.

The spring caribou kill is made during April and May as the animals migrate northward from their winter range, at which time they contain the highest ^{137}Cs concentrations of the cycle. This meat is stored in underground caches for use throughout the summer. Maximum ^{137}Cs levels in the Eskimos usually have occurred during July and August and minimums during January.

The relationship of the ^{137}Cs seasonal cycles in caribou, Eskimos and lichens is shown in Figure 27–3. The graph of Eskimo body burdens prior to January 1964 and after August 1965 is drawn point-to-point and does not show seasonal variations defined by the bi-monthly measurements made during the period January 1964–August 1965. Comparison of ^{137}Cs body burdens from one summer to the next shows a pronounced increase of about 50 percent from 1962 to 1963 and a doubling between 1963 and 1964, followed by a decrease of 30 percent from 1964 to 1965. This is consistent with the pattern of ^{137}Cs concentrations in caribou flesh forming the food base of the people. But, ^{137}Cs concentrations in lichens, the main determinants of the level in caribou flesh, have shown a steady increase with time. Thus, the 30 percent decrease from 1964 to 1965 was

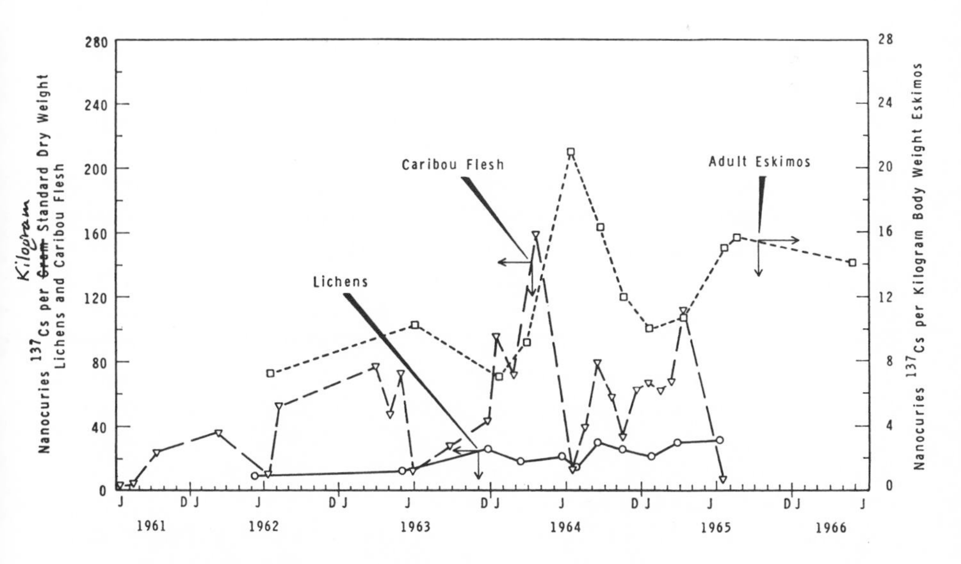

Fig. 27–3. Cesium-137 concentrations in lichens, caribou flesh and Eskimos at Anaktuvuk Pass, Alaska during 1961–1963.

as inconsistent with the ^{137}Cs trend in lichens as was the abrupt increase during the spring of 1964 and emphasizes the probable importance of caribou migration patterns in the determination of ^{137}Cs levels in the Eskimos.

Adult males between the ages of 20 and 50 ordinarily have the highest body burdens of ^{137}Cs, apparently due to their greater proportion of muscle and greater consumption of caribou. Children (3–14 years old) usually contained about half as much ^{137}Cs per kilogram as adults (>21 years old). Minors (15–20 years old) contained about 70 percent as much as adults. Comparison of males and females during the 1964 and 1965 seasonal maxima indicated no important differences in ^{137}Cs body burdens until about the age of 20, males then consistently contained more ^{137}Cs than did females, maximum differences occurred at about the age of 40, and the difference then decreased during advanced age (17).

The biological half-time of ^{137}Cs in adult Eskimos at Anaktuvuk Pass during the summer of 1965 was about 65 days, comparable to that reported for other arctic peoples (18, 19, 20, 21). Nevstrueva *et al.* (5) have reported that measurements of 60 reindeer breeders yielded biological half-times of 97 $\pm$ 50 days in summer and 57 $\pm$ 35 days in winter: but details of the study are not yet available to explain the differences.

Concentrations of ^{137}Cs in various arctic peoples in Finland, Norway, Sweden, the Soviet Union and Alaska are presented in Figure 27–4. These data show that in the three countries where measurements have been performed regularly since 1962, the Finnish Lapps have consistently exhibited the highest values and Alaskan Eskimos have shown the lowest values until recent months. The Swedish investigators are presently defining the seasonal cycle of ^{137}Cs in their northern peoples by frequent counting begun during mid-1965, and initial results appear to follow the same general pattern observed in Alaska during 1964–1965. Values from the three countries were quite comparable during 1962 and 1963, and then diverged during 1964 and 1965. Now a decreasing trend has apparently begun in the Finnish and Swedish Lapps, while the Alaskan Eskimos continue to increase. This contrast may be explained by changes in reindeer herding practices in Scandinavian countries, that may result in the animals regrazing ranges where the top parts of the lichens containing the greater concentrations of ^{137}Cs have previously been removed. The Alaskan trend is consistent with the pattern of steadily increasing ^{137}Cs concentrations in lichens and caribou, which do not occupy such specific herd areas as do reindeer. The single points for Norwegian and Russian reindeer breeders have been included to provide reference to other northern populations.

Wolf flesh obtained from animals of the Anaktuvuk Pass region, presumably dependent upon caribou for their main diet during the period September–March, contained about twice the ^{137}Cs concentration in

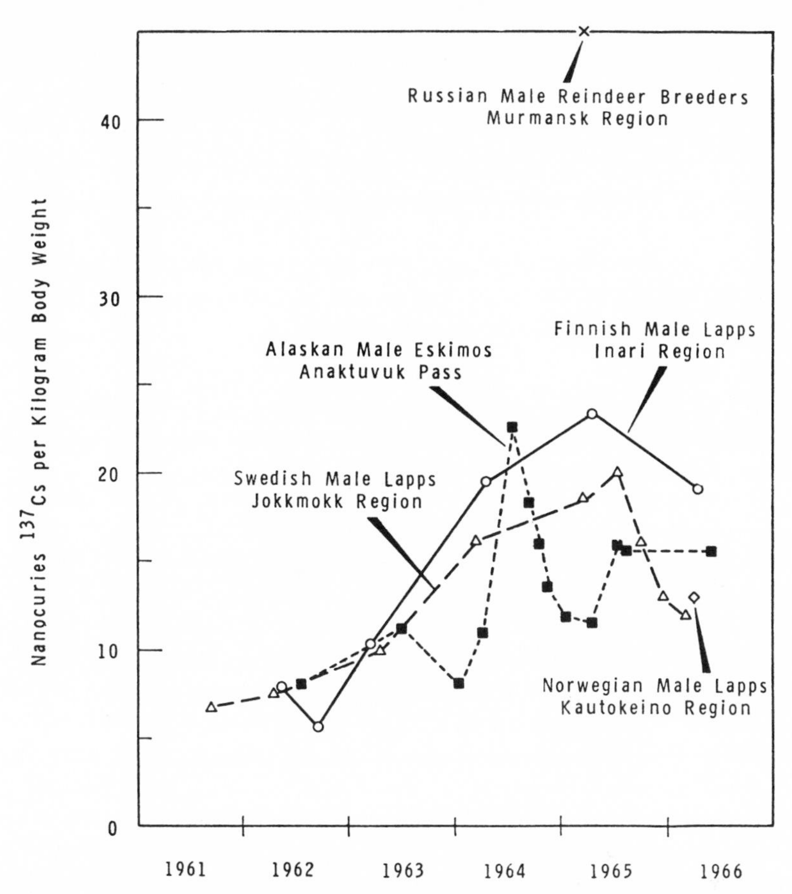

Fig. 27–4. Cesium-137 body burdens in northern human populations of Sweden,(3) Finland,(4) USSR,(5) Norway(22) and Alaska (U.S.A.) during 1961–1966.

caribou flesh. Estimates of ^{137}Cs concentrations in Eskimo muscle were about the same as those for the caribou flesh used for food. Because about half the Eskimo's food was caribou meat, it is reasonable to assign a concentration factor of two to the caribou-carnivore (including man) link of the Alaskan food chain.

SUMMARY AND CONCLUSIONS

Studies of ^{137}Cs in arctic ecosystems have been conducted to provide information about transfer routes and rates of this important fallout radionuclide in a region where wildlife resources are necessary to human welfare. The lichen-caribou (reindeer)-man food chain has been observed to constitute an important ^{137}Cs concentration process. This results from

(1) the unusual capacity of lichens to absorb and retain fallout materials, especially ^{137}Cs; (2) the utilization of lichens for food in winter by caribou and reindeer; and (3) the dependence upon caribou and reindeer for food by several northern populations. Although there are differences in radionuclide concentrations in plants and animals, including man, from various locations within the arctic region, the same general ecological processes have been found to efficiently transfer appreciable amounts of ^{137}Cs through the food chain. Climatic factors, animals behavior, food habits, physiological parameters and human cultural practices have important effects upon ^{137}Cs accumulation and retention.

TABLE 27–1. Cumulative ^{137}Cs fallout in Alaska (nCi/m^2) derived from soil and precipitation samples at Barrow, Fairbanks and Palmer [6,7] and from lichens at Anaktuvuk Pass during 1962–65

	Soil and Precipitation			Lichens
Date	Palmer	Fairbanks	Barrow	Anaktuvuk Pass
July 1962	23	20	8	18
July 1963	32	29	13	28
July 1964	38	33	16	41
July 1965	45	38	16	48

Cesium-137 concentrations increased by a factor of about two at each successive trophic level of the Alaskan lichen-caribou-man(wolf) food chain. The trend with time is one of steady increase, and contrasts with a recent decrease of ^{137}Cs body burdens in Finland and Sweden. Recent measurements in Alaska suggest that ^{137}Cs whole body burdens will continue to increase during the summer of 1966.

REFERENCES

1. D. C. Smith, *Biol. Rev.* 37, 537 (1962).
2. W. C. Hanson, D. G. Watson, and R. W. Perkins, *Proceedings of the International Symposium on Radioecological Concentration Processes*, Stockholm, 1966. Pergamon Press, Oxford (1967).
3. K. Lidén and M. Gustafsson, *Proceedings of the International Symposium on Radioecological Concentration Processes*, Stockholm, 1966. Pergamon Press, Oxford (1967).
4. J. K. Miettinen and E. Häsänen, *Proceedings of the International Symposium on Radioecological Concentration Processes*, Stockholm, 1966. Pergamon Press, Oxford (1967).

5. M. A. Nevstrueva, P. V. Ramsaev, A. A. Moiseev, M. S. Ibatullin, and L. A. Teplykh, *Proceedings of the International Symposium on Radioecological Concentration Processes*, Stockholm, 1966. Pergamon Press, Oxford (1967).
6. E. P. Hardy, Jr. and J. Rivera, HASL Fallout Program—Quarterly Summary Report for June 1, 1964 through September 1, 1964, p. 29, HASL-149. Health and Safety Laboratory, U.S. AEC New York Operations Office (1964).
7. Health and Safety Laboratory, Fallout Program—Quarterly Summary Report, p. 5, HASL-165. U.S. AEC New York Operations Office (1966).
8. G. K. Svensson and K. Lidén, *Health Phys.* 11, 1393 (1965).
9. L. E. Eberhardt, *Nature, Lond.* 204, 238 (1964).
10. L. Ekman and U. Griest, *Proceedings of the International Symposium on Radioecological Concentration Processes*, Stockholm, 1966. Pergamon Press, Oxford (1967).
11. W. C. Hanson and H. E. Palmer, *Health Phys.* 11, 1401 (1965).
12. J. K. Miettinen, *Proceedings of the Third International Conference on the Peaceful Uses of Atomic Energy*, Geneva, 1964, Vol. 14, p. 122. United Nations, New York (1965).
13. C. A. Heller, *J. Am. Dietetic Assoc.* 45, 425 (1964).
14. W. C. Hanson, H. E. Palmer, and B. I. Griffin, *Health Phys.* 10, 421 (1964).
15. W. C. Hanson and H. E. Palmer, *Trans. N. Am. Wildlife and Nat. Resources Conf.* 29, 215 (1964).
16. H. E. Palmer, W. C. Hanson, B. I. Griffin, and L. A. Braby, *Science* 147, 620 (1965).
17. W. C. Hanson, *Proceedings of a Conference on the Pediatric Significance of Peacetime Radioactive Fallout*, San Diego, 1966. American Academy of Pediatrics, Chicago. In press.
18. J. K. Miettinen, A. Jokelainen, P. Roine, K. Lidén, Y. Naversten, G. Bengtsson, E. Häsänen, and R. C. McCall, *Ann. Acad. Sci. Fennicae AII. Chemica* 120, 35 (1963).
19. L. G. Bengtsson, Y. Naversten, and K. G. Svensson, *Assessment of Radioactivity in Man*, Vol. II, p. 21. I.A.E.A., Vienna (1964).
20. K. Lidén, *Assessment of Radioactivity in Man*, Vol. II, p. 33. I.A.E.A., Vienna, (1964).
21. Y. Naversten and K. Lidén, *Assessment of Radioactivity in Man*, Vol. II, p. 79, I.A.E.A., Vienna (1964).
22. K. Madshus, T. Berthelsen, and E. Westerlund, *Proceedings of the International Symposium on Radioecological Concentration Processes*, Stockholm, 1966. Pergamon Press, Oxford (1967).

PROBLEMS OF ENVIRONMENTAL POLLUTION

AIR POLLUTION

28

Damage to forests from air pollution

George H. Hepting

The Clean Air Act (10), passed by the 88th Congress in December 1963, states "that the growth in the amount and complexity of air pollution brought about by urbanization, industrial development, and the increasing use of motor vehicles, has resulted in mounting dangers to the public health and welfare, including injury to agricultural crops and livestock, damage to and the deterioration of property, and hazards to air and ground transportation."

This recognition by Congress, of the hazards imposed by the various forms of air pollution to our health and welfare indicate that they can no longer be ignored and no longer be considered local problems with which public agencies need not become involved. Until recent years most of our streams were treated as common sewers into which virtually any liquid waste could be dumped. This situation resulted in a multitude of health, wildlife, and other water use problems that have led to the widespread adoption of strong measures for stream pollution abatement. The public has been slower to recognize the dangers of polluting the atmosphere. So long as acute impacts remained isolated cases, dealt with locally, settled by agreement or litigation, or considered the result of meteorological acts of God, the air has been widely used in the "common sewer" sense with respect to gaseous emissions.

In recent years, however, there has been a slow but steady increase in public awareness of the chronic buildup of pollution levels in many

Reprinted with permission from the *Journal of Forestry*, 62:630–634 (1964). The author is currently principal research scientist, Forest Disease Research, U.S. Forest Service, Asheville, N.C.

parts of the world. This has been due in part from episodes like the Donora Pennsylvania, fumigation of 1948, the London acute smog of 1952, the smog effects of the Los Angeles area, and from the clear evidence of chronic damage, not only to health but to property, to products of many kinds including paint and rubber, and to agriculture and forestry.

The current literature on the causes and effects of air pollution is enormous and new periodicals on the subject have been appearing at a rapid rate. Good summaries of our air pollution problems were presented by McCabe in 1952, who assembled the contributions to the deliberations of the United States Technical Conference on Air Pollution (24), by Stern, who edited an excellent compendium (35), by several authorities in the monograph "Air Pollution," released by the World Health Organization in 1961 (40), and by the Proceedings of the two National Air Pollution Conferences (37, 38). The possible effects of ionizing radiations on plants are not included in the present review.

Information on the many types of air pollution damage to economic plants has been brought together recently by Thomas (36), Middleton (25), and others (1, 20). In the past the principal pollutants have been oxides of sulfur from industrial sources and from London-type smog, and fluorine mainly from ore reduction and the preparation of phosphate fertilizers. Today many additional constituents of polluted air are known to contribute to plant damage, particularly ozone and peroxidized compounds such as peroxyacetyl nitrate (PAN), which are the main elements of Los Angeles-type smog that are toxic to plants and animals. Ozone and PAN result from photochemical reactions between oxides of nitrogen and organic vapors mostly derived from the incomplete combustion of petroleum. In addition to the four principal pollutants already mentioned (sulfur dioxide, fluoride, ozone and PAN), plant damage has been caused by many other gases including ethylene, chlorine, ammonia, hydrogen chloride, hydrogen sulfide, and others (36). While nitrogen oxides are essential to the formation of photochemical smog, and can themselves be toxic to plants (26), some consider it questionable that they occur in the atmosphere in concentrations high enough to cause injury directly (36).

Virtually all of the principal types of agricultural crops have suffered important damage from air pollution. A few examples among nonforest crops might be mentioned (36). Alfalfa, cotton, and lettuce have been readily injured by sulfur dioxide at field levels; gladiolus, azalea, and vaccinium are among the most sensitive plants to fluorine; ozone produces weather fleck in tobacco (13); ethylene ruins orchid blooms; and a recent report of the University of California states that the photochemical smog from Los Angeles has reduced the production of citrus fruit in the main California citrus area south of that city by 20 to 25 percent in the past 15 years (4).

PAST AIR POLLUTION IMPACTS TO THE FOREST

Scheffer and Hedgcock (32), in their bulletin on injury to northwestern forest trees by sulfur dioxide from smelters, give a brief literature review of air pollution damage to forests and state that sulfur dioxide discharged from smelters had been the major cause of gas injury to forest trees. The National Research Council of Canada in 1939 released a 447-page report (29) on the effects of sulfur dioxide on vegetation, and damage to coniferous trees from smelting was a major feature of this report. Katz (19), in Canada, further described damage to coniferous forests from sulfur dioxide, and later Linzon (21) measured and documented heavy damage to white pine from sulfur dioxide released in connection with ore smelting in the Sudbury, Ontario, area. Gordon and Gorham (12) reported on extensive forest changes attributed to sulfur dioxide around an iron-sintering plant at Wawa, Ontario, where white pine failed to appear on test plots leeward of the plant for a distance of over 30 miles, indicating that this species was the most readily injured of the 30 woody plants recorded. The sensitivity of this tree to the fumes was again pointed up by these authors when white pine showed up for the first time on plots 16 miles leeward from another source, thus making it also the most sensitive of the woody plants recorded in connection with this source.

The Scheffer and Hedgcock study (32) cites the losses from mortality, growth reduction, lack of reproduction, and other impacts to many species of forest trees in the upper Columbia River valley south of the Trail, British Columbia, smelter before corrective measures were taken in 1931. Appreciable damage to conifers extended over 40 miles down the river from the smelter. These authors also cite the heavy damage to timber around the Washoe smelter in the vicinity of Anaconda, Montana, revealed by field studies made in 1910 and 1911.

Before corrective measures were instituted, the smelting operations in the Copper Hill, Tennessee, area, early in the century, had laid waste an area of 17,000 acres and damaged another 30,000 acres of timberland (18), mostly hardwoods. Much of the area is still virtually bare and eroding severely.

In 1949 an intensive investigation was started of the browning and dying of ponderosa pine within a 50-square-mile area around an aluminum ore reduction plant near Spokane, Washington. In this case the characteristic browning and banding was related to fluoride (34). Adams and his coworkers (2, 3) have given us much information on the susceptibility of ponderosa pine foliage, particularly when immature, to fluoride, and their work indicates a higher sensitivity of ponderosa pine and lodgepole pine to fluoride than western white pine, Douglas-fir, or Engelmann spruce.

These acute cases of smelter fume injury, while severe, have been fairly local and not numerous. For the most part highly toxic stack emissions from ore reduction have been controlled by various engineering devices and a major bonus of byproduct recovery has followed air pollution reduction measures in many cases. Yet sulfur dioxide and fluoride damage are still problems around a great many sources.

PRESENT AIR POLLUTION DAMAGE TO THE FOREST

Since the rate of growth of forest stands and the vigor and appearance of individual trees are influenced by so many site factors, including soil type, moisture, temperature, drainage, competition, etc., alien impacts not identifiable with known diseases or insects, can go unrecognized unless very severe damage is done. Partly for this reason we have recognized some measure of air pollution damage in the past only where the cause of such damage was obvious. Within recent years, however, injury to and death of trees over large areas, have been attributed to atmospheric insults not related to smelting (16). The pollution scientist's term, "atmospheric insult," is used advisedly here, instead of "air pollution," because it is not at all certain that the high oxidant levels that result in some injuries, such as the burning of the tips of eastern white pine needles result, mainly, from man's activities.

It is interesting to note that in four of the cases of forest tree damage recently either proven or considered likely to be caused by atmospheric constituents; namely, thermal power plant stack gas injury to white pine (7), ozone injury to white pine (emergence tipburn [6]), chlorotic dwarf of white pine (9), and chlorotic decline of ponderosa pine (30), pathologists spent years eliminating other, more conventional, possible causes before new knowledge and new techniques led the investigators to a consideration of aerological factors.

The loss of individual shade trees is seldom blamed on the atmosphere unless a direct relation of the source of the pollutant is almost self-evident. Cases of killing obviously related to such sources include the blighting of trees in close proximity to automobile exhaust, toxic gas leaks, waste burners, or burning municipal dumps. No one knows the extent of loss of city and highway trees resulting from urban smogs, whether of the reducing (London) type or the oxidizing (Los Angeles) type, but when such data become available they will undoubtedly show that many of our tree species have a low tolerance to urban air. In a general way such tolerance in the East ranges from high, as with the rugged Norway maple and the exotic gingko, to very low as with the sensitive balsam fir (8).

Recent research on the four pine problems already alluded to in connection with atmospheric impacts illustrates that the techniques and fields of knowledge employed in the solution of such problems are quite different from those conventionally used in working with parasites.

CHLOROTIC DECLINE OF PONDEROSA PINE

A condition called chlorotic decline, or X-disease, first noted in the early 1950's has affected ponderosa pine over thousands of acres in the San Bernardino Mountains of California (30). Other intermixed conifers of several genera and species have been unaffected. The decline is characterized by reduction in growth; loss of all but the current year's needles; yellowing, mottling, and stunting of needles; and death of trees. Fumigation experiments and air sampling data in the affected area (28) suggest that photochemical smog, aggravated by drought, is probably the principal cause of the decline of ponderosa pine in this area, in spite of the elevation (4,000 to 5,000 feet above the valley floor) and a distance of over 50 miles from Los Angeles.

Lodge (23) discusses our country's tendency to move from the reducing sulfur dioxide type of urban pollution toward the oxidizing "ozone" pollution as we move from solid to liquid and gas fuels. The particulate matter in smoke tends to reduce photochemical smog through the reduction in light intensity reaching the polluted air. Thus, an increase in oil and gas consumption plus a reduction in smoke can bring us new high levels of oxidant emanating from cities and certain types of industrial activities.

In southern California PAN has apparently been the most destructive smog constituent, while in the East ozone appears to be causing most of the oxidant damage to plants (31). With respect to the recent changes in types of pollution, Rodenhiser (31) emphasizes that whereas fluoride and sulfur dioxide, which are still major pollutants, are usually traceable to a limited number of large industrial plants, smog comes from "millions" of sources throughout our highly mechanized society.

REACTION OF WHITE PINE TO CERTAIN ATMOSPHERIC CONSTITUENTS

Recently Berry and Ripperton (6) reported on studies indicating that one of the long-known needle blights of white pine, investigated by many pathologists since the turn of the century, is due to atmospheric oxidant, probably ozone. Berry (5) first attributed the term "emergence tipburn" (ET) to this trouble, and pointed out that in 1961, when an unusual wave of ET hit the Southeast, an unusual wave of tobacco weather fleck, also an ozone-induced disorder, was reported by North Carolina State College.

When field oxidant concentrations, expressed as ozone, reached 6.5 p.p.h.m. (parts per hundred million) tipburn occurred on susceptible white pine clones. When the ambient air was filtered through an activated carbon filter, thus removing any oxidant, no tipburn occurred. Finally, when susceptible clones were fumigated with ozone at 6.5 p.p.h.m., on the basis of the Mast recorder, tipburn occurred apparently identical in symptoms with the tipburn occurring naturally in the field.

Heggestad and Menser (14) produced tobacco weather fleck at concentrations of ozone, as measured by the Mast recorder, and exposure times almost the same as those used by Berry and Ripperton (6) in tipburning white pine. Thus ET appears to be the pine analogue of tobacco weather fleck, and an explanation is afforded to a perennial pine problem of previously undetermined cause.

Emergence tipburn is probably one of the commonest white pine troubles that have long been known to us, that occurs in increasing intensity going northward from the southern Appalachian Mountains into New England and Canada, and that is seemingly unrelated to pathogen attack (22). A relation to weather, as brought out by Berry (5) in connection with ET, by Linzon (22) for the same or a similar trouble, and as implied in the term weather fleck of tobacco, is a common denominator in these troubles. The rise in atmospheric ozone, leading to these troubles, whether resulting from natural sources (36) or from polluted air (39), appears to be related to certain weather patterns.

Another baffling stunting of white pine, called chlorotic dwarf, that occurs widely in the Northeast and Central States is being investigated by Dochinger and Seliskar (9). Their recent report, based on grafting experiments and other studies, suggested that the trouble results "from a causal agent that acts directly on the foliage," and no virus, fungus, or other pathogen appeared to be implicated. Regarding the cause, some atmospheric impact has been suggested as a strong possibility.

A third decline of white pine related to one or more atmospheric constituents has occurred within about a 20-mile radius of some soft-coal-burning power plants. Berry and Hepting (7) have shown that while fume damage symptoms of the acute sulfur dioxide type, together with elevations in foliar content of sulfate in white pine needles, may occur immediately around a plant of this type, damage of a different kind may extend for 20 or more miles, depending on wind, terrain, and other local conditions. The latter, more extensive type of injury, is at least temporarily being called post-emergence chronic tipburn (PECT) to differentiate it from emergence tipburn (ET). While ET, a type of oxidant (probably ozone) injury, starts and ends during the period of shoot elongation and needle growth, PECT may start any time of year, typically showing up first in the winter. PECT is also characterized by a gradual change from a brown tip to a green base, often with mottling or banding in between.

The separation of these two troubles from some fungus diseases of white pine is described by Hepting and Berry (17).

PECT results in needle blight, a casting of older needles, growth reduction, and often in early death. In the case of all three of the eastern white pine troubles described here, as well as the smog damage to ponderosa pine in the San Bernardino Mountains of California, and fluoride damage to ponderosa pine in Washington (2), there is striking tree-to-tree variability in susceptibility. Normal trees and trees in the last stages of decline may occur side by side, indicating a genetic difference in resistance. These characteristics of resistance or susceptibility are retained in scions after they are grafted on stock trees of the opposite susceptibility tendency. Steps have already been taken to establish a seed orchard of eastern white pine clones resistant to the PECT type of stack gas injury.

When ramets (vegetative reproductions) of a PECT-susceptible clone of white pine were taken from an unpolluted area to an area eight miles from a coal-burning power plant, growth was checked. They lost needles and the remaining needles developed typical PECT symptoms. Ramets of the same clone, in plastic buckets of the same soil, that were left in the unpolluted area remained normal. The trees exposed to pollution for seven months that survived the exposure developed normal foliage within two years after being returned to the unpolluted area (7).

PECT of eastern white pine was observed around several thermal power plants in the Appalachian region although the offending gas or gases are still unknown. Although only white pine showed obvious symptoms once away from the immediate environs of the stacks, we have reasons for being concerned with the effects of the increasing output of power plant and industrial stack gases on our forest acreage (16).

TRENDS IN AIR POLLUTION AS THEY AFFECT TREES

An impressive list of crops damaged by photochemical smog from urban pollution can be compiled (26, 36), including those in the categories of field, flower, fruit, and vegetable crops, as well as forest trees. Unless measures now being taken by various public and private agencies, and spearheaded on a national basis by the U.S. Public Health Service, can successfully combat this problem in the near future, we can expect increasing damage to orchard, forest, and shade trees. As brought out during the last five years, we will also likely be recognizing certain kinds of damage to trees as caused by air pollution that we have not known the cause of before.

Damage from urban smog can take many forms and extend considerable distances. Parmeter, Bega, and Neff (30) point out that the San Bernardino ponderosa pine case involves thousands of acres of land important not only for timber, but for valuable watershed needs, and as a recreational area that attracts more than 4 million visitors each year. Scurfield (33) relates how the British National Pinetum had to be moved from Kew, near London, to an area in Kent because of urban air pollution.

Changes in our climate that have been taking place in the past 70 years have been described as probably affecting the incidence of many of our forest diseases (15), including both those caused by parasites and by physiogenic influences. Certainly we would expect that climate effects would not only influence the concentration or dissipation of man-made smog and the amount of light energy available for photochemical reactions but also, through subsidence, turbulence, or other meteorologic phenomena, could affect the amount of stratospheric ozone that reaches the troposphere, which is the zone in which we and our plants live. In investigating the possible causes of decline in a number of forest tree species in the Northeast, notably sugar maple, ash, black walnut, oak, and birch, the influence of climatic changes must be considered (15). The possible effects of high oxidant levels and other atmospheric impacts are additional influences related to weather that may play a part in these unresolved problems. We must study our trees in relation to their total environment, and polluted air can be an important, and often ignored, part of the environment.

Another upward pollution trend, in addition to urban pollution, is related to thermal power production and it, also, is a source of some concern. Frankenberg (11) depicts the growth in large power units of this type. Up to 1954 we had no plants with a boiler capacity of 2 million pounds of steam per hour. By 1962 over 60 percent of our power was generated in such huge new plants. More plants and larger plants, many using low-grade soft coal, mean far greater stack gas emissions. Since the problem of controlling these complex stack gases, which include among other gases sulfur dioxide, fluoride, and oxides of nitrogen, has only been partly solved, we must learn much more than we now know about the impact of the increasing number of such large-capacity, soft-coal-burning industrial units on the surrounding forest vegetation.

Most people today read and hear much about our own and British urban air pollution problems, especially with regard to human health. From our point of view as foresters, it is interesting to note a 1962 Associated Press dispatch from Rome, Italy, that "a special study commission says many of Rome's pine trees are drying up (sic) largely because of gases in the air." It urged city officials to take steps to limit air contamination caused by industrial fumes, smoke from homes and automobile exhaust fumes.

Scurfield (33) presents, under the title "Air Pollution and Tree Growth," an impressive compendium, including 258 references, of information on sources of air pollution around the world, and gives examples of damage to different tree species by these pollutants in parts of Germany, England, Russia, the United States, Portugal, Tasmania, South America, South Africa, India, and New Zealand. The more important earlier investigations demonstrating air pollution damage to farm and forest crops was done in Germany as an outgrowth of their industrial expansion starting about a century ago. How individual trees or individual species react to different gases is interesting and important, but it is also important that we determine the expected impact of the many major sources of pollution on our forest and shade tree resources in terms of how much they reduce timber and other forest production, and recreational and civic improvement values. I have tried to show; first, that we have had notable but scattered cases of severe air pollution damage to forests in the past; second, that we are being subjected to new forms of air pollution, as part of our urban and industrial growth, which have already severely damaged certain tree species over considerable areas in the United States; and third, that as our research facilities, knowledge, and techniques improve, we are finding tree declines due to air pollution that were either erroneously ascribed to other causes or to no cause at all.

Air pollution authorities point out that we have gone beyond strictly urban problems, in the sense of Pittsburgh, St. Louis, or Los Angeles, to regional problems as in southern California, the East Coast, and other industrialized areas. We are finding that we must manage our air as we do our land, water, and forests.

LITERATURE CITED

1. Adams, D. F. 1956. The effects of air pollution on plant life. Amer. Med. Assoc. Arch. Indus. Health 14:229–245.
2. ———, C. G. Shaw, and W. D. Yerkes, Jr. 1956. Relationship of injury indexes and fumigation fluoride levels. Phytopathology 46:587–591.
3. ———, J. W. Hendrix, and H. G. Applegate. 1957. Relationship among exposure periods, foliar burn, and fluorine content of plants exposed to hydrogen fluoride. Agric. and Food Chem. 5:108–116.
4. Anonymous. 1964. Citrus study under way: smog answer sought. Air in the News 1(9):12.
5. Berry, C. R. 1961. White pine emergence tipburn, a physiogenic disturbance. U.S. Forest Service, Southeastern Forest Expt. Sta. Paper 130, 8 pp.
6. ———, and L. A. Ripperton. 1963. Ozone, a possible cause of white pine emergence tipburn. Phytopathology 53:552–557.
7. ———, and G. H. Hepting. 1964. Injury to eastern white pine by unidentified atmospheric constituents. Forest Sci. 10:2–13.
8. Collingwood, G. H., and W. D. Brush. 1947. Knowing your trees. The American Forestry Assoc., Washington, D.C.
9. Dochinger, L. S., and C. E. Seliskar. 1963. Susceptibility of eastern white pine to chlorotic dwarf. (Abst.) Phytopathology 53:874.

10. Eighty-Eighth Congress. 1st Session. 1963. An act to improve, strengthen, and accelerate programs for the prevention and abatement of air pollution. Public Law 88-206, 88 Cong., H. R. 6518. 10 pp. Dec.
11. Frankenberg, T. T. 1963. Air pollution from power plants and its control. Combustion 34:28-31.
12. Gordon, A. G., and E. Gorham. 1963. Ecological aspects of air pollution from an iron-sintering plant at Wawa, Ontario. Canadian Jour. Bot. 41:1063–1078.
13. Heggestad, H. E., and J. T. Middleton. 1959. Ozone in high concentrations as cause of tobacco leaf injury. Science 129:208–209.
14. ———, and H. A. Menser. 1962. Leaf spot-sensitive tobacco strain Bel W-3, a biological indicator of the air pollutant ozone. (Abst.) Phytopathology 52:735.
15. Hepting, G. H. 1963. Climate and forest diseases. Ann. Rev. Phytopath. 1:31–50.
16. ———. 1963. [Statement in discussion of agricultural, natural resource, and economic considerations.] *In:* Proceedings of the National Conference on Air Pollution of 1962. U.S. Public Health Service. p. 200.
17. ———, and C. R. Berry. 1961. Differentiating needle blights of white pine in the interpretation of fume damage. Internatl. Jour. Air and Water Pollut. 4:101–105.
18. Hursh, C. R. 1948. Local climate in the Copper Basin of Tennessee as modified by the removal of vegetation. U.S. Dept. Agric. Cir. 774. 38 pp.
19. Katz, M. 1952. The effect of sulfur dioxide on conifers. *In:* U.S. Tech. Conf. on Air Pollution Proc. 1952:84.
20. ———, and V. C. Shore. 1955. Air pollution damage to vegetation. Jour. Air Pollut. Control Assoc. 5:144.
21. Linzon, S. N. 1958. The influence of smelter fumes on the growth of white pine in the Sudbury region. Ontario Dept. Lands and Forests and Ontario Dept. Mines, Toronto, Ontario. 45 pp.
22. ———. 1960. The development of foliar symptoms and the possible cause and origin of white pine needle blight. Canadian Jour. Bot. 38:153–161.
23. Lodge, J. P. 1957. [Discussion statements.] *In:* Proceedings of the Air Pollution Research Planning Seminar of 1962. U.S. Public Health Service. pp. 32–33.
24. McCabe, L. C. 1952. Air pollution, U.S. Technical Conference on Air Pollution Proceedings 1950. McGraw-Hill Book Co., New York.
25. Middleton, J. T. 1961. Photochemical air pollution damage to plants. Ann. Rev. Plant Physiol. 12:431–448.
26. ———, and A. O. Paulus. 1956. The identification and distribution of air pollutants through plant responses. Amer. Med. Assoc. Arch. Indus. Health 14:526–532.
27. ———, D. F. Darley, and R. F. Brewer. 1957. Damage to vegetation from polluted atmospheres, Amer. Petrol. Inst. Sect. III, Refining, Proc. 8 pp.
28. Miller, P. Robert, J. R. Parmeter, Jr., O. C. Taylor, and E. A. Cardiff. 1963. Ozone injury to the foliage of Pinus ponderosa. Phytopathology 53:1072–1076.
29. National Research Council of Canada. 1932. Effect of sulfur dioxide on vegetation. Ottawa. Publ. 815.
30. Parmeter, J. R., Jr., R. V. Bega, and T. Neff. 1962. A chlorotic decline of ponderosa pine in southern California. U.S. Dept. Agric. Plant Dis. Rptr. 46:269–273.
31. Rodenhiser, H. A. 1962. Effects of air pollution on crops and livestock. *In:* Proc. of the National Conf. on Air Pollution of 1962. U.S. Public Health Service. pp. 175–178.
32. Scheffer, T. C., and G. G. Hedgcock. 1955. Injury to northwestern forest trees by sulfur dioxide from smelters. U.S. Dept. Agric. Tech. Bull. 1117. 49 pp.
33. Scurfield, G. 1960. Air pollution and tree growth. Forestry Abst. 21:1–20.
34. Shaw, C. G., G. W. Fisher, D. F. Adams, and M. F. Adams. 1951. Fluorine injury to ponderosa pine. (Abstr.) Phytopathology 41:943.

35. Stern, A. C., Editor. 1962. Air pollution. 2 volumes. Academic Press, New York.
36. Thomas, M. D. 1961. Effects of air pollution on plants. *In:* Air Pollution. World Health Organ. Monog. Series 46. pp. 233–278.
37. U.S. Public Health Service. 1958. Proceedings of the National Conference on Air Pollution of 1958. 526 pp.
38. ———. 1963. Proceedings of the National Conference on Air Pollution of 1962. 436 pp.
39. Wanta, R. C., W. B. Moreland, and H. E. Heggestad. 1961. Tropospheric ozone: an air pollution problem arising in the Washington, D.C., metropolitan area. Monthly Weather Rev. 89:289–296.
40. World Health Organization. 1961. Air pollution. Monog. Series 46. 442 pp.
41. Zimmerman, P. W. 1952. Effects on plants of impurities associated with air pollution. *In:* Air Pollution. U.S. Tech. Conf. on Air Pollution Proc. 1950:127–139.

29

Effects of air pollutants on apparent photosynthesis and water use by citrus trees

C. Ray Thompson
O. C. Taylor
M. D. Thomas
J. O. Ivie

Air pollutants, especially photochemical smog and fluorides, cause major damage to agricultural crops when certain levels are exceeded (Darley and Middleton, 1966; Middleton, 1961; Thomas, 1961; Middleton *et al.*, 1965). Commercial production of some leafy vegetables has virtually ceased in the Los Angeles Basin because of oxidant lesions which reduce quality or render the crop totally unsalable. Fluorides can cause visible damage to crops if levels are high enough, but where no outward symptoms occur, environmentally controlled studies are required to determine the extent of injury. The present studies were initiated in 1960 under a unique cooperative effort (Richards and Taylor, 1960) supported financially by agriculture, industry, local and national governments, various private organizations, and the University of California with an avowed purpose to "measure under field conditions the effect of the various atmospheric phytotoxicants on agricultural crops growing in the Upper Santa Ana Drainage Basin."

Reprinted from *Environmental Science and Technology*, 1:644–650 (August 1967). C. Ray Thompson is affiliated with the Statewide Air Pollution Research Center, University of California, Riverside.

There was a serious question as to whether photochemical smog and/or fluorides were responsible for major economic losses by the citrus industry and, if so, how much. Thus, the present study reports the effects of ambient pollutants on apparent photosynthesis and water use of citrus trees because they show definitive results. Design and performance of the greenhouses were reported earlier (Thompson and Taylor, 1966). Details of the systems for supplying controlled levels of nitric oxide and hydrogen fluoride to the trees also have been published (Thompson and Ivie, 1965). Data showing effects on leaf drop, fruit drop, and fruit yield are still being recorded, but preliminary indications are that all three responses are being adversely affected by photochemical smog. Fluorides seem to have no measureable effect. Details will be reported in a subsequent manuscript.

METHODS

The work was begun on lemons because these trees grow rapidly and set several flushes of leaves during the year, and experimental effects would probably appear sooner than with other, slower growing citrus species. Two groves were leased where both photochemical smog and atmospheric fluoride occur. These groves had uniform trees and were six and nine years of age, respectively. Later a navel orange grove nine years of age was added to the study. At each location, 24 experimental trees were selected and divided according to a randomized block design into six treatments with four replications each (Table 29–1). All trees were

TABLE 29–1. Experimental field installations on Citrus, Upland and Cucamonga, Calif.

Tree Atmosphere	Treatment of Atmospheres	Toxicant Remaining
Filtered air	Activated carbon, limestone	....
Ambient air	....	Fluoride, ozone, PAN[a]
Low fluoride air	Limestone	Ozone, PAN
Low ozone air	Nitric oxide	Fluoride, NO, NO_2, PAN
Filtered air + fluoride	Limestone, activated carbon, hydrogen fluoride	Fluoride
Low ozone, low F^- air	Limestone, nitric oxide	PAN, NO, NO_2

[a] Peroxyacyl nitrates.

Fig. 29–1. Plastic covered greenhouses installed over bearing lemon trees near Upland, Calif.

enclosed in individual aluminum-framed plastic covered greenhouses (Thompson and Taylor, 1966) equipped with squirrel-cage blowers (Fig. 29–1). Total volume of the houses was 41.5 cu. meters. Air was supplied continuously to the top of the houses at about 80 cu. meters per minute in an effort to change the air twice per minute. Temperatures rose 4° to 6° C. inside the houses over outside readings on hot days. The blowers were enclosed in steel cabinets. In treatments requiring filters, three Filterfold cannisters (Barneby-Cheney Co., Los Angeles, Calif. 90063) were filled with activated coconut charcoal and/or six-mesh limestone and were interposed in the cabinets ahead of the blower which was rated for 85 cu. meters per minute but inside fiberglass dust filters. Dust filters were changed on a 4-week schedule or more often if accumulations became too heavy.

Commercial furrow-type irrigation practices were followed by the grove owner but when trees approached a soil suction of 0.5 atm., supplemental water was applied. Pruning was done as required to keep the trees from abrading the plastic sides and allow circulation of air.

The "filtered" treatment (Table 29–1) consisted of both activated carbon and limestone filtration. The carbon was used to remove ozone, and peroxyacyl nitrates plus some of the hydrocarbons. Limestone removed acid gaseous fluorides. Typical data showing total oxidant reduction are given in Table 29–2. Oxidant was measured with a Mast ozone

TABLE 29–2. Total oxidant levels in greenhouses after activated carbon filtration or NO treatment[a]

July 5, 1965 Orange 1		July 6, 1965 Lemon 1	
Begun 1350 o'clock		Begun 1605 o'clock	
Outside air	0.16	Outside air	0.25
Filtered air + F^-	0.05	Filtered air	0.05
Filtered air	0.05	Filtered air + F^-	0.05
Low ozone air	0.05	Low ozone air	0.03
Low ozone, low F^- air	0.05	Outside air	0.25
Low fluoride air	0.13	Ended 1525 o'clock	
Outside air	0.16		
Ended 1410 o'clock			

[a] All data are in parts per million by volume.

meter (Mast Development Co., Davenport, Iowa). This instrument gives a partial response to nitrogen dioxide which could account for a small part of the residual oxidant, but small filter voids and leaks also allow some of these compounds to get into the houses.

Several dozen determinations of residual oxidants in "filtered" greenhouses showed a mean reduction of 80 percent. "Ambient air" as it occurred in the area was drawn into the houses through wirecloth screens. If dusty tillage operations occurred, fiberglass dust filters were used. "Low fluoride air" consisted of ambient air which was passed through dust and limestone filters. "Low ozone air" consisted of ambient air to which nitric oxide was added to react with the ozone. This addition was begun when the total oxidant level of the ambient atmosphere of the grove reached 0.10 p.p.m. (Thompson and Ivie, 1965). At that time, 0.20 p.p.m. of NO on a volume basis was added to the intake of the blowers. The NO was diluted with 50 volumes of cylinder nitrogen before injection into the blowers to prevent premature oxidation by oxygen of the air. This 2-to-1 ratio of NO to the total oxidant was maintained as the photochemical smog increased, often necessitating the addition of 0.8 to 1.0 p.p.m. of NO.

To find out what was happening to the oxides of nitrogen both NO and NO_2 were measured at the outlet of greenhouses to which NO was added. Nitric oxide and nitrogen dioxide were determined by aspirating two air samples, one through dichromate coated paper which oxidized the NO to NO_2 and a second untreated sample through Saltzman reagent and measuring color development at 550 mμ (Saltzman, 1960). NO values were obtained by difference. Total oxidant outside and that in air issuing from the houses was correlated with the NO to NO_2, Table 29–3. Because

TABLE 29–3. Total oxidant reduction, nitric oxide, and nitrogen dioxide content of "low ozone air" in greenhouses treated with NO[a]

Date	Outside Total Oxidant	NO Added	At Greenhouse Outlet: Total Oxidant	NO_2	NO	NO plus NO_2
July 20	0.26	0.50	...	0.26	...	...
July 27	0.18	0.30	0.07	0.11	...	...
July 29	0.12	0.20	0.05	0.20	0.11	0.31
August 2	0.20	0.40	0.05	0.28	0.10	0.38
August 16	0.28	0.60	0.02	0.32	0.26	0.58
August 17	0.25	0.50	...	0.31	0.25	0.56
August 25	0.28	0.50	...	0.41	0.12	0.53

[a] All data in parts per million by volume.

of the large volumes of air processed, it was questionable as to whether addition of NO was 100 percent effective in removing ozone. Also, even if ozone removal was quantitative, both NO_2 and PAN react to give some response with the total oxidant analyzer, thus leaving a question as to the actual amounts of residual ozone. Despite these interferences, 0.2 to 0.07 p.p.m. of the total oxidant as measured by the Mast ozone meter was all that remained.

Comparison of the NO_2 produced from the ozone plus NO showed a reasonably quantitative conversion of NO to NO_2. The sum of the two values, NO + NO_2, was also well correlated with the NO added showing that the stoichiometry of the reaction was being measured by the analytical determinations.

"Low ozone, low fluoride air" combined the use of nitric oxide addition and limestone filtration as described previously. "Filtered air plus fluoride" presented a special case because extended studies were unsuccessful in removing oxidants while still allowing all of the atmospheric fluoride to enter the houses. All adsorbents, catalysts, etc., which were tried for taking up or destroying oxidants, removed substantial amounts of fluoride. Accordingly both activated carbon and limestone filters were used as in "filtered air" and hydrogen fluoride was added back to the airstream at ambient levels but on a 24-hour delayed basis. Details are published elsewhere (Thompson and Ivie, 1965).

Tensiometers were installed in all houses about 1½ meters from the tree at a depth of ½ meter to monitor soil moisture. One tree in each block had a tensiometer at ¾ meter to determine moisture at the greater depth. Tabulations of soil suction were made daily. Total oxidant levels

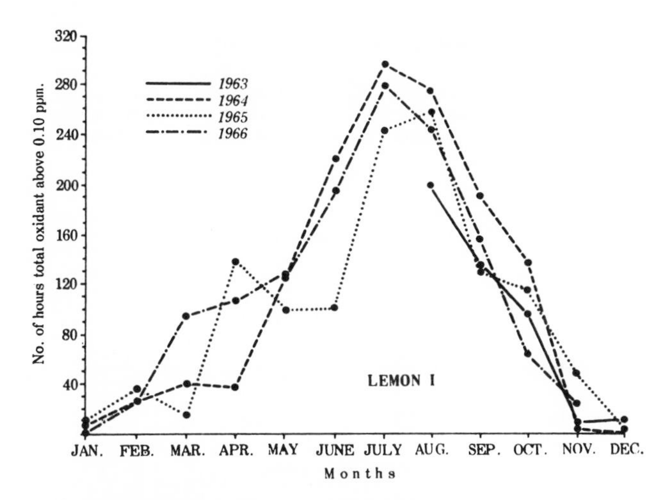

Figure 29–2. Total number of hours per month during which total oxidant levels were above 0.10 p.p.m. at Upland, Calif.

in air were monitored continuously by a Mast ozone meter. Typical results showing the number of hours during which the levels were above 0.10 p.p.m. at Lemon Division 1 are given in Figure 29–2. Atmospheric hydrogen fluoride was measured continuously by a fluorometric analyzer (Ivie *et al.*, 1965), and total fluorides were determined daily by titration or colorimetrically from impinger samples (Bellack and Schouboe, 1958; Willard and Winter, 1933). Typical data for total daily fluoride levels are given in Table 29–4. Checks on the amount of total fluoride inside of the greenhouses were made periodically and leaf accumulations were determined at 3-to 4-month intervals.

The degree of removal of gaseous fluoride, principally hydrogen fluoride, by the limestone filters was difficult to determine precisely because the fluorometric analyzer available (Ivie *et al.*, 1965) had a lower limit of about 0.10 μg. per cu. meter of hydrogen fluoride. With ambient levels of only 0.1 to 0.5 μg. per cu. meter of hydrogen fluoride and with laboratory trials showing 90 to 97 percent removal by the limestone filters, the lower limit of the analyzer was exceeded. Several simultaneous checks of total fluorides from outside and within houses by the impinger method having limestone filters showed a reduction in total fluoride of about 75 percent. Average leaf accumulation in trees receiving limestone filtered air after 8 months of exposure was 10 p.p.m. on a dry weight basis, while trees receiving ambient air had 27 p.p.m. Some of this amount is considered "background" possibly accumulated by absorption and translocation from soil through the tree.

TABLE 29–4. Average total fluoride levels for 24 hours at experimental locations (Micrograms per cubic meter)

(Micrograms per cubic meter)

	Lemon 1 (21st-Euclid, Upland)	Lemon 2 (Ramona-Church Sts.) (Cucamonga)	Orange 1 (23rd-Euclid, Upland)
	JULY 1964		
High	0.80	0.91	0.63
Low	0.30	0.30	0.13
Mean	0.41	0.55	0.32
	AUGUST 1964		
High	0.85	0.93	0.62
Low	0.19	0.34	0.23
Mean	0.55	0.70	0.54
	SEPTEMBER 1964		
High	0.92	1.80	0.56
Low	0.21	0.27	0.25
Mean	0.46	0.77	0.39

EXPERIMENTAL RESULTS

Effect of air pollutants on water use of trees

Within a few weeks after the experimental treatments began in 1962, water use by trees in the various groups in Lemon Division 1 differed appreciably. The owner irrigated on a 15-day schedule, but some trees required "extra" irrigations more frequently than others. "Extra" irrigations were those required when tensiometers buried at 50-cm. depth in soil under the skirts of the trees reached soil suction of 45 centibars. This extra water was applied so that the suction never went above 50 centibars. The tentative conclusions of 1962 were confirmed during the succeeding years (Table 29–5) and showed that trees receiving "filtered air" required more extra irrigations than those with "ambient air" during 1963, 1964, and 1965. In 1963 and 1964, this was statistically greater at the 1 percent level; in 1965, at the 5 percent level. "Ozone low air" trees and "ozone, fluoride low air" trees lost less water than "filtered air" in 1964 and 1965 (5 percent level). All other treatments were not different from "filtered air" statistically but all average values were numerically less.

Typical rates of increase in soil suction are shown in Figure 29–3 for "filtered air" and "ambient air." Lemon 2 did not show the differences in water use because the grove operator irrigated the trees weekly. The orange grove showed a greater numerical frequency of extra irrigations for trees in "filtered air," but these were not statistically different.

TABLE 29–5. Total number of "extra" irrigations required by trees in different treatments during the year

Treatment	Lemon 1				
	1962[a]	1963	1964	1965	1966
Filtered air	23	43	36	24	20
Ambient air	3	7[b]	5[b]	4[c]	6
Low fluoride air	16	37	26	14	18
Low ozone air	7	19	12[c]	10	7
Filtered air plus fluoride	14	39	28	15	16
Low ozone, low fluoride	12	37	22	5[c]	9

[a] Three months of the year.
[b] Significantly less than filtered air at 1%.
[c] Significantly less than filtered air at 5%.

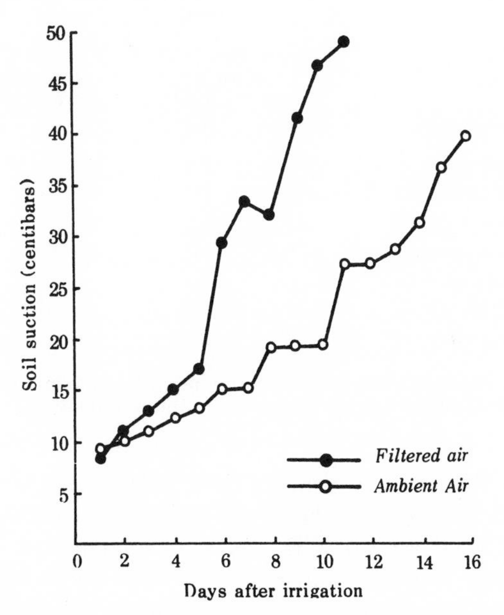

Figure 29–3. Typical record of soil suction increase in tree root zone following irrigation at Lemon Division 1.

Effect of air pollutants on apparent photosynthesis of entire trees

To measure another effect of pollutants on the metabolism of the lemon trees, apparent photosynthesis was measured on the entire trees of Lemon 1 during 1961 and 1962 (Ivie *et al.*, 1963). In this method, doors were installed on the intake blowers which when closed reduced airflow.

This was necessary to allow enough CO_2 reduction so that it could be measured accurately. Speed of the blower motors was also reduced from 1750 to 1140 r.p.m., resulting in a decreased air movement from 80 to 20 cu. meters per minute during the first period of measurement of apparent photosynthesis. Air samples from the intake and outlet of each house were drawn through aluminum tubes to a central control room where they were stored in vinyl plastic bags until analyzed sequentially by an infrared analyzer. Carbon dioxide changes in one set of 12 houses comprising two enclosed trees from each treatment were measured for one week after which the remaining 12 trees were monitored.

Before the actual air treatments were started at Lemon 1 but after erection of the houses, the rate of apparent photosynthesis of all trees was determined as a base line (Table 29–6). These measurements were

TABLE 29–6. Apparent photosynthesis of entire trees as per cent of trees with filtered air

Atmosphere	Initial Period before Treatment, 1961		Treatment Period, 1962	
	AM[a]	PM[b]	AM[a]	PM[b]
Filtered air	100	100	100	100
Ambient air	95	118	61	78
Low fluoride air	127	105	81	97
Low ozone air	125	111	68	93
Low ozone, low fluoride air	104	106	94	95
Filtered air + HF	88	107	120	96

[a] 0700–1000 with reduced airflow (20 cu. meters/min.) through houses.
[b] 1100–1500 with full airflow (80 cu. meters/min.) through houses.

made from September to November 1961. Treatments were then begun in January 1962. The results shown as 1962 treatment were taken from April–June 1962 and represent the average amounts of carbon dioxide absorbed by each tree as compared with trees receiving "filtered air." Because the trees varied in size the pretreatment values varied from the "filtered air" base, but after the treatments of 3 months, the "filtered air" and "filtered air plus fluoride" had the highest average rate of apparent photosynthesis.

A more precise evaluation was made by taking six successive periods in January, June, and November of 1962 during which apparent photosynthesis was measured. The amount of CO_2 absorbed was determined, and the grams of dry matter assimilated per tree per hour were calculated. This assumes 44 percent carbon in vegetation and requires 1.63 grams of CO_2 to produce 1 gram of dry matter (Franck and Loomis,

TABLE 29–7. Apparent photosynthesis of lemon trees during January, June, and November 1962

Date of Treatment	Adjusted Grams Dry Matter per Hour[a]					
	Filtered	Ambient	Filtered + F	Low O_3	Low O_3, F	Low F
1-15-62 to 1-21-62	5.9	2.2	5.3	2.6	4.2	3.8
	2.4	1.3	4.4	2.8	3.2	2.9
1-21-62 to 1-31-62	6.1	3.0	5.1	2.9	3.2	4.1
	3.4	0.9	6.9	2.2	2.5	3.6
Av.	4.5	1.9	5.4	2.6	3.3	3.6
Treatments greater than ambient	b		b			c
6-6-62 to 6-11-62	5.1	3.5	5.0	4.3	5.3	2.8
	5.0	4.3	5.1	3.8	5.2	5.5
6-13-62 to 6-18-62	2.4	2.4	3.6	2.6	2.7	3.9
	3.8	2.8	4.7	2.5	4.5	3.1
Av.	4.1	3.2	4.6	3.3	4.4	3.8
Treatments greater than ambient			c			
11-1-62 to 11-13-62	8.6	8.3	11.5	8.8	10.9	6.7
	12.5	9.2	13.4	9.1	10.0	10.3
11-14-62 to 11-26-62	8.7	6.0	9.8	5.9	5.4	7.1
	10.3	7.0	10.8	5.4	9.3	8.3
Av.	10.0	7.6	11.4	7.3	8.9	8.1
Treatments greater than ambient	c		b			

[a] Adjusted for pretreatment grams dry matter/hour.
[b] Greater than ambient at 1% level of significance.
[c] Greater than ambient at 5% level of significance.

1949). The results (Table 29–7) show how much CO_2 was absorbed during the morning hours with the blower door closed and reduced air flow through the houses. The individual values were adjusted by the pretreatment CO_2 absorption values. The "ambient air" treatment is compared statistically with all other treatments by the procedure of Dunnett (1955) and shows that in January "filtered air" and "filtered air plus fluoride" are significantly greater than "ambient air" at the 1 percent level. "Low fluoride air" is greater at the 5 percent level. In June, "filtered air plus fluoride" was greater than "ambient air" at the 5 percent level. In November, the results showed that "filtered air" and "filtered air plus fluoride" are greater than "ambient air" at the 5 percent level, and in the latter treatment, are greater at the 1 percent level.

Effect of air pollutants on apparent photosynthesis of an isolated branch

In an attempt to obtain a measurement of apparent photosynthesis which could be related directly to leaf area, six leaf chambers were constructed, and the measurement procedure of Taylor *et al.* (1965) was used. In this method five to six leaves on one branch are enclosed in a plastic chamber fitted with an artificial light source. The chambers have inlet and outlet connections for obtaining simultaneous CO_2 samples. These chambers were installed in six greenhouses, one in each of the six treatments, and the amount of CO_2 in the particular greenhouse *vs.* that coming from each leaf chamber was measured for one week; a second set of leaves was measured on the same tree for a second week. The chambers were then moved to a second replicate of six houses for two weeks and successively on to the third and fourth replicates, making an 8-week measurement period in all. Leaf area in each chamber was measured by tracing and planimetry. Carbon dioxide chart readings were converted to parts per million and then to milligrams of CO_2 per sq. decimeter per hour. Data from four 8-week periods of measurement are shown in Table 29–8. For the two periods in 1963, three covariates, tem-

TABLE 29–8. Apparent photosynthesis of selected lemon branches (adjusted average Mg. $CO_2/Dm.^2/Hr.$[a])

Group	Date	Filtered	Ambient	Filtered $+F^-$	Low O_3	Low O_3 F^-	Low F^-
I	8-5-63 to 8-16-63	6.2	3.3	4.6	4.6	5.9	6.1
II	8-19-63 to 8-30-63	5.6	3.0	4.3	5.1	4.9	4.7
III	9-9-63 to 9-20-63	6.7	3.5	4.4	5.8	4.5	4.6
IV	9-23-63 to 10-4-63	7.3	4.9	5.8	5.5	3.7	5.4
Grand av.		6.4	3.7	4.8	5.3	4.7	5.2
Treatments greater than ambient		[b]			[c]		[c]
I	10-7-63 to 10-18-63	6.9	5.1	7.9	3.8	6.7	5.2
II	10-21-63 to 11-1-63	6.4	4.9	7.4	7.0	8.2	4.6
III	11-4-63 to 11-15-63	5.4	3.2	5.6	6.0	4.0	3.1
IV	11-18-63 to 1-31-64	3.7	1.8	2.2	2.5	2.0	3.3
Grand av.		5.6	3.8	5.8	4.8	5.2	4.1
Treatments greater than ambient		[c]		[c]			
IV	6-24-64 to 7-7-64	2.6	1.4	0.9	1.5	2.3	0.6
III	7-8-64 to 8-4-64	2.0	1.1	3.1	1.1	2.5	1.6
I	8-5-64 to 8-19-64	4.8	1.8	2.0	2.4	2.2	4.7
II	8-20-64 to 8-31-64	2.4	1.7	4.0	2.7	2.0	1.3
Grand av.		2.9	1.5	2.5	1.9	2.3	2.1
No treatments greater than ambient							
IV	9-14-64 to 9-26 64	4.0	4.9	4.1	4.8	5.0	3.9
III	9-28-64 to 10-11-64	4.1	1.2	3.8	3.0	4.3	2.2
I	10-12-64 to 10-25-64	6.5	4.4	3.5	2.8	5.0	2.0
II	10-26-64 to 11-8-64	5.4	3.5	2.9	4.8	3.5	2.4
Grand av.		5.0	3.5	3.6	3.9	4.4	2.6
No treatments greater than ambient							

[a] Adjusted for covariates.
[b] Greater than ambient at 1% level of significance.
[c] Greater than ambient at 5% level of significance.

perature, total leaf area, and average area per leaf, were tested for significance. As temperature was the only significant covariate (Fig. 29–4), a mean temperature value was calculated, and the CO_2 values were adjusted for this covariate. These results showed that during August and September "filtered air" had greater CO_2 absorption than "ambient air" at the 1 percent level (Table 29–8). Also "low ozone air" and "low fluoride air" were greater than "ambient air" at the 5 percent level. In October and November, "filtered air" and "filtered air plus fluoride" exceeded "ambient air" at the 5 percent level. The grand averages of all treatments compared were numerically greater than those of "ambient air."

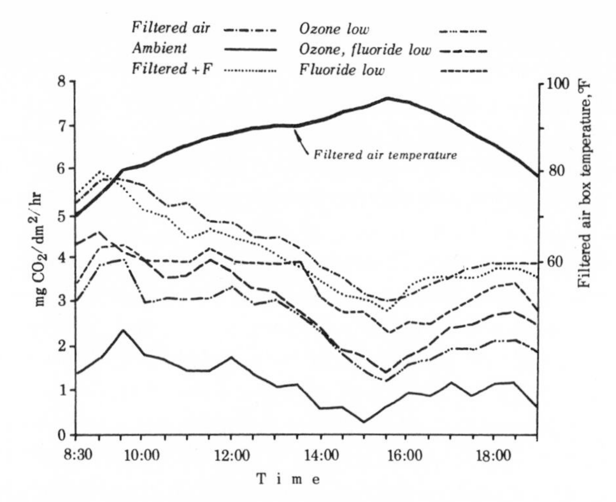

Figure 29–4. Correlation of air temperatures and rates of apparent photosynthesis of lemon branches exposed to the various experimental atmospheres.

In June–October 1964, a similar measurement was made. Evaluation of the data adjusted by the covariates of branch diameter and leaf area showed no treatments were statistically greater than ambient. All grand averages except "low fluoride air" in the last period were numerically greater than "ambient air."

Because the trees which received "filtered air" required more frequent irrigations and were thus assumed to be transpiring more moisture than the other plants, an attempt was made to measure the amount of moisture given off directly. The same air samples used for the apparent photosynthesis measurement in 1964 were exhausted from the infrared CO_2 analyzer over wet bulb thermocouples in an apparatus similar to the equipment designed originally by Slatyer and Bierhuizen (1964). Thus a measure of the water content of the intake and outlet air of the leaf chamber was obtained, and the milligrams of water transpired per square decimeter of leaf surface per hour were determined (Table 29–9). The

TABLE 29–9. Rate of transpiration of selected lemon branches (adjusted average Mg. $H_2O/Dm.^2/Hr.$[a])

Treatment	Filtered	Ambient	Filtered $+F^-$	Low O_3	Low O_3 F^-	Low F^-
Grand av. 6-24-64 to 8-31-64	494	302	389	312	316	399
Grand av. 9-14-64 to 11-8-64	488	492	366	439	383	362
No significant treatment differences						

[a] Adjusted for covariates.

values were adjusted for covariates of temperature and branch diameter. Statistical evaluation showed no differences between the different treatments. Variations were very large, and some unknown or extraneous factors may have affected these measurements to cause such large differences between individual values and thus invalidate the results.

The increased water use by the lemon trees receiving filtered air as compared with ambient air is indicative of an increased transpirational rate. Attempts to examine stomatal openings directly with the microscope or by silicone impressions to make visual comparisons were unsuccessful because the stomates in citrus are deeply recessed. Attempts to confirm the increased rate of transpiration by determining the water content of air samples coming from the leaf chambers could have encountered interference by condensation of water in the aluminum sampling lines which were as much as 40 meters in length. If this occurred during the night and a slow evaporation of liquid droplets occurred during the measuring period, the results could be meaningless. Also transpirational rate of some leaves proceeds by regular cycles (Ehrler *et al.*, 1965), and because the air samples were only drawn from a given sampling chamber for 2 minutes during each 30-minute period, the rates could have varied rather widely depending upon which part of the cycle the leaves were in at the time of sampling.

DISCUSSION

The higher rate of apparent photosynthesis of the entire trees receiving filtered air as compared with ambient air could be caused by several factors. Increased senescence and early abscission of leaves reported with other plants (Darley, 1959) could reduce effective photosynthetic capacity. Probably other metabolic changes such as those described by Dugger *et al.* (1966) are more important. They observed that lemon seedlings, fumigated with ozone, had less starch and more reducing sugars than controls, increased permeability to exogenous sugars and an increased rate of respiration.

Why the apparent photosynthesis of isolated leaves receiving "filtered air" and "filtered air plus fluoride" was not statistically greater than ambient air during 1964 as it was in 1963 is unexplained. Individual variation in photosynthesis between leaves is great, and the cycling in rates of CO_2 absorption observed with cotton by Pallas *et al.* (1967) also occurs with lemons. This phenomenon coupled with the short sampling periods could give such great variability as to invalidate the results during this period even though the average value of the filtered air was highest.

Another possible explanation of the reduced water use and lower apparent photosynthesis of trees in the four groups which did not have carbon filtered air—i.e., ambient, low ozone, low ozone-low fluoride, and low fluoride air—is that peroxyacyl nitrates were causing the effect. These compounds occur in concentrations from 3 to 12 percent of that of the ozone in this area.

In planning the study, the authors assumed that the addition of nitric oxide to react with ozone when total oxidant reached 0.10 p.p.m. would give a beneficial effect. Nitrogen dioxide was thought to be less phytotoxic than ozone. However, the results show little improvement and suggest that either the nitrogen dioxide is as deleterious as ozone or that the treatment with NO should have been started at lower levels of total oxidant—i.e., enough damage was caused by oxidant at 0.10 p.p.m. that additional oxidant had a relatively minor effect. Additional information on the effects of NO_2 on citrus is needed badly to aid in interpreting these results.

Atmospheric fluoride levels in the experimental areas seemed to have little deleterious effect on the trees. Average accumulations in 8-month-old leaves were 50 p.p.m. on a dry weight basis in 1961. Accumulations declined in similar aged leaves to 40 p.p.m. by the close of 1964.

ACKNOWLEDGMENTS

The authors acknowledge the technical aid and advice of W. M. Dugger, Jr., M. J. Garber, E. A. Cardiff, Earl Hensel, and Carol J. Adams.

LITERATURE CITED

Bellack, Ervin, and P. J. Schouboe. 1958. *Anal. Chem.* 30:2032–4.

Darley, E. F. (Pub. 1960). *Proc. Intern. Clean Air Conf. London*, 1959, p. T28.

Darley, E. F., and J. T. Middleton. 1966. *Ann. Rev. Phytopath.* 4:103–18.

Dugger, W. M., Jr., Jane Koukol, and R. L. Palmer. 1966. *J. Air Pollution Control Assoc.* 16(9):467–71.

Dunnett, Charles W. 1955. *J. Am. Statistical Assoc.* 50:1096–1121.

Ehrler, W. L., F. S. Nakayama, and C. H. M. Van Bavel. 1965. *Physiol. Plant.* 18: 766–75.

Franck, J. E., and W. E. Loomis. 1949. "Photosynthesis in Plants," p. 38, Iowa State College Press, Ames, Iowa.

Ivie, J. O., M. D. Thomas, and L. F. Zielenski. 1963. *Biomed. Sci. Instrumentation* 1:45–52.

Ivie, J. O., L. F. Zielenski, M. D. Thompson, and C. R. Thompson. 1965. *J. Air Pollution Control Assoc.* 15:195–7.

Middleton, J. T. 1961. *Ann. Rev. Plant Physiol.* 12:431–48.

Middleton, J. T., L. O. Emik, and O. C. Taylor. 1965. *J. Air Pollution Control Assoc.* 15(10):476–80.

Pallas, J. E., Jr., B. E. Michel, and D. G. Harris. 1967. *Plant Physiol.* 42:76–89.

Richards, B. L., and O. C. Taylor. 1960. *J. Air Pollution Control Assoc.* 11(3):125–8.

Saltzman, B. E. 1960. *Anal. Chem.* 32:135.

Slatyer, R. O., and J. F. Bierhuizen. 1964. *Plant Physiol.* 39:1051–6.
Taylor, O. C., E. A. Cardiff, and J. D. Mersereau. 1965. *J. Air Pollution Control Assoc.* 15:171–3.
Thomas, M. D. 1961. *World Health Organ. Monograph Ser.* MD46:233–78.
Thompson, C. R., and J. O. Ivie. 1955. *Intern. J. Air Water Pollution* 9:799–805.
Thompson, C. R., and O. C. Taylor. 1966. *Trans. Am. Soc. Agr. Eng.* 9(3)338, 339, and 342.
Willard, H. H., and O. B. Winter. 1933 *Ind. Eng. Chem., Anal. Ed.* 5:7–10.

Received for review March 27, 1967. Accepted July 31, 1967. This investigation was supported in part by Grant AP 00270-03 from the National Center for Air Pollution Control. U.S. Public Health Service.

PROBLEMS OF ENVIRONMENTAL POLLUTION

EUTROPHICATION OF AQUATIC ECOSYSTEMS

30

Basic concepts of eutrophication

Clair N. Sawyer

Eutrophication is a new word in the vocabulary of many sanitary engineers and scientists and is destined to become a part of the normal complement of words used by everyone concerned with the broad concept of water resources. The term is not a new one. It has been used by limnologists for nearly 50 years to describe the change in biological productivity which all lakes and reservoirs undergo during their life history.

LIFE HISTORY OF LAKES AND RESERVOIRS

Lakes and reservoirs are not permanent features of the landscape. Geologically speaking, they are only water-filled natural or man-made depressions in the earth's crust that are destined to become filled with soil and organic deposits as time passes. Fundamentally, they are giant sedimentation basins which not only serve to remove suspended matter from tributary waters but also act as giant reaction vessels for biological phenomena involving production of both plants and animals. From a strict conservationist viewpoint, the function of lakes is to retain on the land areas of the world those matters of value to the land and to prevent their being carried to the oceans.

The life span of lakes is normally reckoned in millenniums or even eons of time; however, there are exceptions. Lake Mead, the impound-

Reprinted with permission from *Journal Water Pollution Control Federation*, Vol. 38, pp. 737–744 (May, 1966), Washington, D.C. 20016.

ment above Boulder Dam, is predicted to have a life of less than 150 years because of the tremendous silt load carried by the Colorado River (1). Lake Constance on the Rhine is expected to have a life of only 12,000 years, one-half already spent (2).

Historically, young lakes are relatively barren bodies of water in terms of the amount of biological life which they support. In this phase they are referred to as being oligotrophic. As aging progresses the mate-

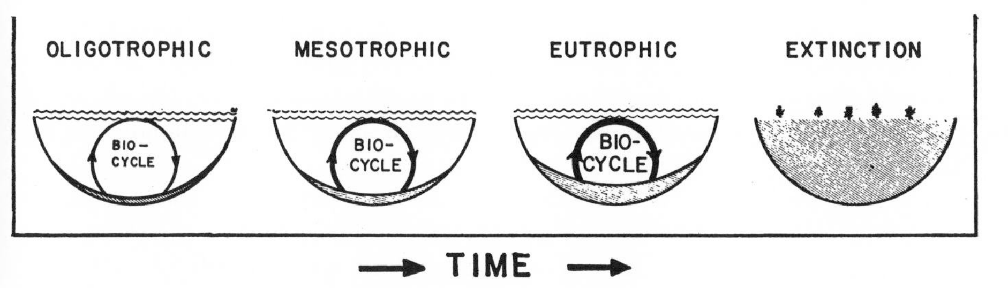

Figure 30–1. Natural transition of a lake through various stages of productivity eventually resulting in extinction.

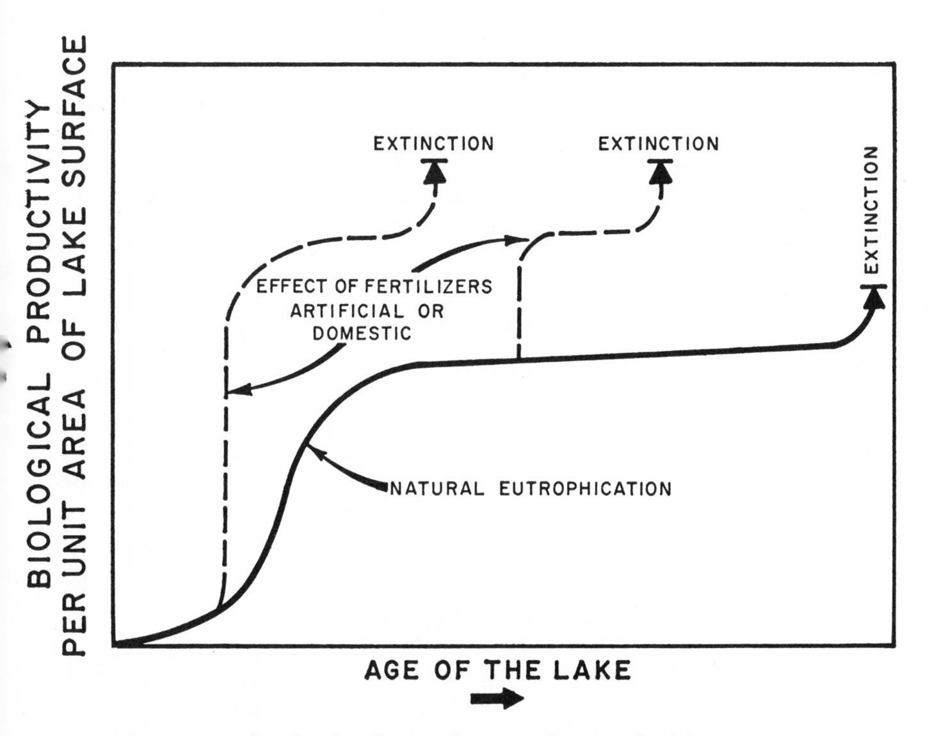

Figure 30–2. Natural and induced eutrophication [from Hasler (3)].

rials retained by the lake gradually increase in the bottom sediments and, through bacterial and other decomposition of the sediments, the lake waters become richer and richer in nutrient materials on which phytoplankton thrive. Concurrent with the increase in phytoplankton, the population of zooplankton and higher animal forms responds accordingly, as the food supply gains in amount.

With the increase in biological productivity of a lake, major changes occur in both the surface and deeper waters. The lake passes from the oligotrophic phase through the mesotrophic and finally into the eutrophic phase. It continues in this until deposits from biological activity, both organic and inorganic, plus materials settled from tributary waters, fill the basin to the extent that rooted aquatic plants take command of the situation and gradually convert the area to marsh land. These changes are illustrated in Figure 30–1.

The aging process of lakes and the effect of fertilization have been described admirably by Hasler (3) as shown in Figure 30–2. It also shows the effect of artificial fertilization on hastening the onset of eutrophication and shortening the life span of lakes. In a nutshell, it explains why we must consider the subject of eutrophication and what we can do about it.

EUTROPHICATION

Is eutrophication good or bad? That is a question to which there is no "pat" answer, since so much depends on the purposes for which the lake or reservoir is used and the degree of eutrophication. Many oligotrophic lakes are "gems of beauty" which reflect the azure blue of the sky beyond description, serve as admirable water supplies, but provide little recreational opportunity beyond swimming and boating. In some instances (4) oligotrophic lakes have been fertilized purposely to increase the production of phytoplankton and indirectly the population of fish life, the desired product.

Some of our eutrophic lakes are extremely valuable because of their ability to provide excellent fishing and serve as general purpose recreational areas. Others are notorious because of their ability to produce highly obnoxious blooms of nuisance algae which are aesthetically objectionable to recreationists and to residents on the shore.

Although there are many factors such as depth, size, shape, and geographical location which determine the degree of eutrophication which would be classed as beneficial or detrimental, in general the basic factor involved is the algae nutrient budget of the lake as shown in Figure 30–3. It is conceded generally that the biological productivity of a lake, in terms of phytoplankton, is related directly to the nutrients available per unit volume of water in the trophic zone. Of the nutrients entering a lake only

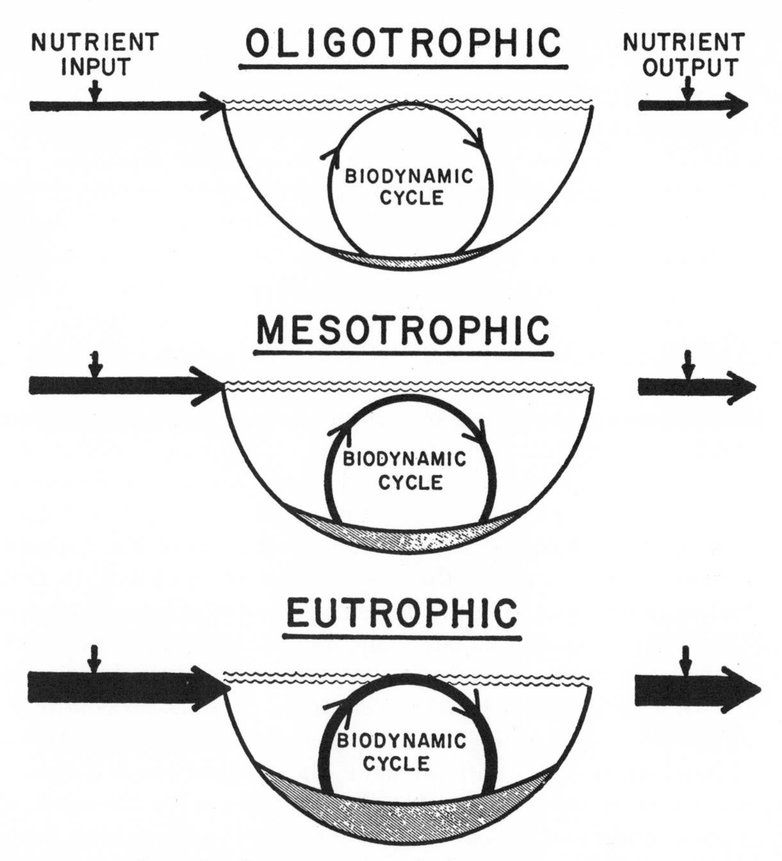

Figure 30–3. Relationship between nutrient budget and the magnitude of the biodynamic cycle in various types of lakes.

a fraction escape in the effluent. The fraction depends on a great many factors of which retention time, location of outlet with respect to prevailing winds, and nature of dominant blooms are perhaps most significant.

Of the nutrients which enter the lake, a major part becomes incorporated into algae and other forms of life which eventually die and settle to the lake bottom. There they are digested by bacteria, protozoa, worms, etc., with much of the nutrient materials solubilized. They then are free to leach back into the waters above and eventually become available to support further phytoplankton growths. The amount of nutrients which recycle from the bottom muds, of course, is proportional to the amount of material which drops to the mud from above.

There are many points of view on the question of eutrophication. Considering extreme positions, one might argue that the natural order of things is to let lakes convert to dry land and do nothing to stay the transition. Such a person has never had the experience of residing near a highly eutrophic lake or lived in an area devoid of lakes. On the other extreme, there are some who take the position of Hasler, who is reported to have said "it is clear that any increase in the rate of eutrophy, even if this involves only the acceleration of a natural and inevitable process is, from a human point of view, thoroughly undesirable." The latter seems somewhat idealistic. Somewhere between the two extremes should lie the "Happy Fishing Lake."

PARAMETERS OF EUTROPHICATION

There are many parameters of eutrophication. In its advanced stages it can be detected by visual observation and often by the sense of smell. When conditions become repulsive and drastic steps are required to control obnoxious growths of algae, then there is no longer need to marshall experts or scientific equipment to explain what has happened.

It is the responsibility of the scientist and the engineer to devise ways and means of measuring the degree of eutrophication and of developing methods of arresting its onslaught. We no longer must be satisfied with methods which simply measure trends toward eutrophication and regard them as irresistible changes over which man has no control. We must recognize as did Strom (2) over 40 years ago as he wrote "When a lake has reached the real eutroph stage, the changes toward a greater degree of eutrophy are very speedily effected, indeed; the character of a lake can be materially changed within a generation" and we must be prepared to act.

The parameters of eutrophication may be placed into two broad classifications. One group may be classed as indirect indicators as they in themselves play no part in increasing the biological productivity on which eutrophication depends. Most of these are chemical in nature and have been called on for lack of more definitive data in many cases. Beeton (5) in his paper discussing eutrophication of the Great Lakes, resorts to the increases of total solids, calcium, sodium, potassium, sulfate, and chloride as supporting evidence of the growing eutrophication of all the lakes except Superior. None of these elements are prime factors in biological productivity. If they were, the oceans of the world would know no limit. The mounting amounts of these materials are concrete evidence of increasing amounts of human, industrial, and agricultural wastewaters, all of which carry prime nutrients and possibly stimulants for biological growths. Beeton made no mention of the nitrogen and phosphorus associated with these cations and anions, an oversight that has been perpetuated time and time again by limnological investigators in spite of the fact that it has been known for over 40 years (6) that the principal components of all aquatic life are carbon, hydrogen, oxygen, nitrogen, and phosphorus.

The direct methods of evaluating eutrophication may be subdivided into qualitative and quantitave types.

Qualitative The qualitative type is well illustrated by the information presented in Table 30–1 which shows the nature of plankton in oligo-

TABLE 30–1. Plankton of oligotrophic and eutrophic lakes

Parameter	Oligotrophic	Eutrophic
Quantity	Poor	Rich
Variety	Many species	Few species
Distribution	To great depths	Trophogenic layer
Diurnal migration	Extensive	Limited
Water—blooms	Very rare	Frequent
Characteristic algal groups or genera	Chlorophyceae Desmids *Staurastrum* Diatomaceae *Tabellaria* *Cyclotella* Chrysophyceae *Dinobryon*	Cyanophyceae *Anabaena* *Aphanizomenon* *Microcystis* Diatomaceae *Melosira* *Fragilaria* *Stephanodiscus* *Asterionella*

trophic and eutrophic lakes according to Rawson (7). Use of the parameters listed for eutrophic lakes and recognition of their presence come much too late on the time scale and drastic action is required to overcome the damage done.

Certain organisms also have been proposed as indicators of the onset of eutrophication. A favorite one has been *Oscillatoria rubescens.* Others have been *Anabaena* of several species, *Aphanizomenon flos-aquae*, and *Microcystis aeruginosa.* All of these indicators signal that eutrophic conditions have arrived and that great damage may have occurred already.

In the case of deep lakes which stratify, salmonid fishes have served as indicators. Their presence is used as evidence that the lake is oligotrophic; their absence and the presence of other coarser forms is considered evidence of eutrophic conditions.

Quantitative A considerable variety of methods exists for the evaluation of the degree of biological productivity of a lake or reservoir. Fundamentally, we are concerned with all gradations of conditions between oligotrophic and eutrophic waters. Since the transition from one form to the other is gradual, the intermediate phase commonly is referred to as mesotrophic and is of great interest to those concerned with multiple uses of water resources.

The methods of quantitative assessment of the degree of eutrophi-

TABLE 30–2. Methods of quantitative assessment of eutrophication

Hypolimnetic oxygen
Dissolved
Rate of consumption
Biological productivity
Standing crop
Volume of algae
Transparency
Chlorophyll in epilimnion
Oxygen production
Carbon dioxide utilization
Nutrient levels
Nitrogen
Phosphorous
Nitrogen—phosphorous ratios

cation are listed in Table 30–2. Of these methods of measurement, oxygen determinations were the first to be used. Observations in the least productive lakes showed little change with depth while the more productive lakes showed a marked decrease in the hypolimnion, during the summer stagnation period, as illustrated in Figure 30–4. The changes in oxygen concentration were associated with the amount of organic matter in the form of dead algae and other life which entered the hypolimnion and

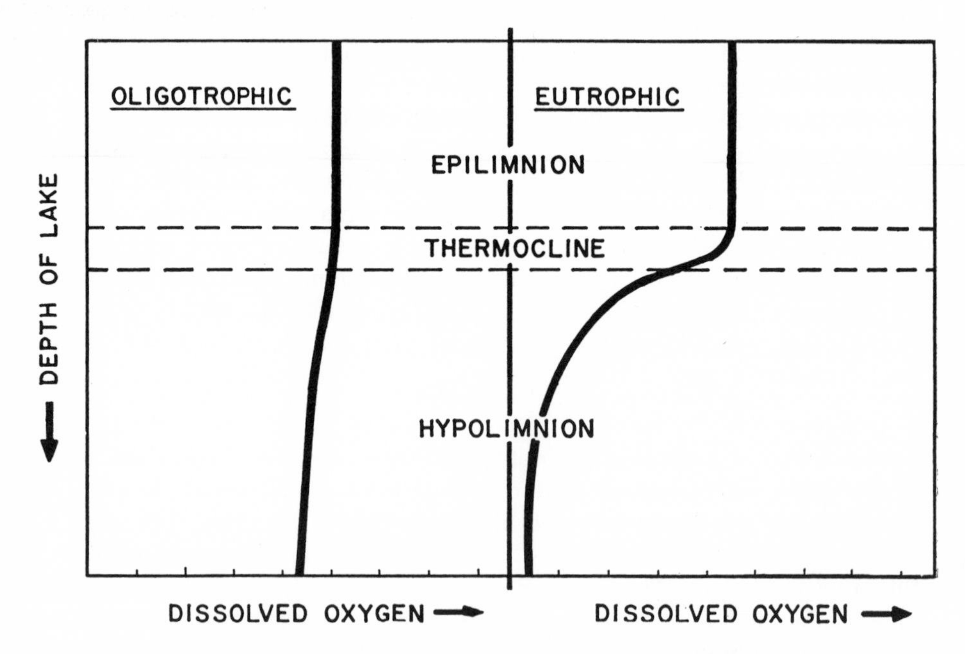

Figure 30–4. Typical DO curves in oligotrophic and eutrophic lakes during summer stagnation.

exerted an oxygen demand in proportion to their quantity due to bacterial decomposition. From these observations, oligotrophic lakes were identified as those in which DO was not depleted seriously during the summer stagnation period and eutrophic lakes vice versa.

The use of DO as a parameter of eutrophication was refined further by Hutchinson (8) when, through observations of the rate of oxygen depletion, he concluded that the rate of loss of hypolimnetic oxygen in oligotrophic lakes ranged from 0.004 to 0.033 mg/day/sq cm and the rate of loss in eutrophic lakes ranged from 0.05 to 0.14 mg/day/sq cm. Mortimer (8) has suggested 0.025 mg/day/sq cm as the upper limit for oligotrophy and 0.055 mg/day/sq cm as the lower limit for eutrophy. The range between 0.025 and 0.055 is reserved for mesotrophic lakes.

Measurements of biological productivity are, of course, the most direct way of evaluating eutrophication. Seven different methods have been used.

1. Standing Crop Perhaps the oldest method of determining the extent of biological productivity of a body of water was to make observations on the concentrations of individual organisms of the various types at periodic intervals over a growing season. Under normal circumstances the results obtained produced a reasonable picture of the biological productivity which could be related to lake type but it entailed a great deal of time and effort.

2. Volume of Algae Because of the great amount of labor involved in evaluating the standing crop many investigators have turned to a gross measurement of algae in terms of volume of algae. The results reported by Anderson (9) to show the increasing eutrophication of Lake Washington as shown in Figure 30–5 are typical.

3. Transparency A favorite and very simple way of measuring eutrophication is by measuring transparency with the Secchi disc. This method also measures gross biological productivity.

4. Chlorophyll in Epilimnion Perhaps the easiest method of all of measuring biological productivity and thereby the state of eutrophication is to determine the chlorophyll *a* content of near surface waters at periodic intervals over a growing season. Since the chlorophyll *a* content of lake waters is related directly to the amount of phytoplankton present, it furnishes indisputable evidence.

5. Oxygen Production The measurement of photosynthetic oxygen by the light- and dark-bottle technique has been used quite extensively to

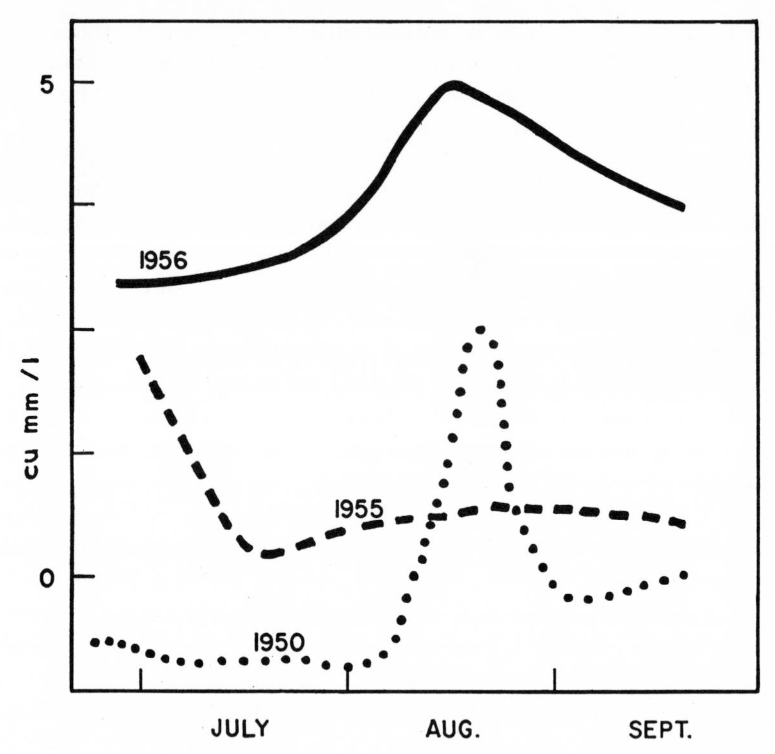

Figure 30–5. Volume of algae as a measure of progressive eutrophication of Lake Washington (9).

measure primary biological productivity but interpretation of data is a much disputed matter. Considerable doubt exists as to the reliability of this method but, in general, it yields quite valuable information.

6. Carbon Dioxide Utilization One of the newer methods of measuring primary biological productivity is by determining carbon dioxide utilization. The technique is simplified somewhat by using $C^{14}O_2$ as a tracer.

7. Nutrient Levels As our knowledge of the primary biological productivity of surface waters has increased there has been a growing awareness of the role of the major nutrients, carbon dioxide, nitrogen, and phosphorus. Since there is little possibility of controlling carbon dioxide levels in natural waters, attention has been focused on the fertilizing elements, nitrogen and phosphorus, which are such significant components of human wastes, certain industrial wastes, and drainage waters from agricultural lands.

The author (10), after studying 17 lakes in southeastern Wisconsin, was so brash as to suggest limits of inorganic phosphorus as a means of differentiating between well-behaved and nuisance lakes. Well-behaved lakes were those whose inorganic phosphorus content did not exceed 0.015 mg/ and whose inorganic nitrogen (NH_3, NO_2^-, NO_3^-) did not exceed 0.30 mg/ at the time of the spring overturn.

Protein is a major component of plankton organisms. Catches of net plankton from the Madison, Wisconsin, lakes showed protein contents ranging from 42 to 58 percent on an ash free basis and nitrogen content ranging from 6.8 to 10 percent (6). It might be reasonable to say that each part of nitrogen will give rise to 12 parts of plankton organisms. Thus, it seems logical to include nitrogen determinations as a means of evaluating biological productivity and, thereby, eutrophication. There are two precautions, however, to keep in mind in using nitrogen as a parameter. In the first place, the evaluations of inorganic nitrogen should be performed at the time the spring turnover is complete but before any significant growth or conversion of inorganic nitrogen to organic forms has occurred. The second relates to the fact that certain blue-green algae are capable of fixing nitrogen from the dissolved nitrogen in the water, provided other necessary nutrients are available. Evaluations of biological productivity on the basis of nitrogen alone are not necessarily reliable, unless phosphorus is included in the considerations.

Phosphorus is also one of the major nutrients required by phytoplankton. It appears to be much more labile or less firmly fixed within the cell tissue than nitrogen and the amounts required do not appear to be established very firmly. However, restrictions on phosphorus can be shown to limit and alter the nature of plankton growths. On the other hand, it has been demonstrated (11) (12) that adequate supplies of phosphorus are stimulatory to nitrogen fixing blue-green algae. For this reason phosphorus has been considered by many investigators as a key element in the control of biological productivity. There is ample reason to believe that it will be used more and more as one of the parameters of eutrophication.

Some investigators believe that nitrogen to phosphorus ratios are significant in predicting biological productivity. It is based on the premise that nutrients should be abstracted from water in the same ratio as that in which they occur in the living organisms. Although the nitrogen to phosphorus ratio varies in individual plankton forms from about 6 to 1 to about 25 to 1, a normal or average ratio for gross plankton is considered to be about 15 to 1. On this basis, it is argued that waters which contain nitrogen and phosphorus in ratios greater than 15 to 1 will show productivities which are nitrogen dependent and waters which have ratios less than 15 to 1 will have productivities that are phosphorus dependent. The latter, because of the possibility of containing excess phosphorus, would be stimulatory toward the development and growth of blue-green algae.

SUMMARY

Lakes and reservoirs are classified normally as oligotrophic, mesotrophic, or eutrophic based on biological productivity or parameters related to it. In the normal aging process natural lakes pass from oligotrophic to eutrophic bodies of water and eventually revert to marshes or moors and then dry land.

The parameters for measuring eutrophication are many. The basic problem concerning sanitary scientists and engineers, however, is how to arrest or control eutrophication.

The natural conversion of oligotrophic lakes to eutrophic lakes has been accelerated greatly by agricultural, human, and industrial wastes carrying fertilizing matters, principally nitrogen and phosphorus, stimulatory to phytoplankton.

It is axiomatic, therefore, that a major factor in the control of eutrophication depends on limiting the amount of nutrients entering a lake.

REFERENCES

1. Stevens, J. C. 1946. "Future of Lake Mead and Elephant Butte Reservoir." *Trans. Amer. Soc. Civil Engr.*, 111:1231.
2. Welch, P. S. 1952. "Limnology." 2nd Ed., McGraw-Hill Book Co., New York, N.Y., 353.
3. Hasler, A. D. 1947. "Eutrophication of Lakes by Domestic Drainage." *Ecology*, 28:383.
4. Hasler, A. D., and W. G. Einsele. 1948. "Fertilization for Increasing Productivity of Natural Inland Waters." *Trans. 13th N. Amer. Wildlife Conf.*, 527.
5. Beeton, A. M. 1965. "Eutrophication of the St. Lawrence Great Lakes." *Limnol. and Oceanog.*, 10:240.
6. Birge, E. A., and C. Juday. 1922. "The Inland Lakes of Wisconsin. The Plankton. I. Its Quality and Chemical Composition." *Wis. Geol. Nat. History Survey Bull.*, 64.
7. Rawson, D. S. 1956. "Algal Indicators of Trophic Lake Types." *Limnol. and Oceanog.*, 1:18.
8. Hutchinson, G. E. 1957. "A Treatise on Limnology." John Wiley & Sons, New York, N.Y., Vol. 1:64.
9. Anderson, G. C. 1961. "Recent Changes in the Trophic Nature of Lake Washington, Algae and Metropolitan Wastes." *Trans. 1960 Seminar*, Robert A. Taft San. Eng. Center, U.S.P.H.S., Cincinnati, Ohio.
10. Sawyer, C. N. 1947. "Fertilization of Lakes by Agricultural and Urban Drainage." *Jour. N. E. Water Works Assn.*, 61:109.
11. Sawyer, C. N., and A. F. Ferullo. 1961. "Nitrogen Fixation in Natural Waters under Controlled Laboratory Conditions." *In:* "Algae and Metropolitan Wastes," Robert A. Taft San. Eng. Center, Cincinnati, Ohio.
12. Dugdale, R. C., and J. C. Nees. 1961. "Recent Observations on Nitrogen Fixation in Blue-Green Algae." *In:* "Algae and Metropolitan Wastes." *Trans. 1960 Seminar*, Robert A. Taft San. Eng. Center, Cincinnati, Ohio.

31

Eutrophication of the St. Lawrence Great Lakes

Alfred M. Beeton

INTRODUCTION

Evidence of appreciable change in the biota and physicochemical conditions in Lake Erie (Beeton 1961) and speculation on possible changes in the other lakes, stemming from increases in total dissolved solids (Ayers 1962; Rawson 1951), have directed attention to eutrophication of the Great Lakes. The question does not concern the existence of eutrophication, because all lakes are aging and there is no reason to believe that the Great Lakes are exceptional. Of greatest consequence is the possibility of detection and perhaps even measurement of the rate of eutrophication of these large lakes. Important also is the effect of mankind on the normal rate of eutrophication.

Accelerated rates of aging due to man's activity have been detected in a number of lakes (Hasler 1947). The classic example is the Untersee of Lake Zürich, which urban effluents have changed from an oligotrophic to an eutrophic lake in a relatively short time. Recently, studies of Lake Washington at Seattle (Edmondson, Anderson, and Peterson 1956), and Fure Lake, Denmark (Berg *et al.* 1958) have demonstrated that these relatively large lakes are undergoing accelerated eutrophication due to man's influence. None of these lakes, however, is large in comparison with any of the St. Lawrence Great Lakes.

Reprinted with permission from *Limnology and Oceanography*, 10:240–254 (1965).

PRESENT TROPHIC NATURE OF THE GREAT LAKES

The meaning of eutrophication seems to vary according to the special interests of the individual. Limnologists agree that eutrophication is part of the aging of a body of water and implies an increase in the nutrient content of the waters. Most lakes change gradually from a nutrient-poor, oligotrophic, to a nutrient-rich, eutrophic, condition. At this point agreement ends in arguments on lake classification. Although the terms eutrophic, mesotrophic, and oligotrophic are used freely by limnologists and are well established in the literature, it is difficult to determine precisely what is meant by them. Various investigators have attached considerable significance to one or more of the following criteria in classifying lakes: abundance or species of plankton or both; benthic organisms; chemical characteristics; sediment types; distribution of dissolved oxygen; productivity; fish populations; and morphometry and morphology of the lake basin. Lake classification has been closely related to regional limnology and has been developed primarily from observations on small lakes. It is not surprising, then, to find that the Great Lakes do not fit readily into the various classification schemes that have been proposed. Rawson (1955, 1956) reviewed the problem of classifying large lakes and considered certain characteristics of the St. Lawrence Great Lakes.

Despite the troublesome problems of lake classification, an attempt should be made to classify the Great Lakes for discussion of their eutrophication and to facilitate comparison with other lakes. A better classification surely is needed since Lake Erie, for example, has been called oligotrophic, mesotrophic, and eutrophic by various investigators over the past 30 years.

Physicochemical conditions

Morphometry, transparency, total dissolved solids, conductivity, and dissolved oxygen content of the water appear to be useful in lake classification. On the basis of these five physical and chemical factors, two of the lakes would be classified as oligotrophic, one as tending toward the mesotrophic, and two as eutrophic. The low specific conductance and total dissolved solids, and the high transparency and dissolved oxygen of Lakes Huron and Superior agree with the commonly accepted characteristics of oligotrophy (Table 31–1). Lakes Erie and Ontario have the high specific conductance and total dissolved solids and, of special significance, the low concentration of dissolved oxygen in the hypolimnion characteristic of eutrophic lakes. Data were not available on the oxygen content of the hypolimnetic waters of Lake Ontario, but it is even more significant that the deep waters had a low percentage saturation of dissolved oxygen under essentially isothermal conditions. The transparencies of Lakes Erie

TABLE 31–1. Physical and chemical characteristics of the Great Lakes*

Lake	Mean depth (m)	Transparency (average Secchi disc depth, m)	Total dissolved solids (ppm)	Specific conductance (μmhos at 18C)	Dissolved oxygen
Oligotrophic	>20	High	Low: around 100 ppm or less	<200	High, all depths all year
Superior	148.4	10	60	78.7	Saturated, all depths
Huron	59.4	9.5	110	168.3	Saturated, all depths
Michigan	84.1	6	150	225.8	Near saturation, all depths
Eutrophic	<20	Low	High: >100	>200	Depletion in hypolimnion: <70% saturation
Ontario	86.3	5.5	185	272.3	50 to 60% saturation in deep water in winter
Erie, average for lake	17.7	4.5	180	241.8	
Central basin	18.5	5.0	—	—	<10% saturation, hypolimnion
Eastern basin	24.4	5.7	—	—	40 to 50% saturation, hypolimnion

*Criteria designating lake types are based primarily on factors considered important by Rawson (1960); specific conductance limits from Dunn (1954); dissolved oxygen criteria, Thienemann (1928). Data from Bureau of Commercial Fisheries except transparency and dissolved oxygen for Lake Ontario (Rodgers 1962).

and Ontario are not especially low in comparison with many small lakes, but in comparison to Lakes Huron and Superior, their average transparencies are indeed low. Lake Michigan falls between the other lakes from the standpoint of specific conductance, total dissolved solids, and transparency. The dissolved oxygen content of Lake Michigan water is near saturation, however, at all depths. On occasion, water in the deeper areas has a saturation between 70 and 80 percent, but these low values are infrequent.

Biological characteristics

The biological characteristics of all the Great Lakes, except Lake Erie, may place them in the oligotrophic category (Table 31–2). This classification surely holds for Lakes Huron, Michigan, and Superior, but is uncertain for Lake Ontario, where recent and detailed information is lacking on benthos and plankton. The types of bottom fauna, especially the dominant midge larvae, have been recognized for many years as good criteria for classifying lakes (Elster 1958); Brundin (1958) held this basis had world-wide application. Lakes Huron, Michigan, and Superior are all of the *Orthocladius-Tanytarsus* (*Hydrobaenus-Calopsectra*) type. *Hydrobaenus* and related genera are the only midges in the deeper parts of these lakes. *Calopsectra* occurs in shallower areas with a variety of other midges. In Lake Erie, *Procladius* and *Tendipes* spp. (*T. plumosus* group) are the dominant midges. *Cryptochironomus* and *Coelotanypus* also are abundant in the western basin. *Calopsectra* and a variety of midge larvae become progressively more abundant toward the eastern end of the lake and only a few *Tendipes* appear in samples from the deep (maximum, 64 m) eastern basin.

The crustaceans, *Mysis relicta* and *Pontoporeia affinis*, are characteristic of oligotrophic lakes, although they occur in lakes classed as mesotrophic. Both organisms require fairly high dissolved oxygen concentrations and cold water. Consequently, they are absent from highly eutrophic lakes. Both crustaceans are found in all of the Great Lakes, but in Lake Erie the major population is restricted to the eastern basin.

Oligotrophic lakes of the north have been considered salmonid lakes. It is not implied that salmonids do not occur in eutrophic lakes, but, when present, they are not the dominant fishes and usually they have a restricted bathymetric distribution. Salmonids dominate the fish populations in all of the lakes except Lake Erie, although in Lake Ontario ictalurids and percids are almost as important in the commercial fishery as salmonids (Table 31–2). Three native species (formerly four) of salmonids occur in Lake Erie, but during most of the year they are restricted to the eastern basin. Warmwater fishes dominate the fish fauna of Lake Erie now, including the eastern basin.

Considerable controversy exists over the value of plankton in lake classification. Rawson (1956) pointed out that the number of species present, as well as the ecological dominants, should be considered in characterizing plankton. He stressed that the dominant species of plankters in the oligotrophic lakes of Canada, as well as the Great Lakes, were not the species usually associated with oligotrophic lakes. Furthermore, a number of plankters commonly accepted as indicative of eutrophy are dominant species in these lakes. These observations are supported by the more recent plankton data from the Great Lakes. *Fragilaria crotonensis* is a dominant diatom species in Lakes Huron and Michigan as well as Lake Erie, and *Melosira granulata* is a dominant plankter in Lake Superior. Both species are considered to be indicators of eutrophy by Teiling (1955). On the other hand, plankters that are accepted as indicators of oligotrophy (*Dinobryon, Tabellaria, Cyclotella*) are also dominant in the Great Lakes (Table 31–2). It is agreed, nevertheless, that the Cyanophyceae, *Aphanizomenon, Anabaena,* and *Microcystis,* which are among the dominant plankters in Lake Erie, are good indicators of eutrophy. Little purpose could be served here, however, by further discussion of the problem of plankton indicators of lake types. Consequently, Rawson's (1956) ranking of algae in order of their occurrence from oligotrophy to eutrophy has been used, in part, in Table 31–2, since he worked on lakes similar in many ways to the St. Lawrence Great Lakes. On the basis of plankton abundance and the dominant species of phytoplankton, Lakes Huron, Michigan, and Superior would be considered oligotrophic and Lake Erie eutrophic.

TABLE 31–2. Biological characteristics of the Great Lakes*

Lake	Bottom fauna and dominant midges	Dominant fishes	Plankton abundance	Dominant phytoplankton†
Oligotrophic	*Orthocladius-Tanytarsus* type (*Hydrobaenus-Calopsectra*)	Salmonids	Low	*Asterionella formosa* *Melosira islandica* *Tabellaria fenestrata* *Tabellaria flocculosa* *Dinobryon divergens* *Fragilaria capucina*
Superior	*Pontoporeia affinis* *Mysis relicta* *Hydrobaenus*	Salmonids	Very low	*Asterionella formosa* *Dinobryon* *Synedra acus* *Cyclotella* *Tabellaria fenestrata* *Melosira granulata*
Huron	*Pontoporeia affinis*‡ *Mysis relicta* *Hydrobaenus* *Calopsectra*	Salmonids	Low	*Fragilaria crotonensis* *Tabellaria fenestrata* *Fragilaria construens* *Fragilaria pinnata* *Cyclotella kutzingiana* *Fragilaria capucina*
Michigan	*Pontoporeia affinis*§ *Mysis relicta* *Hydrobaenus*	Salmonids	Low	*Fragilaria crotonensis* *Melosira islandica*‖ *Tabellaria fenestrata* *Asterionella formosa* *Fragilaria capucina*
Ontario	*Pontoporeia affinis* *Mysis relicta*	Salmonids, ictalurids, percids	—	—
Eutrophic	*Tendipes plumosus* type	Yellow perch, pike, black bass	High	*Microcystis aeruginosa* *Aphanizomenon* *Anabaena*
Erie Central basin	*Tendipes plumosus*	Yellow perch, smelt, freshwater drum	High	*Melosira binderana* *Stephanodiscus* *Cyclotella* *Fragilaria crotonensis*
Eastern basin	*Pontoporeia affinis* few *Calopsectra*	Yellow perch, smelt, few salmonids	High	*Microcystis* *Aphanizomenon*

* Criteria for lake types as follows: tendipedid larvae, Brundin (1958); fish and plankton abundance, Welch (1952); dominant phytoplankton, Rawson (1956).
† Data for Lake Michigan from Bureau of Commercial Fisheries; Lake Superior from Putnam and Olson (1961); Lake Erie from Davis (1962); Lake Huron from Williams (1962).
‡ Data from Teter (1960).
§ Data from Merna (1960).
‖ Refers to *Melosira islandica-ambigua* type.

The combined biological, chemical, and physical characteristics of Lakes Huron and Superior clearly are those of oligotrophy. The biota and the high dissolved oxygen content of the deep hypolimnetic waters characterize Lake Michigan as oligotrophic but contrariwise, the high content of total dissolved solids indicates a trend toward mesotrophy. Lake Ontario, as a mesotrophic lake, retains the biota of an oligotrophic lake, but the physicochemical characteristics are those of eutrophy. The chemical content of the waters of Lake Ontario is closely similar to that of

Lake Erie, since the main inflow to Lake Ontario is from Lake Erie via the Niagara River. The trophic nature of Ontario has been determined to a large extent by the chemical history of Lake Erie waters. Lake Ontario, and perhaps Lake Michigan, would be eutrophic except for the large volumes of deep waters. Even in Lake Erie, the eastern basin has components of a fauna associated with oligotrophy and sufficient deep, cold, oxygenated water to maintain this fauna. These conditions exist despite the highly eutrophic nature of the central basin (flow through the lake is from west to east). The evident ability of the total dissolved oxygen content of the deep hypolimnetic waters to meet the oxygen demand of the organic production of the epilimnetic waters, as well as the oxygen demand of allochthonous materials, makes Lakes Michigan, Ontario, and eastern Lake Erie in some measure oligotrophic (or mesotrophic) because of their morphometry.

EVIDENCE OF EUTROPHICATION

The present trophic nature of the Great Lakes is to a considerable degree the result of their gradual aging since formation. Evidence is accumulating, however, which indicates that human activity is greatly accelerating the eutrophication of all of the lakes but Lake Superior. This evidence is most spectacular for Lake Erie. A difficult problem is one of finding acceptable indices of change.

Various criteria have been used by different investigators to demonstrate eutrophication. Hasler (1947) compiled information on 37 lakes affected by enrichment from domestic and agricultural drainage. Among the changes in many of these lakes were: the dramatic decline and disappearance of salmonid fishes and increases in populations of coarse fish; changes in the species composition of plankton; and blooms of blue-green algae. Special significance has been attached to blooms of *Oscillatoria rubescens.* As the Untersee of Lake Zürich changed from a salmonid to a coarse-fish lake, plankton abundance increased, different species became dominant in the plankton, transparency decreased, and the dissolved oxygen content of the deep waters decreased. At the same time, the concentrations of chlorides and organic matter increased (Minder 1918, 1938, 1943). Minder (1938) attributed the increase and changes in the plankton to the growing amount of phosphorus and nitrogen from domestic sources. *Oscillatoria rubescens* appeared explosively in 1898 and replaced the formerly dominant *Fragilaria capucina.* The cladoceran *Bosmina longirostris* replaced *B. coregoni* after 1911. Blooms of *Oscillatoria rubescens,* declines in the hypolimnetic oxygen, decrease in transparency, and increases in the abundance of plankton were cited by Edmondson, Anderson, and Peterson (1956) as evidence of eutrophication of Lake

Washington. They held this increased productivity to be the result of growing discharges of treated sewage into the lake. Similar changes were observed in Fure Lake by Berg *et al.* (1958). Species composition of the phytoplankton changed, transparency decreased, dissolved oxygen concentrations became low in the hypolimnion, and conductivity rose. These changes have occurred during the last 40 to 50 years. Berg (Berg *et al.* 1958, p. 176) stated, "The cause is an increased introduction of material with the sewage."

Our knowledge of the limnology of the Great Lakes in earlier years is seriously deficient. Observations useful for tracing changes in the Great Lakes are mostly limited to water-quality data, commercial fishing records, and a few observations on plankton. The fishing records and chemical data have the longest history and are the most reliable.

Chemical characteristics

Chemical data representative of the lakes proper were compiled from many sources (Table 31–3) ranging from isolated samplings and water-intake data to extensive lake-wide sampling. Consequently, records for a particular year may represent an average of hundreds, thousands, or only a few determinations. Data on magnesium were available from most of the sources plotted in Figures 31–2 and 31–3, but they are not included because no significant change in concentration could be detected in any of the lakes. For example, magnesium concentrations in Lake Erie averaged 7.6 p.p.m. in 1907 (Dole 1909), 8.0 p.p.m. in 1934 (Mangan, Van Tuyl, and White 1952), and 8.0 p.p.m. in recent years (Bureau of Commercial Fisheries data). Broadly speaking, there has been no significant change in Lake Superior. Other lakes in order of increasing chemical change are Huron, Michigan, Erie, and Ontario.

Lake Superior The indicated slight downward trend in total dissolved solids in Lake Superior is not significant and concentrations have remained at approximately 60 p.p.m. throughout the years (Fig. 31–1). Calcium, chloride, and sulfate concentrations also have remained the same since 1886 (Fig. 31–2). The close agreement among analyses of Lake Superior water by various individuals using different methods and techniques is unusual. The slight decrease in the sodium-plus-potassium content of the water probably is not real, because present analytical methods differ substantially from former ones. The uniformity in the chemical analyses here lends confidence to the reliability of the chemical data for the other lakes.

Lake Huron The slight increase in total dissolved solids in this lake is probably real, since about 30 percent of the inflow to Lake Huron is from Lake Michigan, where dissolved solids have risen significantly (Fig. 31–1). The sodium-plus-potassium content has remained about the same over

TABLE 31–3. Sources of data used in preparing Figs. 1, 2, and 3

Source and date	Data
Allen (1964)	Major ions, south-central Lake Huron, 1956.
Bading (1909)	Chlorides, Lake Michigan, 1909.
Barnard and Brewster (1909)	Chlorides, Lake Michigan, 16 Sept 1908, table 29.
Bartow and Birdsall (1911)	Major ions, average concentrations in 10 open Lake Michigan samples collected about 1910.
Beeton, Johnson, and Smith (1959)	Major ions, Lake Superior, 1953.
Birge and Juday (no date)	Total dissolved solids, Lake Erie, 1928–1930; data on analyses for total dissolved solids by L. A. Youtz for Lakes Erie, Huron, and Superior, 1928.
Bowles (1909)	Total dissolved solids, Lake Michigan, April 1908.
Collins (1910)	Major ions, Lake Michigan, 1885, from J. H. Long; total dissolved solids and chlorides, Chicago, average of weekly values 1897–1900.
Clarke (1924)	Major ions, Lake Michigan, Milwaukee, 1877; Kenosha and Racine, Wis., 1911.
Dole (1909)	Major ions and total dissolved solids, Lakes Erie, Huron, Michigan, Ontario, and Superior, averages of monthly determinations 1906–07.
Eddy (1943)	Major ions and total dissolved solids, Lake Superior, 1934.
Erie, Pa., Bur. Water (1956, 1957, 1959)	Major ions and total dissolved solids, Lake Erie, 1956, 1957, 1959.
Fish (1960)	Chlorides, Lake Erie, 1929.
Hunt (1857)	Major ions in water collected at Pointe des Cascades, Vandreuil, Que., Lake Ontario, 1854.
International Joint Commission (1951)	Chlorides and total dissolved solids, open Lake Erie near Detroit River mouth, sampling ranges P-1-W and LC, table N-17; open Lake Huron above Port Huron; open Lake Ontario, sampling locations 4 miles (6.4 km) or more from Niagara River, 1946–48.
Jackson (1912)	Chlorides and total dissolved solids, Lake Erie, 1910, tables 69, 70; 1911, tables 71, 72; 1912, table 68.
Kramer (1961)	Major ions, except potassium, Lake Erie, 1961.
Kramer (1962)	Major cations, western Lake Ontario, 1959.
Lake Michigan Water Commission (1909)	Chlorides and total dissolved solids, Lake Michigan, October 1908.
Lane (1899)	Sodium chloride, Lake Huron, 12 miles (19.3 km) above Port Huron and Alpena, Mich., 1895; Lake Michigan, 5 miles (8 km) off Milwaukee, Wis., 1895–99; total solids, Chicago, Ill., 1895; major ions and total dissolved solids, open Lake Superior 50 miles (80 km) from Keweenaw Pt., 1886.
Lenhardt (1955)	Major ions and total dissolved solids, Lake Michigan, 1954.
Leverin (1942)	Major ions and total dissolved solids, Lake Huron, Pt. Edward, Ont., 1934–37; Lake Ontario, average of 6 analyses, Kingston, Ont., 1934–38 and 1940.
Leverin (1947)	Major ions and total dissolved solids, Lake Erie, average of 6 determinations, Fort Erie, Ont., 1934–38; Lake Ontario, average of 7 determinations, Toronto, Ont., 1934–38; Lake Superior, average of 5 determinations, Sault Ste. Marie, Ont., 1936–38, one sample taken in open lake midway between Fort William and Sault Ste. Marie, 1942.
Lewis (1906)	Chlorides and total dissolved solids, Lake Erie, average of 6 analyses, Erie, Pa., 1901–1903.
Mangan, Van Tuyl, and White (1952)	Major ions and total dissolved solids, Lake Erie, 1934, 1945, and 1951.
Michigan Water Resources Commission (1954)	Major ions and total dissolved solids, Lake Huron at Alpena, East Tawas, and Harbor Beach, Mich.; Lake Michigan at Muskegon, St. Joseph, and Traverse City, Mich.; Lake Superior at Calumet, Mich.
Ohio, State of (1953)	Major ions, Lake Erie, Lorain, O., average values, 1950–52.
Reade (1903)	Major ions, Lake Ontario, water sample collected in the St. Lawrence River opposite Montreal, Que., 1884.
Thomas (1954)	Major ions and total dissolved solids, Lake Erie, at Chippawa, Ont.; Lake Huron at Goderich and Sarnia, Ont.; Lake Ontario at Gananoque and Port Hope, Ont.; Lake Superior at Sault Ste. Marie, Ont., averages of monthly analyses, 1948–49.
U.S. Geological Survey (1960)	Major ions and total dissolved solids, Lake Erie, Niagara River at Buffalo, N.Y.; Lake Ontario, St. Lawrence River at Cape Vincent, N.Y., analyses for August 1957.
U.S. Public Health Service (1961)	Chlorides, sulfates, and total dissolved solids, Lake Erie at Buffalo, N.Y.; Lake Huron at Port Huron, Mich.; Lake Michigan at Milwaukee, Wis.; Lake Ontario at Massena, N.Y.; Lake Superior at Duluth, Minn.; average values for Oct 1960–30 Sept 1961.
Wright (1955)	Chlorides, western Lake Erie, 1930.

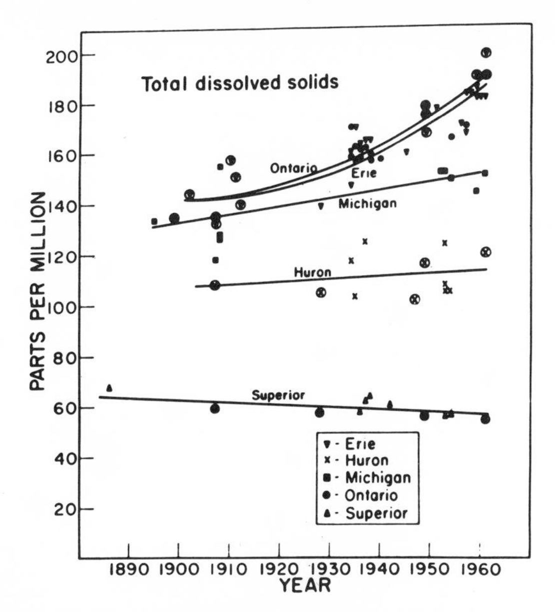

Fig. 31–1. Concentrations of total dissolved solids in the Great Lakes. Circled points are averages of 12 or more determinations. Data are from sources presented in the bibliography and Table 3.

the years of record. Some rather low values were reported for these ions during the 1930's, but the 1890–1910 data agree with recent determinations (Fig. 31–2). An increase of 3 p.p.m. in chloride appears to have occurred during the past 30 years. The major source of chlorides within the Lake Huron watershed is in the Saginaw Valley, where considerable quantities of brine are pumped to the surface in the oil fields and for use in the chemical industry. The increased influx of brine during the past 30 years may account for most of the increase in chloride in the lake. Sulfate concentrations have increased 7.5 p.p.m. in the past 54 years.

Lake Michigan Total dissolved solids have increased about 20 p.p.m. since 1895 (Fig. 31–1). Calcium has remained constant (Fig. 31–2). The greatest increase in any ion in Lake Michigan has been that of sulfate, which has risen 12 p.p.m. since 1877. The chloride concentrations have risen slowly but steadily by 4 p.p.m. The sodium-plus-potassium content has not changed since 1907 but it exceeds that extant in 1877–1900. The determinations before 1907 may be too low, since they are below those reported for Lakes Huron and Superior. If, however, these early determinations are reasonably accurate, the increase that occurred between 1877 and 1907 may be attributed to population growth in the Chicago area. The population of Chicago exceeded 1 million in 1890 and the

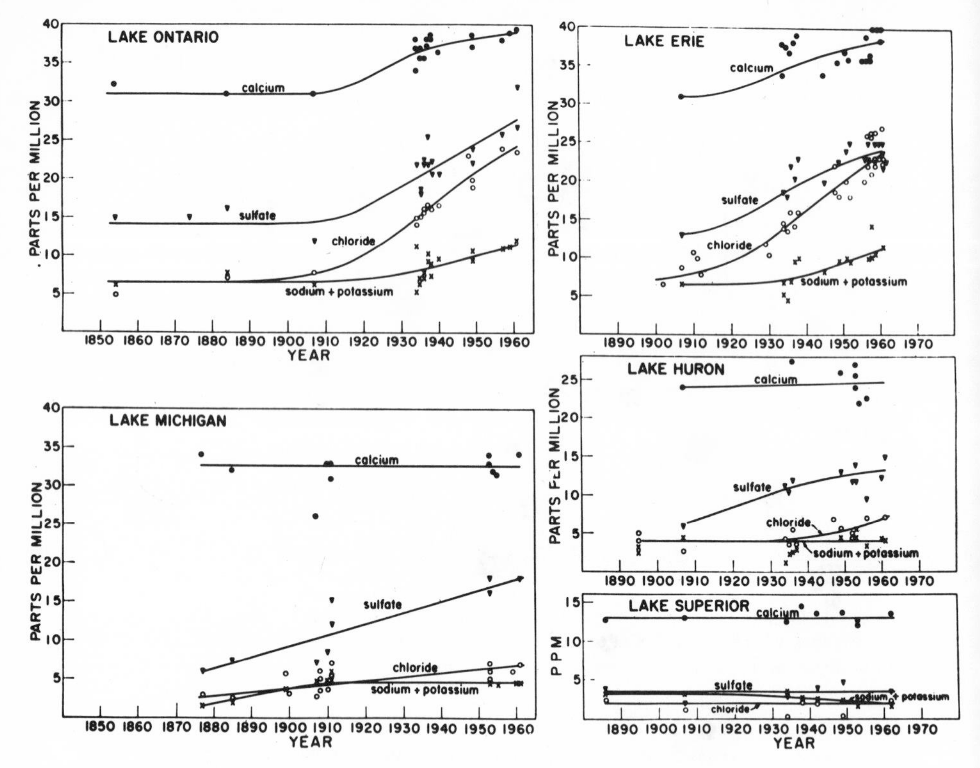

Fig. 31–2. Changes in the chemical characteristics of Great Lakes waters. Data for Lake Erie, 1958; Lake Huron, 1956; Lake Michigan, 1954, 1955, 1961; Lake Ontario, 1961; Lake Superior, 1952, 1953, 1961, 1962 are from the Ann Arbor Biological Laboratory, U.S. Bureau of Commercial Fisheries. Other data are from sources presented in the bibliography and Table 3.

Chicago Sanitary Canal, to divert sewage from the lake, was not completed until 1900. Consequently, during these early years considerable amounts of raw sewage entered the lake at Chicago.

Lake Erie Total dissolved solids, calcium, chloride, sodium-plus-potassium, and sulfate all increased significantly in Lake Erie during the past 50 years. Total dissolved solids have risen by almost 50 p.p.m. (Fig. 31–1). Increases of approximately 8, 16, 5, and 11 p.p.m. have taken place in the concentrations of calcium, chloride, sodium-plus-potassium, and sulfate, respectively (Fig. 31–2).

Lake Ontario The rate of increase in total dissolved solids in Lake Ontario has been the same as in Lake Erie. This rate was similar to that occurring in Lake Michigan prior to the late 1920's but has been higher

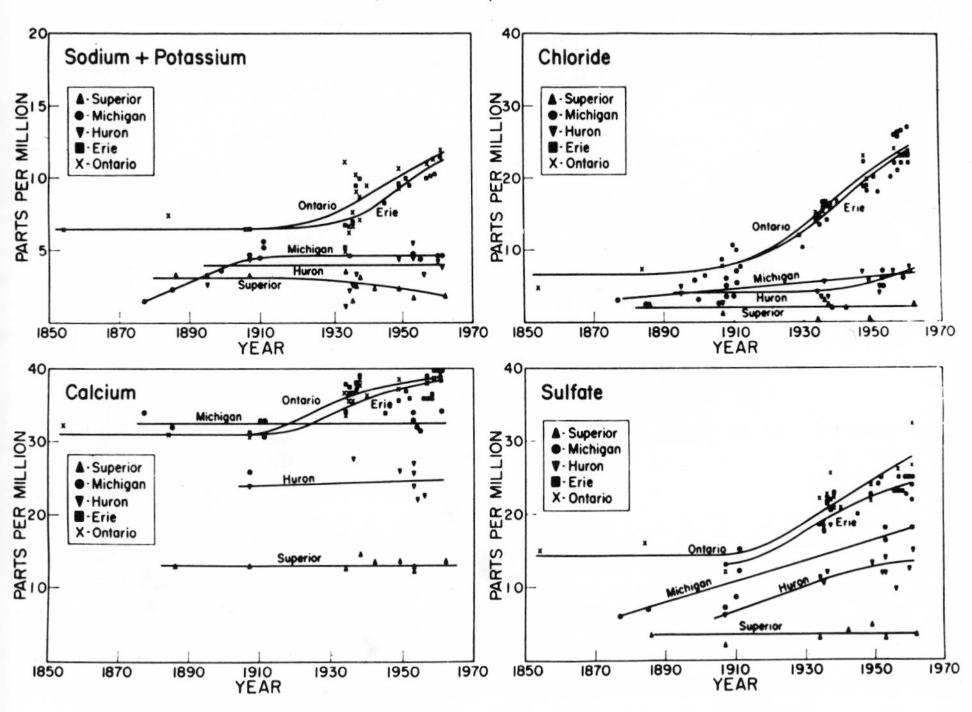

Fig. 31–3. Changes in the concentrations of calcium, chloride, sodium-plus-potassium, and sulfate in each of the Great Lakes. Sources of data same as for Fig. 2.

than in Lake Michigan since about 1930 (Fig. 31–1). Close agreement of chemical data for Lake Ontario for 1854 and 1884 with those for 1907 indicates that the chemical characteristics of the water were altered little during this period (Fig. 31–2). Calcium, chloride, and sodium-plus-potassium increased to the same extent as noted in Lake Erie since 1910. Sulfate concentrations increased by 13 p.p.m., which is somewhat higher than in Lake Erie.

Summary of chemical changes The extent of change in total dissolved solids in Lakes Erie, Michigan, and Ontario has not been as great as that indicated by Ayers (1962) for Lake Michigan or by Rawson (1951) for Lakes Erie and Ontario; both used the observations of Dole (1909) in 1906–1907 as their base. Dole's estimates of total dissolved solids (and of various ions as well) for these lakes were 9 to 16 p.p.m. lower than those of several other investigators during this period. The 1907 data for Lake Michigan on sulfate, chloride, and especially calcium probably were all low because Dole collected his samples in the Straits of Mackinac where Lake Michigan water enters Lake Huron and the mixing of the water from these two lakes, as well as the occasional inflow of Lake Superior water into this area, could produce low concentrations of ions.

Changes in the chemical characteristics of Lake Ontario have closely paralleled those in Lake Erie (Fig. 31–3). Prior to 1910 the chemical characteristics of the two lakes were similar and conditions in Lake Erie were probably the same as indicated by the 1854 and 1884 analyses of Lake Ontario water. Concentrations of calcium, chloride, sodium-plus-potassium, and sulfate have been somewhat higher in Lake Ontario than in Lake Erie during the past 50 years. The greater concentrations of salts in Lake Ontario probably can be attributed to growth of the Toronto, Hamilton, and Rochester metropolitan areas and the industrial expansion along the upper Niagara River.

Lakes Erie and Ontario are the only lakes in which calcium increased materially (Fig. 31–3). Increases in sulfate have been significant in all of the lakes except Superior. The 11 p.p.m. and 13 p.p.m. increases in Lake Erie and Ontario have taken place in 30 years, whereas the rise of 12 p.p.m. in Lake Michigan has been more gradual over a period of 84 years. The sulfate change in Lake Huron parallels that in Lake Michigan and may have resulted largely from the inflow of Lake Michigan waters. The degree of change in the chloride content of Lakes Erie and Ontario is similar to that for sulfate, but chloride has not increased as much as sulfate in Lakes Huron and Michigan.

Plankton

Few plankton data are available for Lakes Huron, Ontario, and Superior, especially for earlier years. Some rather extensive plankton data do exist, however, for Lakes Erie and Michigan.

Lake Michigan Studies of Lake Michigan phytoplankton by Briggs (1872), Thomas and Chase (1886), Eddy (1927), Ahlstrom (1936), Damann (1945), and Williams (1962) show that the diatom species dominant 90 years ago have maintained their importance. The relative abundance and occurrence of individual species give no evidence of change in the phytoplankton. (Some confusion exists over the identification of certain species and some species are listed by one investigator and not by another.)

The best information of change of plankton abundance probably comes from Damann's (1960) publication on 33 years of plankton data from the Chicago water intake which showed an average increase of 13 organisms per ml per year in the standing crop of the total plankton. Damann's work, and indeed most of the past work on plankton, has been in the extreme southern end of the lake except for the open-lake data of Ahlstrom (1936). Damann (1945, p. 771) pointed out that the data he published ". . . are an expression of the plankton activity in only a small portion of the southwestern corner of Lake Michigan."

Recent plankton collections during July 1960 to July 1961 from Gary, Indiana, at the southern end of the lake, and Milwaukee, Wisconsin, near the middle of the lake (Williams 1962), show considerable differences between these two localities. The average phytoplankton count at Milwaukee was 975 plankters/ml. This figure is close to the annual average of 952/ml for the Chicago water intake 1926–1942 (Damann 1960). The average phytoplankton count at Gary was 1,914 plankters/ml, well above the annual average of 1,222/ml reported at Chicago for 1943–1958 (Damann 1960). Local differences in the relative importance of the major diatom species at Gary and Milwaukee also are apparent. Sampling by the Bureau of Commercial Fisheries, in 1960, yielded further evidence of wide variability. The relative abundance of the plankton species was different from that at the Gary and Milwaukee water intakes, and a much higher abundance of plankton was obtained with nets towed vertically. Plankton counts in the open-lake samples ranged from about 450 to 12,000 plankters/ml and averaged around 4,500/ml.

Two changes in the species composition of the zooplankton of Lake Michigan may be of some consequence. Apparently the cladoceran, *Bosmina longirostris*, has replaced *B. coregoni.* A similar change was noted as evidence of eutrophication in Lake Zürich (Minder 1938). Wells (1960) did not find *B. coregoni*, although *B. longirostris* was present in all of his 1954 and 1955 samples. *Bosmina coregoni* (*B. longispina*) was the most abundant cladoceran in samples collected in 1887–1888 and 1926–1927, and *B. longirostris* was listed as rare (Eddy 1927). *Bosmina coregoni* is an important component of the Lake Superior plankton (Putnam and Olson 1961), whereas *B. longirostris* has been important in Lake Erie plankton for many years (Fish 1960; Davis 1962).

The other possible change in the Lake Michigan plankton is the increased prominence of *Diaptomus oregonensis*. This species was not found by Eddy (1927), but Wells (1960) reported it to be present on all collection dates and it was a dominant diaptomid in the fall of 1955.

Lake Erie Some significant changes have been observed in the plankton of Lake Erie. Evidently copepods and especially cladocerans have shown a marked increase in abundance since 1939 (Bradshaw 1964). Bradshaw attached some importance to the "recent" occurrence of the cladocerans *Eurycerus lamellatus*, *Chydorus sphaericus*, and *Ilyocryptus* sp., since they had not been found by Chandler (1940) in 1938–1939. *Eurycerus lamellatus*, *Chydorus sphaericus*, and two species of *Ilyocryptus* were present, however, in 1929 (Fish 1960). A copepod, *Diaptomus siciloides*, which was reported as "incidental" in Lake Erie plankton in 1929 and 1930 (Wright 1955), is now one of the two most abundant diaptomids (Davis 1962). Marsh's (unpublished manuscript) account of plankton in 1929 and 1930 included the statement, "The occurrence of *Diapto-*

mus siciloides in Lake Erie is a matter of decided interest, as it has never before been found in any of the Great Lakes." Later he continued, "*D. siciloides* may be considered an accidental intruder in the lake plankton . . ."

CHANGES IN LAKE ERIE

Several important changes in Lake Erie have not been detected in the other lakes; all indicate an accelerated rate of eutrophication.

Fish populations

The abundance of several commercially important fishes in Lake Erie has changed markedly during the past 40 years.[1] The fish populations of all of the lakes have changed but most changes, except in Lake Erie, have been the direct or indirect consequence of the buildup in sea lamprey populations (Smith 1964). The sea lamprey has not been important in Lake Erie, where few of the tributaries offer suitable spawning conditions.

The lake herring or cisco contributed around 20 million pounds (9 million kg) annually and as much as 48.8 million pounds (22.1 million kg) to the commercial catch (U.S. and Canada) prior to 1925. In 1925, the production declined to 5.8 million pounds (2.6 million kg) and continued to decrease. The take has amounted to a fraction of that since the early 1930's, except for landings of more than 2 million pounds (about 1 million kg) in 1936 and 1937 and a production of 16.2 million pounds (7.4 million kg) in 1946. The total production was only 7,000 pounds (3,200 kg) in 1962.

The whitefish fishery has been at an all-time low since 1948. The 1962 catch was 13,000 pounds (6,000 kg), whereas production had been 2 million pounds (1 million kg) or more for many years.

The sauger was contributing 1 million pounds (500,000 kg) or more to the commercial production prior to 1946. The catch has not reached 0.5 million pounds (0.22 million kg) since 1945 and has declined progressively since 1953. Production has been only 1 to 4 thousand pounds (450–1,800 kg) in recent years.

The walleye production increased during the 1940's and 1950's to reach 15.4 million pounds (7 million kg) in 1956. Production has decreased to less than 1 million pounds (450,000 kg).

The commercial catch of blue pike has dropped disastrously. The production fluctuated around an average of about 15 million pounds

[1] Statistical data cited are from Baldwin and Saalfeld (1962).

(6.8 million kg) for many years. The landings dropped to 1.4 million pounds (640,000 kg) in 1958, to 79,000 pounds (36,000 kg) in 1959, and to only 1,000 pounds (450 kg) in 1962. Of the few blue pike caught in 1963, most were more than 10 years old.

The total production of all species in Lake Erie continues to be around 50 million pounds (22.7 million kg), but only because more fresh-water drum (sheepshead), carp, yellow perch, and smelt are being caught than in the past. The major factor in the decline in the commercial catch of the more desirable species has been their failure to reproduce.

Bottom fauna

The changes in the species composition of the bottom fauna of the area west of the islands in Lake Erie have been sweeping. Carr and Hiltunen (unpublished manuscript)[2] have shown that few of the formerly abundant mayfly nymphs (*Hexagenia*) now inhabit this area and that tubificids are far more abundant now than 30 years ago (Fig. 31–4). Midge larvae and tubificids have increased and mayflies have decreased also among the islands and in the western part of the central basin (Beeton 1961).

Dissolved oxygen

Synoptic surveys of Lake Erie in 1959 and 1960 have demonstrated that low dissolved oxygen concentrations (3 p.p.m. or less) appear in about 70 percent of the hypolimnetic waters of the central basin during late summer (Beeton 1963). Scattered observations of some relatively low dissolved oxygen concentrations have been made during the past 33 years. The information we have indicates that the severity of depletion is more frequent and greater now than in the past and probably affects a more extensive area (Carr 1962).

CONCLUSION

The chemical content of the water in all of the Great Lakes except Lake Superior has changed in some measure. The biota also has changed in Lake Michigan and especially Lake Erie. These changes, remarkable for such large lakes, are those characteristic of eutrophication in smaller lakes and have come about over the relatively short time of 50 to 60 years.

[2] Carr, J. F., and J. K. Hiltunen. Changes in the bottom fauna of Lake Erie, west of the islands, 1930–1961. Paper presented at the 11th annual meeting Midwest Benthological Society, Murfreesboro, Tenn., 18 April 1963.

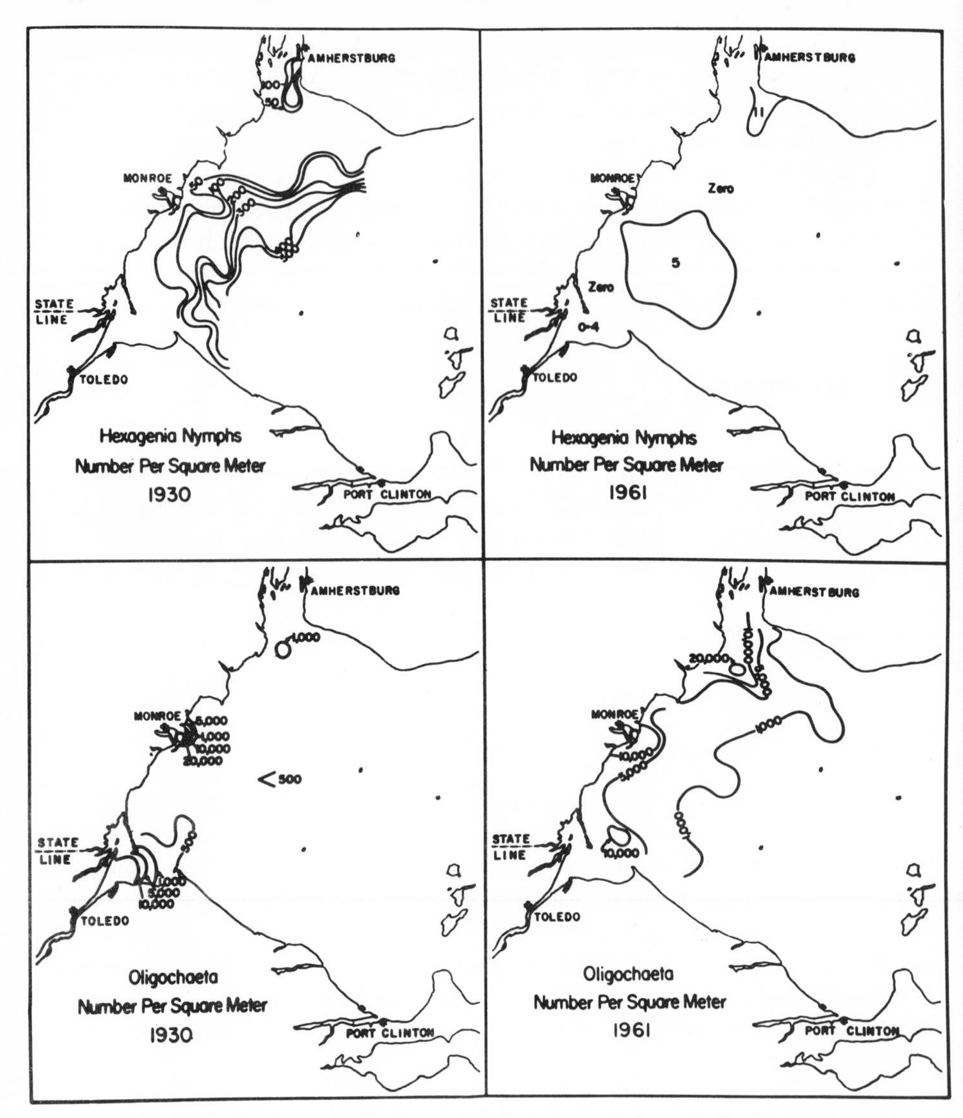

Fig. 31–4. Distribution and abundance of *Hexagenia* nymphs and oligochaetes in western Lake Erie, 1930 and 1961.

Man's activities clearly have accelerated the rate of eutrophication. This rate has been greatest in Lakes Erie, Ontario, and Michigan and these lakes have had the largest population growth within their drainage areas. An indication of the growth of population comes from census data for the northeast central states; the population there increased from 4.5 million to 16 million between 1850 and 1900, and by 1960 the population was 36.3 million. The rate of population growth increased sharply after 1910.

The substantial increases in the chemical content of the waters of Lakes Erie and Ontario also have appeared since 1910. The increases have been greatest for chloride and sulfate, both of which are conspicuous in domestic and industrial wastes, whereas magnesium concentrations have not changed measurably. Most of the magnesium comes from the limestones in the Lake Michigan basin; the stability of magnesium concentrations, therefore, indicates no appreciable change in the erosion of these deposits. The population along Lake Superior always has been sparse. The population along Lake Huron has been far less than on Lake Michigan, Erie, and Ontario. Most changes in the open-lake waters of Lake Huron have resulted from the inflow of Lake Michigan water. Undoubtedly, Saginaw Bay, Lake Huron, has been changed appreciably by the extensive growth of industry and agriculture within the Saginaw River valley during the past 50 years. High chloride concentrations have been measured in the bay (Adams 1937); no attempt has been made, however, to assess the extent of change.

REFERENCES

Adams, M. P. 1937. Saginaw Valley report. Mich. Stream Control Commission. 156 pp.

Ahlstrom, E. H. 1936. The deep-water plankton of Lake Michigan, exclusive of the Crustacea. Trans. Am. Microscop. Soc. 55:286–299.

Allen, H. E. 1964. Chemical characteristics of south-central Lake Huron. Great Lakes Res. Div., Inst. Sci. and Tech., Univ. Mich., Publ. No. 11, pp. 45–53.

Ayers, J. C. 1962. Great Lakes waters, their circulation and physical and chemical characteristics. Am. Assoc. Advan. Sci., Publ. No. 71, pp. 71–89.

Bading, G. A. 1909. Water conditions at Milwaukee. Lake Michigan Water Comm., Rept. No. 1, pp. 36–39.

Baldwin, N. S., and R. W. Saalfeld. 1962. Commercial fish production in the Great Lakes 1867–1960. Great Lakes Fish. Comm., Tech. Rept. No. 3. 166 pp.

Barnhard, H. E., and J. H. Brewster. 1909. The sanitary condition of the southern end of Lake Michigan, bordering Lake County, Indiana. Lake Michigan Water Comm., Rept. No. 1, pp. 193–266.

Bartow, E., and L. I. Birdsall. 1911. Composition and treatment of Lake Michigan water. Lake Michigan Water Comm., Rept. No. 2, pp. 69–86.

Beeton, A. M. 1961. Environmental changes in Lake Erie. Trans. Am. Fisheries Soc., 90:153–159.

———. 1963. Limnological survey of Lake Erie 1959 and 1960. Great Lakes Fish. Comm., Tech. Rept. No. 6. 32 pp.

———, J. H. Johnson, and S. H. Smith. 1959. Lake Superior limnological data. U.S. Fish Wildlife Serv., Spec. Sci. Rept. Fisheries No. 297. 177 pp.

Berg, K., K. Andersen, T. Christensen, F. Ebert, E. Fjerdingstad, C. Holmquist, K. Korsgaard, G. Lange, J. M. Lyshede, H. Mathiesen, G. Nygaard, S. Olsen, C. V. Otterstrøm, U. Røen, A. Skadhauge, E. Steemann Nielsen. 1958. Investigations on Fure Lake 1950–54. Limnological studies on cultural influences. Folia Limnol. Scandinavica, 10(1958). 189 pp.

Birge, E. A., and C. Juday. No date. The organic content of the water of Lake Erie. Supplemental data to "A limnological survey of western Lake Erie with special reference to pollution," by Stillman Wright. Ohio Div. Wildlife. Unpublished manuscript. 281 pp.

Bowles, J. T-B. 1909. Investigation of typhoid fever epidemic at Sheboygan, Wisconsin. Lake Michigan Water Comm., Rept. No. 1, pp. 90–95.

Bradshaw, A. S. 1964. The crustacean zooplankton picture: Lake Erie 1939–49–59; Cayuga 1910–51–61. Verhandl. Intern. Ver. Limnol., 15:700–708.

Briggs, S. A. 1872. The Diatomaceae of Lake Michigan. The Lens, 1:41–44.

Brundin, L. 1958. The bottom faunistical lake type system. Verhandl. Intern. Ver. Limnol., 13:288–297.

Carr, J. F. 1962. Dissolved oxygen in Lake Erie, past and present. Great Lakes Res. Div., Inst. Sci. and Tech., Univ. Mich., Publ. No. 9, pp. 1–14.

Chandler, D. C. 1940. Limnological studies of western Lake Erie. I. Plankton and certain physical-chemical data of the Bass Islands Region, from September, 1938, to November, 1939. Ohio J. Sci., 40:291–336.

Collins, W. D. 1910. The quality of the surface waters of Illinois. U.S. Geol. Surv., Water Supply Papers, 239. 94 pp.

Clarke, F. W. 1924. The composition of the river and lake waters of the United States. U.S. Geol. Surv., Profess. Papers, No. 135. 199 pp.

Damann, K. E. 1945. Plankton studies of Lake Michigan. I. Seventeen years of plankton data collected at Chicago, Illinois. Am. Midland Naturalist, 34:769–796.

———. 1960. Plankton studies of Lake Michigan. II. Thirty-three years of continuous plankton and coliform bacteria data collected from Lake Michigan at Chicago, Illinois. Trans. Am. Microscop. Soc., 79:397–404.

Davis, C. C. 1962. The plankton of the Cleveland Harbor area of Lake Erie in 1956–1957. Ecol. Monographs, 32:209–247.

Dole, R. B. 1909. The quality of surface waters in the United States. Part I. Analyses of waters east of the one hundredth meridian. U.S. Geol. Surv., Water Supply Papers, 236. 123 pp.

Dunn, D. R. 1954. Notes on the bottom fauna of twelve Danish lakes. Vidensk. Medd. Dansk Naturhist. Foren., 116:251–268.

Eddy, S. 1927. The plankton of Lake Michigan. Illinois Nat. Hist. Surv. Bull., 17:203–232.

———. 1943. Limnological notes on Lake Superior. Proc. Minn. Acad. Sci., 11:34–39.

Edmondson, W. T., G. C. Anderson, and D. R. Peterson. 1956. Artificial eutrophication of Lake Washington. Limnol. Oceanog., 1:47–53.

Elster, H.-J. 1958. Das limnologische Seetypensystem, Rückblick und Ausblick. Verhandl. Intern. Ver. Limnol., 13:101–120.

Erie, Pennsylvania, Bureau of Water. 1956. Ninetieth annual report, 1956. 63 pp.

———. 1957. Ninety-first annual report, 1957, 63 pp.

———. 1959. Ninety-third annual report, 1959. 56 pp.

Fish, C. J. 1960. Limnological survey of eastern and central Lake Erie, 1928–1929. U.S. Fish Wildlife Serv. Spec. Sci. Rept. Fisheries, No. 334. 198 pp.

Hasler, A. D. 1947. Eutrophication of lakes by domestic drainage. Ecology, 28:383–395.

Hunt, T. S. 1857. The chemical composition of the waters of the St. Lawrence and Ottawa Rivers. Phil. Mag., Ser. 4, 13:239–245.

International Joint Commmission. 1951. Report of the International Joint Commission United States and Canada on pollution of boundary waters. Washington and Ottawa. 312 pp.

Jackson, D. D. 1912. Report on the sanitary conditions of the Cleveland water supply. Cleveland. 148 pp.

Kramer, J. R. 1961. Chemistry of Lake Erie. Great Lakes Res. Div., Inst. Sci. and Tech., Univ. Mich., Publ. No. 7, pp. 27–56.

———. 1962. Chemistry of western Lake Ontario. Great Lakes Res. Div., Inst. Sci. and Tech., Univ. Mich., Publ. No. 9, pp. 21–28.

Lake Michigan Water Commission. 1909. Comparative analysis of samples of water from Lake Michigan. Rept. No. 1, pp. 103–105.

Lane, A. C. 1899. Lower Michigan waters: a study into the connection between their chemical composition and mode of occurrence. U.S. Geol. Surv. Water-supply Irrigation Papers, 31. 97 pp.

Lenhardt, L. G. 1955. Water quality and water usage of the Great Lakes public water supplies. The Great Lakes and Michigan. Great Lakes Res. Inst., Univ. Mich., pp. 13–15.

Leverin, H. A. 1942. Industrial waters of Canada. Can. Dept. Mines Resources, Mines Geol. Branch, Rept. 807. 112 pp.

———. 1947. Industrial waters of Canada. Can. Dept. Mines Resources, Mines Geol. Branch, Rept. 819. 109 pp.

Lewis, S. J. 1906. Quality of water in the upper Ohio River basin and at Erie, Pa. U.S. Geol. Surv. Water-supply Irrigation Papers, 161. 114 pp.

Mangan, J. W., D. W. Van Tuyl, and W. F. White, Jr. 1952. Water resources of the Lake Erie shore region in Pennsylvania. U.S. Geol. Surv. Circ. 174. 36 pp.

Marsh, C. D. No date. The Crustacea of the plankton of western Lake Erie. Supplemental data to "A limnological survey of western Lake Erie with special reference to pollution," by Stillman Wright. Ohio Div. Wildlife. Unpublished manuscript. 31 pp.

Merna, J. 1960. A benthological investigation of Lake Michigan. M.S. Thesis, Michigan State Univ. 74 pp.

Michigan Water Resources Commission. 1954. Great Lakes water temperatures. Unpublished manuscript. 50 pp.

Minder, Leo. 1918. Zur Hydrophysik des Zürich u. Walensees, nebst Beitrag zur Hydrochemie u. Hydrobakteriologie des Zürichsees. Arch. Hydrobiol., 12:122–194.

———. 1938. Der Zürichsee als Eutrophierungsphänomen. Summerische Ergebnisse aus fünfzig Jahren Zürichseeforschung. Geol. Meere Binnengewässer, 2:284–299.

———. 1943. Neuere Untersuchungen über den Sauerstoffgehalt and die Eutrophie des Zürichsees. Arch. Hydrobiol., 40:279–301.

Ohio, State of. 1953. Lake Erie pollution survey, supplement. Ohio Div. Water, Final Rept., Columbus, Ohio. 39 tables, 125 pp.

Putnam, H. E., and T. A. Olson. 1961. Studies on the productivity and plankton of Lake Superior. School Public Health, Univ. Minn., Rept. No. 5. 58 pp.

Rawson, D. S. 1951. The total mineral content of lake waters. Ecology, 32:669–672.

———. 1955. Morphometry as a dominant factor in the productivity of large lakes. Verhandl. Intern. Ver. Limnol., 12:164–174.

———. 1956. Algal indicators of trophic lake types. Limnol. Oceanog., 1:18–25.

———. 1960. A limnological comparison of twelve large lakes in northern Saskatchewan. Limnol. Oceanog., 5:195–211.

Reade, T. M. 1903. The evolution of earth structure. Longmans, Green, New York. 342 pp.

Rodgers, G. K. 1962. Lake Ontario data report. Great Lakes Inst., Univ. Toronto, Prelim. Rept. No. 7. 102 pp.

Smith, S. H. 1964. Status of the deepwater cisco population of Lake Michigan. Trans. Am. Fisheries Soc., 93:209–230.

Teiling, E. 1955. Some mesotrophic phytoplankton indicators. Verhandl. Intern. Ver. Limnol., 12:212–215.

Teter, H. E. 1960. The bottom fauna of Lake Huron. Trans. Am. Fisheries Soc., 89: 193–197.

Thienemann, A. 1928. Der Sauerstoff im eutrophen und oligotrophen See. Die Binnengewässer, Band 4, Schweizerbart, Stuttgart, 75 pp.

Thomas, F. J. F. 1954. Industrial water resources of Canada. Upper St. Lawrence River-central Great Lakes drainage basin in Canada. Can. Dept. Mines Tech. Surv., Water Surv. Rept. No. 3, Mines Branch Rept. 837. 212 pp.

Thomas, B. W., and H. H. Chase. 1887. Diatomaceae of Lake Michigan as collected during the last sixteen years from the water supply of the city of Chicago. Notarisia, Commetarium Phycologium, 2:328–330.

U.S. Geological Survey. 1960. Quality of the surface waters of the United States. U.S. Geol. Surv., Water Supply Papers, 1520. 641 pp.

U.S. Public Health Service. 1961. National water quality network. Annual compilation of data October 1, 1960–September 30, 1961. U.S. Public Health Serv. Publ. 663. 545 pp.

Welch, P. S. 1952. Limnology. McGraw-Hill, New York, 538 pp.

Wells, LaRue. 1960. Seasonal abundance and vertical movements of planktonic Crustacea in Lake Michigan. U.S. Fish Wildlife Serv., Fishery Bull., 60(172):343–369.

Williams, L. G. 1962. Plankton population dynamics. U.S. Public Health Serv. Publ., 663. 90 pp.

Wright, Stillman. 1955. Limnological survey of western Lake Erie. U.S. Fish Wildlife Serv., Spec. Sci. Rept. Fisheries, 139. 341 pp.

32

Control of eutrophication

R. T. Oglesby
W. T. Edmondson

Eutrophication is a natural process in the development of a lake. The structure of impoundments and lakes causes them to act as concentrators of inorganic salts which in turn are utilized in the production of aquatic plants. It is usually difficult to halt the sequence of events inherent in this process. However, in many cases it is feasible to decrease the rate at which eutrophication proceeds and this is particularly true where the activities of man have caused substantial acceleration by the addition of fertilizing materials. Depending on the methods used, decreases in eutrophication rates may be very rapid but temporary, as when toxic chemicals are applied to kill off aquatic plants, or they may be of a more permanent nature in cases where nutrient inflow is limited for prolonged periods. The latter method of combating eutrophication will be emphasized and illustrated by examples in the following discussion.

TREATING THE SYMPTOMS

With some lakes it is feasible to alleviate temporarily the nuisance of excessive aquatic plant growths by periodic direct killing of the plants. This practice is particularly common in the midwest, where copper sul-

Reprinted with permission from *Journal Water Pollution Control Federation*, Vol. 38, pages 1452–1460 (September, 1966), Washington, D.C. 20016.

fate generally is used for algal control and sodium arsenite is applied to kill rooted vegetation (1) (2). The efficiency of $CuSO_4$ as an algicide decreases with increasing water hardness; hence, the applicability of this control method is limited. Both copper and arsenic are quite toxic to aquatic animals as well as plants, and oxidized compounds of both these elements are highly insoluble and hence accumulate in bottom sediments where they may exert undesirable effects on resident invertebrates. Organic herbicides have been marketed widely in recent years but are relatively expensive and some question exists about their possible longterm effects on aquatic fauna.

Temporary control of nuisance aquatic plants also has been carried out by excluding appreciable light from the water for limited time periods. Carbon black is the usual agent employed for light exclusion. The difficulty in maintaining a film on the water surface under conditions of any appreciable wind action makes this approach feasible only on very small, protected bodies of water.

Both chemical treatment and light exclusion as methods of lake rehabilitation are not only lacking in general applicability but, most importantly, provide only temporary solutions to the problem. Such procedures are comparable to a physician treating the symptoms of a disease rather than removing the cause.

PRINCIPLES OF CONTROLLING EUTROPHICATION

Odum (3) has outlined the theoretically possible approaches for seeking permanent solutions to the problems of excessive aquatic plant production. There are two general methods which seem to offer the greatest chance of success. These are: (*a*) harvesting of the plants and (*b*) limiting nutrient supplies.

Harvesting might be accomplished either directly or through cropping by secondary producers such as herbivorous fish. Although direct harvesting of algae by such processes as foam fractionation (4) and chemical precipitation (5) appears to be practical for very concentrated cultures, this would be infeasible in a natural body of water because of the large volumes involved. Production and harvesting of secondary producers may seem attractive from the standpoint of gaining usable crops of herbivorous fish or invertebrates. However, the degree of algal control obtained by this method would not be great since a relatively dense crop of plants would have to be maintained to support significant populations of the secondary producers.

The most discussed and widely employed method of controlling algal nuisance blooms is the limitation of dissolved nutrients. Nitrogen and phosphorus generally are considered to be the required elements most often limiting plant growth. A common ratio of these in organisms is about 16 nitrogen atoms to each atom of phosphorus. Nitrogen compounds generally are much more abundant in soils and water than are those of phosphorus, and under aerobic conditions they are much more soluble. In addition, many nuisance species of blue-green algae have the ability to fix atmospheric nitrogen and hence may be relatively independent of dissolved nitrogen compounds. For these reasons, phosphorus is considered to be the element most often limiting the level of algal population growth in regions where most nuisance conditions exist.

With the increasing knowledge of algal nutrient requirements obtained from bioassays and culture studies, it is evident that under normal conditions many other chemical elements may be limiting in some instances. For example, Goldman (6) (7) found indications that zinc, manganese, and especially iron may control the level of primary productivity in Lake Tahoe, and Provasoli (8) has shown that organic micronutrients may be vital to many algal species. However, in most cases phosphorus and nitrogen are probably the more important factors in determining plant production. Certainly most instances of man-induced eutrophication can be traced directly to a superabundance of these elements.

METHODS OF NUTRIENT LIMITATION

The principal methods used for controlling nutrient concentrations in lakes have been: (*a*) the removal of nitrogen and phosphorus at their source, (*b*) the diversion of nutrient-rich waters from receiving bodies, and (*c*) the dilution of these elements in lakes by the controlled addition of water low in nutrients. Once the primary source of fertilizing elements has been controlled, reversal of eutrophication may be enhanced by decreasing or preventing the leaching of phosphorus and nitrogen into the water from the bottom sediments. This could be done by bottom sealing, as Sylvester and Seabloom (9) have shown, or by dredging to remove nutrient-bearing sediments. The increasingly popular practice of aerating small lakes in order to prevent anaerobiosis in the hypolimnion also may help by holding phosphorus and iron in their insoluble, oxidized forms.

EXAMPLE OF NUTRIENT DIVERSION FOR CONTROLLING EUTROPHICATION

A common cause of artificial eutrophication has been the delivery of large volumes of wastewater treatment plant effluent to lakes. The few lakes from which wastewater effluent has been diverted are small and fairly productive for natural reasons, as mentioned in a summary by Edmondson (10). Lake Washington, a large, relatively unproductive lake, has shown distinct symptoms of enrichment and therefore has been under study to provide new information on the relation between nutrient income and productivity. The lake has shown a progressive deterioration in the sense of producing denser populations of phytoplankton, especially floating types of blue-green algae, showing reduced transparency and causing a certain amount of public disturbance. The increased algal population evidently results from increased productivity caused by a large increase in the nutrients provided by wastewater. Accounts of these changes and references to the limnological literature on Lake Washington are found in several publications (11) (12) (13) (14) (15).

Public funds were voted for a sewerage system which will divert all the effluent and raw wastewater from the lake. At this time (summer 1965), about half the effluent has been diverted and the project is scheduled for completion during the summer of 1966 (Table 32–1).

TABLE 32–1. The delivery to Lake Washington in 1957 of total phosphorus and effluent by wastewater treatment plants*

Diversion Date	Annual Delivery of Total P		Annual Volume of Effluent	
	(kgm $\times 10^{-3}$)	(%)	(cu m $\times 10^{-3}$)	(%)
Diverted in March 1963	13.97	33.1	2,270	26.4
Diverted in May 1965	5.56	13.2	896	10.4
To be diverted in 1966	22.54	53.4	5,441	63.2
Total 1957	42.07	99.8	8,607	100.0

* The portion that is affected by each diversion is shown. Between 1957 and 1965 the rate of delivery by each plant increased and another plant was added which will be diverted in 1966. Data from City of Seattle Department of Engineering (Hollis M. Phillips).

Obviously, it is too early to make a genuine evaluation of the effect of diversion because the amount diverted in 1963 was a small fraction of the whole and only a few months have passed since the second diversion. Further, lake conditions can be expected to vary from year to year in response to many factors other than nutrient input. Nevertheless, the condition of the lake in late summer 1965 is suggestive. The spring and summer of 1965 were very favorable for algal growth because the weather was unusually sunny. The income of total radiation was 68,210 cal/sq cm for the period March–July 1965 as compared to 62,800 in 1963 and 57,430 in 1964. In fact, the plankton growth started earlier and proceeded more rapidly in 1965 than in 1964 or 1963 as measured by seston (dry weight of total particulate matter), chlorophyll, and transparency (Fig. 32–1).

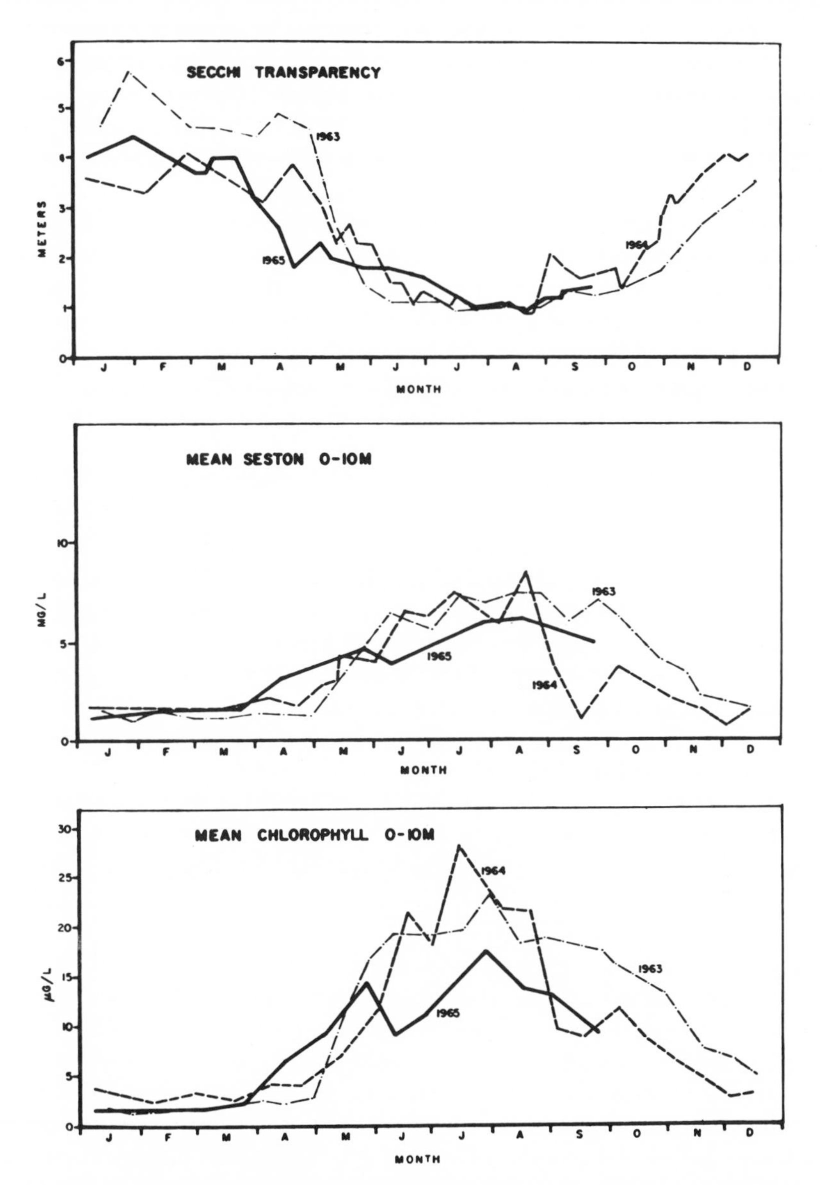

Figure 32–1. Conditions in Lake Washington, 1963–1965. Measurements were taken at a station near the center of the lake; transparency was measured with a Secchi disc; seston, or total particulate matter, was that retained by Millipore HA filter, mean top 10 m; chlorophyll is mean for top 10 m.

Some of the major differences in plankton growth between 1963 and 1964 correspond to differences in radiation. Despite the rapid start in 1965 the lake did not reach the same low transparency and high seston and chlorophyll values as it had in 1963 and 1964. Further, the maximum value of total phosphorus measured in the upper 10 m of the lake was 63 μ/1 in 1965, as compared to 66 in 1964 and 70 in 1963. These differences cannot be attributed to lack of light in midsummer 1965.

The interpretation of these recent results is not entirely clear at the moment because of interactions between temperature, light, and nutrients; however, it seems reasonable to suggest that if nutrients had continued to be fed into the epilimnion at the same rate as in previous years, a denser crop of plankton would have built up in the second half of the summer of 1965 and the lake would have been less transparent. Secchi-disc transparency is strongly affected by the presence of particles and is therefore a useful measure in this situation.

It may be worthwhile to point out a few features of Lake Washington that lead us to expect it to show a prompt and rapid recovery after diversion. Perhaps the most important factor is that the basic water supply is poor in dissolved materials, including nutrients; the calcium content of the Cedar River is about 8 mg/l, TDS 76 mg/l, and the total phosphorus content is usually less than 5 μg/l. Before enrichment with wastewater, the lake did not produce dense crops of algae for prolonged periods. The annual inflow of water is about a third of the volume of the lake. Thus, while the rate of dilution is not unusually large, it is adequate for effective flushing of the lake.

One of the processes controlling productivity is the regeneration of nutrients from the sediments. In Lake Washington, most of the lake bottom drops off very rapidly from the shoreline (16). Thus, soft sediments rich in decomposing matter are deposited in deep water where they are not easily disturbed during periods of mixing and where they are not exposed to an intensity of illumination high enough to permit algae to grow in contact with them. The lake has not been enriched long enough to develop thick deposits of rich sediments. To fill in a lake 65 m deep, principally by artificial enrichment, would be a major achievement. Further, there is relatively little area for the development of rooted plants.

Thus, Lake Washington, which was able to absorb considerable enrichment before deterioration to a publicly noticeable degree, cannot be expected to maintain a high productivity without continued fertilization.

Not all lakes, of course, would match Lake Washington; some would be much more sensitive to enrichment and would maintain high production with less fertilization than Lake Washington. It seems reasonable to expect lakes with water supplies rich in nutrients for natural reasons, especially in regions of hard water, to be more responsive to a given amount of enrichment, other things being equal. Some lakes are naturally productive. It is useful to think of productivity of lakes as being set by the simultaneous operation of a number of controlling factors (10) (17). To evaluate the effect of a given change requires knowledge of other relevant features. When a deep, unproductive lake has been made productive by a rather specific change, such as wastewater disposal, it can be expected to return toward its original condition when relieved of the wastewater. The question of irreversibility of eutrophication needs to be considered, but one should not take a mystical attitude toward the problem, as sometimes seems to be done. The reason that artificial eutrophication may have seemed to be irreversible is that, in some cases of enrichment of naturally productive lakes, an unrealistically high rate of recovery was expected, and in others, shallow lakes in the last stages of filling-in were under consideration. Even Lake Monona has responded favorably to wastewater diversion, but the effects were not immediately recognizable. Interpretation of this example may be confused by the fact that the original condition seems not to be really known. [For summary see Figure 2 of Edmondson (10).]

AN EXAMPLE OF NUTRIENT DILUTION FOR REHABILITATING A EUTROPHIC LAKE

Green Lake, located in the northcentral part of Seattle, is a naturally eutrophic body of water plagued by heavy blooms of blue-green algae during the warmer months. Despite this nuisance, the Lake has extensive and diversified recreational uses, including bathing and trout fishing. In an effort to find means for improving such an important recreational resource, the City of Seattle sponsored a study in 1959 by Sylvester and Anderson (18). After computing nutrient budgets, these workers concluded that phosphorus was the key element controlling the extent of algal blooms. It was recommended that, as a means of keeping algae below nuisance levels, large amounts of nutrient-poor city water be diverted into Green Lake for dilution purposes. The average inflow rate in 1959 was about 2.6 mgd (9,841 cu m/day) and the volume of the lake at normal elevations was about 1,088 bil gal (4 mil cu m). Sylvester and Anderson proposed that up to 10 mgd (37,850 cu m/day) of city water be added to the normal inflow, thereby reducing the mean rate of phosphorus addition by about 70 percent.

The Seattle Park Board accepted this recommendation and, following design and construction of facilities, began adding dilution water in June 1962. Addition rates have been considerably lower than those proposed, ranging from about 0.1 to 6 mgd (380 to 22,710 cu m/day) with 2 periods of several months each when no dilution water was added. To date, enough city water has been added to flush the lake about three times.

During 1965 a three-year study was begun by Oglesby (19) to evaluate the effectiveness of this program in reducing the undesirable effects of eutrophication. Those parameters reflecting both bloom conditions and algal growth potential show trends which appear to reflect at least a partial control of the lake's eutrophication. In the lower part of Figure 32–2 the differences between the phosphorus concentrations for 1965 and 1959 are shown. A reduction, presumably resulting from the intervening addition of dilution water, was especially marked during the spring.

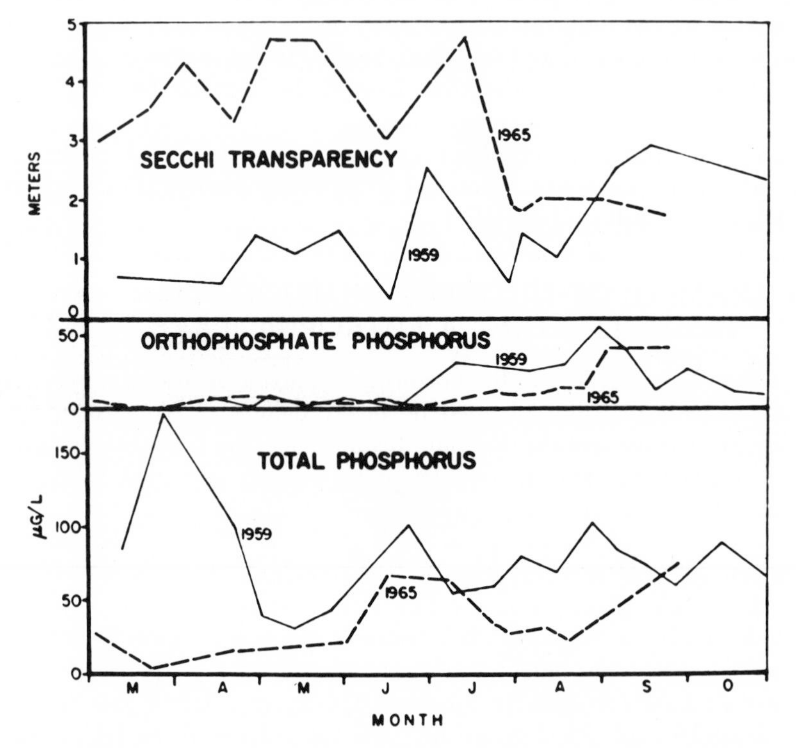

Figure 32–2. Comparison of conditions in Green Lake for the years 1959 and 1965. Phosphorus values represent approximate averages and transparencies were measured with a Secchi disc.

Early in June, water near the bottom suddenly decreased in dissolved oxygen content and this was accompanied by a release of phosphorus into the overlying water. Even after this occurred, concentrations were generally lower in 1965 than in 1959 except during the month of September. During this month there was appreciably more sunlight (Table 32–2) than in September 1959 and consequently conditions for algal growth were more favorable. The higher values in 1965 are thus thought to rep-

TABLE 32–2. Percent of possible sunshine by month for 1959 and 1965*

Year	Month						
	Mar.	Apr.	May	June	July	Aug.	Sept.
1959	38	39	48	41	67	45	22
1965	68	39	42	57	64	55	54

* U. S. Weather Bureau Climatological Data Section, Seattle, Washington.

resent particulate phosphorus in the form of plant cell material and soluble phosphorus resulting from microbial decomposition of algal cells.

A comparison of transparencies observed for the two years is shown in the upper part of Figure 32–2. Until late summer, measurements indicated a one-to-threefold improvement during 1965 despite comparable amounts of sunshine during the spring and summer. The more favorable algal growth conditions in September 1965 are reflected in the lower transparencies compared to those in 1959 when cloudy conditions prevailed.

The species composition and growth patterns of Green Lake phytoplankton show some interesting differences between the two years under comparison. Most noticeable was the complete absence of *Aphanizomenon* in 1965. In 1959 this alga formed unsightly scums on the surface from the latter part of June through the remainder of the warm months. Conversely, *Gleotrichia*, a less objectionable blue-green alga because of its more uniform vertical distribution, was much more common in 1965 than in 1959 and occurred during midsummer rather than in early autumn as in 1959. In general, the algal population in 1965 was more varied in makeup and never reached the nuisance levels observed during the earlier survey.

Care must be taken in drawing conclusions from the above information. Green Lake has been studied only during the years 1959 and 1965 and the natural variability of its characteristics from year to year is unknown. However, climatic conditions in 1965 were very favorable for algal growth and it is to be expected that in the future primary productivity will not increase above the levels currently observed so long as adequate dilution with city water is continued.

Controlled nutrient dilution as a means of lake rehabilitation is feasible only when a relatively plentiful source of water low in nutrients is available for the purpose. Such instances might seem to be quite rare. However, Sketelj and Rejic (20) have proposed a diversion of the River Radovna through Lake Bled in Yugoslavia in order to prevent water quality deterioration in this subalpine lake, and several cities in the

United States and Canada have expressed an interest in using dilution as a method for improving water quality in urban recreational lakes.

Lakes showing only mild symptoms of eutrophication may achieve adequate water quality improvement by periodic additions of dilution water. In the case of a shallow, well-mixed body of water, such as Green Lake, which has extensive organic bottom deposits rich in phosphorus and nitrogen and which receives large quantities of fertilizing elements via subsurface springs and runoff, a permanent reduction of nuisance plant blooms will require a continuing, controlled addition of dilution water.

CONCLUSIONS

Several approaches to rehabilitating eutrophic lakes have received considerable attention in recent years. Instances where these principles have been applied are increasing in number but thus far insufficient time has elapsed to permit conclusive statements about the effectiveness of the various projects. However, in the examples cited above, both diversion and dilution of nutrients give early indications of success as means for significantly decreasing eutrophication rates. Tertiary treatment of a wastewater effluent is currently being practiced to lower phosphorus input into Lake Tahoe (21) and this project also should provide an interesting test of existing hypotheses on eutrophication. Whatever the results of current attempts at lake rehabilitation, it is not likely that any simple formulae will be developed that are generally applicable to all lakes, since each natural body of water is unique in the balance of properties that control productivity and the population density of algae. The principles of lake productivity are reasonably well understood, but we need to establish specifically the quantitative relationships among the various processes to permit evaluation of individual cases and prediction of the effects of particular changes. Obviously, thorough, well-organized limnological studies of appropriate features of lakes will be necessary to provide data for evaluation of the engineering problems.

ACKNOWLEDGMENTS

The authors wish to acknowledge the financial assistance of the U.S. Public Health Service in providing a demonstration project grant (WPD-38-01 (R1)-65) for the evaluation of the Green Lake project and the National Science Foundation grants G 6167 and G 24949 for supporting the limnological study of Lake Washington.

REFERENCES

1. Kanneberg, A., *et al.* 1946. "Aquatic Nuisance Control in Wisconsin." Committee on Water Pollution, Madison, Wis.

2. Mackenthun, K. M. 1961. "Use of Algicides." *In:* "Algae and Metropolitan Wastes." R. A. Taft Sanitary Engineering Center, Cincinnati, Ohio, Tech. Rept. W61-3, 148.
3. Odum, E. P. 1961. "Factors Which Regulate Primary Productivity and Heterotrophic Utilization in the Ecosystem." *In:* "Algae and Metropolitan Wastes." R. A. Taft Sanitary Engineering Center, Cincinnati, Ohio, Tech. Rept. W61-3, 65.
4. Levin, G. B., and J. M. Barnes. 1965. "Froth Flotation for Harvesting Algae and Its Possible Application to Sewage Treatment." *Proc. 19th Ind. Waste Conf.*, Purdue Univ., Ext. Ser. 117:421.
5. Golueke, C. G., and W. J. Oswald. Apr. 1965. "Harvesting and Processing Sewage-Grown Algae." This Journal, 37:4, 471.
6. Goldman, C. R. 1964. "Primary Productivity and Micronutrient Limiting Factors in Some North American and New Zealand Lakes." *Verh. Intl. Ver. Limnol.* (Germany), 15:365.
7. Goldman, C. R., and R. C. Carter. July 1965. "An Investigation by Rapid Carbon-14 Bioassay of Factors Affecting the Cultural Eutrophication of Lake Tahoe, California-Nevada." This Journal, 37:7, 1044.
8. Provasoli, L. 1961. "Micronutrients and Heterotrophy as Possible Factors in Bloom Production in Natural Waters." *In:* "Algae and Metropolitan Wastes." R. A. Taft Sanitary Engineering Center, Cincinnati, Ohio, Tech. Rept. W61-3, 48.
9. Sylvester, R. O., and R. W. Seabloom. 1965. "Quality of Impounded Water as Influenced by Site Preparation." Dept. of Civil Engineering, Univ. of Washington, Seattle, Wash.
10. Edmondson, W. T. (in press). "Water Quality Management and Lake Eutrophication: The Lake Washington Case."
11. Anderson, G. C. 1961. "Recent Changes in the Trophic Nature of Lake Washington." *In:* "Algae and Metropolitan Wastes." R. A. Taft Sanitary Engineering Center, Cincinnati, Ohio, Tech. Rept. W61-3, 27.
12. Brown and Caldwell (Civil and Chemical Engineers). 1958. "Metropolitan Seattle Sewerage and Drainage Survey." Report for the City of Seattle, King County, and the State of Washington.
13. Edmondson, W.T. 1961. "Changes in Lake Washington Following an Increase in the Nutrient Income." *Proc. Intl. Assn. Theoret. and Appl. Limnol.*, 14:167.
14. Edmondson, W. T. 1963. "Pacific Coast and Great Basin." Ch. 13 in "Limnology in North America." Univ. of Wisconsin Press, Madison, Wis.
15. Edmondson, W. T. (in press). "Changes in the Oxygen Deficit of Lake Washington." *Proc. Intl. Soc. Theoret. and Appl. Limnol.*
16. Gould, H. R., and T. F. Budinger. 1958. "Control of Sedimentation and Bottom Configuration by Convection Currents, Lake Washington, Washington." *Jour. Marine Res.*, 17:183.
17. Sawyer, C. N. Mar. 1954. "Factors Involved in Disposal of Sewage Effluents to Lakes." *Sewage and Industrial Wastes*, 26:3, 317.
18. Sylvester, R. O., and G. C. Anderson. 1964. "A Lake's Response to Its Environment." *Jour. San. Eng. Div., Proc. Amer. Soc. Civil Engr.*, 90:SA1, 1.
19. Oglesby, R. T. Unpublished.
20. Sketelj, J., and M. Rejic. 1966. "Pollutional Phases of Lake Bled." *In:* "Advances in Water Pollution Research." Proc. 2nd Intl. Conf. Water Poll. Res., Pergamon Press, Ltd., London, England, Vol. 1, 345.
21. Culp, R. L. 1964. "Tertiary Treatment of Wastes at Lake Tahoe." *Newsletter Pac. Northwest Poll. Control Assn.*, 10:1, 26.

SUSTAINED USE MANAGEMENT

AQUATIC ECOSYSTEMS AND WATER RESOURCES

33

Human water needs and water use in America

Charles C. Bradley

The current rapid rise in population poses many problems, among them the question, Where are the limits, if any? More carefully stated for America, the question seems to be, how many people can we sustain at what standard of living?

My purpose in this article is to examine one vital resource, water—(i) to show the minimum amount necessary to sustain human life, (ii) to show the amount we are now using in the United States to maintain our standard of living, and (iii) to indicate from these figures when we may expect to find certain ceilings imposed on the crop of human beings in this country.

While water economics is admittedly important in the complex problem of water supply, no discussion of this aspect of the problem is attempted in this article.

WATER NEEDS OF MAN

The 2 quarts or so of water which a man needs daily for drinking is a requirement obvious to anyone. Less obvious is the equally vital but much larger volume of water needed to sustain a man's food chain from soil to stomach. This is the water necessary to raise the wheat for his

Reprinted with permission from *Science*, 138(3539):489–491 (26 October 1962).

daily bread and the vegetables that fill his salad bowl. This is also the still larger volume required to raise alfalfa to feed a steer from which a man may get his daily slice of meat. All this water represents a rather rigid requirement for human life, and it is water which is consumed, in the sense that it is removed from the hydrosphere and returned to the atmosphere.

An adult human has a daily food requirement of about 2½ pounds, dry weight. If he is strictly a vegetarian, an illustrative approximation of the water requirements for his food chain can be made by assuming man *can* "live by bread alone."

Wheat has a transpiration ratio of 500 (1); that is, ideally it takes 500 pounds of water circulating through the wheat plant from the soil to the air to bring 1 pound (dry weight) of wheat plant to maturity. If grain to be milled represents half the weight of the wheat plant, we can say that it takes 1000 pounds of water to make 1 pound of milling wheat, or (simplifying again) 1000 pounds of water to make 1 pound of bread. Therefore, it takes 2500 pounds of water, or approximately 300 gallons, to make 2½ pounds of bread. Three hundred gallons per day per person is, therefore, probably not far from the theoretical minimum water requirement to sustain human life.

The introduction of animal protein to a man's diet lengthens the food chain, thereby greatly increasing the water requirement. To illustrate, let us assume what might be called a simplified but generous American diet of 1 pound of animal fat and protein (beef) and 2 pounds of vegetable foods (bread) per day. It takes about 2 years to raise a steer. If butchered when it is 2 years old, the animal may yield 700 pounds of meat. Distributed over the 2 years, this is about 1 pound of meat per day. It may be seen, therefore, that this diet requires a steady-state situation of about one steer per person.

A mature steer consumes between 25 and 35 pounds of alfalfa a day and drinks about 12 gallons of water (2). Alfalfa has a transpiration ratio of 800 (1), hence 20,000 pounds of water are required to bring 25 pounds of alfalfa to maturity. In other words, a little over 2300 gallons per day per man are required to introduce 1 pound of beef protein and fat into a person's diet. Add to this the 200 gallons necessary to round out his diet with 2 pounds of vegetable matter and we have a total water requirement of about 2500 gallons per day per person for a substantial American diet.

It should be remembered that these are conservative figures, because transpiration ratios are derived from carefully controlled laboratory experiments and not from data collected in the field, where perhaps half the total rainfall is lost directly by evaporation and does not pass through the plant body. It should be noted, too, that the water cost of a pound of meat is about 25 times that of a pound of vegetable. We should anticipate a similar ratio for the water cost of wool to that of cotton or for the water cost of butter to that of margarine. In any case, somewhere between 300 and 2500 gallons per day is the bare subsistance water cost for one naked human being.

WATER USE IN THE UNITED STATES

When we talk about "use" we have to add to the foregoing figures the water requirements for all our fibers, lumber, and newsprint, as well as the water needed to process steel, to run the washing machine, to flush the toilet, and to operate our air conditioning and our local laundries, and especially that required to sweep our sewage to the sea. It is therefore pertinent, at this point, to digress slightly in order to clarify our concept of the American standard of living, or at least that portion sustained by water use. The American standard of living is not a wholly unmixed blessing. In achieving such luxuries as the flush toilet, synthetic detergents, cheap newspapers, and atomic power we find ourselves also achieving polluted streams, sudsy well-water, radioactive milk, and poisoned oysters.

Underlying and supporting our standard of living are powerful industrial centers and a mass production scheme which creates inexpensive commodities. This scheme rests firmly upon certain prodigal wastes, polluted streams being a prime example. To clean up the streams would take a tremendous amount of money which might otherwise be spent on cheap commodities. On the basis of some standards of values, this could be construed as a lowering of the standard of living.

To illustrate the magnitude of the practical problems we have created for ourselves, we note that if river-disposal of waste were suddenly denied the city of St. Louis, the city fathers would have to decide what else to do with the daily discharge of 200,000 gallons of urine and 400 tons of solid body-wastes, to say nothing of all the industrial wastes. River disposal of human waste, though cheap, involves a double loss of resources. On the one hand there is the polluted river; on the other, the depleted soil. So long as these losses are deemed less important than the production of inexpensive commodities which they support, we will have to accept our befouled streams and depleted soil as part of the cost of our standard of living. In addition to waste disposal we can see the water power, river transportation, fisheries, and water recreation are all well-established items in our standard of living. Therefore, as we move into a discussion of water use, especially future use of surface waters, we must remember that most of our runoff is already committed to our living standard and is working hard to support it.

A figure for water use in the United States can be obtained by subtracting that water which we are *not* using from the total water available.

Thirty inches of annual rainfall on the surface area of the United States (exclusive of Alaska and Hawaii) gives us theoretically nearly 5000 billion gallons per day, a figure which represents the total water available for our use (3). Of this 5000 billion gallons, about 1300 billion gallons a day, or about one-fourth of the rainfall, is discharged by our rivers (4). It may also be said that this discharge figure contains the groundwater increment, since stream flow is largely maintained by effluent seepage from the ground.

It can be seen that 75 percent of our rainfall is returned to the atmosphere through evaporation and transpiration. It is difficult to assess the relative contributions of these two factors. A ration of 50:50 is probably not far from the truth. From a utilitarian standpoint, evaporation constitutes pure waste, and it may be that here some significant gains in water conservation can be made. But until this is done, we have to reckon this loss, too, as part of the price being paid for our standard of living.

Very little of the area of the United States which could produce crops for man is not actually doing so. The largest nonproducing area is, of course, our desert, and even here we are irrigating, using stream water exported from regions of water surplus. Additionally, we are forcing the desert to raise crops through the use of ground water. But in many such areas we have considerable evidence that the annual draft from the ground-water reservoir exceeds the annual recharge. Consequently, some of these operations will be short-lived and perhaps socially and economically catastrophic for the people involved.

About 2 percent or more of the surface of the United States is "paved" with cities and roads and will probably remain agriculturally unproductive until some far-sighted city planners provide for extensive roof gardens. Another 2 or 3 percent of the land in this country is devoted to wilderness and national parks. While these do not directly produce crops for man, we do include them and their waters in our standard of living. Finally, we can say that bad agricultural management has reduced the productivity of a fraction of our arable land, and that this percentage must be added to our total for unproductive lands. Let us make a quasi-educated guess and say that as much as 10 percent of our land in areas of abundant rainfall is, at the moment, non-productive.

Three-fourths of the nation's rain (3700 billion gallons per day) falls on about half the nation's area, and it is this three-fourths, largely unmetered, that does the big job of raising crops for America. As concluded previously, perhaps one-tenth of this rain falls on unproductive areas. Hence we may say that 3300 billion gallons per day are productive of crops or surplus water. Of this, about one-fourth is unconsumed runoff, giving a remainder of about 2500 billion gallons per day which we are *consuming*, though perhaps wastefully, to raise our crops. In a population of 180 million people, this amounts to approximately 13,800 gallons per day per person. In addition, 240 billion gallons per day are metered out of our streams, lakes, and ground-water reservoirs to serve industry, municipalities, and rural areas (4); over half of it is consumed in irrigation and other processes. This 240 billion gallons per day is almost 1400 gallons per day per person, a figure which now must be added to the 13,800 gallons for a grand total of 15,200 gallons per day per person. Thus we find that the per capita daily use of water in the United States is in excess of 15,000 gallons, 95 percent of which is consumed.

SOME POPULATION LIMITS IN THE UNITED STATES

How many people could we feed if all the rainfall in the United States were completely utilized? Since 300 gallons per person per day is needed for a vegetarian diet, we could, in theory, sustain about 17 billion people, or approximately 8 times the present world population. If, on the other hand, we decided to feed people on the "generous American" diet, we discover, by the same sort of calculation, that we could feed about 2 billion people, or somewhat less than the present world population. If we admit that loss of water through evaporation is unavoidable, as discussed earlier in this article, we must cut these figures to 8 billion and 1 billion, respectively.

Assuming a population of 180 million and a rainfall of 5000 billion gallons per day, we discover that each person today theoretically has about 28,000 gallons per day for his use. We are now using 15,000 gallons per day per person, 95 percent of it consumptively. We might, therefore, conclude that if we could use every drop of rain that falls we could almost double our population with no decrease in the standard of living. But this is far from possible because there would then be no surface water to generate power, float ships, raise fish, and carry away the national sewage and waste.

The extent to which we can consume our runoff before our standard of living suffers is difficult to foresee. Involved are not only the waste-disposal and commercial uses of rivers but the fact that river water is generally most abundant and most available where it is least needed for agriculture.

Let us guess that we might safely and profitably use one-third of our remaining river water, or 400 billion gallons per day, for future development without expecting a resultant drop in our standard of living. Add to this figure the amount of water that falls on unproductive areas which might rather easily be made productive. We now have a total of about 750 billion gallons per day for future development. At 15,000 gallons per day per person we seemingly can accommodate 50 million more people, or a total population of 230 million, before our standard of living starts to suffer. There is little doubt that America will have reached that population figure well before the year 2000. The evidence of the moment suggests, then, that young Americans alive today will see a significant deterioration in their standard of living before they are much past middle age. Improved cropping, mulching, and other conservation practices could, of course, extend the grace period by a few years.

How far deterioration in the American standard of living will progress depends, of course, upon what action Americans choose to take on their own numbers problems—upon *what* action, and especially upon *when* they take it. Fortunately we have at our disposal human intelligence and considerable time in which intelligence can function. At present rates of rainfall and of population growth we should have almost 200 years before the American standard of living drops to subsistence level and Malthusian controls eliminate the necessity for intelligent action.

REFERENCES AND NOTES

1. Maximov, N. A. 1929. *The Plant in Relation to Water.* Macmillan, New York.
2. Woll, F. W. 1921. *Productive Feeding of Farm Animals.* Lippincott, New York.
3. It is doubtful whether artificial conversion of salt water will ever make a significant difference in this total, although it may be of great significance to certain communities.
4. Leopold, L. B., and W. B. Langbein. 1960. *A Primer on Water.* U.S. Government Printing Office, Washington, D.C.

34

Effects of decreased river flow on estuarine ecology

B. J. Copeland

With rapidly expanding industrialization along the waterways of the world, water usage has become a legal and economic problem. The phenomenon of low flow occurs naturally in times of drought and many studies during drought conditions have shown that the effects of reduced river flow are diverse and far-reaching. The practice of decreasing river flow by the installation of reservoirs is frequently followed without consideration or prior knowledge of the effects downstream, particularly in estuaries.

Estuaries are important in many ways. Many important commercial and sport fishes and invertebrates use estuaries during some stage of their development. Estuaries probably are the greatest producing areas accessible to man for available protein. Many people use these waters for recreation, and industry uses them as factory sites because water is available for transportation, processing, and waste disposal.

As used in this paper, the word "estuary" means a body of water between the mouth of a river and the sea. Since lagoons are really "old estuaries," the term estuary, as used here, includes lagoons, bays, and sounds.

Reprinted with permission from *Journal Water Pollution Control Federation*, Vol. 38, pages 1831–1839 (November, 1966), Washington, D.C. 20016.

CURRENTS AND HYDROGRAPHY

An intricate current system within estuaries is produced by the balance of river discharge and the contribution of the sea. With decreased river flow, the current system can be so altered that shoaling and scouring can set up complete foreign physical conditions.

The most important hydrobiological parameter is salinity. Under normal conditions, salinity is variable in some estuaries. Collier and Hedgpeth (1) showed a direct correlation between river flow and estuarine salinity in Corpus Christi Bay, Texas. If river flow is restricted by upstream reservoirs, the salinity level in the receiving estuary may increase to the detriment of estuarine biological communities.

Closely connected to river flow is the maintenance of natural inlets between barrier islands, connecting estuaries to the sea. Simmons and Hoese (2) suggested the necessity of river flow for maintaining discharge through Cedar Bayou, a natural inlet on the Central Texas coast. Lack of river flow during the early 1950's resulted in the closing of Cedar Bayou. Passes can be kept open only at great expense if river flow is reduced below the level required to maintain them. The passes are important passageways for fish and invertebrates between nursery grounds and feeding areas (2) (3) (4) (5) (6) (7).

OYSTER PRODUCTION

Oysters flourish in a wide range of salinity, and the production of oyster is exclusively estuarine. Oyster production in 1962 amounted to about $16,000,000 in Chesapeake Bay and $6,000,000 on the Gulf Coast. Production by weight was about the same, however, amounting to 20×10^6 and 19×10^6 lb (8.6×10^6 and 9.1×10^6 kg) respectively (8).

Under normal estuarine conditions, occasional flushing with fresh water from rivers helps rid oyster populations of damaging parasites and eliminates species that compete with oysters for food (9) (10). With reduced river flow, however, salinities may remain near or above that of seawater with the result that damaging parasites and competing species thrive.

Wells (11) reported a decrease in number of species of oyster parasites and scavengers in the Beaufort, N.C., area as salinity decreased. The abundance of boring sponges, major oyster parasites, decreases as salinity decreases (12) (13) (14). Most of the boring sponges in the high salinity areas of Newport River, N.C., were killed when the salinity was lowered during heavy flooding of the river (14).

Dermocystidium marinum, a parasitic fungus, is often a serious threat to oyster populations and may be controlled by dilution with fresh water (15) (16) (17) (18) (19) (20) (21) and (many others). *D. marinum* is common on oysters in moderate salinities, nearly always absent in salinities below 10–15 ppt, and only occasionally found in salinities approaching 35 ppt (18). Although the average salinities may be similar in various estuaries, areas with the greatest fluctuation in freshwater inflow (salinity fluctuation) have the least *D. marinum* infestation. A sudden drop in salinity as a result of flooding with fresh water may be enough to check the fungal growth. Apparently it is necessary that freshwater inflow occur in spring to check the infestation effectively.

Another widespread and destructive oyster parasite is the oyster drill, *Thais haemastoma*, which also is sensitive to salinity changes and is usually eliminated by moderate to low salinities (22). The Atlantic oyster drills, *Urosalpinx* and *Eupleura*, are sometimes quite destructive to oysters (23). They, too, are controlled in moderate to low salinities and/or with periodic flushing with fresh water.

Franklin (24), in a well-documented newspaper article, discussed the new, yet unnamed, oyster disease MSX. Apparently, MSX is limited to waters with over 15 to 20 ppt salinity, thus controlled by freshwater contributions.

In spite of the tremendous need for fresh water by the oyster industry, an over abundance will upset the balance. Andrews *et al.* (25) reported oyster kills when the salinity was lowered in the James River, Va. Butler (26) and Gunter (27) reported oyster kills over large areas of the Mississippi Sound during floods of fresh water. However, if flooding occurs for only a short period, oysters are capable of remaining closed for several days and avoiding the harmful effects.

Galtsoff (10) discussed the decrease in the oyster population of Texas bays during the drought of the 1950's. He suggested that the resulting increase in salinity in coastal waters during that time allowed the influx of parasites and diseases and the higher salinities (above 40 ppt) restricted gonad development. Surveys by the Texas Parks and Wildlife Department (28) (29) indicate little or no oyster harvest in Texas bays that receive little fresh water.

SHRIMP

Commercially important shrimp (*Peneidae*) are spawned offshore and the young migrate into estuarine nursery grounds, generally as postlarvae, to complete the life cycle. There, utilizing the higher productivity and abundant dissolved organic material, they grow rapidly, sometimes more than one millimeter per day. Along the Gulf coast shrimp is the most valuable of all fishery products, amounting to more than $60 million in 1962 (8).

The entrance of postlarval penaeid shrimp through the Aransas Pass Inlet, Texas, corresponds to high flow of the rivers of the area in spring and fall (Fig. 34–1) (30). Undoubtedly, this coupling of peak migration and increased river flow (accomplished through years of natural selection) is essential for the propagation of penaeid shrimp.

Important fluxes occur in estuarine ecosystems during the high flows of spring and fall, including flows of vitamins and other dissolved organic compounds, nutrients, lowered salinity by the addition of fresh water, and flushing and mixing influences. Burkholder and Burkholder (31) reported a greater concentration of vitamin B_{12} in the mud and estuarine waters of Georgia than in the adjacent seawater. Starr and Sanders (32) found similar results in other areas, and postulated that productivity of the nearshore sea was greatly dependent on suspended solids brought into the estuary by river flow. In attempts at raising shrimp from the egg to juvenile stage in laboratory experiments, it was found that it was necessary to add vitamin B_{12} to the sea water aquarium, according to a personal communication from the Bureau of Commercial Fisheries, Galveston, Texas.

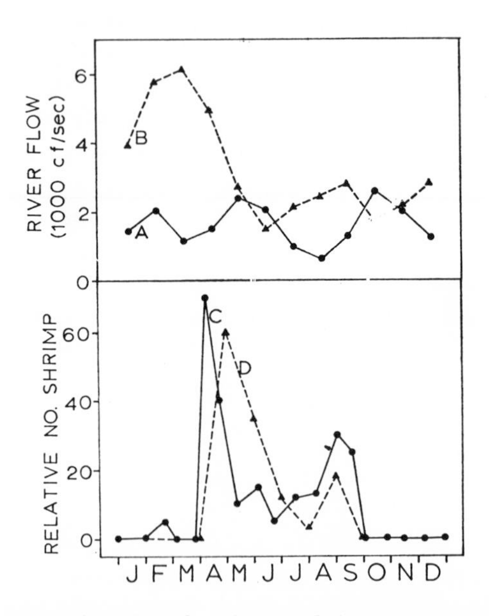

Figure 34–1. Relative number of postlarval penaeid shrimp entering the bays, and river flow of coastal rivers. A. River flow for Guadalupe River near Victoria, Tex.; 10-yr. avg. 1953–1962; B. river flow for Neuse River, near Kingston, N. C.; 10-yr. avg. 1955–1964; C. postlarval *Penaeus aztecus* and *P. duorarum* entering Texas bays through the Aransas Pass Inlet (30); D. postlarval *P. aztecus* and *P. duorarum* entering the Brunswick-Onslow bay area of North Carolina (59).

Gunter (33) found that postlarval penaeid shrimp were most abundant in waters of moderate salinities, although other features were the same. At least 2 species of *Penaeus* are hypoosmotic to seawater and hyperosmotic to low-salinity waters, with isotonicity occurring between 25 and 30 ppt (34). At lowered temperatures, however, the ability to regulate osmosis is slightly impaired, which may account for movement to deeper (and more saline) waters.

Gunter and Hildebrand (35) showed a correlation between catch of white shrimp on the Texas coast and average rainfall for the state. Their correlation coefficients were significant to the one-percent level when rainfall of the two previous years was correlated with an annual catch. This method of analysis was considered valid since the reproductive cycle of shrimp is one year or more, large bodies of water change their salinity slowly, and dry land absorbs more water (less runoff) following droughts.

A plot of shrimp catch and average rainfall for Texas is shown in Figures 34–2. An increase or decrease in rainfall is followed by similar fluctuations in shrimp catch, generally after a two-year period. After the drought during the early 1950's, white shrimp populations never fully recovered, in spite of technological advances and increased fishing effort. Presumably, the addition of several reservoirs in Texas during this time

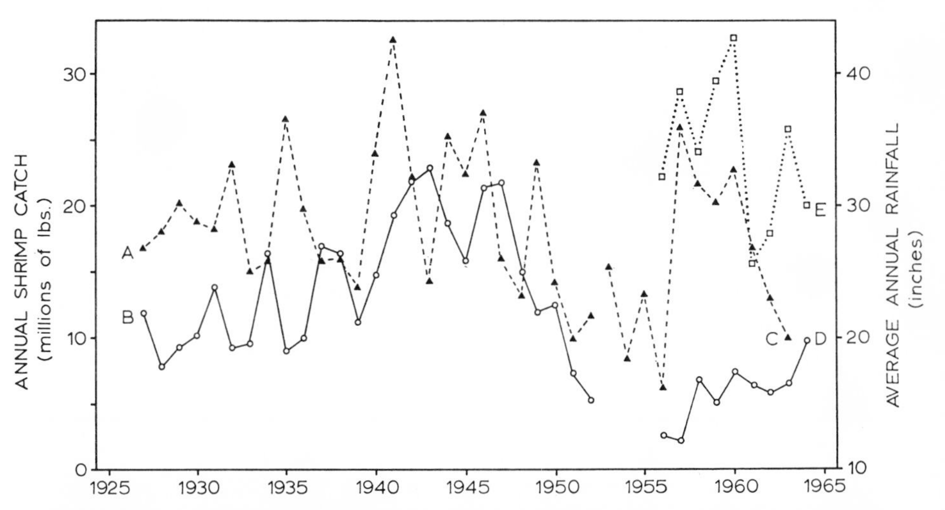

Figure 34–2. Annual shrimp catch and average rainfall for Texas, 1927–1964. A. Avg. rainfall for Texas, 1927–1952 (35); B. annual catch of white shrimp (*Penaeus setiferus*) from Texas waters, 1927–1952 (35); C. avg. rainfall for Texas, 1953–1964 (U. S. Weather Bureau, Climatological Data, Annual Summary, Vol. 58–70); D. annual catch of white shrimp (*P. setiferus*) from Texas waters, 1956–1964 (from U. S. Bureau of Commercial Fisheries, Gulf Coast Landings, 1956–1964). E. annual catch of brown shrimp (*P. aztecus*) from Texas waters, 1956–1964 (from U. S. Bureau of Commercial Fisheries, Gulf Coast Landings, 1956–1964).

prevented river runoff from approaching the level of previous years.

Gunter (36) reported a general decline of white shrimp along the Texas coast and suggested that it be attributed to the drought. He further postulated that the appearance of larger white shrimp populations in Galveston Bay and smaller populations in the bays to the south be attributed to the decrease in freshwater contribution to Texas bays in a southerly direction.

BLUE CRAB

The blue crab (*Callinectes* sp.) fisheries ranked sixth in value in the United States, with a value of $1.5 million on the Gulf coast in 1962 (8). The life cycle of blue crabs is closely connected to estuaries, and estuarine ecology is a vital factor in the maintenance of a continuous crab fishery.

There have been pronounced fluctuations in crab production through the years, according to U.S. Fisheries statistics. There is a direct correlation between crab reproduction and the salt content of estuarine waters (37). The optimum range of salinity for hatching of blue crab eggs was between 23 and 28 ppt. Hoese (38) noted a decrease in the crab population in Mesquite Bay, Texas, during the Texas drought, when salinities were greater than that of seawater. After the drought was broken and salinities were moderately low, the blue crab population increased to its normal level.

PRIMARY PRODUCTION

Estuaries are among the most productive ecosystems known (39). The maintenance of high productivity in estuaries is due to the "nutrient trap" created by the mixing of river waters with ocean waters. The inflow of nutrient sulfates, carbonates, phosphorus, and nitrogenous compounds via rivers contributes greatly to estuarine productivity.

Considerable amounts of organic detritus are brought into the estuary from adjacent marshes and from upstream (40). Detritus is the principal food of many estuarine organisms (39). This material is decomposed by bacteria and fungi, releasing large amounts of organic and inorganic substances which are absorbed by estuarine organisms.

Estuarine phytoplankton growth is related to terrigenous nutrients in Patuxent River and Chesapeake Bay region (41). Plankton blooms occurred just after peak flows down the Patuxent. Similarly, nutrient concentrations were higher following the peak river flow and prior to the plankton blooms. Hutner and Provasoli (42), in their review of research in algal nutrition, have verified that abundant growth of marine algae requires vitamins and nutrients in greater concentration than those found in ordinary seawater.

Estuaries in the southeastern United States are more productive near the mouths of contributing rivers (43) (44) (45) (46). It is not clear, however, whether higher productivity is because of the contribution of rivers or is because the greater mixing regenerates nutrients in that area.

On the other hand, accumulated nutrients tend to be flushed out of estuarine areas receiving large amounts of flood waters of low nutrient concentration. Such floods also kill many of the larger organisms so that their stored nutrients also are lost (38).

The productivity of Texas estuaries was lower just after heavy floods following prolonged droughts (47) (48). During times of normal river discharge, however, the algal productivity was higher than that in most estuarine areas. Texas estuaries are unique in that they are very shallow and have minimal tidal fluctuations; thus, freshwater flow is important in mixing processes.

GENERAL DISCUSSION

The influx of fresh water is one of the principal sources of dissolved nutrients in estuaries. The relationship of growth in estuarine phytoplankton to terrigenous nutrients is shown by the increase following periods of heavy rainfall and runoff. Organic materials, which are decomposed by bacteria and fungi to release large amounts of organic and inorganic substances that are absorbed by estuarine organisms via the aquatic medium, are brought into the estuaries by river inflow (40).

Estuaries that have proved to be important nursery areas possess a well-defined salinity gradient between river mouth and tidal pass, accommodating a large variety of species. River inflow maintains the salinity gradient to a large extent and without it the entire estuary could become hypersaline, as in the Texas Laguna Madre. On the other hand, too much freshwater inflow may cause the entire estuary to become fresh or near-fresh (Sabine Lake, in east Texas) and destroy the salinity gradient.

The estuarine systems of Texas lie in a broad arc of approximately 375 miles (600 km) and pass through a variety of climatic regions, ranging from almost fresh water in Sabine Lake to hypersaline water in the Laguna Madre. Characteristics of the five estuaries other than Sabine Lake and Laguna Madre show the effects of freshwater contributions. Since 1959, the average annual freshwater contribution to these estuaries has been 20.5 acre-ft per surface acre (62.5×10^3 cu m/ha) for Galveston Bay, 4.4 (13.4×10^3) for Matagorda Bay, 11.5 (35.0×10^3) for San Antonio Bay, 1.4 (4.3×10^3) for Aransas Bay, and 1.9 (5.8×10^3) for Corpus Christi Bay. The large inflow in Galveston and San Antonio Bays, plus rainfall, has contributed to the maintenance of thousands of acres of marshes and bayous that provide habitat for the young of many important animals.

Galveston Bay has continually produced more oysters than other Texas bays, presumably because of the more favorable freshwater conditions. The San Antonio Bay system is the only other bay approaching Galveston Bay in producing oyster reefs. Matagorda, Aransas, and Corpus Christi Bays produce few or no oysters at the present time (27) (28). Oyster production corresponds very closely to the amount of fresh water received by each estuary.

Blue crab production was highest in the Galveston and San Antonio Bay areas, and lowest in the bays receiving the least amount of fresh water (49).

Shrimp are the most important fishery product on the Texas coast. Galveston and Matagorda Bay support the largest shrimp populations and Corpus Christi Bay the smallest. White shrimp are now almost nonexistent in Aransas and Corpus Christi Bays (receiving very little fresh water), where in years past they were quite abundant (33) (50) (51).

The data available for comparison of freshwater input and commercial fishery output are scant. However, in Texas, where freshwater input is at a critical level in most estuarine areas, some startling conclusions can be drawn. The minimum freshwater contribution required to maintain the present commercial fishery is not reached in some years in Matagorda, Aransas, and Corpus Christi Bays (Fig. 34–3). In Galveston and San Antonio Bays, larger commercial fishery yields have been harvested during years of intermediate or below-average freshwater input.

The data presented in Figure 34–3 do not complete the real picture. Perhaps more important than the total commercial fisheries output is the change in species composition that makes up the total. During the year following an above-average freshwater input, shrimp and oysters make up a larger percentage of the total. These products are more valuable economically than are fish products, which make up a larger percentage of the total during years of below-average freshwater input.

Many fish species whose juveniles reside in low-salinity waters depend on estuaries for the completion of their life cycles (52) (53). Freshwater input is important to the propagation of fishes because of its salinity control.

Many of the same effects that have been observed in North American estuaries have been observed in estuaries in other areas. In the St. Lucia estuary system on the east coast of South Africa, serious silting and variations in rainfall, along with man's increasing demand for water, has affected the salinity gradient (54) (55). Most of the formerly rich fauna was

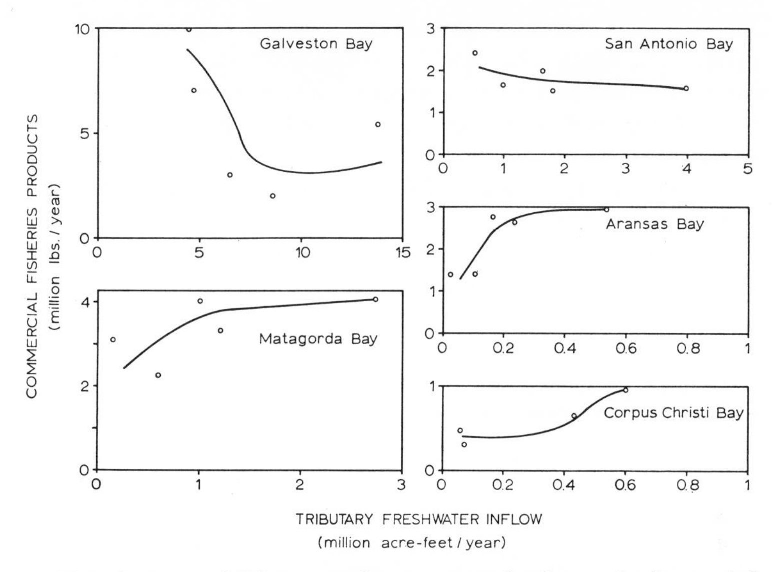

Figure 34–3. Commercial fisheries products vs. annual tributary freshwater inflow for Texas bays, 1959–1964. One year's freshwater contribution was plotted against the following year's commercial harvest. Solid lines indicate trends only. Data on annual commercial fishery yields were obtained from the Bureau of Commercial Fisheries, Biological Laboratories, Galveston, Texas. Data on tributary freshwater contribution to each bay were obtained from the Texas Water Commission.

lost because of the practices of civilization. On the other hand, Richard's Bay, just 36 miles (58 km) south, has not suffered the effects seen so clearly in the St. Lucia system (56). Presumably, this is because the river flow into Richard's Bay has not been changed by man and his enterprises.

As shown by several studies in Russia, the regulation of flow of rivers by the construction of dams led to a considerable accumulation of nutrient salts in reservoirs and to a sharp reduction in the amounts of nutrient salts flowing to the estuaries (57). Following the construction of the Tsimliansk hydroelectric dam, the biomass of the phytoplankton in the Sea of Azov was reduced by 2 to 3 times, and the zooplankton from about 600 mg/cu m to about 50 mg/cu m. Ultimately, this had an adverse effect on the feeding conditions of fishes and their populations.

An important question regarding low river flow is whether the loss of river input and the resulting higher salinities actually result in lessened productivity in the estuary or simply in changes in the productivity channels. Galtsoff (10) reported a general change in the species of oysters in bays of the central Texas coast as salinities increased during the drought of the early 1950's. With the increase in salinity there was a gradual replacement of *Crassostrea virginica*, the commercial oyster, by *Ostrea equestris*, the noncommercial oyster. In 1952 over half of the young oysters (spat) were *O. equestris*, whereas in years of normal salinity the reefs were comprised almost entirely of *C. virginica*. Gunter (58) has discussed this subject and concluded that salinity and freshwater inflow are important factors in limiting the species composition of estuaries.

As has been shown in the previous discussion, freshwater input to estuaries is an important factor. Without it, estuaries become hypersaline and species composition can be altered drastically. With continuation of man's activities in allowing less and less fresh water downstream to the estuary, man may have to pave the estuarine areas and sell them for real estate.

ACKNOWLEDGMENTS

The author acknowledges the courtesy of C. H. Chapman and R. A. Diener of the Bureau of Commercial Fisheries and of F. Masch and C. Urban of the Texas Water Commission in permitting the use of data included in Figure 34–3.

REFERENCES

1. Collier, A., and J. W. Hedgpeth. 1950. "An Introduction to the Hydrography of Tidal Waters of Texas." *Publ. Inst. Marine Sci. Univ. Texas*, 1:2, 121.
2. Simmons, E. G., and H. D. Hoese. 1959. "Studies on the Hydrography and Fish Migration of Cedar Bayou, a Natural Tidal Inlet on the Central Texas Coast." *Publ. Inst. Marine Sci. Univ. Texas*, 6:56.
3. Copeland, B. J. 1965. "Fauna of the Aransas Pass Inlet, Texas. I. Emigration as Shown by Tide Trap Collections." *Publ. Inst. Marine Sci. Univ. Texas*, 10:9.
4. Daugherty, F.M. 1952. "The Blue Crab Investigation, 1949–50." *Texas Jour. Sci.*, 4:77.
5. Hoese, H. D. 1958. "The Case of the Pass." *Texas Game Fish.*, 16(6):18 and 30.
6. Reid, G. K., Jr. 1955. "A Summer Study of the Biology and Ecology of East Bay, Texas." *Texas Jour. Sci.*, 7:316.
7. Reid, G. K., Jr. 1957. "Biological and Hydrographic Adjustment in a Disturbed Gulf Coast Estuary." *Limnol. Oceanogr.*, 2:198.
8. U.S. Fish and Wildlife Service, "United States Fisheries." Annual Summary, Bureau of Commercial Fisheries, Commercial Fisheries Statistics, No. 3471.
9. Collier, A., S. M. Ray, A. W. Magnitzky, and J. O. Bell. 1953. "Effect of Dissolved Organic Substances on Oysters." *Fishery Bull.*, U.S. Fish and Wildlife Serv., 54: 167.
10. Galtsoff, P. S. 1964. "The American Oyster *Crassostrea virginica* Gmelin." *Fish. Bull.*, U.S. Fish and Wildlife Serv., 64:1.
11. Wells, H. W. 1961. "The Fauna of Oyster Beds, with Special Reference to the Salinity Factor." *Ecol. Monogr.*, 31:239.
12. Hopkins, S. H. 1956. "Notes on the Boring Sponges in Gulf Coast Estuaries and Their Relation to Salinity." *Bull. Marine Sci. Gulf Caribb.*, 6:44.
13. Hopkins, S. H. 1962. "Distribution of Species of *Cliona* (Boring Sponge) on the Eastern Shore of Virginia in Relation to Salinity." *Chesapeake Sci.*, 3:121.
14. Wells, H. W. 1959. "Boring Sponges (Clionidae) of Newport River, North Carolina." *Jour. Elisha Mitchell Sci. Soc.*, 75:168.

15. Andrews, J. D. 1955. "Notes on Fungus Parasites of Bivalve Mollusks in Chesapeake Bay." *Proc. Natl. Shellfish. Assn.*, 45:157.
16. Andrews, J. D., and W. G. Hewatt. 1957. "Oyster Mortality Studies in Virginia. II. The Fungus Disease Caused by *Dermocystidium marinum* in Oysters of Chesapeake Bay." *Ecol. Monogr.*, 27:1.
17. Hewatt, W. G., and J. D. Andrews. 1954. "Oyster Mortality Studies in Virginia. I. Mortalities in Oysters in Trays at Gloucester Point, York River." *Texas Jour. Sci.*, 6:121.
18. Hoese, H. D. 1963. "Absence of *Dermocystidium marinum* at Port Aransas, Texas, with Notes on an Apparent Inhibitor." *Texas Jour. Sci.*, 15:98.
19. Mackin, J. G. 1951. "Histopathology of Infection of *Crassostrea virginica* (Gmelin) by *Dermocystidium marinum* Mackin, Owen, and Collier." *Bull. Mar. Sci. Gulf Caribb.*, 1:72.
20. Mackin, J. G. 1955. "*Dermocystidium marinum* and Salinity." *Proc. Natl. Shellfish. Assn.*, 46:116.
21. Mackin, J. G. 1962. "Oyster Disease Caused by *Dermocystidium marinum* and Other Microorganisms in Louisiana." *Publ. Inst. Marine Sci. Univ. Texas*, 7:132.
22. Butler, P. A. 1953. "The Southern Oyster Drill." *Proc. Natl. Shellfish. Assn.*, 44: 67.
23. Carriker, M. R. 1955. "Critical Review of Biology and Control of Oyster Drills *Urosalpinx* and *Eupleura*." Spec. Sci. Rept. U.S. Fish and Wildlife Serv. 148:1.
24. Franklin, B. A. (October 31, 1965). "Once-Plentiful Eastern Oyster Has Become Victim of Drought and Disease." *The N.Y. Times.*
25. Andrews, J. D., D. Haven, and D. B. Quayle. 1959. "Fresh-Water Kill of Oysters (*Crassostrea virginica*) in James River, Virginia, 1958." *Proc. Natl. Shellfish. Assn.*, 49:29.
26. Butler, P. A. 1952. "Effect of Floodwaters on Oysters in Mississippi Sound in 1950." Res. Rept. U.S. Fish and Wildlife Serv. 31:1.
27. Gunter, G. 1953. "The Relationship of the Bonnet Carre Spillway to Oyster Beds in Mississippi Sound and the 'Lousiana Marsh', with a Report on the 1950 Opening." *Publ. Inst. Marine Sci. Univ. Texas*, 3:1, 17.
28. Hefferman, T. L. (mimeo). "Computation, Analysis and Preparation of Coastwide Oyster Population Data." Marine Fisheries Projects Reports for 1961–62, Texas Game and Fish Comm., Austin, Texas.
29. Hofstetter, R. (mimeo). "A Summary of Oyster Studies along the Texas Coast." Coastal Fisheries Projects Reports for 1963, Texas Parks and Wildlife Dept., Austin, Texas, 163.
30. Copeland, B. J., and M. V. Truitt. 1966. "Fauna of the Aransas Pass Inlet, Texas. II. Penaeid Postlarvae." *Texas Jour. Sci.*, 18:65.
31. Burkholder, P. R., and L. M. Burkholder. 1956. "Vitamin B_{12} in Suspended Solids and Marsh Muds Collected Along the Coast of Georgia." *Limnol. Oceanog.*, 1:202.
32. Starr, T. J., F. Sanders. 1959. "Some Ecological Aspects of Vitamin B_{12}-active Substances." *Texas Rept. Biol. Med.*, 17:49.
33. Gunter, G. 1950. "Seasonal Population Changes and Distributions as Related to Salinity, of Certain Invertebrates of the Texas Coast, Including the Commercial Shrimp." *Publ. Inst. Marine Sci. Univ. Texas*, 1:2, 7.
34. Williams, A. B. 1960. "The Influence of Temperature on Osmotic Regulation in Two Species of Estuarine Shrimps (*Penaeus*)." *Biol. Bull. Marine Biol. Lab., Woods Hole*, 119:560.
35. Gunter, G., H. H. Hildebrand. 1954. "The Relation of Total Rainfall of the State and Catch of the Marine Shrimp (*Penaeus setiferus*) in Texas Waters." *Bull. Marine Sci. Gulf Caribb.*, 4:95.

36. Gunter, G. 1962. "Shrimp landings and production of the state of Texas for the period 1956–1959, with a comparison with other gulf states." *Publs. Inst. Marine Sci. Univ. Tex.*, 8:216.
37. Sandoz, M., and R. Rogers. 1944. "The Effect of Environmental Factors on Hatching, Moulting, and Survival of Zoea Larvae of the Blue Crab *Callinectes sapidus* Rathbun." *Ecology*, 25:216.
38. Hoese, H. D. 1960. "Biotic Changes in a Bay Associated with the End of a Drought." *Limnol. Oceanog.*, 5:326.
39. Odum, E. P. 1959. "Fundamentals of Ecology." 2nd Ed., W. B. Saunders Co., Philadelphia, Pa.
40. Diener, R. A. 1964. "Texas Estuaries and Water Resource Development Projects." *Proc. 9th Conf. Water for Texas*, 9:25.
41. Nash, C. B. 1947. "Environmental Characteristics of a River Estuary." *Jour. Marine Res.*, 6:147.
42. Hutner, S. H., L. Provasoli. 1964. "Nutrition of Algae." *Amer. Rev. Plant Physiol.*, 15:37.
43. Marshal, N. 1956. "Chlorophyll *a* in the Phytoplankton in Coastal Waters of the Eastern Gulf of Mexico." *Jour. Marine Res.*, 15:14.
44. Pomeroy, L. R. 1959. "Algal Productivity in Salt Marshes of Georgia." *Limnol. Oceanog.*, 4:386.
45. Pomeroy, L. R., H. H. Haskin, and R. A. Ragotskie. 1956. "Observations on Dinoflagellate Blooms." *Limnol. Oceanog.*, 1:54.
46. Ragotskie, R. A. 1959. "Plankton Productivity in Estuarine Waters of Georgia." *Publ. Inst. Marine Sci. Univ. Texas*, 6:146.
47. Odum, H. T., and C. M. Hoskin. 1958. "Comparative Studies of the Metabolism of Marine Waters." *Publ. Inst. Marine Sci. Univ. Texas*, 5:16.
48. Odum, H. T., and R. F. Wilson. 1962. "Further Studies on Reaeration and Metabolism of Texas Bays, 1958–1960." *Publ. Inst. Marine Sci. Univ. Texas*, 8:23.
49. Childress, U. R. (mimeo). "Coordination of the Blue Crab Studies of the Texas Coast." Marine Fisheries Projects Reports for 1963, Texas Parks and Wildlife Dept., Austin, Texas.
50. Compton, H. (mimeo). "A Study of the Bay Populations of Juvenile Shrimp, *Penaeus aztecus* and *Penaeus setiferus*." Marine Fisheries Projects Reports for 1960–61, Texas Game and Fish Comm., Austin, Texas.
51. Moffett, A. (mimeo). "A Study of the Texas Bay Populations of Juvenile Shrimp." Marine Fisheries Project Reports for 1963, Texas Parks and Wildlife Dept.
52. Gunter, G. 1938. "Seasonal Variations in Abundance of Certain Estuarine and Marine Fishes in Louisiana, with Particular Reference to Life Histories." *Ecol. Monogr.*, 8:313.
53. Springer, V. G., and K. D. Woodburn. 1960. "An Ecological Study of the Fishes of the Tampa Bay Area." Florida State Bd. Conservation Professional Paper Ser. No. 1, 1.
54. Day, J. H. 1951. "The Ecology of South African Estuaries, Part I: A Review of Estuarine Conditions in General." *Trans. Roy. Soc. S. Africa*, 33:53.
55. Day, J. H., N. A. H. Millard, and G. J. Broekhuysen. 1954. "The Ecology of South African Estuaries. Part IV: The St. Lucia System." *Trans. Roy. Soc. S. Africa*, 34:129.
56. Millard, N. A. H., and A. D. Harrison. 1954. "The Ecology of South African Estuaries. Part V: Richard's Bay." *Trans. Roy. Soc. S. Africa*, 34:157.
57. Nikolsky, G. V. 1963. "The Ecology of Fishes." Academic Press, New York.
58. Gunter, G. 1961. "Some Relations of Estuarine Organisms to Salinity." *Limnol. Oceanog.*, 6:182.
59. Williams, A. B. 1955. "A Survey of North Carolina Shrimp Nursery Grounds." *Jour. Elisha Mitchell Sci. Soc.*, 71:200.

35

The water balance and land-use

R. G. Downes

INTRODUCTION

Before civilized man has exercised his influence, an area of land, irrespective of its dimensions, is a dynamically balanced system in which there is a definite relationship between the kind and relative abundance of species and the other features of the environment, namely, climate, topography and soils. Such a system had evolved or was in the process of evolution toward a climax condition, at which there is the maximum sustainable production of plant and animal life possible from the particular array of species available during the period of evolution. The particular characteristics of such an eco-system are the result of the interaction of many features such as parent materials, soils, topography, water balance, plants and animals, under the influence of the climate.

Civilized man strives to make land produce the kinds of plants and animals which are useful to him, and by introducing new species of plants and animals into modified environments, he has achieved some phenomenal successes but has also created some enormous problems. Some introduced plants and animals have been so suited by their environment that they have become noxious weeds and vermin. Modifications of the environment to suit the species to be introduced have seriously affected the stability of some systems and have caused soil erosion and loss of productivity.

Changes in the water balance have been common, and the different

---

Reprinted with permission from *Water Resources, Use and Management*, pp. 329–341 (Melbourne University Press, 1964).

pattern of water distribution within the water cycle has caused a multitude of problems, including waterlogging, aridity, salting, soil erosion and deposition.

For these reasons conservationists consider the only satisfactory systems of land-use are those that not only achieve the objective of higher productivity from more desirable plant and animal species, but also provide a stability similar to that existing in an unimpaired eco-system. In this respect the water cycle is an exceedingly important characteristic, the balance of which must be maintained, or at the least, any possible changes should be forecast and provided for in the changed system of land-use.

When considering the water balance in relation to land-use, it is obvious that climate will determine the amount of water in the system, but it is not so obvious how the other features of the environment exercise a dominating role in the way in which the water is apportioned between run-off, soil-moisture storage, evaporation, transpiration, and percolation to the deep aquifers. The nature of the original disposition of water in the undisturbed eco-system will determine the forms of land-use which may or may not be contemplated. The original character of the water cycle can be significantly altered, not only by the form of land-use, but also by the subsequent management. The stability of the system of land-use and management will depend on the possible effects of the changed water cycle, and whether provision is made to safely dispose of any excess water or make up any deficiency.

A study of the water balance in relation to land-use is vitally important for two reasons: to ensure that the ill effects of wrong systems of land-use do not impair, or lead to less efficient collection and use of a most important natural resource—water; and, to ensure that ill effects of wrong systems of land-use do not upset the water balance in a way which will lead to the destruction of the next most important natural resource—soil.

An examination of the various components of the water cycle, and the environmental characteristics on which they are particularly dependent, enables the deduction of certain principles which should be observed when considering the possible effects of different land-use practices on the water cycle. To some extent these deductions can be substantiated by experimental data but, unfortunately, this is not sufficiently common.

THE WATER BALANCE

There is a convenient equation used by Downes (1958b), which expresses the nature of the water cycle for an area of land:

$$P = E + T + R + I$$

where P = precipitation;
E = evaporation from soil and plant surfaces;
T = plant transpiration of water extracted from the soil;
R = surface run-off;
I = infiltration into the soil.

The soil-moisture increment or infiltration (I) can be disposed of in three ways which are expressed by the following equation:

$$I = U + S + A$$

where U = loss to underground by deep percolation;
S = seepage or laterally moving water;
A = the increment to the possible soil-moisture storage.

The relative magnitude of the individual components of the equation will vary considerably for different kinds of land. They will also vary for similar kinds of land in catchments of different sizes, and they will vary for a single catchment according to whether the values in the equation refer to annual, monthly or weekly totals, or to individual storms. In spite of this, it is possible to deduce the effects of particular changes in land-use on the relative magnitude of the various components of the water cycle and their significance in relation to particular problems.

THE EFFECT OF LAND-USE ON THE COMPONENTS OF THE WATER CYCLE

There are variants of the basic hydrological equation for three broadly differing regions, namely, arid areas where, on an annual basis, the precipitation is small in relation to potential evaporation and transpiration; humid areas where the reverse situation holds; and the intermediate areas where the precipitation about balances the potential evaporation and transpiration.

Arid and semi-arid areas

The precipitation (P) is so low in relation to the potential values which could be attained for evaporation (E) and transpiration (T) if water were available, that the values for run-off (R) and infiltration (I) become insignificant, except when considered for individual storms on relatively small areas. The basic equation becomes

$$P = E + T$$

In terms of land-use, these are areas of water need. This can be satisfied either by importing water for irrigation or by using water-harvesting techniques whereby the water disposed of in R from individual storms on certain areas is collected to increase the value of P and the productivity of much smaller areas lower down in the catchment.

Humid areas

The precipitation (P) is large in relation to the potential values for evaporation (E) and transpiration (T), which in these areas of water sufficiency represent the actual values, so that run-off (R) and infiltration (I) assume significant values. In these areas, the conditions are so suitable for plant growth that the ground surface cover precludes a high value for R from small areas. Of course, in large catchments, R is the stream-flow, which results from the accretion of the excess of water from infiltration (I) after soil moisture storage has been satisfied on many small areas. In terms of land-use, these are areas of water excess from which water should be harvested for use elsewhere.

Sub-humid and semi-arid areas

These are areas in which the precipitation (P) about balances the potential values for evapotranspiration ($E + T$). In drier than normal years, the basic equation is similar to that for arid areas, all the water being disposed of as $E + T$, but in wetter than normal years, the equation is similar to that for humid areas; there is excess water to give values for $R + I$. Because of this balanced condition, the land-use in these areas must ensure efficient use of water and maintenance of the water balance. Any change in the land-use which throws the equation permanently toward either the arid or the humid side has been shown by Downes (1958a, b) to produce undesirable consequences.

It is now appropriate to examine the individual components of the water-cycle equation in relation to forms of land-use and their possible effects.

Precipitation (P) The value for P is primarily determined by the climate, but forms of land-use can modify it. A most significant modification occurs when irrigation water is applied. In these circumstances, the system may be subjected to several times the value of P for which it was in equilibrium. This can do harm if we realize that the original equation, $P = E + T$, has now become $P = E + T + R + I$, and provision is made to dispose of the increased values of $R + I$ by drainage. It is even better and more efficient to limit the value for P by the application of only sufficient water so that the equation $P = E + T$ still holds. In other words, the ideal arrangement in an irrigation system is to apply only the amount of water required to wet the root zone at sufficiently frequent intervals to satisfy the transpiration of the crop. Excess brings its problems of waterlogging and salting, as well as being a waste of water.

The value of P may also be modified by land-use in humid areas by changes of vegetation in specific circumstances. In a recent monograph, Penman (1963) has examined the various arguments and data advanced with respect to the effect of vegetation in inducing rainfall. The conclusion reached is that dense vegetation exists in various parts of the world because of the incidence of rainfall and that there is no evidence that such areas of vegetation induce rainfall. However, there is evidence of a significant proportion of P being due to horizontal interception of fog by vegetation in certain parts of the world. Evidence for increased precipitation at the edge of the intercepting forest with a diminution of the effect with distance inward from the edge is quoted by Penman (1963) from a variety of sources. In Australia, Costin (1961b) has found interception by snow gums in alpine areas to be significant.

Another effect for which evidence is available, both in the United States (Anderson, 1956; Love, 1953) and in Australia (Costin, 1961a), is that of vegetation in affecting the depth of snow-pack and so increasing the precipitation, at least locally. Perhaps this is not really an increase of P when considering a whole water-year, but it certainly changes the effective distribution of P on a monthly basis. The release of stored precipitation late in the season as melting snow provides more useful water, and is preferable to snow being blown away and precipitated as rain at lower altitudes where it provides run-off at a time when the water is of little value.

Evaporation (E) The value for E represents the water in the cycle which is directly evaporated from streams, ponds, bare soil or from the surface of vegetation, following the interception of precipitation. The potential value is determined by the climate, and depends on the amount of solar energy available to evaporate water, but the actual value depends on the nature of the vegetation and the soils, and the amount of water available for evaporation.

Vegetation intercepts rain which would otherwise fall on the soil surface to either infiltrate or run off. The amount of precipitation which is intercepted and is subsequently evaporated, is related to the character of the vegetation, density of foliage, the nature of the trunk and limbs of the trees or shrubs and the orientation of the foliage. The vegetation can only intercept so much water, and the proportion of the total becomes less as the intensity and the duration of any particular rainstorm increases.

Changes in land-use which alter the character of the vegetation so that there is a different amount of interception, and of the water held by the saturated vegetation, cause a significant change in the value for E and, in consequence, changes of the other components of the hydrological equation. For example, the clearing of forested land for cultivation under a system which includes a period of bare fallow is a major change.

Of course, evaporation of water from the foliage of plants conserves the available water in the soil, because until the water on the leaves is evaporated practically no transpiration will take place.

Evaporation from bare soil becomes an important component in arid and semi-arid areas where the vegetative cover is sparse. It is affected to a considerable extent by the nature of the soil. When soils can only absorb water slowly, much of it lies on the surface and can be evaporated. The rate at which water in the soil can be removed by evaporation is influenced by the nature of the soil itself. In arid areas where there is plenty of solar energy to enable evaporation, the loss of water from the surface soil occurs quite rapidly. In sandy soils, this soon provides a mulch of air-dry soil above the wetted zone, and subsequent loss of moisture from the soil can only be the result of vapour pressure and temperature gradients. This movement of water as vapour is much slower than the initial rate of loss due to soil-moisture tension gradients causing movement of water to the surface as liquid. In clay soils, the latter form of loss persists for much longer and since there is more water available for evaporation from saturated clay soils, the value of E can be large.

Forms of land-use which tend to improve soil structure and increase the infiltration can effectively reduce evaporation losses. On the other hand, management techniques, such as the deep ploughing of soils to leave a very cloddy surface, can encourage the complete air-drying of the ploughed layer during the long dry summer and so create a condition requiring a much larger amount of water at the beginning of the next season to bring the moisture content of the surface soil above wilting point.

Transpiration (T) As for evaporation, the potential value for T is a function of the climate and depends on the solar energy being received by plants. However, the actual value of T depends on the amount of water available for transpiration which in turn depends on two things. First, the amount of water in the soil, and second, the ability of the vegetation to make use of it.

In humid areas, where availability of water is nonlimiting, the actual value of T can be computed from meteorological data by using one of several well-known methods, such as those of Blaney and Criddle (1950), Penman (1956), Thornthwaite (1954), Turc (1953), but in sub-humid and arid areas, the actual value for T will be below the potential value, and can only be evaluated by applying these physical concepts to calculate daily water balances while water is available for plants.

Some plants appear to be able to take an active part in determining the transpiration by reducing the rate of using water under stress.

Downes (1958a, b) has shown how forms of land-use, by changing the vegetation, can affect the structure of the soils and the infiltration capacity and so affect the amount of water entering the soil and becoming available in the root zone for transpiration by plants; and also how a change of vegetation, which encourages less use of water, can lead to undesirable excesses of water in the system which can cause soil erosion and a decline in fertility.

For example, the change of vegetation from a deep-rooted perennial tree or shrub vegetation to that of an annual crop only once in each two or three years, with either bare fallow or shallow-rooted annual pastures in the other years, will reduce the water use considerably. Although this water conservation enables a commercial crop to be grown it can cause troubles due to excess water. Problems of this origin are evident in the light-textured soils of the rises in north-western Victoria.

The change from trees to perennial grass to annual pastures on solodic soils in central Victoria has also provided problems because of the change in the basic hydrological equation, $P = E + T$. When T is reduced, this becomes $P = E + T + R + I$. Special measures are required to ensure that the increased run-off and infiltration do not cause problems such as gully erosion, tunnel erosion and salting, and in some places mass movement of soils by landslips and earth-flows.

There are many circumstances when a change in transpiration can create problems and so, when introducing new forms of land-use on areas where under natural conditions P is approximately equal to $E + T$, it is important to ensure that the introduced vegetation has the same water-using characteristics and a rooting habit which enables it to explore and take water from the same volume of soil as the original vegetation.

Surface run-off (R) Surface run-off in small catchments is mainly from overland flow, but in large catchments more frequently the accretion of both overland flow and sub-surface seepage creates streams.

In humid areas, the dense vegetation cover precludes much overland flow, and the run-off from a catchment is largely the result of accretion of underground seepage to create springs and streams.

In sub-humid and arid areas, overland flow becomes important, its magnitude depending on the amount and kind of vegetative cover, the infiltration capacity of the soil, the topography and the climatic characteristics, particularly the intensity and duration of rainstorms.

Some of these characteristics can be greatly influenced by land-use, particularly within small catchments; for example, subterranean clover can improve infiltration capacity, excessive grazing can reduce vegetative cover and the trampling by animals can compact soils.

Infiltration (I) On small catchments the value for I is complementary to the value for R. Water which does not infiltrate is run-off. However, in large catchments the run-off is an accumulation of that portion of I which is not lost to deep aquifers and taken outside the catchment or that required to replenish soil-moisture storage.

Infiltration is disposed of in bringing the soil-moisture storage up to field capacity, after which there are two possible paths. First, seepage past the root zone to lower levels either to create springs within the catchment or to be lost to other catchments. The second is by lateral seepage through the soil on sloping land.

The proportion of water disposed of in lateral seepage is determined by soil characteristics and slope, and also by storm characteristics. Soil having a light-textured surface layer which absorbs water readily will not provide much opportunity for surface run-off until this layer is saturated. How soon it becomes saturated depends on the nature of the subsoil. If the surface soil is shallow over an impeding horizon such as impermeable clay or rock, then the soil is soon saturated above this layer and lateral seepage (S) will occur before surface run-off begins.

It follows, therefore, that forms of land-use which improve the infiltration capacity of the surface soil could lead to an increase of water occurring as lateral seepage, and this would happen more readily if the form of land-use reduced the use of water from a clay subsoil so that it remained moist and therefore a more effective barrier to downward movement of water. A change from perennial to annual pasture or bare fallowing will have just such an effect and in some circumstances, the results can be disastrous.

WHAT DO WE KNOW ABOUT THE WATER BALANCE AND LAND-USE?

From these basic concepts of the mechanics of the hydrological cycle and the way in which the various features of the environment interact, it is possible to consider the advisability of any proposed changes of land-use in relation to important catchment characteristics.

When considering a catchment from which water is to be harvested as the main form of production from the land, certain characteristics need to be maintained.

The important characteristics of a catchment are its ability to yield water of good quality, and the way in which the yield is distributed throughout the year. Consequently the systems of land-use and management should not diminish the yield nor reduce the quality of the water, neither should they seriously alter the seasonal distribution of the flow. Any changes of land-use which can improve the water yield without increasing the sediment load or turbidity will increase the value of a catchment, particularly if a good distribution of water yield throughout the year, a characteristic which can be easily destroyed by changes in land-use, can be maintained.

When considering land as the major asset, the productivity of which must be maintained, the following characteristics of the water cycle are important.

(a) The relative proportions of water in the various components of the basic equation.

(b) The possible result of changes of the value in one or other of the components, and whether such a result can be avoided by adjustments in the proposed land-use.

Although effects of land-use may be deduced, it would be an advantage to have much more experimental confirmation.

There is a wide range of research into particular aspects of the hydrological cycle. Most of this work is being done not because of its value to catchment hydrology but because of its value to agriculture, forestry, or engineering. Studies of infiltration and movement of water through soils, the physics of evaporation, the physical chemistry of evaporation suppression, the effect of water stress in plants and stream geometry in relation to flow characteristics, are all important investigations in their own right. However, there is insufficient investigation into the whole hydrological cycle in which the knowledge from these various investigations is integrated into an understanding of the whole problem.

The main value of more detailed study of the hydrological cycle in catchments is twofold. First, it will enable the determination of land-use to ensure that the catchment will be maintained as an efficient water producing unit, to be based on experimental evidence and not deductions. Second, certain management operations to increase water yield safely may become more readily apparent.

Penman (1963) has summarized some complete catchment studies which are becoming more common now in many parts of the world. It is only during the last decade that any of these kinds of investigations have been attempted in Australia, a strange fact in view of the importance of water in this continent. With three stations now established at the Parwan on pastoral land having a 21-in. average annual rainfall, at Stewart's Creek on forest land having a 40-in. average annual rainfall, and at Reefton on forest land having a 55-in. average annual rainfall, the Soil Conservation

Authority of Victoria has initiated the basis of a useful programme of whole catchment study in an attempt to overcome a deficiency of information in that State.

These experiments have been established to determine the effects on the hydrological cycle of different kinds of land-use which seem appropriate for each area. It is only at the Parwan station that the investigations have advanced sufficiently for results to be forthcoming (Dunin and Downes, 1962) but these, although as yet only preliminary, appear to confirm some previous deductions concerning the possible effects of land-use.

Studies on a complete catchment basis are expensive, long term, and have some inherent difficulties. The condensation of continuous measurements recorded on charts to readily available and easily used sets of figures is a problem. So also is the interpretation of any derived data, because whole catchments cannot be replicated to allow the application of the more usual statistical techniques. Finally, there is the problem of applicability. The results obtained from one catchment do not necessarily apply for another which may appear to be similar, but is located in another district. Furthermore, the expected quantitative data from a catchment of one size is almost certainly not linearly related to measured results from a catchment of a different size. A significant investigation into this aspect of the problem of apply experimental data is that of Harrold (1957) at Coshocton in the United States, and, as Harrold himself remarks, "in physiographic provinces other than this, like some Arizona areas, the effect of watershed size may have exactly the opposite effect."

Catchment experiments are of value because they provide information about the relative values for the different components of the hydrological cycle, and in this respect they can correct completely erroneous ideas. In the experiments at the Parwan station on badly-eroded catchments, the average annual run-off was found to be only about 5 percent of the rainfall (21 in.), a much smaller proportion than had been imagined. Even for high-intensity storms, the run-off only amounted to a maximum of 60 percent of the total rain.

The study of the whole hydrological cycle provides basic information about catchment mechanics, the behaviour under a range of seasonal and storm conditions following a variety of antecedent conditions. For this reason, the information is of considerable value as background information for the design and subsequent interpretation of data from short-term *ad hoc* experiments.

In experiments established in Israel recently, Shachori *et al.* (1962) have used a plot technique in an attempt to measure the relative water-use of pines, maquis and grass on limestone mountain catchments. Dr. Pereira of Southern Rhodesia and the author were F.A.O. consultants in the planning of these experiments, and the decision to adopt such a technique was made because of the nature of the data already obtained from a study of the hydrology of whole catchments.

Plot experiments of short duration to determine some particular point may constitute the main line of future hydrological research in relation to the effect of land-use practices. But the data will need to be interpreted always in the light of information obtained from whole catchment studies. Intensive investigations of the water balance for particular vegetative types, such as that reported by Slatyer (1959), could be the most useful approach in arid and semi-arid areas.

From the studies already made in various parts of the world, it is difficult to derive any universal principles concerning the effect of land-use on the hydrological cycle. This is because the effect is so dependent on the size of catchment being considered. Some land-use practices can be tremendously significant in their effect on single paddocks, whole farms or even minor catchments, but apparently not particularly significant in their total effect for whole river basins.

Land-use changes which reduce the ability of the catchment to absorb water are detrimental. The pattern of water-yield distribution is upset and a serious load is imposed on the watercourses, so causing erosion and an increased sediment movement. More arid conditions are created for the remaining vegetation and this tends to hasten further deterioration.

Land-use changes which reduce the volume of soil from which water is being extracted by vegetation leads to an excess of water in the system. The redistribution of this water can cause considerable deterioration of the area both as a catchment to yield water and as land having a productive potential.

This is not the occasion for an exhaustive review of research into effects of land-use on catchment hydrology. Reference to a few examples will serve to indicate the individual nature of the investigations and how difficult it would be to translate the results obtained in one area to another. However, they provide information which can be a useful guide for research in other places.

Shachori and Michaeli (1962) have reviewed the available experimental evidence from many parts of the world with respect to the water-use of forested, woodland and shrub-covered areas in relation to that from grassed and barren areas. They concluded that in all studies the total run-off from bare or grassed areas was greater than that from forested, woodland or scrub-covered areas, and that there was a tendency for the magnitude of the difference to increase with increasing rainfall. However, in considering this fact in relation to catchment management, an increased total water yield need not be the significant attribute to be achieved. Unless any increased yield is of the right quality and available when required, it is of no real value.

Humphrey (1959) in a study of evidence available concerning the management of range vegetation in relation to water-use stated: "Water use varies approximately in proportion to the length of time a plant is actively growing or is producing a crop of green leaves, the depth and extent of the root system and the amount of water available."

Liacos (1962), in a recent study of the effect of grazing on water yield from grassland catchments in California, showed from plot studies on areas subjected to heavy, light and no grazing for long periods, significant differences in water yield. The differences were great enough for him to remark that "range managers have a very good tool in their hands to affect water production in quantity and quality."

Storey (1959), in a review of the effects of forests on run-off, points out the significant effect of a forest cover on interception, infiltration and the effect of these on storm run-off, and so on water yield. He refers to possible water-yield management by forest harvesting techniques, and the removal of trees from along watercourses.

Patric (1961), in discussing results from the San Dimas large lysimeters, each of which has an area of 0.005 acre and 6-ft. depth soil, makes these comments:

> During a series of dry seasons 1952-1956, the average annual rainfall being 20½ inches, the water yield was entirely from surface run-off, except for a small amount of seepage from the grass plot. Woody plants used all available moisture during each dry season, while about 10 inches of available water remained under the bare and grass covered plots. During winter and spring, pine and grass dried the soil more rapidly than scrub oak.
>
> During a very wet year 1957-58, the rainfall was 48.4 inches—more than 20 inches above the normal average annual rainfall. Regardless of the amount of rainfall, the bare plot only absorbed 9.3 inches and the seepage yields were greater under grass than any other cover. This is a significant fact for catchment management, because good grass cover will give clean water if there is a higher yield.

A most interesting set of catchment experiments for which data are available are those organized by Pereira (1962) in East Africa. The experiments concern the effect of land-use practices on the hydrological cycle in a range of environments. The first concerns the development of tea estates in areas normally carrying tall rain forest, the second concerns grazing control in semi-arid catchments, and the third, the conversion of bamboo forest to pine plantations. A further investigation was into the effect of peasant cultivation in steep stream-source valleys.

In the first of these experiments, there is clear evidence of the efficiency of a tall rain forest as a regulator of storm flow. Although only one-third of the catchment has been cleared for tea plantations, there are significant departures of catchment behaviour following storms.

In the second, there is data to support the contention that in semi-arid catchments water yield is dependent on excess of storm rainfall over the amount required to satisfy surface-storage capacity, and it was "remarkably insensitive to rainfall intensities."

In the third experiment, direct measurement of water use from 10 ft. depth of soil showed equal consumption by 120-ft.-high 26-year-old *Pinus radiata*, 50-ft.-high 16-year-old Monterey cypress, 30-ft.-high 10-year-old Patula pine, and by an undisturbed indigenous bamboo thicket.

In the fourth experiment, in a period of two years the continuous arable cropping on steep slopes has doubled the suspended sediment load under steady flow conditions, indicating that soil stability is inadequate for the form of land-use being followed.

In commenting on these experiments, Penman (1962) states: "This group of studies is almost alone in its recognition that the behaviour of a catchment must be considered as a single problem, whose conventional components have interactions that are often more important than their individual characters."

FURTHER RESEARCH REQUIREMENTS

Soil conservationists are concerned with the establishment of an ecological balance with the system of land-use and management devised for each different kind of environment. In determining whether there is a balance, the effect on the hydrological cycle can often be the significant indicator. It will not be a balanced system unless the hydrological cycle remains unchanged, or if it is changed, the new circumstances are provided for in the system so that no permanent damage to productivity will result.

A study of the parts of the system as separate components is not enough, because the interactions of the components are important and can be even more important than the components themselves. For these reasons, study of whole catchments or whole eco-systems is essential if the necessary information about water resources, their conservation, management and use is to be obtained. Furthermore, there is considerable need for research into the manner of distribution and use of water when it is taken into areas of water need.

Research at present is too unco-ordinated, possibly because it is a sideline for many organizations, but not the prime responsibility of any.

There are many useful investigations with respect to various components of the hydrological cycle, but few concerned with the whole cycle within various kinds of environments.

Whole catchment studies are expensive and long range, and it would be unwise for a single organization to be responsible for their establishment and direction throughout Australia. Nevertheless, it would be desirable to have some guiding and co-ordinating centre to advise and assist in co-operative programmes. The East African experiments organized by Dr. Pereira appear to have much to commend them in the manner in which various organizations were able to provide a co-ordinated effort to produce effective results.

The water research problems of Australia in relation to land-use sort themselves into three categories on a geographical basis:

(a) Problems of the water-producing areas where precipitation greatly exceeds potential evaporation and transpiration.

(b) Problems of the water-balance areas in which local conservation and use becomes important because precipitation about equals the potential evaporation and transpiration.

(c) Problems of the areas of water need—where precipitation is much less than the potential evaporation and transpiration and water must be imported for improved production from the land.

Of these three sets of problems, the only one being pursued with any vigour is that in relation to the semi-arid and arid areas, namely, irrigation research.

There appears to be a need for some organization responsible for hydrological research which should initially concentrate on the fundamental biological aspects of catchment hydrology, and act as a centre for devising techniques of study and suitable instruments. In this way necessary studies of the whole water cycle by interested bodies throughout the country would be encouraged and assisted.

For a continent so dry as Australia, with a good record of scientific achievement in many aspects of the development, conservation and use of natural resources, it is incredible that so little has been attempted with respect to water.

REFERENCES

Anderson, H. W. 1956. "Forest-cover effects on snow pack accumulation and melt, Central Sierra Snow Laboratory," *Trans. Amer. Geophys. Un.*, 37:307–12.

Blaney, H. F., and W. D. Criddle. 1950. *Determining water requirements in irrigated areas from climatological and irrigation data*, U.S. Dept. Agric., Soil Conserv. Service Tech. Paper 96.

Costin, A. B. 1961. *Studies of catchment hydrology in the Australian Alps: III. "Preliminary snow investigations"; "IV. Interception by trees of rain, cloud and fog"*; C.S.I.R.O. Div. Plant Industry Tech. Papers Nos. 15, 16.

Downes, R. G. 1958a. "Land management problems following disturbance of the hydrologic balance of environments in Victoria, Australia," paper presented at 7th Tech. Meeting Intern. Union for Conservation of Nature—Natural Resources, Athens.

——. 1958b. "Soil salinity in non-irrigated arable and pastoral land as the result of unbalance of hydrologic cycle," paper presented at UNESCO Arid Zone Symposium, Teheran.

Dunin, F. X., and R. G. Downes. 1962. "The effect of subterranean clover and Wimmera ryegrass in controlling surface runoff from four-acre catchment areas, Bacchus Marsh," *Aust. J. Exp. Agric. and Anim. Husb.*, 2:148–52.

Harrold, L. 1957. "Minimum water yield from small agricultural watersheds," *Trans. Amer. Geophys. Un.*, 38:201.

Humphrey, R. R. 1959. "Forage and water," *J. Range Management*, 12:164–70.

Liacos, L. G. 1962. "Water yield as influenced by degree of grazing in the California winter grasslands," *J. Range Management*, 15:34–42.

Love, L. D. 1953. "Watershed management in the Colorado Rockies," *J. Soil and Water Cons.*, 8:107–12.

Patric, J. H. 1961. "The San Dimas large lysimeters," *J. Soil and Water Cons.*, 16:13–17.

Penman, H. L. 1956. "Estimating evaporation," *Trans. Amer. Geophys. Un.*, 37:43–50.

———. 1963. *Vegetation and Hydrology*, Tech. Comm. No. 53 (Commonwealth Agric. Bureaux, Farnham Royal).

Pereira, H. C. *et al.* 1962. "Hydrological effects of changes in land-use in some East African catchment areas," *E. African Agric. and For. J.*, XXVII, Special Issue.

Shachori, A. Y., and A. Michaeli. 1962. "Water yields of forest, maquis and grass covers in semi-arid regions," paper presented at UNESCO Arid Zone Symposium, Montpellier.

———, G. Stanhill, and A. Michaeli. 1962. "The application of integrated research approach to the study of different cover types on rainfall disposition in the Carmel Mountains–Israel," paper presented at UNESCO Arid Zone Symposium, Montpellier.

Slatyer, R. O. 1959. "Methodology of a water balance study conducted on a desert woodland community in Central Australia," paper presented at UNESCO Arid Zone Symposium, Madrid.

Storey, H. C. 1959. "Effects of forests on runoff," *J. Soil and Water Cons.*, 14:152.

Thornthwaite, C. W. 1954. 'The determination of potential evapotranspiration,' Appendix XVI in "The Measurement of Potential Evapotranspiration" by J. R. Mather, *Publications in Climatology*, VII:218 (Johns Hopkins University, Baltimore).

Turc, L. 1953. "Inter-relations of rainfall, evaporation and runoff," *African Soils*, III: 139–72.

SUSTAINED USE MANAGEMENT

TERRESTRIAL ECOSYSTEMS AND SOILS RESOURCES

36

Nutrient cycling

F. H. Bormann
G. E. Likens

Life on our planet is dependent upon the cycle of elements in the biosphere. Atmospheric carbon dioxide would be exhausted in a year or so by green plants were not the atmosphere continually recharged by CO_2 generated by respiration and fire (1). Also, it is well known that life requires a constant cycling of nitrogen, oxygen, and water. These cycles include a gaseous phase and have self-regulating feed-back mechanisms that make them relatively perfect (2). Any increase in movement along one path is quickly compensated for by adjustments along other paths. Recently, however, concern has been expressed over the possible disruption of the carbon cycle by the burning of fossil fuel (3) and of the nitrogen cycle by the thoughtless introduction of pesticides and other substances into the biosphere (4).

Of no less importance to life are the elements with sedimentary cycles, such as phosphorus, calcium, and magnesium. With these cycles, there is a continual loss from biological systems in response to erosion, with ultimate deposition in the sea. Replacement or return of an element with a sedimentary cycle to terrestrial biological systems is dependent upon such processes as weathering of rocks, additions from volcanic gases, or the biological movement from the sea to the land. Sedimentary cycles are less perfect and more easily disrupted by man than carbon and nitrogen cycles (2). Acceleration of losses or, more specifically, the disruption

Reprinted with permission from *Science*, 155(3761):424–429 (27 January 1967).

of local cycling patterns by the activities of man could reduce existing "pools" of an element in local ecosystems, restrict productivity, and consequently limit human population. For example, many agriculturalists, food scientists, and ecologists believe that man is accelerating losses of phosphorus and that this element will be a critical limiting resource for the functioning of the biosphere (1, 5).

Recognition of the importance of these biogeochemical processes to the welfare of mankind has generated intensive study of such cycles. Among ecologists and foresters working with natural terrestrial ecosystems, this interest has focused on those aspects of biogeochemical cycles that occur *within* particular ecosystems. Thus, information on the distribution of chemical elements and on rates of uptake, retention, and release in various ecosystems has been accumulating (6). Little has been done to establish the role that weathering and erosion play in these systems.

Yet, the rate of release of nutrients from minerals by weathering, the addition of nutrients by erosion, and the loss of nutrients by erosion are three primary determinants of structure and function in terrestrial ecosystems. Further, with this information it is possible to develop total chemical budgets for ecosystems and to relate these data to the larger biogeochemical cycles.

It is largely because of the complex natural interaction of the hydrologic cycle and nutrient cycles that it has not been possible to establish these relationships. In many ecosystems this interaction almost hopelessly complicates the measurement of weathering or erosion. Under certain conditions, however, these apparent hindrances can be turned to good advantage in an integrated study of biogeochemical cycling in small watershed ecosystems.

It is the function of this article (i) to develop the idea that small watersheds can be used to measure weathering and erosion, (ii) to describe the parameters of watersheds particularly suited for this type of study, and (iii) to discuss the types of nutrient-cycling problems that this model renders susceptible to attack. Finally (iv), the argument is developed that the watershed ecosystem provides an ideal setting for studies of ecosystem dynamics in general.

ECOSYSTEM DEFINED

Communities such as fields and forests may be considered as ecological systems (7) in which living organisms and their physical and biological environments constitute a single interacting unit. These ecosystems occupy an arbitrarily defined volume of the biosphere at the earth-atmosphere interface.

Lateral boundaries of an ecosystem may be chosen to coincide with those of a biological community, such as the edges of a forest, or with the boundary of some pronounced characteristic of the physical environment, such as the shoreline of a small island. Most often, however, the continuous nature of vegetation and of the physical environment makes it difficult to establish exact lateral boundaries on the basis of "community" or "environmental discontinuity" (8). Often the investigator arbitrarily selects an area that may be conveniently studied.

From a functional point of view it is meaningless to include within the vertical limits of an ecosystem *all* of the column of air above and of soil and rock below the laterally defined ecosystem. For a working model of an ecosystem, it seems reasonable to include *only* that part of the column where atoms and molecules may participate in the chemical cycling that occurs with the system (see the "intrasystem cycle" of Fig. 36–1). When the biological community is taken as a determinant, the vertical extensions of the terrestrial ecosystem will be delimited by the top of the vegetation and the depth to which roots and other organisms penetrate into the regolith (9). Vertical dimensions, defined in the manner, can expand or contract depending on the growth potential of present

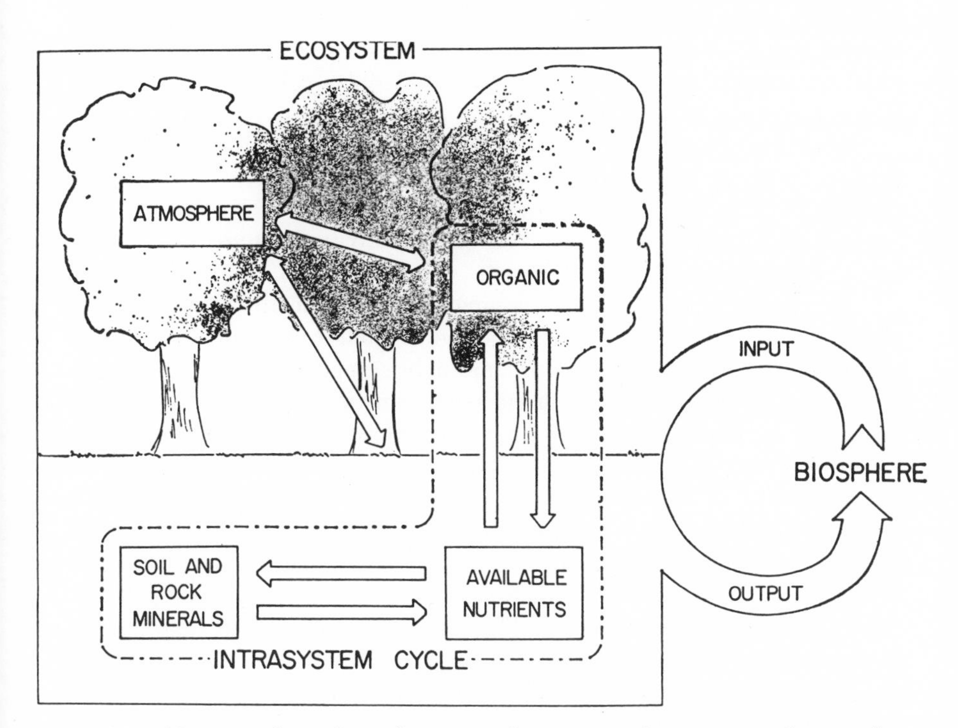

Fig. 36–1. Nutrient relationships of a terrestrial ecosystem, showing sites of accumulation and major pathways. Input and output may be composed of geologic, meteorologic, and biologic components, as described in the text.

or succeeding communities. Thus, volumetric changes with time can be considered—for example, those associated with primary and secondary succession or with cliseral changes.

THE ECOSYSTEM AND BIOGEOCHEMICAL CYCLING

The terrestrial ecosystem participates in the various larger biogeochemical cycles of the earth through a system of inputs and outputs. Biogeochemical input in forest or field ecosystems may be derived from three major sources: geologic, meteorologic, and biologic. Geologic input is here defined as dissolved or particulate matter carried into the system by moving water or colluvial action, or both. Depositions of the products of erosion or mass wasting and ions dissolved in incoming seepage water are examples of geologic input. Meteorologic input enters the ecosystem through the atmosphere and is composed of additions of gaseous materials and of dissolved or particulate matter in precipitation, dust, and other wind-borne materials. Chemicals in gaseous form fixed by biologic activity within the ecosystem are considered to be meteorologic input. Biological input results from animal activity and is made up of depositions of materials originally gathered elsewhere; examples are fecal material of animals whose food was gathered outside the system, or fertilizers intentionally added by man.

Chemicals may leave the ecosystem in the form of dissolved or particulate matter in moving water or colluvium, or both (geologic output); through the diffusion or transport of gases or particulate matter by wind (meteorologic output); or as a result of the activity of animals, including man (biologic output).

Nutrients are found in four compartments within the terrestrial ecosystem: in the atmosphere, in the pool of available nutrients in the soil, in organic materials (biota and organic debris), and in soil and rock minerals (Fig. 36–1). The atmospheric compartment includes all atoms or molecules in gaseous form in both the below-ground and the above-ground portions of the ecosystem. The pool of available nutrients in the soil consists of all ions adsorbed on the clay-humus complex or dissolved in the soil solution. The organic compartment includes all atoms incorporated in living organisms and in their dead remains. (The distinction between living and dead is sometimes hard to make, particularly in the case of woody perennial plants.) The soil-rock compartment is comprised of elements incorporated in primary and secondary minerals, including the more readily decomposable minerals that enter into equilibrium reactions with available nutrients.

The degree to which a nutrient circulates within a terrestrial ecosystem is determined, in part, by its physical state. Gases are easily moved by random forces of diffusion and air circulation; consequently, nutrients with a prominent gaseous phase tend not to cycle within the boundaries of a particular ecosystem but, rather, to be continually lost and replaced from outside. On the other hand, elements without a prominent gaseous phase may show considerable intrasystem cycling between the available-nutrient, organic, and soil-rock compartments (Fig. 36–1). This internal cycle results from (i) the uptake of nutrients by plants, (ii) the release of nutrients from plants by direct leaching, (iii) the release of nutrients from organic matter by biological decomposition, and (iv) equilibrium reactions that convert insoluble chemical forms in the soil-rock compartment to soluble forms in the available-nutrient compartment, and vice versa.

Available nutrients not only enter the ecosystem from outside but are added by the action of physical, chemical, and biological weathering of rock and soil minerals already within the system. Although some ions are continually withdrawn from the available-nutrient compartment, forming secondary minerals in the soil and rocks, for most nutrient elements there is a net movement out of the soil-rock compartment. As the ecosystem is gradually lowered in place by erosion or by the downward growth of roots, new supplies of residual rock or other parent material are included; in some systems these materials may also be added as geologic input.

HYDROLOGIC-CYCLE, NUTRIENT-CYCLE INTERACTION

At many points, nutrient cycles may be strongly geared to the hydrologic cycle. Nutrient input and output are directly related to the amounts of water that move into and out of an ecosystem, as emphasized by the "leaching" and "flushing" concepts of Pearsall (10), Dahl (11), and Ratcliffe (12), while temporal and absolute limits of biogeochemical activities within the system are markedly influenced by the hydrologic regime. Biologic uptake of nutrients by plants and release of nutrients by biological decomposition are closely related to the pattern of water availability. Potential levels of biomass within the system are determined in large measure by precipitation characteristics. Similarly, the nature and rate of weathering and soil formation are influenced by the hydrologic regime, since water is essential to the major chemical weathering processes [ion exchange, hydrolysis, solution, diffusion, oxidation-reduction, and adsorption and swelling (13)].

BIOGEOCHEMICAL STUDIES OF ECOSYSTEMS

Although study of nutrient input, nutrient output, and weathering is necessary for an understanding of field and forest ecosystems (6), ecologists, foresters, and pedologists have generally focused attention on the internal characteristics alone. Thus, considerable information has been accumulated on uptake, retention, and release of nutrients by the biota of ecosystems, and on soil-nutrient relationships (see, for example, 6, 14). Rarely are these internal characteristics of the ecosystem correlated with input and output data, yet all these parameters are necessary for the construction of nutrient budgets of particular ecosystems, and for establishing the relationship of the smaller system to the biosphere.

Quantitative data on input-output relationships are at best spotty. There are many data on nutrient output due to harvesting of vegetation, but for particular natural ecosystems there are only sporadic data on nutrient input in precipitation, or on nutrient output in drainage waters (6). Recently, small lysimeters have been used successfully in the measurement of nutrient dynamics within the soil profile and in the measurement of nutrient losses in drainage water (15). The lysimeter technique seems to be well suited for studying ecosystems characterized by coarse-textured soils with a relatively low field capacity (16), high porosity, and no surface runoff. For most ecosystems, however, lysimeters are probably of limited value for measuring *total* nutrient output because (i) they are of questionable accuracy when used in rocky or markedly uneven ground, (ii) they cannot evaluate nutrient losses in surface waters, and (iii) their installation requires considerable disturbances of the soil profile.

The lack of information on the nutrient-input, nutrient-output relationships of ecosystems is apparently related to two considerations: (i) integrated studies of ecosystems tend to fall into an intellectual "no man's land" between traditional concepts of ecology, geology, and pedology; (ii) more important, the measurement of nutrient input and output requires measurement of hydrologic input and output. Unquestionably this lack of quantitative information is related to the difficulties encountered in measuring nutrients entering or leaving an ecosystem in seepage water or in sheet or rill flow, and to the high cost, in time and money, of obtaining continuous measurements of the more conventional hydrologic parameters of precipitation and stream flow. In many systems the problem is further complicated by the fact that much water may leave by way of deep seepage, eventually appearing in another drainage system; direct measurement of loss of water and nutrients by this route is virtually impossible.

SMALL-WATERSHED APPROACH TO BIOGEOCHEMICAL CYCLING

In some ecosystems the nutrient-cycle, hydrologic-cycle interaction can be turned to good advantage in the study of nutrient budgets, erosion, and weathering. This is particularly so if an ecosystem meets two specifications: (i) if the ecosystem is a watershed, and (ii) if the watershed is underlain by a tight bedrock or other impermeable base, such as permafrost. Given these conditions, for chemical elements without a gaseous form at biological temperatures, it is possible to construct nutrient budgets showing input, output, and net loss or gain from the system. These data provide estimates of weathering and erosion.

If the ecosystem were a small watershed, input would be limited to meteorologic and biologic origins. Geologic input, as defined above, need not be considered because there would be no transfer of alluvial or colluvial material between adjacent watersheds. Although materials might be moved within the ecosystem by alluvial or colluvial forces, these materials would originate within the ecosystem.

When the input and output of dust or windblown materials is negligible (this is certainly not the case in some systems), meteorologic input can be measured from a combination of hydrologic and precipitation-chemistry parameters. From periodic measurements of the elements contained in precipitation and from continuous measurements of precipitation entering a watershed of known area, one may calculate the temporal input of an element in terms of grams per hectare. Noncoincidence of the topographic divide of the watershed and the phreatic divide may introduce a small error (17).

Losses from this watershed ecosystem would be limited to geologic and biologic output. Given an impermeable base, geologic output (losses due to erosion) would consist of dissolved and particulate matter in either stream water or seepage water moving downhill above the impermeable base. Although downhill mass movement may occur within the system, the products of this movement are delivered to the stream bed, whence they are removed by erosion and stream transportation.

Geologic output can be estimated from hydrologic and chemical measurements. A weir, anchored to the bedrock (Fig. 36–2), will force all drainage water from the watershed to flow over the notch, where the volume and rate of flow can be measured. These data, in combination with periodic measures of dissolved and particulate matter in the outflowing water, provide an estimate of geologic output which may be expressed as grams of an element lost per hectare of watershed.

The nutrient budget for a single element in the watershed ecosystem may be expressed as follows: (meteorologic input + biologic input) − (geologic output + biologic output) = net loss or gain. This equation may be further simplified if the ecosystem meets a third specification—if it is part of a much larger, more or less homogeneous, vegetation unit. Biological output would tend to balance biological input if the ecosystem contained no special attraction or deterrent for animal populations moving at random through the larger vegetation system, randomly acquiring

Fig. 36–2. A weir showing the v-notch, recording house, and ponding basin. [Courtesy of the Northeastern Forest Experiment Station]

or discharging nutrients. On this assumption, the nutrient budget for a single system would become: (meteorologic input per hectare) − (geologic output per hectare) = net gain or loss per hectare. This fundamental relationship provides basic data for an integrated study of ecosystem dynamics.

SMALL WATERSHEDS FOR ECOSYSTEM RESEARCH

The relationship of the individual terrestrial ecosystem to biogeochemical cycles of the biosphere can be established by the small-watershed approach. Data on input and output of nutrients provide direct measurements of this relationship, while data on net loss provide, as explained below, an indirect measure of weathering rates for soil and rock minerals in relatively undisturbed ecosystems.

The small watershed may be used for experiments at the ecosystem level. This has been shown by numerous experiments concerned with hydrologic relationships (see, for example, 18). Thus, it is possible to test the effects of various experimental treatments on the relationship of the individual ecosystem to the biospheric nutrient cycles. Experiments can be designed to determine whether logging, burning, or use of pesticides or herbicides have an appreciable effect on net nutrient losses from the system. This information is not generally available at the ecosystem level.

The small watershed, with its measured parameters of hydrologic and chemical input, output, and net change, is an excellent vehicle for the study of interrelationships within a single ecosystem. Nutrient output may be related to hydrologic parameters such as seasonal and diurnal variations in stream flow, seasonal patterns of precipitation, individual rainstorms, and variations in evapotranspiration. Characteristics of the nutrient cycle may also be related to phenological events occurring within the ecosystem, such as leaf development, initiation of root growth, leaf fall, and litter turnover. In combination with current methods of biomass and nutrient analysis (see, for example, 6), the small-watershed approach provides a comprehensive view of the status and behavior of individual elements within an individual ecosystem.

Weathering, or the rate at which an element bound in soil and rock minerals is made available, can be estimated from net losses of that element as calculated by the nutrient-budget method. Within the ecosystem (watershed), atoms of an element (one that lacks a gaseous form at ecosystem temperatures) may be located in (i) soil and rock minerals, (ii) the biota and organic debris, and (iii) the pool of available nutrients (Fig. 36–1). There is an intense intrasystem cycling between categories (ii) and (iii) as large quantities of ions are taken up by the vegetation each year and released by direct leaching or stepwise decomposition in the food chain. Ions are continually released to the intrasystem cycle by weathering of soil and rock material. Some of these ions, however, are reconstituted as secondary minerals. If an ecosystem is in a state of dynamic equilibrium, as the presence of climax forest would suggest (19), ionic levels in the intrasystem cycle must remain about the same for many years. Thus, in the climax ecosystem, net ion losses (output minus input) must be balanced by equivalent additions derived from weathering of soil and rock minerals. Thus, net ionic losses from an undisturbed, relatively stable terrestrial ecosystem are a measure of weathering within the system. In a successional ecosystem (in which nutrients are accumulating in biomass and organic debris over the course of years), the rate at which an ion is released by weathering must equal its rate of net loss from the ecosystem plus its rate of net accumulation in the biota and organic debris (Fig. 36–1).

The watershed model allows comparison in relative importance of solution and suspended bed load in removing nutrients from an ecosystem. Nutrient matter can be removed from an ecosystem by three forms of transportation in streams: in solution in the stream water, as inorganic and organic suspended load kept in motion by turbulent flow, and as inorganic and organic bed load slid or rolled along the stream bottom (20). Solution losses may be measured, as described above, from stream-flow data and periodic measurements of dissolved substances in the stream. Part of the losses of suspended matter may be estimated from stream-flow data and periodic measurements of particulate matter obtained by straining or filtering stream water as it comes over the weir. The remaining suspended matter and all of the bed load may be measured above the weir, where these materials collect in the ponding basin (Fig. 36–2). These comparative measurements should be of interest not only to the ecologist concerned with ecosystem dynamics but also, since stream transportation is one of the important aspects of fluvial denudation, to geologists.

The small-watershed approach provides invaluable baseline information for the investigation of stream biology. Life-history studies of stream organisms, population studies, and shifts in community structure and diversity might be correlated with the measured physical and chemical parameters of drainage streams. Analyses of uptake, release, and transport of various nutrients by stream organisms could be made. Moreover, the vegetation of a watershed and the stream draining it are an inseparable unit functionally, and it would be of great interest to obtain information on the biological interaction between them.

SITES FOR WATERSHED STUDIES

Small watersheds meeting the conditions outlined above are probably common. However, even if the desired conditions are met, the investigator studying nutrient cycling is faced with the task of initiating a hydrologic study before he can attack his major problem. This is a time-consuming and expensive procedure, involving construction and maintenance of weirs, establishment of a precipitation network, and continuous collection of records, as well as land rental fees and possible road construction costs. A practical solution to this problem is inauguration of nutrient cycling studies at established hydrologic laboratories, where the required conditions exist and where hydrologic parameters are being measured and data are available.

The feasibility of this approach is demonstrated by our study at the Hubbard Brook Experimental Forest in West Thornton, New Hampshire. There, with the support of the National Science Foundation and the excellent cooperation of the Northeastern Forest Experiment Station, we are studying nutrient cycling and ecosystem dynamics on six small monitored watersheds. We have accumulated data on weathering rates, input,

TABLE 36–1. Budgets for dissolved cations in watershed No. 3 (42.4 hectares) for the period June 1963 to June 1964

Cation	Input (kg/ hectare)	Output (kg/ hectare)	Net change (kg/ hectare)
Calcium	3.0	7.7	−4.7
Sodium	1.0	6.3	−5.3
Magnesium	0.7	2.5	−1.8

output, and the annual budget of several ions in this northern hardwood ecosystem. Also, studies of biomass, phenology, productivity, annual rates of nutrient turnover, and other factors are being made in one undisturbed watershed and in one in which conditions are being experimentally modified.

Preliminary data on input, output, and net change for three cations are presented in Table 36–1. These results allow us to add some numerical values to our ecosystem model (Fig. 36–3). For the calcium cycle, for example, input would be about 3 kilograms per hectare, while output

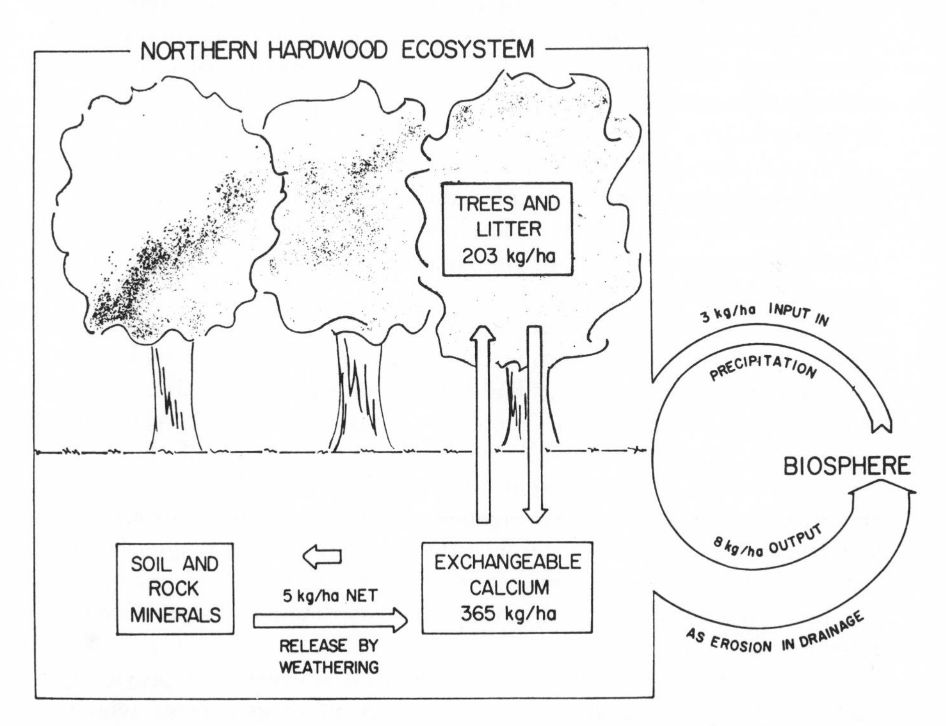

Fig. 36–3. Estimated parameters for the calcium cycle in an undisturbed northern hardwood ecosystem in central New Hampshire. [Data on trees, litter, and exchangeable calcium from Ovington (*6*)]

(erosion) is estimated to be about 7.9 kilograms per hectare. Of this latter amount, 98 percent (7.7 kilograms per hectare) is lost in the form of dissolved substances in the stream water, while first approximations indicate that 2 percent is lost as calcium incorporated in organic matter flushed out of the ecosystem. On the basis of assumptions discussed above, it is estimated that the net amount of calcium lost, approximately 5 kilograms per hectare, is replaced by calcium released from soil and rock minerals by weathering. Hence 5 kilograms of calcium per hectare is added to the system each year by weathering.

As yet we have not measured the calcium content of the soil and vegetation or the annual uptake and release of calcium by the biota. From Ovington's data (6), for a beech forest in West England, which must be of about the same magnitude as values for our forest in New Hampshire, we see that 203 kilograms of calcium per hectare are localized in the trees and litter, while 365 kilograms per hectare represent exchangeable calcium in the soil. This gives a total of 568 kilograms of calcium per hectare in organic matter or as available nutrient. Assuming that our forest (Fig. 36–3) contains a similar amount of calcium, we estimate that a net annual loss of 5 kilograms per hectare would represent only nine-tenths of 1 percent of the total. This suggests a remarkable ability of these undisturbed systems to entrap and hold nutrients. However, if these calculations were based on actual amounts of calcium circulated each year rather than on the total, the percentage losses would be higher.

On its completion, the Hubbard Brook study will have yielded estimates, for individual elements, of many of the parameters and flux rates represented in the nutrient cycle shown in Fig. 36–1. These data will increase our understanding of fundamental nutrient relationships of undisturbed northern hardwood forests, and they will provide baseline information from which we can judge the effects on nutrient cycling of such practices as cutting, burning, and the application of pesticides.

Studies similar to these at Hubbard Brook could be established elsewhere in the United States. There are thousands of gaged watersheds operated by private and public interests (17), and some of these must meet the proposed requirements. On selected watersheds, cooperative studies could be made by the agencies or organizations controlling the watershed and university-based investigators interested in biogeochemical cycling. Just such cooperation, between federal agencies and universities, has been urged by the Task Group on Coordinated Water Resources Research (21).

Cooperative studies of this type have the advantage of providing a useful exchange of ideas between scientists in diverse fields who are working on the same ecosystem. The studies would provide a larger yield of information on a single system, the prospect of new concepts arising from the available information, and a greater scientific yield per dollar invested. Finally, cooperative studies would make available, for interpretation from the standpoint of nutrient cycling, an invaluable record of past hydrologic performance and, in some cases, of the responses of watersheds to experimental manipulation.

CONCLUSION

The small-watershed approach to problems of nutrient cycling has these advantages. (i) The small watershed is a natural unit of suitable size for intensive study of nutrient cycling at the ecosystem level. (ii) It provides a means of reducing to a minimum, or virtually eliminating, the effect of the difficult-to-measure variables of geologic input and nutrient losses in deep seepage. Control of these variables makes possible accurate measurement of nutrient input and output (erosion) and therefore establishes the relationship of the smaller ecosystem to the larger biospheric cycles. (iii) The small-watershed approach provides a method whereby such important parameters as nutrient release from minerals (weathering) and annual nutrient budgets may be calculated. (iv) It provides a means of studying the interrelationships between the biota and the hydrologic cycle, various nutrient cycles, and energy flow in a single system. (v) Finally, with the small-watershed system we can test the effect of various land-management practices or environmental pollutants on nutrient cycling in natural systems.

REFERENCES AND NOTES

1. Cole, L. C. Apr. 1958. *Sci. Amer.* 198:83.
2. Odum, E. P. 1963. *Ecology*. Holt, Rinehart, and Winston, New York.
3. Revelle, R., W. Broecker, H. Craig, C. D. Keeling, and J. Smagorinsky. 1965. *In: Restoring the Quality of Our Environment*. Government Printing Office, Washington, D.C., pp. 111–133.
4. Cole, L. C. May 7, 1966. *Saturday Rev.*
5. Odum, E. P. 1959. *Fundamentals of Ecology*. Saunders, Philadelphia.
6. Ovington J. D. 1962. *In: Advances in Ecological Research*, J. B. Cragg, Ed. Academic Press, New York, Vol. 1.
7. Tansley, A. G. 1935. *Ecology*, 16:284.
8. Slobodkin, L. B. 1961. *Growth and Regulation of Animal Populations*. Holt, Rinehart, and Winston, New York.
9. Buckman, H. O., and N. C. Brady. 1960. *The Nature and Properties of Soils*. Macmillan, New York.
10. Pearsall, W. H. 1950. *Mountains and Moorlands*. Collins, London.
11. Dahl, E. 1956. "Rondane. Mountain vegetation in South Norway and its relation to the environment," *Skrifter Norske Videnskaps-Akad. Oslo 1, Mat. Naturv. Kl. 1956*.

12. Ratcliffe, D. A. 1959. *J. Ecol.*, 47:371.
13. Bould, C. 1963. *In: Plant Physiology, A Treatise*, F. C. Steward, Ed. Academic Press, New York, Vol. 3.
14. Bray, J. R., and E. Gorham. 1965. *In: Advances in Ecological Research*, J. B. Cragg, Ed. Academic Press, New York, 1964, Vol. 2; K. J. Mustanoja and A. L. Leaf, *Botan. Rev.*, 31:151.
15. Cole, D.W., and S. P. Gessel. 1961. *Soil Sci. Soc. Amer. Proc.*, 25:321.
——. 1965. *In: Forest-Soil Relationships in North America,* C. T. Youngsberg, Ed. Oregon State Univ. Press, Corvallis.
16. Cole, D. W. 1958. *Soil Sci.*, 85:293.
17. Wisler, C. O., and E. F. Brater. 1959. *Hydrology.* Wiley, New York.
18. Dils, R. E. 1957. *A Guide to the Coweeta Hydrologic Laboratory.* Southeastern Forest Experimental Station, Asheville, N.C.
19. Whittaker, R. H. 1953. *Ecol. Monographs*, 23:41.
20. Strahler, A. N. 1963. *The Earth Sciences.* Harper and Row, New York.
21. Revelle, R. 1963. *Science*, 142:1027.
22. Financial support was provided by National Science Foundation grants GB 1144 and GB 4169. We thank J. Cantlon, N. M. Johnson, R. C. Reynolds, R. H. Whittaker, and G. W. Woodwell for critical comments and suggestions during preparation of the manuscript.

37

The ecological basis for judging condition and trend on mountain range land

Lincoln Ellison

Since late pioneer days mountain lands of the West have been important in providing summer range for livestock. The history of most of these lands includes a period of severe overgrazing in the latter part of the past century and the early years of this one, resulting in depletion and change in the plant cover and in more or less erosion of the soil. In the past forty years deterioration has continued on many ranges. In part at least, it has continued because range managers were unable to tell how serious was the condition of these ranges, and because they believed, on insufficient evidence, that the trend was toward improvement when in fact it was not.

The range manager's problem is not an easy one. He is assigned the job of managing a piece of land which has been used and abused for from fifty to one hundred years. Ordinarily, he has little idea of what the original vegetation looked like, and hence, in many cases, little idea of the range's true potentialities. The original forage plants—at least the perennial herbs and grasses—died out years ago, and other plants have taken their place. Are they the same species? He has no way of knowing. The soil-protecting qualities of the present vegetation, its crownspread density and litter cover, may or may not be as good as that of the original vegetation; he does not know. When he sees signs of soil erosion, it may be only normal geologic erosion; it may be accelerated above normal. He

Reprinted with permission from the *Journal of Forestry*, 47:786–795 (1949).

does not know for sure. He sees that some plants are reproducing. Is their increase sufficiently aggressive to stop the erosion? And so on, the questions come thick and fast, compounded by variations in elevation, slope, exposure, and soil; but answers founded on fact are few.

The man assigned the management of mountain range land faces problems whose final solutions require years of scientific study, but he is expected to deliver immediate answers that are both correct and practical. With the help of a few ecological principles he must be his own scientist, and by observation ascertain what standards he can use for range in ideal condition, on a variety of sites. He must appraise the condition of each site—the character of soil erosion, vegetal cover, and plant composition in relation to the site's potentialities. Finally, he must weigh the evidences of change, to ascertain whether range trend is toward or away from the kind of plant cover and soil stability that is desired.

The purpose of this article is to help the manager of mountain range land formulate standards of satisfactory condition and develop a firmer conception of what the terms *range condition* and *range trend* mean. To accomplish this purpose certain basic ecological principles are outlined, and their application to judging condition and trend is given in the light of opportunities and practical limitations with which the land manager has to deal in assessing the facts.

THE RANGE COMPLEX

Any given area of range is a *complex*, or, to use the term proposed by Tansley (7), an *ecosystem*. It is made up of many different but closely interrelated parts—biotic, climatic, edaphic, and topographic (4). So closely are these components knit together in their influence upon one another that under natural, undisturbed conditions they can be said to form an integrated whole. Evidence of integration resides in the fact that the complex normally undergoes orderly successional change in which all its parts in some degree modify the others, and are in some degree modified, without any one changing disproportionately to the rest.

Primary succession

A succession beginning on bare talus is probably the primary succession most generally applicable to sloping mountain lands. The one diagrammed in the first four panels of Figure 37–1 occurs on warm, dry, subalpine slopes of the Wasatch Plateau, and will be described in detail elsewhere (3).

In panel 1 the rock has reached an angle of repose. Small fragments broken from larger rocks by weathering and perhaps a little soil from

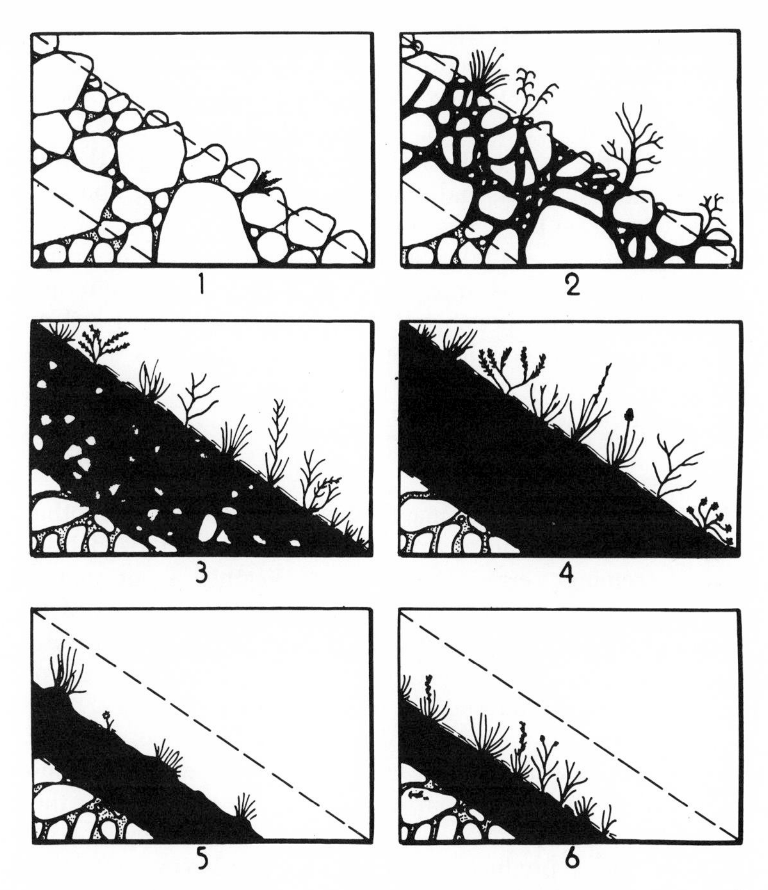

Fig. 37–1. Panels 1 to 4, progressive development of soil mantle from talus rock in the course of primary succession; panel 5, the soil mantle partly stripped away by accelerated erosion; panel 6, the eroded slope subsequently stabilized by vegetation. All highly diagrammatic.

lichen growth and wind-borne dust have accumulated in the interspaces. Where the accumulation is sufficiently near the surface, a fern or flowering plant has been able to establish itself. Often these earliest flowering plants are shrubs or trees rooted deep in the crevices. As disintegration of the rock continues and more and more soil is formed (panel 2), more plants become established and presumably hasten chemical weathering of the rock. In panel 3 the soil is no longer confined to crevices in a matrix of rock; rather, the rock is so far decomposed that its remnants lie in a matrix of soil. Soil maturity, shown in panel 4, is arbitrarily assumed to have been reached when the remnants of rock are essentially all decomposed.

The attempt is made in these diagrams to emphasize two related and important facts. First, soil and vegetation develop concurrently. Second, during the entire successional process soil is practically always under some kind of protection against surficial erosion. When the soil is scanty, occurring only in crevices, it is protected by rocks. When it appears at or near the surface it is also protected by plants, their living crowns, their litter, or both; and when rocks are no longer abundant on the surface, vegetation has the entire protective function.

How long this process takes is not known. Hundreds of years, certainly (old logs, decaying stumps, and mature trees in the intermediate stages suggest as much); possibly thousands. A well-defined soil mantle in the mountains, even though not mature as in panel 3, is something of great antiquity. From the standpoint of practical range management, this is a most important fact.

Integration and change A soil mantle develops, together with other elements of the complex, in an environment that is characterized by cyclic changes and seemingly erratic fluctuations. Weather is not constant; it undergoes a cyclic change within every year. Periods of years of low or high precipitation are the rule, and sudden torrential storms may interrupt periods of calm. The biota is not a constant; cyclic increases in rodents and predators, periodic infestations of insects or fungi, occur. Periodic fires are part of the natural course of events. We have no reason to believe that these fluctuating natural factors have been essentially different during the past few decades from what they have been during the past few centuries, or even, perhaps, the past few millenia. They make differences today in volume production of vegetation and in cover-effectiveness; they have always made differences. Yet soil and vegetation have grown up together on slopes so steep that the soil would fall downhill if it were not held in place by vegetation. This is what is meant when it is said that the components of the complex—plants, animals, and factors of soil, topography, and climate—are knit together in an *integrated whole.*

In normal succession a change in character of one component of the range complex is accompanied by orderly, far-reaching changes in other components. The process resembles orderly growth. Like growth, succession is change—change that is simultaneous in all parts of the organism. As an increment of change in vegetation occurs, for example, it is accompanied by increments of change in the character of soil, of animal life, of microclimate, and of microtopography. This does not mean that a change in vegetation is the cause of these other changes; it means simply that an essential balance exists, an integration of forces, so that, whatever the cause, all the changes occur together and in harmony.

Some ecologists may wish to restrict the idea of *balance* to an ideal climax state where a dynamic equilibrium exists, and it is for that reason that the term *essential balance* is used here. It is needed to direct attention to the interplay and equipoise of forces at any moment in normal succession—a fact of possibly greater significance than the fact of an infinitesimal increment of change. This point of view may be illustrated by considering a healthy child. His health bespeaks a proper balance among the organs and tissues of his body. In a very real and important sense his health reflects an *essential* equilibrium, even though it is true that he is changing, by infinitesimal degrees, into an adult.

Soil an index of integration The presence of a well-developed soil mantle is evidence of successful integration between climatic and topographic factors, vegetation and animal life, over a great period of time. Soil stability in relation to vegetal cover is one of the most practical indices of this integration.[1]

On the other hand, an erosion rate incompatible with the formation of such a soil mantle is evidence of disintegration, of relatively recent change in relationship between components of the range complex that were formerly well integrated. It is an indication that a change of drastic proportions, over and above the normal amplitude of environmental stress, has taken place. Usually this disintegration begins with such reduction of the vegetal cover as to expose the soil to destructive forces.

The concept of range as a complex of integrated parts goes beyond mutual development of vegetation and soil. It includes corresponding development of microclimate, animal life, and topography. However, in judging range condition in the light of present knowledge of indicators, we may restrict consideration to the balance between vegetation and soil.

Secondary succession

Secondary succession is an interruption or modification that may occur at any time during primary succession when the vegetation is destroyed, in whole or in part, so that a different group of species is able to become established. The character of secondary succession is determined largely by the disturbing agent (fire, logging, or grazing), the character of seed source for reinvasion, and the extent of denudation. Compared with primary succession, secondary succession is a short-time affair, and normally becomes submerged in the primary successional process.

As defined here, secondary succession includes those changes resulting from *class overgrazing* where stability of the soil is maintained—e.g., a trend in which certain palatable herbs diminish in a stand grazed by sheep while relatively unpalatable grasses increase; or, with cattle, a

[1] Other practicable indices of integration in the range complex are geomorphic in character. They utilize accordance or discordance as the case may be, between observed and normal stream cutting and deposition, and between the character of observed and normal flood deposits (1).

trend in which grasses are largely eliminated and certain rather unpalatable herbs supplant them. It includes the invasion of thorn shrub and xeric conifers into heavily grazed grassland, and, on the other hand, the elimination of reproduction of hardwoods (e.g., aspen) by heavy grazing of livestock or big game, with the ultimate result that the overstory is eliminated. But secondary succession does not include these changes when they occur in company with accelerated erosion. This is something quite different, beyond the limits of normal integrative adjustment. It will be treated under *destructive change.*

In short, secondary succession, like primary succession is predicated upon internal adjustment and orderly change. In neither are the balance and integration of the range-watershed complex disrupted. The character of the complex may undergo change, in some instances rather marked change, but the essential wholeness of the complex remains.

Destructive change

While secondary succession often presents management problems, a much more difficult and pressing problem is brought about as a result of accelerated erosion. If the overgrazing responsible for it were to cease after the first year—as a fire ordinarily ceases after it has consumed all the fuel available—the problem would hardly exist. What makes it a problem is that overgrazing commonly continues year after year, tending to insure denudation and a material loss of topsoil. Moreover, after much loss has taken place accelerated erosion may continue for many years, even though grazing is discontinued. Invasion of vegetation is not sufficiently aggressive on so greatly deteriorated a soil surface to cope readily with the combined forces of gravity, wind, and rain.

This condition is shown diagrammatically in the fifth panel of Figure 37–1. Contrasting this with the four panels illustrating soil development during the process of normal succession, we see that this state of affairs is abnormal. It is something that has never happened before, at least not for hundreds or thousands of years. Such complete denudation and such severe erosion must be extremely rare, if not nonexistent under natural conditions. If they did occur commonly, soil development certainly could not have proceeded so far and over such extensive areas as it has.

The simple fact is that, when accelerated erosion is taking place, soil is being destroyed. This destruction is not the reverse of normal succession, any more than wasting away with disease is the reverse of growth. Accelerated erosion is not *retrogression* in any practical sense. Retrogression means progress backward along a path that has already been trodden, but the trend followed in accelerated erosion represents a new route not comparable to the successional process of soil development.

This point is stressed to offset a fairly prevalent notion that soil development and soil erosion are counterparts, and that somehow soil can be built up by judicious management about as easily as it can be lost by mismanagement. No fallacy could be more disastrous. Accelerated erosion is, practically, an irreversible process. We can stop it, if it has not gone too far, but we cannot add an appreciable amount in our single lifetime by way of replacement for the soil that has been lost. A soil mantle can be lost in a matter of decades; to rebuild it may take thousands of years.

RANGE CONDITION

Range condition may be defined, adhering in part to a definition agreed upon at the 1944 Methods and Techniques Conference of the Forest Service (8), as follows:

> Range condition is range health; it is the relative position of a range with regard to a standard set up by management objectives within the practicable potentialities of the site. Satisfactory condition requires vegetation both sufficiently dense to maintain soil stability and having a considerable proportion of choice forage species. Unsatisfactory condition involves either vegetation so sparse as to permit erosion rates greater than normal, or, if sufficiently dense to prevent erosion, a preponderance of undesirable species.

Note the stipulation that range condition is to be judged in relation to potentialities of the site. One may not say that a meadow is in better condition than a hillside because it produces more forage or more effective cover. Rather, the meadow must be compared with other meadows, the hillside with other hillsides, of similar steepness, exposure, soil, etc. A corollary following from the phrase "practicable potentialities of the site" is that each site must be judged in relation to its present status in primary succession. Thus, while a talus slope may have the potentiality of becoming a soil-covered hillside some thousands of years hence, its *practicable* potentialities from the land manager's viewpoint are only those of a talus slope, and it must be judged accordingly.

This definition of range condition raises this question: For any given site, how can a person tell what the normal potential is? The most direct way is by comparison with vegetation and soil of similar sites that have never been grazed. Usable areas of this sort are no longer easy to find, even in the mountainous West, and because of this fact their value is so great that those which still exist should be preserved at all reasonable costs. Their preservation is really a great economy, in view of the information they provide which can be obtained otherwise, if at all, only at great expense and years of research effort. Lacking natural areas, the range manager can sometimes synthesize a partial conception of the normal potential by putting together scraps of evidence—remnants of soil or relic species preserved in one way or another—from parts of the grazed range in different degrees of depletion, or from invasion in artificially protected areas and on permanent plots. So detailed an approach, however, is really a research job. Even so, some acquaintance with natural areas is almost essential in reconstructing a true picture of natural conditions.

Returning to the definition, we know that range condition ceases to be satisfactory when accelerated soil erosion sets in, when destructive processes clearly exceed constructive processes. Hence a basic criterion of range condition is degree of soil erosion, and a minimal requirement for satisfactory condition is normal soil stability.

Although to determine precisely what normal erosion *is* may be difficult, even by study of natural areas, to determine what normal erosion *is not* is often easy. Inasmuch as formulation of a soil mantle is an imperceptibly slow process, the process of normal erosion on any widespread scale must also be imperceptibly slow. This inference gives rise to a rough rule: "If you can see it, it's accelerated." Like all rules of thumb this one has its limitations: (1) it is applicable only under local climatic conditions that permit formation of a soil mantle—something generally true in mountain range land; (2) the evidences of erosion must be widespread, not confined to occasional spots a few square feet, or less, in area; and (3) there must be evidence of continuance over several years. These restrictions are not so burdensome as they might seem; and the rule of the thumb is pretty generally useful on mountain range.

Finally, in addition to the prime requisite of a vegetal cover adequate to prevent accelerated erosion, the value of the vegetation as forage must be taken into account. Judgment of condition must weigh the value of present vegetation against that which might grow in its place.

There are thus two bases for judging range condition—soil stability and forage composition. The two cannot well be combined into a single term or figure; they are different kinds of things and are not additive. This difficulty is lessened somewhat by the fact that they are of very unequal weight, the latter being far less important than the former, so that the question of forage composition can be deferred until that of soil stability is evaluated. This is to say, judgment should be based primarily on soil stability, and on forage only after it is certain that vegetal cover is sufficient to preserve the soil.

Weather and condition

Range condition is often attributed to weather or "climate." The belief finds extreme expression in the assertion that deteriorated range will be restored "if and when it rains" (5).

In order to evaluate this element of range condition it is necessary to make a distinction between weather and climate. Climate is an integration or summation of weather over periods to be reckoned in centuries, and perhaps millenia. Compared with weather, which may fluctuate rapidly between one extreme and another, climate is a constant. It changes so slowly—if it is changing at all—that instrumental records for the past two centuries fail to demonstrate a permanent shift (6). It is climate rather than weather that determines the essential nature of the range complex. The succession of plants and animals, the succession of microclimates at the soil surface, the development of soil, the evolution of land forms, is each a characteristic expression of a prevailing overall climate. The period in which these successional processes have continued is sufficiently long to include the entire amplitude of weather fluctuations characteristic of the climate. Hence what a person may consider an abnormal drought or wet spell or freeze may well be infrequent and exceptional in his personal experience, but still only part of the normal climatic pattern.

In formulating a concept of range condition we must know, at least broadly, what effects these variations in weather cause; and what, in and of themselves, they cannot cause. Certainly they cause great variations in volume of forage production from year to year. They cause great differences in species composition whenever there is a great difference between low phase and high phase in the precipitation cycle. These facts are amply documented by ecological studies during and since the latest drought period on the Great Plains.

On the other hand, variations in weather alone cannot account for the incision of a soil mantle by an active gully system; they cannot account for the cutting of torrents through an alluvial fan, carrying coarse materials beyond it. Soil mantle and alluvial fan are themselves evidence of thousands of years of slow, constructive processes under the influence of a particular climate. To attribute recent, severe erosion to prolonged drought or intense rainfall, therefore, one must postulate a recent shift in climate, sufficiently radical to undo the edaphic and geomorphic work of ages. There is no evidence of such a shift. On the contrary, there is ample reason to believe that the present climate, which has existed during the few generations since settlement, with all its extremes in weather, is essentially the same as that climate which has extended over the past few thousand years.

Thus, although variations in weather are certainly accompanied by variations in vegetation (although to a less extent in mountain climates than in the more arid lowlands), and doubtless by variations in rate of erosion, they have only a superficial, not a fundamental, bearing on range condition. Fundamentally, range condition refers to those persisting qualities of the range complex that endure throughout vicissitudes of weather, that—following Webster in definition of condition—characterize "a mode or state of being."

If condition of the range may properly be likened to health of an individual, variations due to weather are analogous to an individual's moods. An invalid may be cheerful, and at times a healthy person is depressed. A mood is hardly a valid indication of the true "mode or state of being," and so it is with weather and range condition.

Condition on eroded soils

The fact has been stated that accelerated erosion has been widespread on mountain range lands of the West, particularly on herbaceous slopes. Panels 5 and 6 of Figures 37–1 illustrate the point diagrammatically. Although the diagrams show that about half the original soil mantle has been lost, and although some sites have lost more than half their mantle and some less, this proportion is not to be interpreted as the average loss. Nobody knows, as a matter of fact, what the average loss has been.

When soil is being lost, as in panel 5, condition is certainly unsatisfactory. But when the slope is stabilized by vegetation, as in panel 6, what then? Should condition be regarded as unsatisfactory because the soil has been deteriorated, as compared with panels 3 and 4? If this convention were to be adopted, the slope would be unsatisfactory for generations, even though it supported the maximum volume and quality of vegetation of which the site is now capable. It would appear wiser to accept the fact of deterioration, and the fact that the lost soil is water under the bridge—or, rather, mud under the bridge—and base management judgments on the "practicable potentialities of the site" as it now exists.

But for sites so deteriorated, what standards of vegetal volume and composition should be adopted? Can sites so deteriorated support a vegetal cover that will stabilize the surface as successfully as the original cover? Nobody knows. As has been pointed out, these eroded surfaces are something new under the sun, and time and study will be required to learn what their true potentialities are.

For the time being there are two estimates of what these eroded soils are likely to support when they are stabilized. One is derived from comparison with natural areas, and is probably an overestimate. The second is derived from comparison with the rest of the more or less deteriorated range, and is very likely to be an underestimate. The true value may be expected to lie somewhere between. Artificial reseeding has possibilities for revealing the true potentiality of eroded soils. These possibilities have hardly been explored as yet. For the time being, if one must err, it is certainly better to err on the side of an overestimate than an underestimate until an accurate appraisal of the true potentiality of the range can be made.

Thus as a practical expedient the natural area remains the most serviceable standard of range potential; and the rule remains valid, for present practical purposes, that visible, widespread erosion on a soil mantle is accelerated erosion. With these criteria many of the errors that have been made in the past, through underestimating what the range should look like and what it should produce, can be avoided.

RANGE TREND

Elaborating on the definition arrived at by the Forest Service Conference in Denver in 1944 (8), we may define range trend as change in the range complex from one condition to another. When sufficient vegetation grows on the range so that the soil is stable, trend may be either upward or downward depending on whether the amount and quality of vegetation is improving or declining. When the soil is unstable, trend must of practical necessity be considered downward.

Succession is involved in range trend, but in practical range management trend has little direct reference to normal, primary succession. The trends with which the range manager is mostly concerned are those involved in loss or stabilization of soil, and decline or improvement in forage value. They are mostly the brief changes that may be expected within a single lifetime. The range manager's most urgent concern is with trends toward or away from soil stability. If he can stabilize the soil, constructive trends can follow. Normal succession, including normal soil development, can proceed, although for reasons already given the progress made during the range manager's lifetime will probably be imperceptible.

Trend is measured between two points in time, usually between the present and a time in the past, either by means of some kind of permanent plot or by means of indicators such as pedestaled plants or an age-class distribution. In any case, the trend in which the range manager is most interested—What will happen in the future?—cannot be examined and measured; by projecting past trends forward it can only be inferred. The more accurately past trends can be measured, therefore, and the more nearly up-to-date they can be brought, the more accurate will predictions of future trend be.

Range trend is often difficult to evaluate, especially by inspection unaided by permanent-plot records, because quantitative measurements can seldom be used. It is easy enough, when shrubs or trees make up part of the cover, to tell whether the woody plants have been increasing or decreasing by counting growth rings and numbers of individuals of different sizes. But with herbaceous species, classification into age classes is very rough, and the age classes are bounded by loose qualitative limits —very old, mature, young, very young, for example—rather than by precise quantitative limits. Moreover, there is at present no means, capable of application on a wide scale at least, for evaluating rate of soil loss. Except when extreme changes are under observation it is practically impossible to tell whether the rate of loss has recently increased or lessened. Finally, in appraising trend, one must make allowance for merely apparent trends due to current fluctuations in weather which may obscure the true, long-time trend. This means that often a true but recent change in trend cannot be certainly known while it is still recent; time must pass before it can be verified.

These difficulties explain why disagreement as to trend is so common, and why expected trends very often do not materialize. There are too few facts to go on. Clearly, there is need for more and better permanent plots and photographs, and for increased skill in interpreting the changes they reveal. Lacking these records and this skill, we obviously need to use caution and conservatism in basing predictions upon indicators observed on the range. Unfortunately present judgment of range trend must be based almost entirely upon this imperfect record—or, at least, the record we imperfectly understand—written on the range by past events.

Trend when soil is stable

So long as the soil is stable the range manager dependent upon indicators can evaluate trend by the evidence he finds indicating increase or decrease of the various species in the stand. His interpretation will vary with specific objectives of management—for cattle forage certain species may be desirable that would be relatively undesirable for sheep, deer, or elk. And, aside from their relative palatability or their nutritive quality, some species may be highly regarded for their watershed protective value, or for their low consumptive loss of moisture. Again, best use of the land might call for the reproduction, or suppression of reproduction, of certain trees or shrubs. In any case, the range manager attempts to manipulate secondary succession; and indicators that tell him whether or not the stand is changing as he wishes are the age-class distributions of desirable or undesirable plants.

Trend on eroding soil

The question is more complicated when the soil is no longer stable. Trends in vegetation are taking place, either toward more effective cover or less, and at the same time soil is eroding. What is the net trend? This is a problem of widespread and practical urgency on many range-watershed lands (2).

Figure 37–2 is constructed to represent trend in range condition over a past time period *OB* and a future time period *BD*. At time *O*, when the

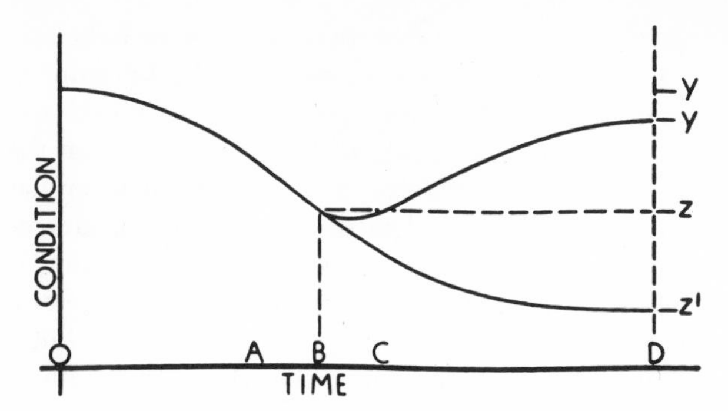

Fig. 37–2. Diagram to represent various elements in range trend. *Oy* is virgin condition, *Bz* is condition today after a certain amount of deterioration, and *Dy'*, *Dz*, and *Dz'* are possible condition levels of vegetation or soil at some future time, *D*.

white man's livestock first began to graze, the range was in virgin condition, *y*. Between time *O* and the present, the range has deteriorated through accelerated soil erosion, and the cover has been depleted by loss of desirable forage species. At present *B*, after the range has deteriorated to a lower condition-level *z*, corrective measures are being taken.

If accelerated soil loss is stopped immediately, amount of soil will remain essentially constant as represented by the horizontal line at level *z*. Soil formation is so exceedingly slow that no measureable increase in depth of soil can be expected during the few decades *BD*. Hence the very most that can be done, so far as amount of soil is concerned, is to hold fast that which remains. Improvement, meaning to make better, is hardly a suitable term in this sense.

To stop accelerated soil loss so quickly is not generally possible without artificial measures. Even if all livestock were to be taken off deteriorated range, accelerated soil loss would not stop immediately everywhere, and it might long continue on some sites. Time is required to build up an adequate protective cover of vegetation upon which soil stability depends. Hence trend in loss of soil will follow some such course as shown by line *z'*. Before full stability is reached some further loss of soil is inevitable.

Since rate of loss at time *C* (say 6 tons per acre annually) is less than at time *A* (say 10 tons), can it be said that improvement has taken place in the interval? Certainly deceleration of erosion represents improvement *in rate*. Yet if maintenance of full soil stability cannot be called improvement of the resource, since deepening of the soil mantle is negligible within the range manager's lifetime, how can deceleration in loss be called improvement? "To improve" means "to make better," not "to make worse at a less rapid rate."

The conclusion to be reached from this reasoning is that, so long as soil is being lost in appreciable amount, trend of the range is downward.

Doubtless it is philosophically true that, as erosion is being checked, a point will be reached where the increment of vegetal gain must counter the increment of soil lost. But, practically, how is one to recognize this point? Very likely it involves a rate of accelerated erosion that is imperceptible. Doubtless, too, some rates of vegetal increase are so great that an eroding soil will be stabilized in the foreseeable future, but how can one be sure of this if he cannot measure either rate? Taking into account the difficulty of field measurement of trend, both in vegetation and soil, and the value of a soil resource which can be destroyed but not practicably restored, we can reach the conclusion that, practically, soil erosion always means downward trend.

This being so, improvement can only begin where soil is stable. It will be reflected in the quality of the soil, in its structure, organic-matter content, and receptivity to rainfall. It will be reflected in amount and kind of vegetation. In either case, trend in soil quality and in amount and quality of vegetation, after soil stability is achieved, may be represented by some such line as y'. Note that this line tends to level off at a condition somewhere below the pristine level y. It fails to reach this level in time period *BD* because soil depth is less than that of the pristine range, and the deteriorated range now probably has a different potential than the original. This point has already been considered.

It is more difficult to determine trend than condition. This is unfortunate because the range manager, anxious for improvement, is more interested in trend. For one thing, in thinking of trend an element of prediction is involved, of looking into the future. For another, recognizable changes are not always a result of livestock management, and skill and judgment are needed to distinguish changes occurring as a result of weather, rodents, insects, game animals, etc., from those that occur as a result of grazing by livestock. The population of one plant species may be changing in one direction while an apparently equivalent species is changing in another, for no known reason; this, of course, is simply a reflection of our own lack of ecological understanding. Finally, vegetal trends and soil trends do not always parallel one another. They may be widely divergent on eroding range (2), and if the range manager is impressed by evidence of change in vegetation and not by evidence of soil erosion, he may be led astray. In short, trend on mountain range land is not readily describable in a word or a single phrase, and it is fair to suspect, when one encounters such an easy description, that it is an inadequate summary of the facts on the ground.

SUMMARY

1. Because the range is a complex of closely interrelated parts normally in essential balance with one another, a standard of integrity is

provided against which range condition may be judged.

2. As a result of these interrelations and because of the reactions of plants and animals on the environment, orderly change (primary succession) takes place without loss of integrity in the complex; this change is much too slow to be manipulated practically by the land manager.

3. Secondary successions result from disturbance of the complex; but by definition changes accompanying long-continued soil erosion are not considered successional. Secondary successions are much more rapid than primary successions, and are capable of being manipulated by the land manager so as to produce desired results within a reasonable time.

4. When disturbance is so extreme as to destroy integrity in the complex, as evidenced by accelerated erosion, the term *succession*, which should imply an integrated complex, is no longer appropriate, and *destructive change* is applied in its place.

5. Range condition must be judged on the basis of the normal potential of the site. What this normal potential is can be ascertained most practicably by comparison with natural areas that have never been grazed.

6. While condition on eroded, but now stabilized, soils properly requires new standards, the best approximations now available are natural areas.

7. In judgment of range condition, soil stability is paramount in importance; forage values are secondary. Where soil is eroding, condition is unsatisfactory. Where soil is stable, condition may be satisfactory or unsatisfactory, depending on the desirableness of the vegetation in relation to the management objective.

8. Where soil is eroding at an accelerated rate, trend is downward. Improvement while soil is eroding is very nearly a logical impossibility, and, from a practical standpoint in judging range trend, should be regarded as impossible.

9. Trend where soil is stable may be either upward or downward. It is to be judged by evidence of change in vegetal composition, toward dominance either by more desirable or less desirable species, as revealed for the most part by age-class distributions.

LITERATURE CITED

1. Bailey, Reed W., C. L. Forsling, and R. J. Becraft. 1934. Floods and accelerated erosion in northern Utah. U.S. Dept. Agric. Misc. Pub. 196, 21 pp.
2. Ellison, Lincoln. 1943. What is range improvement? The Ames Forester, 31:15–22.
3. ———. 1947. Subalpine vegetation of the Wasatch Plateau, Utah. In manuscript.
4. ———, and A. R. Croft. 1944. Principles and indicators for judging condition and trend of high range-watersheds. Intermountain Forest and Range Exp. Sta. Res. Paper 6 (Processed.)

5. Mollin, F. E. 1938. If and when it rains: the stockman's view of the range question. American National Live Stock Association, 515 Cooper Building, Denver, Colorado.
6. Russell, Richard Joel. 1941. Climatic change through the ages. Pp. 67–97 in Climate and Man, 1941 Yearbook of Agriculture, Washington, D.C.
7. Tansley, A. G. 1935. The use and abuse of vegetational concepts and terms. Ecology, 16:284–307.
8. U.S. Department of Agriculture, Forest Service. Proceedings (of) . . . conference . . . on methods and techniques relating to national forest range and wildlife management, December 4–16, 1944. (Processed.)

38

A dangerous game: taming the weather

Frederick Sargent, II

Environment has become a topic of discussion, a focus of research, and a cause of concern among administrators, politicians, planners, scientists, technologists, and citizens. Man has rather recently come to realize that environment is really a vital resource with finite dimensions. If it is wisely managed, it is renewable and bountiful. If it is unwisely manipulated—consciously or inadvertently—the consequences can be serious. Because man is inexorably bound to his environment, his manipulations have both direct and indirect effects on his well-being. Thus, what he actually upsets is a system that involves both himself and his environment—the ecosystem.

Man is now embarking on a new kind of manipulation—"taming" the weather; increasingly large experiments in weather modification are being contemplated, with little understanding of their effects on the ecosystem. There is a danger that man is already too far down this road to be stopped. This is because the issue is not simply scientific; it is also political. We have in conscious weather modification an example of the social disintegration cited by Lord Russell Brain as a characteristic of our times.[1] Politically expedient decisions have been made by men who have not been sufficiently apprised of the consequences of their actions. It is the responsibility of the scientist to inform the politicians of the alternatives that must be weighed in decision-making. Since the costs must be evaluated in terms of the risks—or, to put it another way, the benefits

"A Dangerous Game: Taming the Weather" by Frederick Sargent, II. Reprinted from *Scientist and Citizen*, 9(5):81–88 (May 1967). Used by permission.

must be weighed against the costs—the politicians equally have the responsibility to make informed decisions. Perhaps man is not so far down the road that his course cannot be diverted. Perhaps he can be shown the wisdom of following a path along which the risks are at least anticipated rather than simply ignored.

Attention must be directed to the ecosystem to gain some appreciation of the complex implications and consequences arising from current and planned environmental manipulations.

THE ECOSYSTEM AS A NATURAL RESOURCE

The idea of the ecosystem is one of the fundamental concepts of biology. All living things are related one to another in societies and communities and each is inexorably bound to its habitat. The complex interdependence of environment and organism is an essential characteristic of the ecosystem. This bond developed in the long course of evolution of organisms and environment. As environment evolved, organisms adapted. The properties and processes of the environment were unique, for they provided the ingredients and conditions that met the specific needs of organisms. A mutual adaptation emerged and it is reasonable to speak of it as fitness of the ecosystem.[2]

The energy that drives the ecosystem is solar energy, captured by the green plants in photosynthesis. Energy and nutrients such as water, carbon, hydrogen, and nitrogen flow through the system by way of such consumers as bacteria, herbivores, carnivores, and man himself. These flows may be viewed as the metabolism of the ecosystem. The numbers and interrelations of the organisms are controlled by regulatory processes. For example, the density of small rodents limits the number of predators on these animals, such as hawks and owls. In turn, the numbers of small rodents are limited by weather conditions that favor or discourage the growth of plants providing nuts and seeds on which these species feed. By means of feedback mechanisms, the whole system, a complex interacting web of living things, is maintained in a balanced or steady state.

Although man is himself part of the ecosystem, he looks on it as a natural resource. He extracts from it energy and nutrients for his subsistence and raw materials for his environmental protection. He has learned to manipulate the ecosystem by introducing his own regulatory processes to maximize the growth of some species of plants and animals, and inhibit others. This manipulation of the ecosystem has been most important for man, for it provided the basis for the rise of agriculture, freed him from having to engage in hunting and gathering to provide subsistence and environmental protection for himself and his family and supported the growth of the urban community and the occupational specialization that makes the community function.

THREATS TO THE ECOSYSTEM

In recent decades human activities have posed increasing threats to the ecosystem. Some of the threats derive from inadvertent influences on weather or climate. The use of the word "inadvertent," of course, is man's neat way of apologizing for the fact that when the changes were first instituted, their remote ecological consequences were not appreciated. Indeed, who would have thought that when Prometheus brought fire to man, he laid the foundations for air pollution? Mythology aside, man has known fire for some 500,000 years. Today air pollution—most of it from incomplete combustion—has become a common characteristic of the urban atmosphere and has been implicated in climatological alterations.

First, it has made the urban area a "heat island" where the temperature of the air over cities ranges several degrees centigrade above that of the surrounding country.[3] This heating originates largely from human activities. For example, Hickman observed that the daily energy released by human activities in North America amounts to 0.2 percent of the energy received each day from the sun. This release is, on the average, sufficient to raise the temperature 0.6° C. Since this energy release is concentrated in such areas as the eastern megalopolis, St. Louis-Chicago, and Los Angeles, the local effect is greater than the average. When combined with carbon dioxide and air pollutants, which act as heat traps, the thermal increment might be as large as 3.6° C over 1,000 square miles. He then suggested, "Civilization creates its own drought in proportion to its concentration of people and thus of water needs . . ."[4]

The increase, during the past fifty years, of fifteen to twenty-five percent in the carbon dioxide content of the air, has led meteorologists to speculate that this increment may lead to a general heating of the atmosphere. Carbon dioxide absorbs and radiates back toward the earth outgoing infrared radiation, raising the temperature of the lower air as the glass in a greenhouse does. In an editorial in *Science*, P. H. Abelson points out that if present trends of carbon dioxide release continue until the year 2000, global temperatures could be increased, through a greenhouse effect, by as much as 4° C.[5]

It is true that since about 1890 there has been a gradual and significant increment of about 1.6° C in the mean global temperature.[6,7] Because this increase coincided with a substantial rise of the atmospheric carbon dioxide, it tended to support the theory of the greenhouse effect. However, since the 1940's, the trend of global temperatures has reversed. By the late 1950's, in spite of a continuing increase in atmospheric carbon dioxide, there had accrued a significant drop of about 0.3° C.

The reversal of the trend of the mean global temperature suggests that the atmosphere is now cooling. Cooling might arise from a decrease in solar energy reaching the lower layers of atmosphere. There are processes acting to reduce the heat budget, and these have received little emphasis in recent reports on the deteriorating quality of the environment. The atmosphere has become more hazy due to accumulation of suspended matter (dusts, aerosols, and other particulates) on which moisture can condense to form small droplets of water or ice crystals. These nuclei have increased several-fold.[8] In addition, careful measurements of atmospheric turbidity (haziness) reveal substantial increments in the past several decades.[9] These increments have been detected not only over cities but also in remote places such as the Alps. Furthermore, the jet airplane discharges freezing nuclei and moisture into high layers of the atmosphere and under some conditions the vapor trails of these jets have been observed to evolve into sheets of cirrus clouds (clouds of ice crystals). The net effect of all these processes is to reflect incoming solar radiation.

Now one might be inclined to ascribe the recent lowering of global temperatures to these processes. However, it must be emphasized that long-term swings of global temperatures are not a new phenomena. They have been known for many years. Thus, one must be most cautious about proposing cause and effect relationships and predicting the consequences of the inadvertent changes that man is imposing on the atmosphere. We have neither sufficient temporal perspective nor enough detailed knowledge yet to unravel the complex interrelationships. In the long run we cannot now foretell whether the heating effect of carbon dioxide or the cooling effect of suspended material will dominate. Because both processes are continuing ones, we can but make more intensive measurements and work deliberately and systematically toward a better understanding of the atmosphere.

TAMING THE WEATHER

As far back in the fossil record as the story is clear, there is evidence that weather and climate have exerted significant selective pressures on living things. A major characteristic of most terrestrial animals is their capacity to adapt to the vicissitudes of weather or the changes of climate. Whether the adaptation be physiological or behavioral, the animal manages to achieve some stability in the life processes.

Man has had remarkable technological success in being able to provide for himself, even in the most remote and hostile regions, a comfortable surrounding. He had also developed techniques for protecting other aspects of the ecosystem from the effects of weather. Serious damage to commercially important plants from freezing is controlled by using wind machines, smoke pots, and spraying or flooding. The latter technique, for example, is employed in cranberry bogs to protect the plants from frost.

Second, evaporation from the soil can be reduced through the use of polyethylene mulches. Third, soil erosion from wind can be mitigated by the use of "shelter belts" of trees.

Although these techniques might be viewed as adaptations, definite modification of the microclimate has been achieved. Nevertheless, the burdens imposed by weather continue to be tremendous, although they are no longer so much problems of simple physical protection. Three sets of problems will serve to illustrate the nature of what is now a socio-economic burden: (1) problems associated with storms, (2) problems arising from dependence on water, and (3) problems related to unreliability of seasonal change.

Storms

Storms such as tornados, hurricanes, and blizzards, even when forecast with reasonable precision, cause major damage to crops and property and high casualty rates. The reason that the dimensions of these losses seem to increase is not because the weather has changed but rather because there is more human development in areas where storms prevail. The inconvenience and hazard from fog have intensified because of the accelerating growth of automobile and airplane traffic. The Santa Ana, a hot, dry wind in California, has received recent notoriety not because it blows more often, but because the urban sprawl of Los Angeles pushes homes farther into the dry hills covered with chaparral or brush. When this brush ignites at the time of the Santa Ana, property damage of major proportions accrues. In essence, the dimensions of meteorological disasters expand as the human population and the cities expand.

Water Requirements and Needs

Other problems arise out of the dependence of man's technological and agricultural establishments on water. As industry expands and agricultural productivity rises, the use of water increases. These problems have been intensified with the expansion of agriculture into marginal lands where the reliability of rainfall is less than in arable regions. In many regions, both marginal and arable, the water table has been lowered appreciably. Because the demands for water are now so great, the risk of disaster from prolonged drought has been magnified.

Unreliability of Seasonal Change

Still other problems stem from the unreliability of seasonal changes of weather. The risks involved in planting, growing and harvesting crops arise in large measure from uncertainty that the weather will be propitious during the critical phases of commercial agriculture. Transportation and storage of agricultural produce no longer present crucial problems, at least in the developed countries, for man has learned engineering techniques to maintain proper conditions.

Everywhere, however, unseasonably hot and dry weather, particularly in the early phases of plant growth, can be damaging to the harvest. A reduced harvest threatens human life most severely in countries where subsistence is marginal. Recent experiences in India illustrate this point.

The generalization that emerges from this brief overview of socio-economic problems created by weather is that the human population depends for its existence on the biological productivity of the ecosystem. Although man still behaves as if the resources of the ecosystem were infinite, it has a limited capacity to support the human population. As that limiting carrying capacity is approached, perturbations in the productivity of the system have more and more profound impacts on the dependent populations of living organisms, including man.

Man Plans Deliberate Manipulation

To reduce the hazards and socio-economic problems created by the vagaries of weather, to control the distribution of precipitation over the land and to increase the reliability of seasonal change of weather, man has begun seriously to contemplate conscious weather modification. Conscious weather modification means deliberate human manipulation of atmospheric processes so that they will more nearly fulfill man's needs and requirements. Were such modification successful, storms would be dissipated, mitigated or diverted; clouds would release more water than they would have naturally; and the time and intensity of precipitation would be regulated to fulfill specific agricultural or industrial needs.

Conscious weather modification constitutes one of man's most ambitious attempts deliberately to manipulate the ecosystem to meet his own specific needs and requirements. Already vast sums have been invested in experiments on weather modification conducted primarily in the U.S.S.R. and the U.S. Late last year, both a Panel on Weather Modification of the National Academy of Sciences' Committee on Atmospheric Sciences[3] and a Special Commission on Weather Modification of the National Science Foundation[10] agreed that the success to date warranted greatly increased support for experiments on weather modification. An expenditure of thirty million dollars annually by 1970 was recommended.

In 1961 the UN passed Resolution 1721 (XVI)[11] calling for the development of intensive international research in the atmospheric sciences to advance the knowledge of the physical processes of the atmosphere so that weather modification could be undertaken and weather forecasts improved. The intent of this resolution was reinforced by Resolution 1802

(XVII)[11] passed in 1962 encouraging the World Meteorological Organization to develop in detail its plans for a "World Weather Watch."[12] In 1965, the International Council of Scientific Unions assigned to a Committee on Atmospheric Sciences of the International Union of Geodesy and Geophysics (IUGG) the task of developing the scientific details of a "Global Atmospheric Research Program."[13, 14] This plan will be presented to IUGG in September of this year. With the information accruing, in part, from these studies of atmospheric processes and, in part, from research on cloud physics, man should be in a position in the foreseeable future to determine whether modification of weather can be achieved and on what scale.

What has been accomplished that encourages man to press on in this effort to tame the weather? Three important atmospheric processes seem to be yielding to conscious manipulation.[15] This has excited man's imagination about the possibilities of controlling rather large scale atmospheric processes, processes dominating the weather over several hundred square miles. This is the domain of meteorology which is nowadays identified as mesometeorology.

In the first place, cold fogs over airports can be dispersed by such agents as propane gas, brine, and dry ice. The moisture droplets of cold fogs are super-cooled; when the agents are sprayed into the fog, freezing takes place and snow falls as the air clears. Success here has been sufficiently uniform that the procedure is already operational at several major airports. Warm fogs, fogs which are not supercooled, on the other hand, have so far resisted intervention. In the second place, rainfall from already precipitating clouds can be increased from fifteen to twenty percent by dispersing crystals of dry ice or silver iodide. In the third place, the Russians claim that supersonic sounds will suppress hail. American investigators have not yet confirmed this feat. These achievements support the inference that conscious modification of mesoscale weather systems could be operational on a limited basis within a decade. In other words, man is on the threshold of being able to manipulate the hydrologic cycle of the ecosystem on a scale of the order of hundreds of square miles.

The period since World War II has been one of headlong experimentation by the atmospheric scientists, with encouragement and financial support from government sources. In its budget for fiscal year 1967, the Bureau of Reclamation, for example, has four million dollars to spend on weather modification. Other moneys of comparable dimensions have been allocated to the Environmental Sciences Program of the National Science Foundation (NSF) and to the Environmental Science Services Administration, Department of Commerce.

In spite of the achievements with conscious weather modification, many meteorologists have told the author that they are not as impressed with the evidence as are the learned panels.

ECOLOGICAL RISKS

What do ecologists say? Too few have given this serious matter their attention. But those who have thought deeply about the implications for the biosphere have signaled caution. Because so much is still unknown about how weather influences communities of interacting organisms, the ecological risks cannot yet be fully evaluated.[16] Indeed, those who have considered the human dimensions have likewise recommended that more thought and study be devoted to the remote and indirect consequences of modifying weather. Already claims and counter claims made by those convinced that "cloud seeders" have damaged crops or undermined business have led to legal confusion and premature precedents have been set by courts and legislatures.[17,18]

What are some of the ecological risks in weather modification? One way of thinking about this question is to utilize the concept of the ecosystem. Weather modification may exert an influence by altering the flow of either energy or water through this system. To predict what the consequences of weather modification might be, one needs detailed and quantitative knowledge of the influence of weather and climate not only on individual organisms but also—and more importantly—on societies and communities of organisms. The interrelation and interdependence of groups of organisms in the ecosystem are hierarchically organized and maintained through complex networks of metabolic cycles (e.g., energy and water) and regulatory processes. For this reason, attention must be focused on the susceptibility to environmental change of crucial links—links essential for the sequential flows of energy and water—in the networks. The unfortunate situation, however, is that our present knowledge does not permit detailed prediction of the consequences—immediate and direct or remote and indirect—of weather modification for the ecosystem.

The Weather Working Group of the Ecological Society of America summarizes the state of the art for the NSF Special Commission on Weather Modification as follows:

> Existing theory relating organisms and their climatic environment is far from being adequate to meet the demands these operations are likely to place upon it. Understanding of the ways in which a very small number of species react in isolation to temporary changes in climatic parameters is satisfactory, but when one moves away from concern with a single organism to populations of a single species, communities of many species, or ecosystems, the situation is not nearly so good. It is these more complicated systems that will be affected by intermediate scale operations and we must be prepared to understand the consequences of the operations if climatic manipulation is to be of practical use.[16]

The scale of weather events referred to here includes convective processes that lead to cumulus clouds, cloud systems evolving from air masses being forced upward by mountain barriers, and the familiar mid-latitude low pressure systems depicted on the daily weather map.

Among the consequences of repeated intermediate scale operations foreseen by the Weather Working Group were "shifts of range, local or complete extermination of species, and at least an initial increase of weeds and pests." It was emphasized that not all the changes would be reversible. Short-term alterations would probably have no more influence than natural variations of weather. "The modification of such conditions as tornados, hail, and lightning would probably not be biologically disadvantageous, but the control of larger catastrophes, such as hurricanes, droughts and extremes of temperatures, is likely to produce serious and unpredictable biological consequences." Other problems anticipated were those related to vector-borne diseases and the ecological impact of the "rain-shadow" downwind from the area where cloud seeding operations cause the release of moisture. In the immediate vicinity of these operations rainfall may increase; to the leeward there may be a deficiency of precipitation (the rain-shadow).

Why might modifying a catastrophic storm such as a hurricane have serious biological consequences? Granted such severe storms entail significant hazards and cause appreciable socio-economic losses. These storms are also weather systems that transport moisture from one region to another and provide convective processes whereby that moisture is released as precipitation. Tropical cyclones and hurricanes, for example, account for appreciable portions of the rainfall received by East Coast communities as far north as New England. Mr. G. W. Cry recently collected and carefully analyzed data from 202 selected weather stations that give some insight into the magnitude of this contribution. Since the season of tropical cyclones, including hurricanes, extends from June to October, he restricted his investigations to that period. He studied thirty years of records. Some figures taken from his thesis are presented in Table 38–1. These figures reveal that during the season of tropical cyclones, these storms might, at most, contribute between thirteen and forty-one percent of the total precipitation distributed along the Gulf and Atlantic Coasts. Surprisingly, the highest contributions were made over the middle Atlantic States. Actually, Atlantic City received the largest share of September rainfall from tropical cyclones. "Of the September normal of 3.90 inches of precipitation, an average of 1.61 inches or some 47.5 percent, resulted from tropical cyclones, occurring during only 16 of the 30 years."[20] Inland these contributions declined rapidly.

If tropical storms were dissipated, mitigated, or diverted, from whence would this moisture come? Does the risk of a water shortage on the East Coast outweigh the biological damage and the socio-economic losses from storm and flood? If the hurricane's course could be directed by a given amount in a given direction, who would make the decision to

TABLE 38–1. Contribution of tropical storms to rainfall

Mean (50 per cent probability) monthly precipitation (amount in inches) for representative locations along Gulf and Atlantic Coast, 1931–1960.*

Location	Month of maximum rainfall from tropical storms	Total mean rainfall	Per cent of total mean rainfall from tropical storms
Western Gulf			
Brownsville, Tex.	August	2.13	31
Eastern Gulf			
Homa, La.	September	5.38	29
South Atlantic			
Homestead, Fla.	September	9.87	13
Charleston, S. C.	September	5.57	13
Middle Atlantic			
Atlantic City, N. J.	September	2.54	41
Storrs, Conn.	September	3.09	30

* Figures given were extracted from Figures 9, 10, 14, and 15 of Cry's thesis.[19] The mean precipitation values obtained from gamma-fitted distributions of total precipitation.

change the course so that instead of striking Florida, it would be diverted toward Georgia or the Carolinas? On what basis would the decision be made?

These considerations bring us from the general realm of the ecosystem to specific human, and especially socio-economic, consequences of weather modification. The situation here is really no better than in the case of broad ecological consequences. For a single industry the cost-benefit analysis is straightforward. A group of farmers, for example, can easily calculate the cost of investing in cloud-seeding and the financial benefit accruing from increased agricultural productivity per increment of precipitation. So can a hydroelectric power company. A cost-benefit analysis for a large region with diverse socio-economic activities, however, is as complex as predicting how weather change will act on the ecosystem. There simply are not the data available at the present time to allow a critical systems analysis.[17, 21]

The legal ramifications of weather modification are horrendous. Even before there was any generally accepted body of information specifically relating a modifying operation to a weather change, professional cloud seeders were busy and many people were convinced that these entrepreneurs were affecting alterations in precipitation. Farmer X hired a fellow to make it rain; his crops needed water. His neighbor, Farmer Y, sued because his crops were damaged by the extra rain. A resort owner sued because the cloudy, rainy weather produced by cloud seeders seeking to alleviate a water shortage ruined his business. The courts in some cases favored the plaintiff; in others, the defendant. The precedents set by these actions, taken without clear evidence of cause and effect, may be serious impediments for future operations. Because the confusion has been so great, Maryland has made any form of weather modification a crime and

Pennsylvania has assigned to its counties the option to outlaw weather modification.[22]

In the province of Quebec, during the past few years there has been a twenty-five percent increase in the rainfall resulting in floods, damaged crops, eroded soil, and reduced tourist business. Mothers even came to fear that their children would be growing up without adequate sunshine. Because this augmented rainfall happened to coincide with rain-making operations, the public assigned the cause to the cloud seeders. More than 60,000 women signed a petition requesting governmental intervention. Television and radio advertisements and letters to newspapers threatened the cloud seeders. The government enjoined the weather modifying operations and paid hundreds of thousands of dollars to farmers for flood damage. A federal report stated that cloud seeding had actually reduced the rainfall by five percent; the twenty-five percent increase was attributed to natural causes.[23]

The national legal scene became chaotic before there was any valid evidence that the weather was in fact operationally modifiable. Large scale modification programs would have international legal implications with no body of law or precedent—except the unfortunate national ones—to deal with the problems.

Because weather has no national boundaries, what one country does will assuredly affect its neighbors. Studies of weather modification thus demand international cooperation so that rational procedures may be developed for the peaceful use and legal control of operations aimed at taming the weather.

Surveillance and Models

Because large scale experimentation on weather modification is already in progress, it is essential that ecologists participate in the planning and conduct of these investigations so that there can be a continuing ecological surveillance of the possible impact of weather modification on the metabolism and regulation of the ecosystem. Primarily because ecologists were not involved earlier, there may be insufficient time to make base-line measurements of ecosystems in areas where weather-modifying operations are contemplated. Where time can be devoted to making base-line studies before atmospheric manipulations begin, they should be undertaken. Because of the complexities of ecosystems and the shortage of ecologists, incomplete monitoring may be the only possibility. That would be better than no monitoring! Under these circumstances, attention might be focused on "those species of plant pathogens or insects known or suspected of being capable of damaging crops or causing defoliation of the natural vegetation."[16]

These suggestions will be most difficult to implement for several reasons. First, the techniques of measuring in detail the complex transactions of the ecosystem have not yet been developed. Second, the crucial links in the transactional network of the ecosystem may vary from rain forest to grassland to tundra. Third, there are not enough trained ecologists to divert some from current tasks to monitoring weather-modifying operations.

For these reasons educational programs for training individuals in the measurement and systems analysis of ecosystems must be expanded. Furthermore, more attention should be given to the field study of ecosystems as opposed to work on single species in isolation so that a body of data will begin to accumulate on the perturbations of ecosystems under natural variations of weather and climate and so that crucial links in the transactional networks can be identified. A useful associated investigation would be the examination in depth of the ecological impact of past meteorological extremes and disasters, particularly the long-lasting ones, such as droughts. As yet these have not been systematically studied and they would provide invaluable insights into certain aspects of conscious weather modification.

One procedure that would provide strategic information on the ecological consequences of weather modification is computer simulation of the ecosystem. In these studies a mathematical model of the ecosystem would have to be constructed comprising "a set of functional relationships which mimic the dynamic properties of all relationships between and within soils, plants, animals, site factors and weather, with respect to both changes in variables through time and dispersal of entities through space with the passage of time."[16] With such a model, questions could be asked of the consequences of weather change and the expected economic benefits or losses. Certainly it would be far less expensive to simulate weather modification on the computer than to undertake numerous experiments in the field, experiments that might have serious unforeseen and irreversible biological effects.

A word of caution, however, might be added even here. To construct a mathematical model of a complete ecosystem requires far more detailed knowledge than we now possess. Where knowledge is incomplete, assumptions and educated guesses have to be made. The more assumptions made, the farther the model is from reality. The more idealized that model is, the more caution the experimenter must exercise in evaluating the results of simulated weather modification and then applying them to natural situations.

SUMMARY

Man, the dominant animal of the biosphere, has achieved this role through the exercise of his intelligence and the application of his behavioral products to draw from Nature and to utilize Nature to support his own welfare. The ecosystem is a vital natural resource for him, but his manipulation of it has become increasingly risky. In order to conserve this resource more wisely, it is essential that he develop a resource management strategy. To formulate a realistic strategy, it is important that he comprehend more fully the complex transactions and interactions of the ecosystem else he intervene—even inadvertently—with disastrous biological consequences.

It is urgent that he formulate this strategy soon for already his activities have brought about significant inadvertent weather changes and he has achieved some success with conscious weather-modifying operations. As we have seen, both involve ecological risks that we are not yet prepared to evaluate because of lack of appropriate detailed information.

ACKNOWLEDGMENT

In preparing this manuscript I have received many constructive comments from colleagues both in the atmospheric sciences and in the life sciences. Although I alone assume responsibility for the implications noted herein, I wish to express particular appreciation to Professor L. D. Bliss, Professor R. A. Braham, Jr., Dr. T. F. Malone, Dr. R. A. McCormick, Professor R. A. Ragotzkie, Professor V. J. Schaefer, and Dr. G. E. Sprugel, Jr., for their critical reviews of my manuscript.

REFERENCES

1. Brain, W. R. 1965. "Science and Antiscience," *Science*, 148:192–198.
2. Sargent, F., II, and D. M. Barr. 1965. "Health and the Fitness of the Ecosystem," *In: The Environment and Man*, Travelers Research Center, Hartford, Conn.
3. *Weather and Climate Modifications. Problems and Prospects.* 2 vols. Publication No. 1350. National Academy of Sciences-National Research Council, Washington, D.C. 1966.
4. Hickman, K. 1966. "Oases for the Future," *Science*, 154:612–617.
5. Abelson, P. H. 1967. "Global Weather," *Science*, 155:153.
6. Mitchell, J. M., Jr. 1961. "Recent Secular Changes of Global Temperature," *In: Solar Variations, Climatic Change and Related Geophysical Problems*, R. W. Fairbridge, ed. Ann. N.Y. Acad. Sci. 95(1):235–250.
7. Mitchell, J.M., Jr. 1963. "On the World-Wide Pattern of Secular Temperature Change," *In: Changes of Climate*. Arid Zone Research (UNESCO), 20:161–181.
8. Schaefer, V. J. 1966. "Ice Nuclei from Automobile Exhaust and Iodine Vapor," *Science*, 154:1555–1557.
9. McCormick, R. A. 1967. Personal communication.
10. *Weather and Climate Modification*, NSF 66–3. National Science Foundation, Washington, D.C. 1966.
11. *An Outline of International Programs in the Atmospheric Sciences.* Publication No. 1085, National Academy of Sciences-National Research Council, Washington, D.C. 1963.

12. *The Essential Elements of the World Weather Watch.* World Meteorological Organization, Geneva, Switzerland. 1966.
13. *First Report of IUGG Committee on Atmospheric Sciences.* March 20, 1965.
14. *Second Report of IUGG Committee on Atmospheric Sciences.* September 15, 1966.
15. Wycoff, P. H. 1966. "Evaluation of the State of the Art." *In: Human Dimensions of Weather Modification.* (W. R. D. Sewall, Ed.) Dept. Geog. Res. Paper No. 105. University of Chicago, Chicago, Ill., pp. 27–39.
16. Livingstone, D. A. (Chm.). 1966. "Biological Aspects of Weather Modification." A report from the Ecological Society of America's *ad hoc* Weather Working Group of the Ecological Study Committee to the Special Commission for Weather Modification of the National Science Foundation. *Bull. Ecol. Soc. Am.*, 47:39–78.
17. Sewall, W. R. D. (Ed.), *Human Dimensions of Weather Modification*, op. cit.
18. Morris, E. A. 1965. "The Law and Weather Modification." *Bull. Am. Meteorol. Soc.* 46:618–622.
19. Cry G. W. 1966. *Effects of Tropical Cyclone Rainfall on the Distribution of Precipitation over the Eastern and Southern United States.* M.S. Thesis. Rutgers-The State University, New Brunswick, N.J.
20. Ackerman, E. A. 1958. "Design Study for Economic Analysis of Weather Modification." *Final Report of Advisory Committee on Weather Control.* U.S. Government Printing Office, Washington, D.C. Vol. II, pp. 235–245.
21. Morris, S. A. "Institutional Adjustments to an Emerging Technology; Legal Aspects of Weather Modification." *In: Human Dimensions of Weather Modification,* op. cit.
22. Kane, J. June-July, 1966. New York's Droughts vs. Quebec's Floods. *The Conservationist*, pp. 12–13.

IV
Outlook for the future

39

The conservation ethic

Aldo Leopold

When god-like Odysseus returned from the wars in Troy, he hanged all on one rope some dozen slave-girls of his household whom he suspected of misbehavior during his absence.

This hanging involved no question of propriety, much less of justice. The girls were property. The disposal of property was then, as now, a matter of expediency, not of right and wrong.

Criteria of right and wrong were not lacking from Odysseus' Greece: witness the fidelity of his wife through the long years before at last his black-prowed galleys clove the wine-dark seas for home. The ethical structure of that day covered wives, but had not yet been extended to human chattels. During the three thousand years which have since elapsed, ethical criteria have been extended to many fields of conduct, with corresponding shrinkages in those judged by expediency only.

This extension of ethics, so far studied only by philosophers, is actually a process in ecological evolution. Its sequences may be described in biological as well as philosophical terms. An ethic, biologically, is a limitation on freedom of action in the struggle for existence. An ethic, philosophically, is a differentiation of social from anti-social conduct. These are two definitions of one thing. The thing has its origin in the tendency of interdependent individuals or societies to evolve modes of coöperation. The biologist calls these symbioses. Man elaborated certain advanced symbioses called politics and economics. Like their simpler biological antecedents, they enable individuals or groups to exploit each other in an orderly way. Their first yardstick was expediency.

Reprinted with permission from the *Journal of Forestry*, 31:634–643 (1933).

The complexity of coöperative mechanisms increased with population density, and with the efficiency of tools. It was simpler, for example, to define the anti-social uses of sticks and stones in the days of the mastodons than of bullets and billboards in the age of motors.

At a certain stage of complexity, the human community found expediency-yardsticks no longer sufficient. One by one it has evolved and superimposed upon them a set of ethical yardsticks. The first ethics dealt with the relationship between individuals. The Mosaic Decalogue is an example. Later accretions dealt with the relationship between the individual and society. Christianity tries to integrate the individual to society, Democracy to integrate social organization to the individual.

There is as yet no ethic dealing with man's relationship to land and to the non-human animals and plants which grow upon it. Land, like Odysseus' slave-girls, is still property. The land-relation is still strictly economic, entailing privileges but not obligations.

The extension of ethics to this third element in human environment is, if we read evolution correctly, an ecological possibility. It is the third step in a sequence. The first two have already been taken. Civilized man exhibits in his own mind evidence that the third is needed. For example, his sense of right and wrong may be aroused quite as strongly by the desecration of a nearby woodlot as by a famine in China, a near-pogrom in Germany, or the murder of the slave-girls in ancient Greece. Individual thinkers since the days of Ezekial and Isaiah have asserted that the despoliation of land is not only inexpedient but wrong. Society, however, has not yet affirmed their belief. I regard the present conservation movement as the embryo of such an affirmation. I here discuss why this is, or should be, so.

Some scientists will dismiss this matter forthwith, on the ground that ecology has no relation to right and wrong. To such I reply that science, if not philosophy, should by now have made us cautious about dismissals. An ethic may be regarded as a mode of guidance for meeting ecological situations so new or intricate, or involving such deferred reactions, that the path of social expediency is not discernible to the average individual. Animal instincts are just this. Ethics are possibly a kind of advanced social instinct in-the-making.

Whatever the merits of this analogy, no ecologist can deny that our land-relation involves penalties and rewards which the individual does not see, and needs modes of guidance which do not yet exist. Call these what you will, science cannot escape its part in forming them.

ECOLOGY — ITS ROLE IN HISTORY

A harmonious relation to land is more intricate, and of more consequence to civilization, than the historians of its progress seem to realize.

Civilization is not, as they often assume, the enslavement of a stable and constant earth. It is a state of *mutual and interdependent coöperation* between human animals, other animals, plants, and soils, which may be disrupted at any moment by the failure of any of them. Land-despoliation has evicted nations, and can on occasion do it again. As long as six virgin continents awaited the plow, this was perhaps no tragic matter—eviction from one piece of soil could be recouped by despoiling another. But there are now wars and rumors of wars which foretell the impending saturation of the earth's best soils and climates. It thus becomes a matter of some importance, at least to ourselves, that our dominion, once gained, be self-perpetuating rather than self-destructive.

This instability of our land-relation calls for example. I will sketch a single aspect of it: the plant succession as a factor in history.

In the years following the Revolution, three groups were contending for control of the Mississippi valley: the native Indians, the French and English traders, and American settlers. Historians wonder what would have happened if the English at Detroit had thrown a little more weight into the Indian side of those tipsy scales which decided the outcome of the Colonial migration into the cane-lands of Kentucky. Yet who ever wondered why the cane-lands, when subjected to the particular mixture of forces represented by the cow, plow, fire, and axe of the pioneer, became bluegrass? What if the plant succession inherent in this "dark and bloody ground" had, under the impact of these forces, given us some worthless sedge, shrub, or weed? Would Boone and Kenton have held out? Would there have been any overflow into Ohio? Any Louisiana Purchase? Any transcontinental union of new states? Any Civil War? Any machine age? Any depression? The subsequent drama of American history, here and elsewhere, hung in large degree on the reaction of particular soils to the impact of particular forces exerted by a particular kind and degree of human occupation. No statesman-biologist selected those forces, nor foresaw their effects. That chain of events which in the Fourth of July we call our National Destiny hung on a "fortuitous concourse of elements," the interplay of which we now dimly decipher *by hindsight only*.

Contrast Kentucky with what hindsight tells us about the Southwest. The impact of occupancy here brought no bluegrass, nor other plant fitted to withstand the bumps and buffetings of misuse. Most of these soils, when grazed, reverted through a successive series of more and more worthless grasses, shrubs, and weeds to a condition of unstable equilibrium. Each recession of plant types bred erosion; each increment to erosion bred a further recession of plants. The result today is a progressive and mutual deterioration, not only of plants and soils, but of the animal community subsisting thereon. The early settlers did not expect this, on the cienegas of central New Mexico some even cut artificial gullies to hasten it. So subtle has been its progress that few people know anything about it. It is not discussed at polite tea-tables or go-getting luncheon clubs, but only in the arid halls of science.

All civilization seem to have been conditioned upon whether the plant succession, under the impact of occupancy, gave a stable and habitable assortment of vegetative types, or an unstable and uninhabitable assortment. The swampy forests of Caesar's Gaul were utterly changed by human use—for the better. Moses' land of milk and honey was utterly changed—for the worse. Both changes are the unpremeditated resultant of the impact between ecological and economic forces. We now decipher these reactions retrospectively. What could possibly be more important than to foresee and control them?

We of the machine age admire ourselves for our mechanical ingenuity; we harness cars to the solar energy impounded in carboniferous forests; we fly in mechanical birds; we make the ether carry our words or even our pictures. But are these not in one sense mere parlor tricks compared with our utter ineptitude in keeping land fit to live upon? Our engineering has attained the pearly gates of a near-millennium, but our applied biology still lives in nomad's tents of the stone age. If our system of land-use happens to be self-perpetuating, we stay. If it happens to be self-destructive we move, like Abraham, to pastures new.

Do I overdraw this paradox? I think not. Consider the transcontinental airmail which plies the skyways of the Southwest—a symbol of its final conquest. What does it see? A score of mountain valleys which were green gems of fertility when first described by Coronado, Espejo, Pattie, Abert, Sitgreaves, and Couzens. What are they now? Sandbars, wastes of cobbles and burroweed, a path for torrents. Rivers which Pattie says were clear, now muddy sewers for the wasting fertility of an empire. A "Public Domain," once a velvet carpet of rich buffalo-grass and grama, now an illimitable waste of rattlesnake-bush and tumbleweed, too impoverished to be accepted as a gift by the states within which it lies. Why? Because the ecology of this Southwest happened to be set on a hair-trigger. Because cows eat brush when the grass is gone, and thus postpone the penalties of over-utilization. Because certain grasses, when grazed too closely to bear seed-stalks, are weakened and give way to inferior grasses, and these to inferior shrubs, and these to weeds, and these to naked earth. Because rain which spatters upon vegetated soil stays clear and sinks, while rain which spatters upon devegetated soil seals its interstices with colloidal mud and hence must run away as floods, cutting the heart out of country as it goes. Are these phenomena any more difficult to foresee than the paths of stars which science deciphers without the error of a single second? Which is the more important to the permanence and welfare of civilization?

I do not here berate the astronomer for his precocity, but rather the ecologist for his lack of it. The days of his cloistered sequestration are over:

"Whether you will or not,
You are a king, Tristram, for you are one
Of the time-tested few that leave the world,
When they are gone, not the same place it was.
Mark what you leave."

Unforseen ecological reactions not only make or break history in a few exceptional enterprises—they condition, circumscribe, delimit, and warp all enterprises, both economic and cultural, that pertain to land. In the cornbelt, after grazing and plowing out all the cover in the interests of "clean farming," we grew tearful about wild-life, and spent several decades passing laws for its restoration. We were like Canute commanding the tide. Only recently has research made it clear that the implements for restoration lie not in the legislature, but in the farmer's toolshed. Barbed wire and brains are doing what laws alone failed to do.

In other instances we take credit for shaking down apples which were, in all probability, ecological windfalls. In the Lake States and the Northeast lumbering, pulping, and fire accidentally created some scores of millions of acres of new second-growth. At the proper stage we find these thickets full of deer. For this we naively thank the wisdom of our game laws.

In short, the reaction of land to occupancy determines the nature and duration of civilization. In arid climates the land may be destroyed. In all climates the plant succession determines what economic activities can be supported. Their nature and intensity in turn determine not only the domestic but also the wild plant and animal life, the scenery, and the whole face of nature. We inherit the earth, but within the limits of the soil and the plant succession we also *rebuild* the earth—without plan, without knowledge of its properties, and without understanding of the increasingly coarse and powerful tools which science has placed at our disposal. We are remodelling the Alhambra with a steam-shovel.

ECOLOGY AND ECONOMICS

The conservation movement is, at the very least, an assertion that these interactions between man and land are too important to be left to chance, even that sacred variety of chance known as economic law.

We have three possible controls: Legislation, self-interest, and ethics. Before we can know where and how they will work, we must first understand the reactions. Such understanding arises only from research. At the present moment research, inadequate as it is, has nevertheless piled up a large store of facts which our land using industries are unwilling, or (they claim) unable, to apply. Why? A review of three sample fields will be attempted.

Soil science has so far relied on self-interest as the motive for conservation. The landholder is told that it pays to conserve his soil and its fertility. On good farms this economic formula has improved land-practice, but on poorer soils vast abuses still proceed unchecked. Public acquisition of submarginal soils is being urged as a remedy for their misuse. It has been applied to some extent, but it often comes too late to check erosion, and can hardly hope more than to ameliorate a phenomenon involving in some degree *every square foot* on the continent. Legislative compulsion might work on the best soils where it is least needed, but it seems hopeless on poor soils where the existing economic set-up hardly permits even uncontrolled private enterprise to make a profit. We must face the fact that, by and large, no defensible relationship between man and the soil of his nativity is at yet in sight.

Forestry exhibits another tragedy—or comedy—of *Homo sapiens*, astride the runaway Juggernaut of his own building, trying to be decent to his environment. A new profession was trained in the confident expectation that the shrinkage in virgin timber would, as a matter of self-interest, bring an expansion of timber-cropping. Foresters are cropping timber on certain parcels of poor land which happen to be public, but on the great bulk of private holdings they have accomplished little. Economics won't let them. Why? He would be bold indeed who claimed to know the whole answer, but these parts of it seem agreed upon: modern transport prevents profitable tree-cropping in cut-out regions until virgin stands in all others are first exhausted; substitutes for lumber have undermined confidence in the future need for it; carrying changes on stumpage reserves are so high as to force perennial liquidation, overproduction, depressed prices, and an appalling wastage of unmarketable grades which must be cut to get the higher grades; the mind of the forest owner lacks the point-of-view underlying sustained yield; the low wage-standards on which European forestry rests do not obtain in America.

A few tentative gropings toward industrial forestry were visible before 1929, but these have been mostly swept away by the depression, with the net result that forty years of "campaigning" have left us only such actual tree-cropping as is under-written by public treasuries. Only a blind man could see in this the beginnings of an orderly and harmonious use of the forest resource.

There are those who would remedy this failure by legislative compulsion of private owners. Can a landholder be successfully compelled to raise any crop, let alone a complex long-time crop like a forest, on land the private possession of which is, for the moment at least, a liability?

Compulsion would merely hasten that avalanche of tax-delinquent land-titles now being dumped into the public lap.

Another and larger group seeks a remedy in more public ownership. Doubtless we need it—we are getting it whether we need it or not—but how far can it go? We cannot dodge the fact that the forest problem, like the soil problem, *is coextensive with the map of the United States.* How far can we tax other lands and industries to maintain forest lands and industries artificially? How confidently can we set out to run a hundred-yard dash with a twenty foot rope tying our ankle to the starting point? Well, we are bravely "getting set," anyhow.

The trend in wild-life conservation is possibly more encouraging than in either soils or forests. It has suddenly become apparent that farmers, out of self-interest, can be induced to crop game. Game crops are in demand, staple crops are not. For farm-species, therefore, the immediate future is relatively bright. Forest game has profited to some extent by the accidental establishment of new habitat following the decline of forest industries. Migratory game, on the other hand, has lost heavily through drainage and over-shooting; its future is black because motives of self-interest do not apply to the private cropping of birds so mobile that they "belong" to everybody, and hence to nobody. Only governments have interests coextensive with their annual movements, and the divided counsels of conservationists give governments ample alibi for doing little. Governments could crop migratory birds because their marshy habitat is cheap and concentrated, but we get only an annual crop of new hearings on how to divide the fast-dwindling remnant.

These three fields of conservation, while but fractions of the whole, suffice to illustrate the welter of conflicting forces, facts, and opinions which so far comprise the result of the effort to harmonize our machine civilization with the land whence comes its sustenance. We have accomplished little, but we should have learned much. What?

I can see clearly only two things:

First, that the economic cards are stacked against some of the most important reforms in land-use.

Second, that the scheme to circumvent this obstacle by public ownership, while highly desirable and good as far as it goes, can never go far enough. Many will take issue on this, but the issue is between two conflicting conceptions of the end towards which we are working.

One regards conservation as a kind of sacrificial offering, made for us vicariously by bureaus, on lands nobody wants for other purposes, in propitiation for the atrocities which still prevail everywhere else. We have made a real start on this kind of conservation, and we can carry it as far as the tax-string on our leg will reach. Obviously, though it conserves our self-respect better than our land. Many excellent people accept it, either because they despair of anything better, or because they fail to see the *universality of the reactions needing control.* That is to say their ecological education is not yet sufficient.

The other concept supports the public program, but regards it as merely extension, teaching, demonstration, an initial nucleus, a means to an end, but not the end itself. The real end is a *universal symbiosis with land,* economic and esthetic, public and private. To this school of thought public ownership is a patch but not a program.

Are we, then, limited to patchwork until such time as Mr. Babbitt has taken his Ph.D. in ecology and esthetics? Or do the new economic formulae offer a short-cut to harmony with our environment?

THE ECONOMIC ISMS

As nearly as I can see, all the new isms—Socialism, Communism, Fascism, and especially the late but not lamented Technocracy—outdo even Capitalism itself in their preoccupation with one thing: The distribution of more machine-made commodities to more people. They all proceed on the theory that if we can all keep warm and full, and all own a Ford and a radio, the good life will follow. Their programs differ only in ways to mobilize machines to this end. Though they despise each other, they are all, in respect of this objective, as identically alike as peas in a pod. They are competitive apostles of a single creed: *salvation by machinery.*

We are here concerned, not with their proposals for adjusting men and machinery to goods, but rather with their lack of any vital proposal for adjusting men and machines to land. To conservationists they offer only the old familiar palliatives: Public ownership and private compulsion. If these are insufficient now, by what magic are they to become sufficient after we change our collective label?

Let us apply economic reasoning to a sample problem and see where it takes us. As already pointed out, there is a huge area which the economist calls sub-marginal, because it has a minus value for exploitation. In its once-virgin condition, however, it could be "skinned" at a profit. It has been, and as a result erosion is washing it away. What shall we do about it?

By all the accepted tenets of current economics and science we ought to say "let her wash." Why? Because staple land-crops are overproduced, our population curve is flattening out, science is still raising the yields from better lands, we are spending millions from the public treasury to retire unneeded acreage, and here is nature offering to do the same thing free of charge; why not let her do it? This, I say, is economic reasoning. *Yet no man has so spoken.* I cannot help reading a meaning into this fact. To me it means that the average citizen shares in some degree the intuitive and instantaneous contempt with which the conservationist would regard such an attitude. We can, it seems, stomach the burning or plowing-under of over-produced cotton, coffee, or corn, but the destruction of mother-earth, however "sub-marginal," touches something deeper, some sub-economic stratum of the human intelligence wherein lies that something—perhaps the essence of civilization—which Wilson called "the decent opinion of mankind."

THE CONSERVATION MOVEMENT

We are confronted, then, by a contradiction. To build a better motor we tap the uttermost powers of the human brain; to build a better countryside we throw dice. Political systems take no cognizance of this disparity, offer no sufficient remedy. There is, however, a dormant but widespread consciousness that the destruction of land, and of the living things upon it, is wrong. A new minority have espoused an idea called conservation which tends to assert this as a positive principle. Does it contain seeds which are likely to grow?

Its own devotees, I confess, often give apparent grounds for skepticism. We have, as an extreme example, the cult of the barbless hook, which acquires self-esteem by a self-imposed limitation of armaments in catching fish. The limitation is commendable, but the illusion that it has something to do with salvation is as naive as some of the primitive taboos and mortifications which still adhere to religious sects. Such excrescenses seem to indicate the whereabouts of a moral problem, however irrelevant they be in either defining or solving it.

Then there is the conservation-booster, who of late has been rewriting the conservation ticket in terms of "tourist-bait." He exhorts us to "conserve outdoor Wisconsin" because if we don't the motorist-on-vacation will streak through to Michigan, leaving us only a cloud of dust. Is Mr. Babbit trumping up hard-boiled reasons to serve as a screen for doing what he thinks is right? His tenacity suggests that he is after something more than tourists. Have he and other thousands of "conservation workers" labored through all these barren decades fired by a dream of augmenting the sales of sandwiches and gasoline? I think not. Some of these people have hitched their wagon to a star—and that is something.

Any wagon so hitched offers the discerning politician a quick ride to glory. His agility in hopping up and seizing the reins adds little dignity to the cause, but it does add the testimony of his political nose to an important question: is this conservation something people really want? The political objective, to be sure, is often some trivial tinkering with the laws, some useless appropriation, or some pasting of pretty labels on ugly realities. How often, though, does any political action portray the real depth of the idea behind it? For political consumption a new thought must always be reduced to a posture or a phrase. It has happened before that great ideas were heralded by growing-pains in the body politic, semi-comic to those onlookers not yet infected by them. The insignificance of what we conservationists, in our political capacity, say and do, does not detract from the significance of our persistent desire to do something. To turn this desire into productive channels is the task of time, and ecology.

The recent trend in wild life conservation shows the direction in which ideas are evolving. At the inception of the movement fifty years ago, its underlying thesis was to save species from extermination. The means to this end were a series of restrictive enactments. The duty of the individual was to cherish and extend these enactments, and to see that his neighbor obeyed them. The whole structure was negative and prohibitory. It assumed land to be a constant in the ecological equation. Gun-powder and blood-lust were the variables needing control.

There is now being superimposed on this a positive and affirmatory ideology, the thesis of which is to prevent the deterioration of environment. The means to this end is research. The duty of the individual is to apply its findings to land, and to encourage his neighbor to do likewise. The soil and the plant succession are recognized as the basic variables which determine plant and animal life, both wild and domesticated, and likewise the quality and quantity of human satisfactions to be derived. Gun-powder is relegated to the status of a tool for harvesting one of these satisfactions. Blood-lust is a source of motive-power, like sex in social organization. Only one constant is assumed, and that is common to both equations: the love of nature.

This new idea is so far regarded as merely a new and promising means to better hunting and fishing, but its potential uses are much larger. To explain this, let us go back to the basic thesis—the preservation of fauna and flora.

Why do species become extinct? Because they first become rare. Why do they become rare? Because of shrinkage in the particular environments which their particular adaptations enable them to inhabit. Can such shrinkage be controlled? Yes, once the specifications are known. How known? Through ecological research. How controlled? By modifying the environment with those same tools and skills already used in agriculture and forestry.

Given, then, the knowledge and the desire, this idea of controlled wild culture or "management" can be applied not only to quail and trout, but to *any living thing* from bloodroots to Bell's vireos. Within the limits imposed by the plant succession, the soil, the size of the property, and the gamut of the seasons, the landholder can "raise" any wild plant, fish, bird, or mammal he wants to. A rare bird or flower need remain no rarer than the people willing to venture their skill in *building it a habitat.* Nor need we visualize this as a new diversion for the idle rich. The average dolled-up estate merely proves what we will some day learn to acknowledge: that bread and beauty grow best together. Their harmonious integration can make farming not only a business but an art; the land not only a food-factory but an instrument for self-expression, on which each can play music of his own choosing.

It is well to ponder the sweep of this thing. It offers us nothing less than a renaissance—a new creative stage—in the oldest, and potentially the most universal, of all the fine arts. "Landscaping," for ages dissociated from economic land-use, has suffered that dwarfing and distortion which always attends the relegation of esthetic or spiritual functions to parks and parlors. Hence it is hard for us to visualize a creative art of land-beauty which is the prerogative, not of esthetic priests but of dirt farmers, which deals not with plants but with biota, and which wields not only spade and pruning shears, but also draws rein on those invisible forces which determine the presence or absence of plants and animals. Yet such is this thing which lies to hand, if we want it. In it are the seeds of change, including, perhaps, a rebirth of that social dignity which ought to inhere in land-ownership, but which, for the moment, has passed to inferior professions, and which the current processes of land-skinning hardly deserve. In it, too, are perhaps the seeds of a new fellowship in land, a new solidarity in all men privileged to plow, a realization of Whitman's dream to *"plant companionship as thick as trees along all the rivers of America."* What bitter parody of such companionship, and trees, and rivers, is offered to this our generation!

I will not belabor the pipe-dream. It is no prediction, but merely an assertion that the idea of controlled environment contains colors and brushes wherewith society may some day paint a new and possibly a better picture of itself. Granted a community in which the combined beauty and utility of land determines the social status of its owner, and we will see a speedy dissolution of the economic obstacles which now beset conservation. Economic laws may be permanent, but their impact reflects what people want, which in turn reflects what they know and what they are. The economic set-up at any moment is in some measure the result, as well as the cause, of the then prevailing standard of living. Such standards change. For example: some people discriminate against manufactured goods produced by child-labor or other anti-social processes. They have learned some of the abuses of machinery, and are willing to use their custom as a leverage for betterment. Social pressures have also been exerted to modify ecological processes which happened to be simple enough for people to understand—witness the very effective boycott of birdskins for millinery ornament. We need postulate only a little further advance in ecological education to visualize the application of like pressures to other conservation problems.

For example: the lumberman who is now unable to practice forestry because the public is turning to synthetic boards may be able to sell man-grown lumber "to keep the mountains green." Again: certain wools are produced by gutting the public domain; couldn't their competitors, who lead their sheep in greener pastures, so label their product? Must we view forever the irony of educating our sons with paper, the offal of which pollutes the rivers which they need quite as badly as books? Would not many people pay an extra penny for a "clean" newspaper? Government may some day busy itself with the legitimacy of labels used by land-industries to distinguish conservation products, rather than with the attempt to operate their lands for them.

I neither predict nor advocate these particular pressures—their wisdom or unwisdom is beyond my knowledge. I do assert that these abuses are just as real, and their correction every whit as urgent, as was the killing of egrets for hats. *They differ only in the number of links composing the ecological chain of cause and effect.* In egrets there were one or two links, which the mass-mind saw, believed, and acted upon. In these others there are many links; people do not see them, nor believe us who do. The ultimate issue, in conservation as in other social problems, is whether the mass-mind *wants to* extend its powers of comprehending the world in which it lives, or, granted the desire, *has the capacity to do so.* Ortega, in his "Revolt of the Masses," has pointed the first question with devastating lucidity. The geneticists are gradually, with trepidations, coming to grips with the second. I do not know the answer to either. I simply affirm that a sufficiently enlightened society, by changing its wants and tolerances, can change the economic factors bearing on land. It can be said of nations, as of individuals: "as a man thinketh, so is he."

It may seem idle to project such imaginary elaborations of culture at a time when millions lack even the means of physical existence. Some may feel for it the same honest horror as the Senator from Michigan who lately arraigned Congress for protecting migratory birds at a time when fellow-humans lacked bread. The trouble with such deadly parallels is we can never be sure which is cause and which is effect. It is not inconceivable that the wave phenomena which have lately upset everything from banks to crime-rates might be less troublesome if the human medium in which they run *readjusted its tensions.* The stampede is an attribute of animals interested solely in grass.